Engineering Materials

This series provides topical information on innovative, structural and functional materials and composites with applications in optical, electrical, mechanical, civil, aeronautical, medical, bio- and nano-engineering. The individual volumes are complete, comprehensive monographs covering the structure, properties, manufacturing process and applications of these materials. This multidisciplinary series is devoted to professionals, students and all those interested in the latest developments in the Materials Science field, that look for a carefully selected collection of high quality review articles on their respective field of expertise.

Indexed at Compendex (2021) and Scopus (2022)

Om Prakash Pandey · Piyush Sharma ·
Shagun Kainth
Editors

2D Materials for Electrochemical Water Splitting

Springer

Editors
Om Prakash Pandey
Department of Physics and Materials Science
Thapar Institute of Engineering and Technology
Patiala, Punjab, India

Piyush Sharma
Department of Physics and Materials Science
Thapar Institute of Engineering and Technology
Patiala, Punjab, India

Shagun Kainth
Department of Chemistry and Biochemistry
Thapar Institute of Engineering and Technology
Patiala, Punjab, India

ISSN 1612-1317 ISSN 1868-1212 (electronic)
Engineering Materials
ISBN 978-981-95-2002-2 ISBN 978-981-95-2003-9 (eBook)
https://doi.org/10.1007/978-981-95-2003-9

This Springer imprint is published by the registered company Springer Nature Singapore Pte Ltd.
The registered company address is: 152 Beach Road, #21-01/04 Gateway East, Singapore 189721, Singapore

Preface

The global energy landscape is undergoing a transformative shift, driven by an urgent need to move away from fossil fuel dependence toward clean, sustainable, and renewable energy sources. Amidst rising concerns over climate change, environmental degradation, with finite natural resources, hydrogen has emerged as a compelling vector for decarbonized energy systems. Among various methods for hydrogen production, *electrochemical water splitting* is considered as one of the most promising, environmentally benign, and scalable technologies. This process, driven by electricity—ideally sourced from renewables such as solar or wind—splits water molecules into hydrogen and oxygen gases, thereby offering a direct and efficient route to green hydrogen.

Despite its conceptual simplicity and environmental appeal, electrochemical water splitting faces several formidable scientific and technological challenges. Among these the main reason is the sluggish kinetics of the half-reactions—Hydrogen Evolution Reaction (HER) and Oxygen Evolution Reaction (OER)—and the high cost, instability, and scarcity of traditional noble metal-based catalysts (e.g., Pt, IrO_2, RuO_2). Overcoming these barriers is crucial for the commercial viability of water electrolysis systems and for achieving the broader vision of a hydrogen economy.

In recent years, **two-dimensional (2D) materials** have revolutionized the fields of materials science, nanotechnology, and catalysis. Since the isolation of graphene in 2004, the library of 2D materials has expanded rapidly to include a wide array of classes such as transition metal dichalcogenides (TMDs), 2D trichalcogenides, MXenes, Graphdiyne, metal-organic frameworks (MOFs), and emerging categories like high-entropy 2D materials. These materials exhibit extraordinary properties such as high surface area, tunable electronic structure, atomic thickness, and unique active site exposure, making them ideal candidates for catalyzing electrochemical reactions.

The exploration of 2D materials for **electrochemical overall water splitting**—which integrates both HER and OER in a single system—is currently a frontier area of multidisciplinary research. This book is devoted to unveiling the potential, progress, and future directions of 2D materials in this context. It aims to provide a comprehensive and critical overview of the fundamental mechanisms, synthesis strategies, structural-property relationships, and application prospects of 2D materials in water splitting. While there exists a growing body of literature on HER- or OER-specific catalysis, this book is distinctive in its integrative treatment of overall water splitting systems, emphasizing the synergistic design of bifunctional catalysts, tandem systems, and scalable architectures.

The idea for this book emerged from a recognition of the vast but fragmented nature of current research on 2D materials for water splitting. Numerous review articles, monographs, and research papers address various facets of the topic—from specific materials classes to individual electrochemical reactions—but very few offer a unified, in-depth analysis across the entire water splitting process. This book seeks to fill that gap by incorporating different synthesized materials and technology developed from chemistry, physics, materials science, nanotechnology, and electrochemical engineering into a coherent and accessible volume.

The timing of this book is also significant. The development of 2D materials has reached a level of maturity where researchers are now exploring hybrid, doped, defective, and heterostructured versions tailored for specific catalytic functions. Simultaneously, electrochemical water splitting technologies are being integrated into prototype devices and pilot-scale operations. Bridging these advances is critical for accelerating the translation of laboratory breakthroughs into industrial solutions.

Moreover, as the urgency for climate action increases and nations around the world set ambitious targets for carbon neutrality, the role of green hydrogen will become more central. In this evolving energy ecosystem, understanding how to efficiently and sustainably generate hydrogen through materials engineering is not just a scientific challenge—it is rather a societal imperative.

This book offers a comprehensive overview of the evolving landscape of 2D materials for electrochemical overall water splitting, progressing logically from fundamental principles to advanced applications. It begins with the basics of electrochemical water splitting, including thermodynamics, kinetics, and mechanisms of the hydrogen and oxygen evolution reactions (HER and OER). Readers are then introduced to the design principles of 2D electrocatalysts, exploring concepts such as active site engineering, doping, defect tuning, and charge transfer optimization. Subsequent chapters provide in-depth discussions on specific 2D material families, including transition metal dichalcogenides (TMDs), trichalcogenides, MXenes, graphdiyne, MOFs, and emerging high-entropy materials—detailing their synthesis, structure–performance relationships, and mechanistic roles. The book emphasizes bifunctional catalyst design for integrated HER and OER performance, as well as practical strategies for device integration, addressing challenges in electrode fabrication, stability, and scalability.

This book, authored by prominent specialists, is an invaluable resource for researchers, engineers, and students working to enhance clean hydrogen technologies using next-generation 2D materials. We hope that this book will motivate readers to investigate the fascinating realm of 2D materials for electrochemical water splitting, fostering interdisciplinary collaboration and generating novel research. This collection aims to expand the boundaries of materials science and electrochemical technologies by integrating fundamental research and practical engineering viewpoints, so making a significant contribution to the realization of a sustainable, hydrogen-powered future.

Dr. Om Prakash Pandey
Senior Professor, Department
of Physics and Materials Science
(DPMS)
Thapar Institute of Engineering
and Technology
Patiala, Punjab, India
oppandey@thapar.edu

Dr. Piyush Sharma
Department of Physics and Materials
Science (DPMS)
Thapar Institute of Engineering
and Technology
Patiala, Punjab, India

Dr. Shagun Kainth
Department of Chemistry
and Biochemistry (DCBC)
Thapar Institute of Engineering
and Technology
Patiala, Punjab, India

Acknowledgments

Compiling this book has been both a scientific and collaborative journey. The editors would like to express heartfelt gratitude to all contributing authors for their enthusiasm, expertise, and timely submissions. We are also thankful to the peer reviewers who provided critical feedback to maintain the scholarly standard of the content.

We owe special thanks to the institutions and laboratories that supported this endeavor, and to the funding agencies and consortia that continue to champion research at the intersection of materials science and sustainable energy. Additionally, we appreciate the editorial and production teams at the publisher for their professional guidance throughout the process.

A Forward-Looking Vision

As we look to the future, it is evident that hydrogen will play a transformative role in decarbonizing sectors ranging from transportation to heavy industry. Yet, this transformation will only be possible if the materials that enable water splitting become cheaper, more efficient, and more durable.

Two-dimensional materials, with their vast structural diversity and tunable properties, are at the forefront of this transformation. They are not just the subject of scientific curiosity—they are engines of innovation that could redefine the economics and ecology of hydrogen production.

This book stands as a testament to the remarkable progress that has been made and a blueprint for what lies ahead. It is our hope that this volume will not only inform but also inspire—empowering researchers, educators, and innovators to contribute meaningfully to a cleaner and more sustainable world.

Acknowledgments

Contents

Role of 2D Materials in Electrochemical Water Splitting in the Global Energy Landscape

Shagun Kainth, Piyush Sharma, and O. P. Pandey

Abstract Electrochemical water splitting (EWS) is seen as a critical process of future energy strategies, driven by rising worldwide energy demand and the need to transition from fossil fuels to renewable sources. EWS is the process of producing hydrogen and oxygen via electrolysis of water. In this chapter, we will look at how this technology is influencing the shifting energy landscape. Integrating EWS with renewable energy systems to produce green hydrogen creates a clean, storable, and transportable energy vector with a wide range of uses in industry, transportation, and residential buildings. This chapter discusses the fundamental principles of water splitting, various EWS technologies and 2D materials of interest. It also evaluates the regulatory frameworks, scalability, and techno-economic aspects that influence the adoption of EWS technologies. This chapter discusses in depth how EWS may help decarbonize energy systems while also attaining long-term sustainability and energy security objectives via the combination of research, engineering, and policy.

Keywords Electrochemical water splitting · Green hydrogen · Electrocatalysis · Energy sustainability · Decarbonization

1 Introduction

There has been a recent surge of interest in hydrogen as a viable clean fuel and energy carrier for meeting "net-zero" emissions targets [1]. Interest in hydrogen has emerged in multiple phases in recent times, influenced by surges in oil prices, variations in supply and demand, and the implementation of stricter environmental regulations along with other significant technical and scientific advancements in the

S. Kainth
Department of Chemistry and Biochemistry, Thapar Institute of Engineering and Technology, Patiala, India

P. Sharma · O. P. Pandey (✉)
Department of Physics and Materials Science, Thapar Institute of Engineering and Technology, Patiala, Punjab, India
e-mail: oppandey@thapar.edu

O. P. Pandey et al. (eds.), *2D Materials for Electrochemical Water Splitting*, Engineering Materials, https://doi.org/10.1007/978-981-95-2003-9_1

past several years. This has sparked the hydrogen research and resulted in initiation of multiple projects focussed on the hydrogen fuel's production, storage, transportation, and beyond [2]. As more and more local governments around the globe pass laws requiring "net-zero" emission plans, clean hydrogen technology is being mass-produced as a solution to the climate crisis [3]. The present surge of hydrogen deployment is a direct result of this change. Recent deployments have accelerated dramatic movements toward a hydrogen economy, bringing the possibility of a worldwide clean energy economy one step closer. Many years have passed since the idea of a "hydrogen economy" first arose. However, the primary motivators thus far have been technologies dependent on fossil fuels [4]. Currently, steam reforming of natural gas is the most common and cost-effective method, accounting for about half of global hydrogen production [5]. However, approximately 95% of hydrogen is derived from carbonaceous sources, primarily fossil fuels. Alternative methods include water electrolysis, biomass gasification, photobiological processes, and thermochemical cycles [6].

Electrochemical water splitting (EWS) is emerging as a crucial technology for sustainable energy production and storage, particularly for hydrogen production as a clean fuel [7]. Compared to other hydrogen production technologies, EWS offers high efficiency, low emissions, and compatibility with renewable energy sources [8]. This process involves the hydrogen evolution reaction (HER) and oxygen evolution reaction (OER), with ongoing research focused on developing efficient and cost-effective electrocatalysts to replace expensive noble metals [9]. The global transition to net-zero emissions could require between 1.7 and 13 Terawatt of water electrolysis capacity by 2050, necessitating significant scaling up of the electrolyser industry [10]. Proton Exchange Membrane (PEM) and Alkaline Electrolysers (AE) driven by wind and solar energy are projected to produce hydrogen at USD 2–3/kg, with costs expected to decrease by over 40% by 2030 [11]. These two half-reactions are sluggish, which reduces the water-splitting efficiency significantly. This highlights the critical need for immediate research into and development of improved water-electrolysis catalysts.

Ongoing research aims to enhance the efficiency, durability, and affordability of electrocatalysts for overall water splitting [12]. To conduct electrochemical water splitting, the electrocatalyst must be able to support the evolution of both hydrogen and oxygen. Noble metals (Pt, Ru) and earth-abundant alternatives based on transition metals such as transition metal oxides, nitrides, sulphides, phosphides, selenides, carbides are among the most popular electrode materials [13]. Recent research has also focused on developing nanomaterials with different dimensions for advanced electrocatalysis. Zero-dimensional (0D) nanoparticles, one-dimensional (1D) nanowires/nanotubes, two-dimensional (2D) nanosheets, and three-dimensional (3D) nanostructures have been explored for their unique properties and enhanced catalytic performance [14, 15]. Two-dimensional nanomaterials have emerged as promising candidates due to their exceptional physical, chemical, and electronic properties [16, 17]. Figure 1 shows that 2D electrocatalyst with different morphologies such as sheet, plate, or flake-like forms are viable options for water

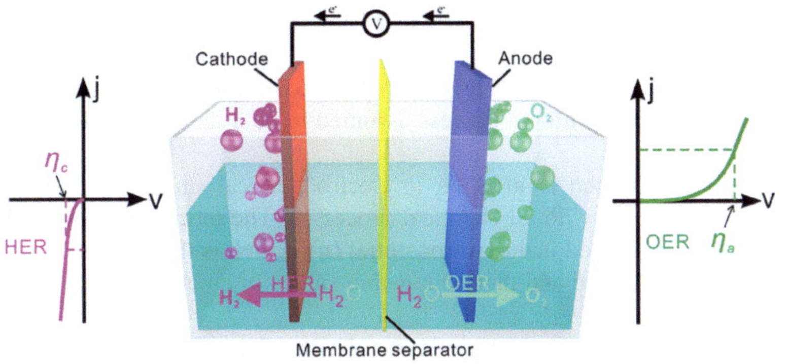

Fig. 1 A schematic representation of HER and OER of electrochemical water splitting. Reprinted (adapted) with permission from [23]. Copyright 2020 American Chemical Society

electrolysis catalysts due to their ability to shorten ion transport distances, which speeds up electron transfer and electrolyte ion diffusion [18, 19].

Additionally, the 2D electrocatalyst can reduce the stress caused during volumetric change and supress the damage of bubbles. Modulating the adsorption/desorption strength using intermediates is another important function of 2D electrocatalyst. For instance, the 2D Mo–Ni–Fe nanobelts significantly improved the OER activity in comparison to Mo–Ni nanobelts in alkaline medium due to weakened binding with OH*, O* and OOH* [20]. These efforts revealed that the shape, size, thickness, and presence of porous defects are strongly correlated with the physicochemical properties of electrocatalysts. This chapter sheds light on the fundamental of electrochemical water-splitting (both HER and OER) as well as the relevance of 2D material structure and properties in electrochemistry will be covered in the present chapter. The chapter also emphasizes on the efficient approaches to enhance electrochemical water splitting. Moreover, new materials and catalysts as well as recent efforts in overcoming durability challenges with original methodologies will be covered in relation to 2D electrode materials for electrochemical water-splitting.

2 Fundamentals of Electrochemical Water-Splitting

Although electrochemical water splitting holds great potential as a method for producing clean hydrogen, the creation of efficient and affordable electrocatalysts remains a significant obstacle. The original design for an electrolytic cell, which had a cathode, an anode, and an electrolyte, dates to 1789 [21]. The process of electrochemically splitting water involves following reaction;

$$H_2O = H_2(g) + \frac{1}{2}O_2(g) \tag{1}$$

In the above reaction the $\Delta G = +237.2$ kJ/mol and $\Delta E = 1.23V$ versus *RHE*.

By electrolyzing 1 mol of H_2O, one can obtain 1 mol of H_2 gas and 0.5 mol of O_2 gas. The thermodynamic parameters (enthalpy and entropy) for the above reaction are shown in Table 1, with the process assumed to be at 298 K and 1 atmospheric pressure [22].

To produce hydrogen from water in electrochemical studies, energy must be applied, as is the case with all chemical processes. The term used to describe this amount of energy needed is the overpotential (η), which is the difference in potential between the required and thermodynamic values of an electrochemical process (Fig. 1).

The process supplies the energy for the dissociation and gas expansion during the reaction. Such phenomena are associated with change in enthalpy (Table 1). Assuming the reaction occurred at 298 K and 1 atmospheric pressure, the equation for pressure–volume-work can be written as:

$$\begin{aligned} W &= P\Delta V = \left(101.3 \times 10^3\right)(1.5\,\text{M})\left(22.4 \times 10^{-3}\ \text{m}^3/\text{mol}\right)(298\ \text{K}/273\ \text{K}) \\ &= 3715\ \text{J} \end{aligned} \tag{2}$$

As the enthalpy can be written as $H = U + PV$, the value of ΔU will be 282.1 kJ. The shift in the internal energy occurred due to the gas expansion. This ultimately reflects the energy needed to finish electrolysis. So, the change in Gibb's free energy can be written as:

$$\Delta G = \Delta H - T\Delta S = 285.83\ \text{kJ} - 48.7\ \text{kJ} = 237.1\ \text{kJ} \tag{3}$$

In general, the EWS comprises of two types of reactions, i.e., hydrogen evolution reaction (HER) and oxygen evolution reaction (OER). The HER reaction occurs at the cathode (Eq. 4), while the OER reaction occurred at the anode (Eq. 5) [24].

$$\text{HER cathode: } 2\text{H}^+(\text{aq}) + 2\text{e}^- \rightarrow \text{H}_2(\text{g}) \tag{4}$$

$$\text{OER anode: } 2\text{H}_2\text{O (l)} \rightarrow \text{O}_2(\text{g}) + 4\text{H}^+ + 4\text{e}^- \tag{5}$$

Furthermore, water is inherently stable and possesses low conductivity. Nevertheless, its conductivity is augmented by electrolytes such as H_2SO_4 (acidic medium), KOH (basic medium), and potassium phosphate buffer (neutral medium). The electrochemical water splitting in these three media follows different reactions [9]:

Table 1 Thermodynamic parameters for electrochemical water splitting

Thermodynamic parameters	H_2O	H_2	0.5 O_2	Change (kJ)
Enthalpy (kJ)	−285.85	0	0	$\Delta H = 285.83$
Entropy (J/k)	69.91	130.68	102.57	$T\Delta S = 48.7$

In acidic medium [25],

$$\text{HER cathode: } 4\text{H}^+(\text{aq}) + 4\text{e}^- \rightarrow 2\text{H}_2(\text{g}),\ \text{E}_0 = 0.0\ \text{V} \tag{6}$$

$$\text{OER anode: } 2\text{H}_2\text{O (l)} \rightarrow \text{O}_2(\text{g}) + 4\text{H}^+ + 4\text{e}^-,\ \text{E}_0 = +1.23\ \text{V} \tag{7}$$

In basic medium,

$$\text{HER cathode: } 4\text{H}_2\text{O(aq)} + 4\text{e}^- \rightarrow 2\text{H}_2 + 4\text{OH}^-(\text{g}),\ \text{E}_0 = -0.828\ \text{V} \tag{8}$$

$$\text{OER anode: } 4\text{OH}^- \rightarrow \text{O}_2(\text{g}) + 2\text{H}_2\text{O(aq)} + 4\text{e}^-, E_0 = +0.401\ V \tag{9}$$

In neutral medium,

$$\text{HER cathode: } 4\text{H}_2\text{O(aq)} + 4\text{e}^- \rightarrow 2\text{H}_2 + 4\text{OH}^-(\text{g}),\ \text{E}_0 = +0.413\ \text{V} \tag{10}$$

$$\text{OER anode: } 2\text{H}_2\text{O(aq)} \rightarrow \text{O}_2(\text{g}) + 4\text{H}^+ + 4\text{e}^-, E_0 = +0.817\ V \tag{11}$$

EWS at extreme pH levels has advantages and downsides. The number of protons available for rapid charge and discharge in an acid increases the efficacy of the hydrogen evolution reaction (HER) [26, 27]. Nevertheless, OER could only be enabled with the help of oxides of the costly and precious metals iridium (Ir) or ruthenium (Ru) at a low overpotential in acid medium. It is essential to recognize that only a limited number of active oxides-based catalyst demonstrate metallic-type conductivity. IrO_2 and RuO_2 exhibit metallic properties due to their high electrical conductivity of 104 cm^{-1}. This results from the lack of a band gap, attributed to the minimal electronegativity difference between oxygen and metal (below 1.3 units). The reduced over-potentials for O_2-evolution observed in both oxides result from weak oxygen binding and somewhat stronger hydroxyl (HO*) binding on the surface relative to other metal anodes. Furthermore, the binding energies of –O, –OH, and –OOH on the (1 1 0) surfaces follow universal linear equations similar to those observed on metallic surfaces such as RuO_2 and IrO_2 [28, 29]. As a result, researchers continue to face challenges while studying the OER in acidic conditions. The efficiency of the OER improves dramatically in alkaline conditions when oxides or hydroxides of cost-effective 3d transition metals are used. However, hydrogen production remains low due to the lack of rapid H^+ availability at high pH levels [30, 31].

Additionally, water splitting happens at elevated cell potentials (1.8–2.0 V) in commercial electrolysers, exceeding the theoretic value (1.23 V) by ~0.55 to 0.77 V. [32]. This additional potential is necessary to surmount the energy barrier of the reaction system, referred to as overpotential. Overpotential can occur because of the electrode and electrolyte's contact and solution resistance, separation, circuit variables, and activation obstacles caused by kinetic processes at both the cathode and anode. A potential exceeding the thermodynamic potential is necessary for the efficient execution of water splitting. Overpotential can be minimized through the utilization of highly efficient, sustainable, and eco-friendly HER and OER catalyst, while other resistances may be diminished by improvising the design of electrolyzer, thereby rendering electrochemical water splitting more environmentally sustainable and economically viable [33, 34].

2.1 Hydrogen Evolution Reaction (HER)

The process of electrochemical HER exhibits three possible fundamental steps that can be taken to reduce proton (acid medium) or H_2O molecules (alkaline medium) to H_2 on the electrode surface at a minimal applied external potential [24]. In step I, a proton and an electron undergo the Volmer reaction (Eqs. 11 and 12) to yield an adsorbed hydrogen atom (H*) on the surface of the electrode material. In acidic medium (Fig. 2a), the proton source is the hydronium cation (H_3O^+), whereas in alkaline medium (Fig. 2b), the proton source is the water molecule.

Afterwards, the production of H_2 might happen through Tafel reaction (Eq. 13) or the Heyrovsky reaction (Eqs. 14 and 15) or their combination. During Tafel reaction, two nearby H* react on the electrode surface to produce H_2. While, in a Heyrovsky process, an additional proton moves to the H* and subsequently becomes H_2 through a reaction with an additional electron. A generalized HER can be expressed as [35]:

Step I: Hydrogen adsorption (Volmer reaction)

$$\text{Acidic medium: } H_3O^+ + M + e^- \rightleftharpoons M - H^* + H_2O \tag{12}$$

$$\text{Alkaline medium: } 2H_2O + M + e^- \rightleftharpoons M - H^* + OH^- \tag{13}$$

Step II: Chemical desorption (Tafel reaction)

$$\text{Both medium: } 2M - H^* \rightleftharpoons H_2 + 2M \tag{14}$$

Or

Step III: Electrochemical desorption (Heyrovsky reaction)

$$\text{Acidic medium: } H^+ + e^- + M - H^* \rightleftharpoons H_2 + M \tag{15}$$

$$\text{Alkaline medium: } H_2O + e^- + M - H^* \rightleftharpoons H_2 + OH^- + M \tag{16}$$

The mechanism of HER could be estimated through evaluating the Tafel slope (b). The Tafel slope represents the change in potential required to increase or decrease the current density via ten-fold [35]. For a rapid Volmer reaction where the chemical desorption reaction estimates the reaction rate, the Tafel slope value of 29 mV/dec

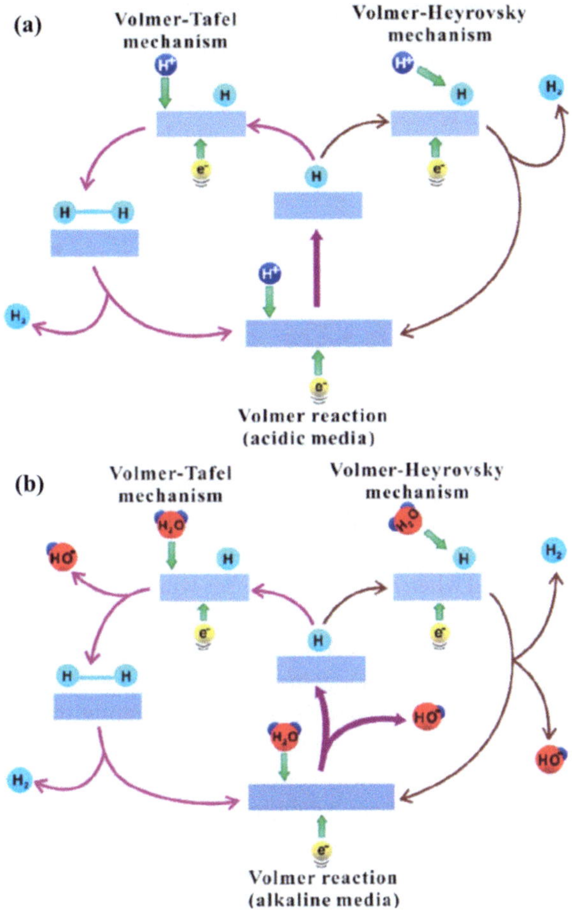

Fig. 2 Mechanism of HER in **a** an acidic and **b** alkaline medium. Reprinted (adapted) with permission from [23]. Copyright 2020 American Chemical Society

must be observed at 25 °C and represented by $b = \frac{2.3RT}{2F} = 0.029\ V/dec$. When there is rapid discharge reaction and the Heyrovsky reaction determines the rate. Whereas the Tafel slope must be 39 mV/dec and can be written as $b = \frac{4.6RT}{3F} = 0.039\ V/dec$ at 25 °C. If the discharge reaction is sluggish, Tafel slope must be 116 mV/dec and can be represented as $b = \frac{4.6RT}{F} = 0.116\ V/dec$ at 25 °C.

Moreover, Sabatier principle states that effective catalyst-reactive intermediate interactions are essential [36]. In a weak interaction, the reaction is slowed down because not enough intermediates attach to the catalyst surface. Overly strong interactions between intermediates prevent reaction products from dissociating and blocks the active site, leading to ineffective activity. By measuring ΔG_{H*} for the HER reaction pathways, one can assess the H* adsorption and H_2 desorption from a physical chemistry perspective. In general, the value ΔG_{H*} must be zero when the maximum value of HER is j_0. Figure 3 shows that Parsons linked the j_0 value to the ΔG_{H*} produced from quantum chemistry using a "volcano-type" plot [37].

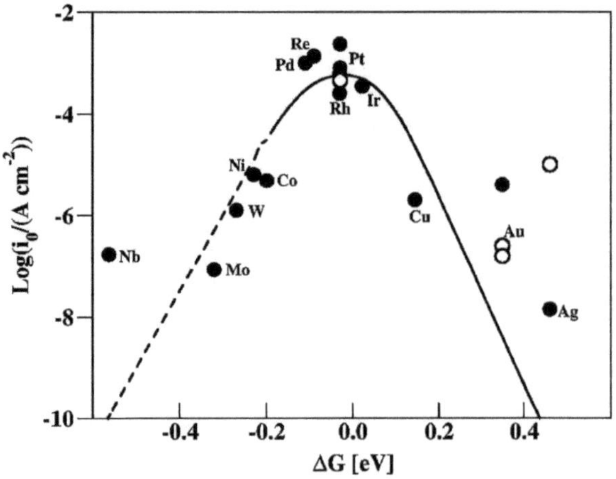

Fig. 3 A plot between j_0 and ΔG_{H^*} according to Langmuir adsorption model. Reprinted (adapted) with permission from [37]. Copyright 2010 American Chemical Society

The link between the observed j_0 value and the ΔG_{H^*} estimated using DFT. This led to identification of the "volcano-type" trend [37–39]. The highest point of the volcano is attained when $\Delta G_{H^*} = 0$. In cases where $\Delta G_{H^*} > 0$, H* adsorption is weak, resulting in an exponential rise of j_0 as ΔG_{H^*} decreases. On the other hand, in cases where $\Delta G_{H^*} < 0$, H* adsorption is comparatively robust, causing j_0 to fall exponentially as ΔG_{H^*} decreases (Fig. 3). Drawing inspiration from Parsons's research, Trasatti presented an alternative "volcano" curve by correlating the log j_0 values of metals with the measured bonding strength between metal and hydrogen [40]. Research by Nørskov et al. [38] shows that as computational science and theoretical calculations advance, DFT calculations are becoming more important for producing ΔG_{H^*} values, which in turn help with building volcanic curves. The ΔG_{H^*} values for different catalysts have been accurately calculated using DFT calculations [41–46].

2.2 Oxygen Evolution Reaction (OER)

The OER can occur in both alkaline as well as acidic mediums. The process of OER is illustrated in Fig. 4 [47]. The initial step involves an electrolyte hydroxyl anion (OH^-) being adsorbed onto the active site (M) via one e^- oxidation, resulting in M-*OH. Here, * stands for O-intermediate adsorption onto M. It is M-*O that is produced via removal of H^+ and e^- from M-*OH. There are three different mechanisms for OER, i.e., adsorbate evolution mechanism (AEM) (Fig. 4a), lattice oxygen mechanism (LOM) (Fig. 4b) and oxide path mechanism (OPM) (Fig. 4c) [48, 49].

The AEM process starts with H_2O adsorption onto catalyst surface, which sets in motion a chain reaction that culminates in the production of molecular oxygen via operations such as charge-proton transfer, chemical bond cleavage/formation, and the adsorption/desorption of an intermediate surface oxygen species [51]. One of the most well-known routes comprises four coordinated proton-electron transport

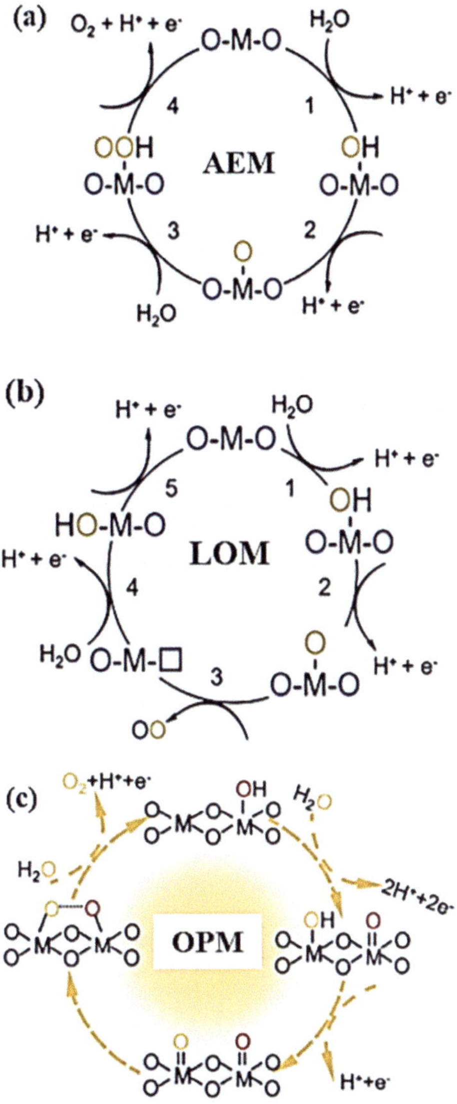

Fig. 4 **a** AEM, **b** LOM and **c** OPM for OER in acidic medium. Reprinted (adapted) with permission from [50]. Copyright 2024 Elsevier

pathways in acidic medium [29], as seen in Fig. 2a and Eqs. (16)–(19).

$$2H_2O\ (l) + * \rightleftharpoons H^+ + HO^* + e^- \tag{17}$$

$$HO^* \rightleftharpoons H^+ + O^* + e^- \tag{18}$$

$$O^* + 2H_2O\ (l) \rightleftharpoons H^+ + HOO^* + e^- \tag{19}$$

$$HOO^* \rightleftharpoons H^+ + O_2(g) + e^- + * \tag{20}$$

$$2H_2O\ (l) \rightleftharpoons 4H^+ + O_2(g) + 4e^- \tag{21}$$

Figure 3a shows the process of proton dissociation and proton removal from HO* and O*, which results from a water molecule adsorbed on the active site. The next step involves the sequential synthesis of HOO*, desorbed O_2, and a released active site after another water molecule performs a nucleophilic assault on oxygen species (O*) [52, 53]. Equation (20) illustrate the total OER in acid medium. In a state of equilibrium, the computed Gibbs free energy (ΔG) of the total process is 4.92 eV, while the theoretic potential of OER is 1.23 V. Each H^+–e^- transfer stage, ΔG can be expressed on the basis of two deposited oxygen intermediates' Gibbs free chemisorption energy. Even though ΔG of the entire process is defined as the summation of ΔG for each individual step, ΔG of each elementary step can nevertheless vary. In thermodynamics, the overpotential is also connected to the ΔG of each individual step. The step where ΔG changes the most, commonly referred to as the potential-determining step [47, 54]. This step is the last one that can happen naturally as the potential increases (U). Equation (21) can be used to determine the theoretic overpotential for OER (η_{OER}) under typical conditions [55].

$$\eta_{OER} = \frac{\Delta G_{max}}{e} - 1.23\ V \tag{22}$$

The OER process may also occur in alkaline medium by following the reaction pathway as presented below:

$$HO^- + * \rightleftharpoons HO^* + e^- \tag{23}$$

$$HO^* + HO^- \rightleftharpoons O^* + H_2O + e^- \tag{24}$$

$$O^* + HO^- \rightleftharpoons HOO^* + e^- \tag{25}$$

$$HOO^* + HO^- \rightleftharpoons H_2O + O_2(g) + e^- + * \tag{26}$$

Many researchers propose alternative reaction pathways, arguing that the mentioned pathways are unnecessary for the OER process [56]. The origin of the adsorbed O species (O*) and the product O_2 is a topic of heated debate among them. Equation (26) can be used to generate O*, while the direct coupling of two O* species (Eq. 27) may produce O_2. It is important to note that H_2O_2 may be generated as a by-product through Eq. (28) in a less vigorous oxidation environment.

$$HO^* + HO^* \rightleftharpoons O^* + H_2O + * \tag{27}$$

$$O^* + O^* \rightleftharpoons O_2(g) + 2* \tag{28}$$

$$HO^* + HO^* \rightleftharpoons H_2O_2 + 2* \tag{29}$$

In AEM, the OER takes place at a single active site and is strongly linked to the O-intermediates adsorption energies [29]. An efficient catalyst must have all the adsorption reactions having ΔG of the same magnitude at $U = 0$ (where U is the external bias) [47, 57]. Since adsorption energies of O-intermediates involved during AEM are linearly associated, the attachment of *OH and *OOH is closely connected with a difference of approximately 3.2 eV (Fig. 4c) [47, 53, 58, 59]. To improve the OER kinetics, it is crucial to disrupt this scaling relationship, which puts the bottom boundary of the OER at $\eta = 370$ mV. Recent research aimed on controlling the structure of catalysts to improve ΔG_{*O}, ΔG_{*OH}, and ΔG_{*OOH}, so that they bind O-intermediates with an optimal strength [57].

Adsorption energy scaling among HO*/HOO* limits OER activity in mechanism I. The lattice oxygen evolution mechanism (LOM) is often known as mechanism II [60]. As the oxide catalyst's lattice oxygen can take part in O–O coupling and OER without being constrained by the AEM pathway's adsorption energy scaling relationship. In this way, the LOM may be able to overcome AEM's limitations. Consequently, LOM-based catalysts may provide significantly enhanced OER activity [60]. Reportedly, the robust metal–oxygen covalency in metal oxides is crucial for LOM. This knowledge provides the groundwork for customizing the electrical structure to regulate OER mechanism of LOM [61]. The O lattice of the metal oxide catalyst contributes to oxygen evolution, and deprotonation of *OH determines the rate. Thus, one thermodynamic strategy to enhance OER activity is to adjust the metal–oxygen covalency [61]. In LOM, non-concentrated proton-electron transfer steps are often used when the oxidation of lattice oxygen (OH^- adsorbed on oxygen sites) commences on the metal oxides surface exhibiting a greater metal–oxygen covalent nature. These steps include $O \rightarrow M^{(n-1)+} - OO \rightarrow O - M^{n+} \rightarrow V_O$ (oxygen vacancy) $+ O_2 + e^-$. Oxides with a stronger covalent bond demonstrated pH dependent OER w.r.t. reversible hydrogen electrode (RHE). Thus, a smaller Tafel slope may indicate the mechanism associated with the OER via the LOM pathway. The LOM pathway (mechanism II) has a smaller Tafel slope of around 2RT/3F (R is ideal gas constant, T is temperature, and F is Faraday's constant), while AEM pathway (mechanism I) exhibit greater Tafel slope of either $\approx$ RT/3F or 2RT/F. For example, a lower value of Tafel slope (~43 mV/dec) was observed for $LaNiO_3$, which happened because of the lattice oxygen contributing in the OER [62]. Thermodynamic calculation revealed that the lattice oxygen involves during OER has negative impact on the stability of the catalyst. Rong and colleagues present a DFT-based reaction pathway that includes lattice oxygen through the reversible production of surface oxygen vacancies (VO) [63]. Subsequently, Yoo et al. provided a more detailed description of this outcome, which is given in Eq. (22)–(25) [64].

$$OH^* \rightleftharpoons \left(V_0 + OO^*\right) + H^+ + e^- \tag{30}$$

$$2H_2O\ (l) + \left(V_O + OO^*\right) \rightleftharpoons O_2(g) + \left(V_O + OH^*\right) + H^+ + e^- \tag{31}$$

$$\left(V_O + OH^*\right) + H_2O\ (l) \rightleftharpoons \left(HO - site^* + OO^*\right) + H^+ + e^- \tag{32}$$

$$\left(HO - site^* + OO^*\right) \rightleftharpoons OH^* + H^+ + e^- \tag{33}$$

Furthermore, the scaling relationship results in insufficient catalytic activity in AEM, and the unstable crystal structure causes poor stability in LOM. Achieving equilibrium in activity and stability proves to be challenging. The identification of the oxide path mechanism (OPM) facilitates the advancement of superior oxygen evolution reaction (OER) electrocatalysts [65]. The OPM pathway facilitates immediate O–O radical coupling without generating oxygen defects or supplementary chemical steps (*OOH). Only *O and *OH serve as intermediates for the oxygen evolution reaction in this mechanism. In these cases, the catalysts are often quite active and stable. Moreover, incorporation of Mo atoms into Co_3O_4 activates the OPM pathway via synergistic interactions involving O_2 vacancies and Mo sites. This led to a highly active and stable catalyst [66]. In a study by Lin et al. [67], α-MnO_2 nanofibers reinforced Ru single-atom array and met the OPM pathway requirements. Theoretical simulations show that the Ru–Ru distance in this structure is much less (2.9 Å) than in commercial RuO_2 (3.1 Å). This structure enhances direct coupling processes between nearby oxygen species while avoiding Ru and Mn termination (Fig. 5a–c). In accordance with the OPM process described, it may conclude that the OPM route is strongly tied with the geographical distribution of active sites in the material. The OPM pathway is more likely to activate when active sites are sufficiently near to work synergistically [67].

3 Current Technologies in Water Electrolysis

Hydrogen generation using water electrolysis has recently gained attention as a viable and environmentally friendly option, facilitating a green approach to stock and use renewable energy. Since it can work well with very modest conditions and is well-matched with intermittent renewable power resources, low-temperature water electrolysis is becoming a popular choice among other ways for producing hydrogen. Alkaline Water Electrolysis (AWE), Proton Exchange Membrane (PEM), and Anion Exchange Membrane (AEM) are the prevailing low-temperature electrolysis methods at present. While each of these technologies brings something unique to the table in terms of operating parameters, materials, and basic principles, they all play a vital part in the developing hydrogen economy. Design concerns for activity and long-term stability are shown in Fig. 6, along with the configurations of these three water electrolyzers [68].

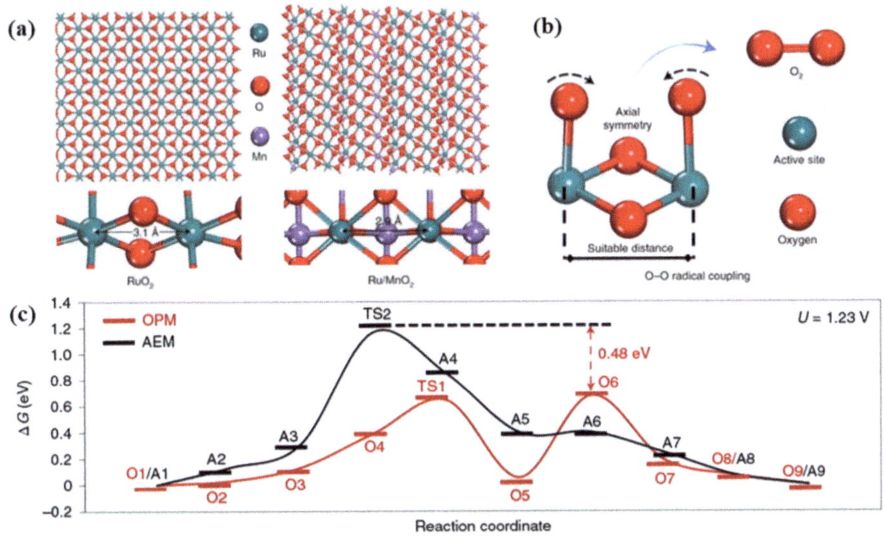

Fig. 5 **a** Structure of RuO_2 and Ru/α-MnO_2 **b** O–O radical coupling; **c** ΔG plots of AEM and OPM at U = 1.23 V versus RHE. Reprinted (adapted) with permission from [67]. Copyright 2021 Springer Nature

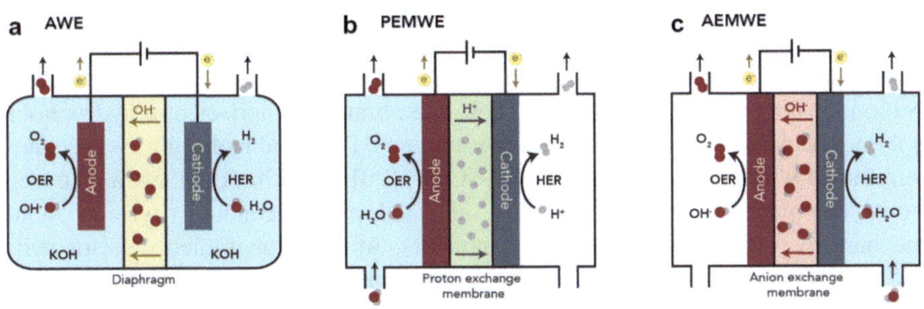

Fig. 6 A graphic representation of **a** AWE, **b** PEM, **c** AEM water electrolyzers. Reprinted (adapted) with permission from [68]. Copyright 2021 Elsevier

3.1 *Alkaline Water Electrolysis (AWE)*

This is the most advanced and well-known technology because it has been used in business for over one hundred years [68]. A standard 20–40% wt% KOH is employed as the electrolyte in alkaline electrolyzers (Fig. 6a). The working electrodes are Ni-based transition metal. In AWE, there is a diaphragm between the electrodes that lets OH^- ions pass through but restrict H_2 and O_2 gases. Usually, AWE devices work in the temperatures between 60 and 90 °C [69]. Although this renders AWE robust, it requires more time than contemporary methods for hydrogen production.

AWE is advantageous because to its low initial cost and reliance on readily available resources. Most alkaline electrolysers require a cell voltage of ~2 V and a current density of around 300 mA/cm^2 for operation [32]. The system usually lasts for more than 20 years because Ni-based catalysts are very stable in alkali medium. In liquid electrolytes, on the other hand, it is hard to keep gas bubbles from establishing on the electrode surface, which limits the current density. The generation of gas bubbles results in a reduced active electrode area, thereby increasing Ohmic loss. Therefore, AWE exhibits sluggish adaptability, possesses a low current density, and presents a risk of gas crossover, rendering it unsuitable for load-following applications with renewable energy sources.

3.2 Proton Exchange Membrane (PEM)

Proton Exchange Membranes (PEMs) are key components in various electrochemical energy conversion and storage systems, most notably Proton Exchange Membrane Fuel Cells (PEMFCs) and electrolyzers. The PEM (Fig. 6b) is a semipermeable membrane engineered to ease the conduction of H^+ ions from the anode to the cathode, while simultaneously serving as a barrier to gases like hydrogen and oxygen, so preventing their crossover and assuring the efficient functioning of the electrochemical cell [68]. This selective ion conduction plays a vital role in the efficiency, stability, and performance of fuel cells and electrolyzers. A typical PEM is composed of perfluorosulfonic acid (PFSA)-based polymers, the most well-known of which is Nafion, developed by DuPont [69]. These membranes comprises of a hydrophobic polymer backbone (like polytetrafluoroethylene, PTFE) and hydrophilic side chains terminated with sulfonic acid groups (–SO_3H) [70]. PEM has demonstrated exceptional chemical and thermal stability, elevated proton conductivity, and robust mechanical properties in humid environments. As the global energy sector shifts towards cleaner and more sustainable alternatives, PEM-based technologies have gathered noteworthy consideration because of their high-power density, low operating temperatures (typically 50–80 °C), environmentally friendly operation and faster hydrogen production [71].

The material design of PEMs must equilibrate numerous essential qualities, including proton conductivity, chemical resistance, dimensional stability, and water management. Proton conduction is predominantly enabled by hydrated sulfonic acid groups (–SO_3H) linked to the polymer backbone, which establish ionic channels for proton mobility by the Grotthuss mechanism and vehicular diffusion [72]. Nonetheless, sustaining sufficient hydration is crucial; PEMs operate most effectively at elevated humidity and temperatures generally ranging from 60 to 80 °C. Severe dehydration can significantly diminish conductivity, whereas excessive hydration may result in swelling, mechanical failure, or inundation of electrodes. To address these problems, researchers have investigated several composite membranes that include inorganic fillers (such as SiO_2, TiO_2, or zirconium phosphate), graphene oxide, or metal–organic frameworks (MOFs) to improve water retention, mechanical strength, and oxidative stability [73]. These solutions sustain proton conductivity even under low-humidity or high-temperature settings, enhancing the operational versatility of PEM systems in practical environments.

A vital element of PEM design is its durability and longevity, particularly in fuel cell applications that anticipate prolonged operation. Nafion and other perfluorinated substances exhibit exceptional chemical resistance; yet they remain vulnerable to deterioration from chemical assaults by radicals (e.g., hydroxyl and superoxide radicals) generated during oxygen reduction processes at the cathode. This deterioration results in membrane thinning, loss of mechanical integrity, and ultimate failure. To alleviate such consequences, antioxidant additions, radical scavengers, or stabilizing cross-linking agents are incorporated into the membrane matrix. Moreover, investigations have explored alternative hydrocarbon-based proton exchange membranes (PEMs), including sulfonated poly(ether ether ketone) (SPEEK), polybenzimidazole (PBI), and various aromatic copolymers, which present reduced costs and adjustable characteristics but frequently necessitate compromises regarding conductivity or chemical stability [74].

Nevertheless, extensive implementation encounters obstacles including elevated material expenses, particularly for perfluorinated membranes and platinum catalysts, alongside issues related to scalability and recyclability. Substantial initiatives are in progress to diminish or substitute costly components with more sustainable options. Research is actively conducted on non-precious metal catalysts, bio-derived polymers, and recyclable PEM materials [74]. Lifecycle analysis and techno-economic assessments facilitate the optimization of PEM systems for commercial feasibility and environmental effect mitigation. Progress in membrane chemistry, fabrication, and integration consistently enhances the efficiency, durability, and cost-effectiveness of PEM-based devices, advancing the pursuit of a sustainable energy future. As research advances and economies of scale are achieved, PEM technologies are set to assume a fundamental role in global initiatives aimed at carbon neutrality, energy independence, and environmental stewardship.

3.3 Anion Exchange Membrane (AEM)

This is a developing technology that aims to integrate the benefits of both AWE and PEM systems. AEM electrolyzers function with a solid anion-conducting membrane in an alkaline milieu, enabling the utilization of non-noble metal catalysts while capitalizing on the compactness and efficiency of membrane-based devices. AEMs (Fig. 6c) are engineered to selectively facilitate the passage of anions while obstructing cations and other charged entities, functioning as essential elements in numerous electrochemical devices, including fuel cells, water electrolyzers, redox flow batteries, and electrodialysis systems. Their essential composition generally consists of a polymer matrix modified with quaternary ammonium or other cationic groups that promote the movement of anions such as hydroxide (OH^-), carbonate (CO_3^{2-}), or chloride (Cl^-) ions. Recent interest in AEMs has escalated owing to their capacity to facilitate alkaline electrochemical environments that diminish necessity on noble metal, enhance reaction kinetics for hydrogen evolution and oxygen reduction reactions, and improve cost-efficiency in clean energy technologies [74]. Unlike their counterparts, cation exchange membranes (CEMs) utilized in PEMs, AEMs facilitate operation in alkaline environments, permitting a wider array of cost-effective, abundant catalysts, including nickel, manganese, and cobalt-based materials [74].

The foundation of AEMs generally consists of durable polymers like poly(arylene ether), polysulfone, or polyethylene, which are chemically modified with cationic groups, predominantly quaternary ammonium moieties. These functional groups not only facilitate anion conductivity but also dictate the membrane's chemical stability and ion transport efficacy. The incorporation of cationic groups into the polymer matrix can be accomplished through direct polymerization of functionalized monomers or post-functionalization methods, both presenting advantages and disadvantages for synthetic control and membrane efficacy. A vital design factor in AEMs is attaining a compromise between elevated ionic conductivity and mechanical, chemical, and dimensional stability. A high ion exchange capacity (IEC) frequently correlates with enhanced conductivity; nevertheless, it may also lead to excessive water absorption, swelling, and mechanical degradation, eventually diminishing membrane durability [75].

A significant hurdle in the development of durable AEMs is their chemical instability in alkaline environments. In contrast to acidic PEM, AEM are subjected to hydroxide ions that might engage in nucleophilic substitution and Hofmann elimination processes, thereby compromising the cationic sites and polymer backbones. This results in the deterioration of functional groups and a gradual decline in conductivity over time. Benzyltrimethylammonium, a frequently utilized functional group in AEMs, is susceptible to degradation via SN2 and E2 reactions, especially under increased temperatures and high pH conditions [76]. Recent improvements have investigated other cationic groups, including imidazolium, phosphonium, sulfonium, and guanidinium, alongside sterically hindered or cyclic quaternary ammonium structures that provide enhanced alkaline stability [77]. Similarly, the advancement of stable polymer backbones, i.e., poly(phenylene oxide), poly(arylene ether sulfone), and poly(benzimidazole) has reduced degradation concerns, enabling extended operational durations and enhanced thermal stability in practical applications.

A vital performance indicator of AEMs is their ionic conductivity, which is directly affected by membrane shape, ion exchange capacity, hydration level, and the characteristics of the mobile anions. Hydroxide conductivity generally exceeds that of other anions like bicarbonate or acetate, rendering it the favored ion for fuel cell and electrolyzer applications. Conductivities between 10 and 100 mS/cm are deemed effective for optimal device performance, with newer materials achieving or exceeding this standard by incorporating well-connected hydrophilic domains that enhance anion transport efficiency [78].

Considerations of sustainability have becoming progressively significant in AEM development. Initiatives are underway to integrate recyclable polymers, minimize fluorinated substances, and employ environmentally friendly synthetic methods. The advancement of circular materials, especially bio-derived polymers or renewable feedstocks, is accelerating in AEM research. Moreover, the life-cycle performance, recyclability, and cost-effectiveness of AEMs are incorporated into techno-economic evaluations and life-cycle assessments to assess their commercial feasibility and environmental consequences.

Each of these methods has proved a key roleplay in the present and future landscape of hydrogen generation. AWE remains a dependable and low-cost solution for centralized, steady-state hydrogen generation, especially in industrial applications. PEM electrolysis is better suited for applications requiring high responsiveness and integration with intermittent renewable sources due to its superior dynamic performance and system compactness. Meanwhile, AEM electrolysis, though still in the early stages, holds potential as a affordable, efficient substitute that could transform the economics of green H_2 if current research challenges are overcome.

Alongside its technological dimensions, the environmental and economic ramifications of these technologies are as significant. The carbon footprint associated with H_2 production via electrolysis is affected by the type of electrical energy utilized. When utilizing renewable energy, these three methods provide a means to achieve genuinely green hydrogen, hence facilitating global decarbonization objectives. Nonetheless, the expense of hydrogen generated through electrolysis continues to exceed that resulting from fossil fuels, particularly steam methane reforming (SMR), highlighting the necessity for additional innovation and expansion. Governmental backing, public–private collaborations, and enhancements in electrolyzer efficiency, longevity, and cost-effectiveness are essential for bridging this gap.

Furthermore, progress in catalyst development, membrane technology, and system integration is anticipated to yield substantial enhancements in all three technologies. In PEM systems, initiatives are being implemented to diminish reliance on precious metals by innovating catalyst supports and alloy formulations. AWE is leveraging advancements in electrode design and electrolyte management to improve performance and longevity. Research in AEM focuses on creating high-performance membranes and affordable catalysts that work efficiently and stably in alkaline environments. Figure 7 shows the important parameters to be considered for designing an electrolyzer. Table 2 presents a detailed evaluation of the key features of alkaline, PEM, and AEM electrolyzers.

Ultimately, the future of low-temperature water electrolysis technologies will likely entail a diversified strategy wherein each type—PEM, AWE, and AEM—occupies its specific role within the hydrogen value chain. For instance, PEM may prevail in mobility and distributed energy systems, AWE in large-scale industrial hydrogen production, and AEM in next-generation, cost-sensitive applications as the technology advances. Collectively, these technologies constitute the foundation of an advancing green hydrogen infrastructure that facilitates the worldwide shift towards a low-carbon economy. The increasing demand for sustainable hydrogen in sectors like transportation, energy storage, chemical production, and steel manufacturing necessitates ongoing investment in and advancement of electrolysis technologies to meet climate objectives and promote a resilient, decarbonized energy future [75].

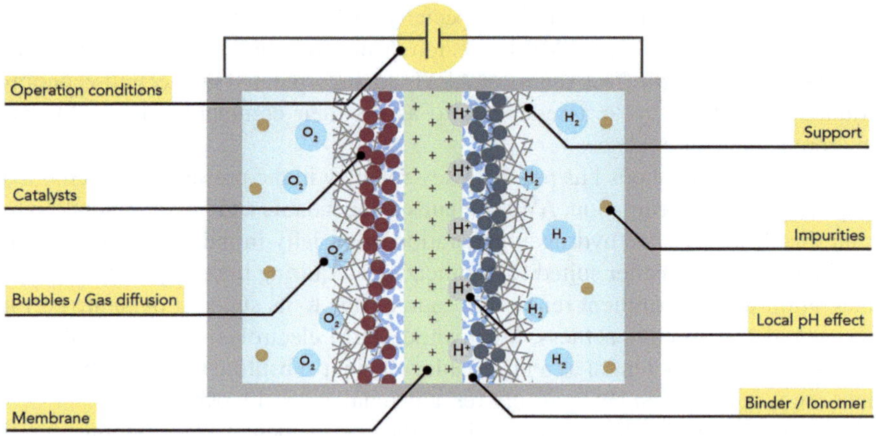

Fig. 7 Important aspects to consider while developing electrolyzers with high activity and long-term stability. Reprinted (adapted) with permission from [68]. Copyright 2021 Elsevier

Table 2 Various features of Alkaline, PEM and AEM electrolyzers

Characteristics	Alkaline water electrolyzer	PEM water electrolyzer	AEM water electrolyzer
Separator	Diaphragm (Zirfon Perl 500 μm)	Nafion 117	AEM (20–100 μm)
Catalysts	Anode: Ni, Ni–Co alloys Cathode: Ni, Ni–Mo alloys	Anode: Ru/IrO_2 Cathode: Pt, Pt–Pd	Anode: Ni, Fe, Co oxides Cathode: Ni and Ni alloys
Electrolyte	KOH (20–40 wt%)	High-purity deionized water (18 MΩcm)	Alkaline solution or deionized water
Ion transport	OH^-	H^+	OH^-
Temperature (°C)	50–80	50–80	50–70
Pressure (bar)	1–30	30–76	1–30
Current density (A cm^{-2})	0.2–0.6	0.6–2.0	0.2–0.4
Cell voltage (V)	1.8–2.4	1.8–2.2	1.8–2.2
Efficiency (%)	60–70	70–80	
Gas purity (%)	>99.5	>99.9999	>99.99
Technology status	Mature technology	Mature for small scale	Under development
Advantages	Non-noble metal catalysts, great durability, affordable	Compact design, fast response/start-up, high-purity production	Combined advantages of alkaline and PEM water electrolysis
Disadvantages	Slow dynamics, corrosive electrolyte, further purification stage	Noble metal catalysts, poor durability, high capital cost	Low OH^- conductivity in polymeric membranes

4 2D Materials of Interest

Recently, the advent of 2D materials has created great interest in energy conversion owing to their distinctive structural, electrical, and physicochemical properties. The ultrathin characteristics of these materials, sometimes minimized to monolayer or few-layer structures, provide a high surface area and many active sites, which are crucial for catalysis. Moreover, 2D materials have adjustable electrical configurations, varied defect chemistries, and the potential for heteroatom doping and phase engineering, enabling precise modulation of their catalytic efficacy [38].

Although graphene was the first extensively studied 2D material, its inherent catalytic activity is constrained by its inert basal planes. This constraint has prompted the investigation of several 2D material families with inherently enhanced electrocatalytic characteristics. This encompasses transition metal dichalcogenides (TMDs) are recognized for their edge-active hydrogen evolution reaction (HER) characteristics; transition metal trichalcogenides (TMTs) featuring distinctive quasi-one-dimensional chain architectures; MXenes, an extensive category of 2D transition metal carbides/nitrides exhibiting exceptional metallic conductivity; MBenes, 2D transition metal borides; metal–organic frameworks (MOFs) and their two-dimensional derivatives that merge porosity along with adjustable redox activity; graphdiyne, a carbon-dense two-dimensional material characterized by intrinsic sp–sp^2 hybridization; and, more recently, high entropy two-dimensional materials, which leverage entropic stabilization to develop multicomponent systems with synergistic catalytic properties [76].

Several important factors determine how well these 2D materials perform in water splitting reactions. These include the energy they adsorb toward important intermediates (e.g., H*, OH*), the efficiency with which electrons are transferred, the stability of their structure under operating conditions, and the thermodynamic overpotential needed to drive HER or OER. Their electrocatalytic efficacy has been enhanced to the point that it can compete with or even outperform that of conventional noble-metal catalysts by the modification of their chemical composition, shape, and surface functionalities [76]. The integration of 2D materials into water electrolysis technologies also aligns with the SDG 7 and SDG 13, targeting affordable green energy and climate action. Moreover, green hydrogen generated through electrolysis can serve as a key vector in the sectoral decarbonization of transportation, power generation, and industrial processes such as ammonia production and steelmaking. As countries develop hydrogen strategies to achieve net-zero goals, the search for cost-effective, high-performance electrocatalysts based on 2D materials becomes a central scientific and technological priority.

Despite remarkable advances, several scientific and engineering challenges remain. These include the need for mechanistic understanding of catalytic pathways at the atomic level, achieving long-term operational stability in corrosive electrolytes, and developing universally accepted metrics to compare catalyst performance across different systems. The role of DFT and machine learning in guiding rational design of 2D materials is also becoming increasingly prominent, enabling the rapid screening of materials with desirable catalytic descriptors such as low ΔG_{H*} or favourable electronic band structures [38].

In this part, we have described the function of several 2D material families in electrochemical water splitting. Each section delves into the structural characteristics, electrocatalytic mechanisms, performance metrics, and integration strategies of a particular material class, followed by a discussion on its relevance to the global energy transition. By evaluating these materials from both a scientific and energy systems perspective, this work strives to offer a knowledge of how emerging 2D materials are reshaping the landscape of sustainable hydrogen production.

4.1 Transition Metal Dichalcogenides (TMDs)

TMDs are characterized by generic formula MX_2 (where M is a transition metal and X is a chalcogen), have arose as a highly promising category of 2D materials for electrocatalytic water splitting. Their inherent stratified architecture, diverse edge chemistry, and adjustable electrical characteristics provide them appealing options for both the HER and OER in many electrolytic settings [77]. TMDs comprise hexagonal layers having a metal atom interposed between two chalcogen atoms, resulting in a sandwich-like X–M–X arrangement. The layers are adhered through weak van der Waals interactions, facilitating facile exfoliation into few-layer or monolayer nanosheets. TMDs may manifest in several polymorphic phases, particularly the 2H (trigonal prismatic) and 1T (octahedral) phases. The 2H phase is semiconducting, while the 1T phase exhibit metal like characteristics, which are preferable for catalytic activity owing to superior electrical conductivity.

Number of techniques are utilized to improve the catalytic response of transition metal dichalcogenides (TMDs), including phase engineering, doping, defect and strain engineering [77]. Although TMDs are conventionally linked to HER, certain TMDs (e.g., $NiSe_2$, $CoSe_2$) have exhibited significant activity for OER, particularly when doped or hybridized. Metallic transition metal dichalcogenides provide avenues for stabilizing these intermediates, particularly when influenced by defect chemistry or heterostructuring with conductive substrates. To address the inherent limitations of TMDs, including limited conductivity in the 2H phase and inadequate stability, various hybrid materials have been developed. Hybrid topologies augment the active surface area, enhance electrolyte accessibility, and promote charge separation, hence improving overall water splitting efficacy.

The practical implementation of TMDs in water electrolysis depends on various parameters, including stability, scalability, and integration with PEM and alkaline electrolyzers. Furthermore, attaining exact control over flaws, edges, and doping at the atomic scale continues to pose a synthetic challenge. The stability during prolonged operation and across varying pH conditions must be considered. A comprehensive knowledge of multi-electron transfer processes and active site dynamics under biased settings necessitates more in situ/operando investigations. Novel methodologies, such as machine learning-assisted materials discovery, in situ spectroscopy (e.g., XAS, Raman, FTIR), and atomic-scale characterisation (e.g., aberration-corrected TEM), are utilized to enhance the comprehension of TMD catalytic behavior.

4.2 Transition Metal Trichalcogenides (TMTs)

A type of low-dimensional materials in electrocatalysis that has received little attention but has great promise are transition metal trichalcogenides (TMTs), which have the generic formula MX_3 (where M = Ti, Zr, Nb, etc., and X = S, Se, Te). Anisotropic charge transport and unique catalytic characteristics are produced by TMTs, which differ from the well-established layered structures of TMDs (e.g., MoS_2) by commonly exhibiting quasi-one-dimensional (1D) chain-like or ribbon-like motifs inside a layered framework. Their metallic or semimetallic conductivity, exposed edge and defect sites, and modifiable surface chemistry give rise to their emerging role in electrochemical water splitting, specifically in the hydrogen evolution reaction (HER) and, to a lesser degree, the oxygen evolution reaction (OER) [78].

The two-dimensional nanosheets and nanoribbons can be fabricated by exfoliating TMTs from their chain-based or ribbon-like crystal structures. As an example, TiS_3 is a type of TMT that has quasi-1D chains that are oriented along the b-axis and create weak van der Waals contacts between the layers. The presence of a high density of unsaturated edge atoms, which serve as catalytically active sites, allows for efficient in-plane charge transport due to this unique anisotropy [79]. In addition to improving electrocatalytic efficiency, the electronic anisotropy caused by these structural motifs allows for rapid electron migration along the conductive chains. Based on the combination of transition metal and chalcogen, TMTs can have either narrow-gap semiconductor or semimetal band structures. Sulfur vacancies, oxygen substitution or doping, heteroatom doping, and strain engineering are edge and surface modification procedures that greatly improve the catalytic activity of transition metal tetrahedra (TMTs).

Considering the materials processing aspect, TMTs can be produced using cost-effective and scalable processes such chemical vapor transport (CVT), liquid phase exfoliation (LPE), hydrothermal, and solvothermal techniques. Decentralized hydrogen systems, alkaline water electrolyzers, PEM electrolyzers, and TMTs are all potential applications for TMTs because of their stability and promising activity. Despite the potential of TMTs, they encounter various obstacles on the path to widespread use, such as a lack of investigation into relevant materials, erratic synthesis, and an inadequate grasp of the underlying mechanisms. In the future, scientists want to apply machine learning to optimize processes, find new TMTs through high-throughput computational screening, and further integrate their systems with photonic and bioinspired ones. TMTs are poised to transition from being mere theoretical interests to being essential components of sustainable hydrogen technologies, thanks to these breakthroughs.

4.3 MXenes for Electrochemical Water Splitting

MXenes, a substantial family of 2D transition metal carbides, nitrides, or carbonitrides, have swiftly emerged as adaptable and efficient materials for electrochemical applications, encompassing energy storage, sensors, and catalysis. Since their discovery in 2011 by Naguib et al. [80], MXenes have shown significant potential in water splitting owing to their distinctive amalgamation of metallic conductivity, adjustable surface chemistry, hydrophilicity, and structural adaptability. MXenes are generally produced by the selective etching of the "A" element from MAX phases ($M_{n+1}AX_n$), yielding layered structures of the form $M_{n+1}X_nT_x$, where M denotes an early transition metal (Ti, Mo, Nb, etc.), X represents C and/or N, and T_x indicates surface functional groups such as –OH, –F, and –O. These functional groups enhance the hydrophilicity of MXenes and are crucial in regulating their electrical characteristics and catalytic activity.

The ΔG_{H*} of the terminal functional groups attached to MXenes determines their hydrogen evolution reaction (HER) activity. Due of its strong hydrogen adsorption, Pristine Ti_3C_2 performs less than optimally in hydrogen evolution reactions. However, by functionalizing the surface with O or OH, the ΔG_{H*} value may be significantly adjusted towards thermoneutral values (~0 eV), which improves the kinetics of HER. Essential for the Volmer, Tafel, and Heyrovsky processes in both acidic and basic electrolytes are intermediate complexes (e.g., H_{ads}, OH_{ads}) that can be formed by the active terminal groups. The improved catalytic capabilities of bimetallic MXenes are thought to be the result of a synergistic interaction between several transition metals, which increases conductivity and diversifies active sites [81]. Composite systems, including MXene/oxide hybrids, have shown bifunctionality with low overpotentials and prolonged performance in alkaline medium [82]. Pure MXenes, on the other hand, have decreased efficiency in OER due to considerable oxygen adsorption. Doping, surface termination control, hybridization, and defect engineering are some of the engineering tactics that have been used to improve MXene-based water splitting. Additionally, MXenes may be easily transformed into inks, films, and foams, making them suitable for a variety of applications such as flexible electrolyzers, electrocatalytic membranes, wearable hydrogen generators, conductive catalyst supports for PEM and alkaline cells, and electrocatalytic membranes.

4.4 MBenes: 2D Transition Metal Borides

Two-dimensional (2D) MBenes—recently emerged transition metal borides derived from MAX phases—have garnered increasing interest in the field of electrochemical water splitting due to their unique combination of high electrical conductivity, tunable surface chemistry, and intrinsic chemical stability. Structurally analogous to MXenes, MBenes are exfoliated from MAB phases (where M = transition metal, A = IIIA or IVA group element, B = boron), yet they offer a distinct advantage due to the incorporation of boron, which imparts greater chemical and thermal stability [83, 84]. Unlike MXenes, which are terminated with fluorine or hydroxyl groups post-etching, MBenes often exhibit clean, non-terminated surfaces, allowing for more straightforward surface modification strategies and preserving active sites essential for hydrogen evolution reaction (HER) and oxygen evolution reaction (OER).

The intrinsic metallicity of MBenes such as Mo_2B, Ti_2B, and Cr_2B ensures excellent charge transport, a crucial criterion for efficient electrocatalysts. Recent density functional theory (DFT) calculations have demonstrated that monolayer Mo_2B exhibits a low Gibbs free energy of hydrogen adsorption ($\Delta G\_H^* \approx 0.06$ eV), placing it near the ideal range for HER catalysis. Furthermore, Mo_2B and Ti_2B have been shown to possess favorable band structures for proton reduction and water oxidation, with high electrochemical stability under acidic and alkaline conditions [85]. The presence of boron also allows for strong M–B covalent bonding, which resists corrosion and enhances structural integrity during prolonged electrolysis.

Functionalization and heteroatom doping strategies have been employed to further enhance MBene activity. For instance, nitrogen-doped Ti_2B and Mo_2B monolayers exhibit improved electronic conductivity and optimized adsorption energies for water splitting intermediates [86]. Co-doping with transition metals such as Ni or Fe into MBenes has also been shown to modulate their d-band center, thereby increasing catalytic turnover [87]. Surface engineering approaches, including defect formation and hybridization with conductive carbon frameworks or conductive polymers, have been explored to address the inherently low surface area of some pristine MBenes.

Experimentally, MBenes remain relatively underexplored compared to MXenes, largely due to challenges in selective etching of the A element from MAB phases without damaging the boron framework. However, recent advances in etching methods using molten salts or electrochemical routes have opened new opportunities for scalable synthesis [88]. Liu and Gong [89] synthesized α-MoB_2 nanosheets with remarkable HER performance, achieving low overpotentials of 124 and 141 mV at 10 mA cm^{-2} in alkaline and acidic media, respectively. The same authors previously reported β-MoB_2 nanosheets with promising HER activity in acidic medium.

In conclusion, 2D MBenes represent a frontier class of electrocatalysts for water splitting, especially HER, owing to their excellent conductivity, stability, and tunable catalytic properties. As synthetic challenges are gradually overcome, and more experimental studies validate theoretical predictions, MBenes may become key materials in next-generation clean energy technologies.

4.5 Metal–Organic Frameworks (MOFs) and 2D MOF Derivatives

MOFs provide unparalleled structural and porosity tunability. MOFs comprised of metal ions/clusters coordinated to organic linkers. The ultrathin 2D MOF nanosheets, derived from bulk MOFs, facilitate electrochemical water splitting due to their extensive surface area, readily available active sites, and diverse catalytic tunability. Many MOFs have low conductivity when they are in their purest state. The 2D nanosheets of MOFs may be synthesized by exfoliation or bottom-up synthesis, doped with redox-active metals like Co, Ni, or Fe, and then enhanced in conductivity and catalytic activity by adding conductive linkers like tetrathiafulvalene. For example, metal–organic frameworks (MOFs) may undergo pyrolysis to become porous carbon frameworks doped with metal or metal oxide nanoparticles, resulting in carbon with a coordination ion (Co@C) or cobalt oxide (CoOx@N) doping, or bimetallic MOFs becoming dual-function HER/OER catalysts. These composites made of MOFs are perfect for integrated electrolyzers since they are stable, conductive, and durable [90].

4.6 Graphdiyne for Electrochemical Water Splitting

Graphdiyne (GDY) is an innovative carbon allotrope comprising of sp and sp^2 hybridized carbon atoms interconnected by diacetylenic links (–C≡C–). Its distinctive π-conjugated structure and inherent nanopores provide elevated charge mobility and many active sites, establishing it as an exceptional non-metallic HER/OER material. GDY exhibits an optimal bandgap (~0.5–1.2 eV), inherent cavities for reactant transport, and significant structural flexibility. GDY-metal hybrids (e.g., GDY/Ni, GDY/Mo) provide synergistic interactions, with GDY facilitating rapid electron transfer and stabilizing nanoparticles. Doped GDY (e.g., N, B, S) can precisely adjust the Fermi level and enhance intermediate adsorption for HER and OER [91].

4.7 High-Entropy 2D Materials

High-entropy materials (HEMs) consist of multiple principal elements (≥ 5) in near-equimolar ratios, enabling entropic stabilization and vast compositional diversity. Their 2D analogs—HE oxides, nitrides, or chalcogenides—exhibit synergistic catalytic behavior and excellent thermal and electrochemical stability [92]. HEMs allow tailored d-band centers, multi-site catalysis with diverse $\Delta G_{H*}/\Delta G_{OER}$ and intrinsic robustness to corrosion. For instance, high-entropy transition metal oxides have shown OER overpotentials as low as 270 mV and HER activity comparable to MoS_2, with excellent cycling stability.

2D materials represent a transformative group of electrocatalysts in electrochemical water splitting, addressing twin challenges of performance and sustainability in the global hydrogen economy. As we progress toward scalable green hydrogen systems, each material class—TMDs, TMTs, MXenes, MBenes, MOFs, GDY, and HEMs—offers distinct advantages in terms of conductivity, catalytic site density, tunability, and cost-effectiveness. A multi-pronged approach combining material innovation, system-level engineering, and policy integration could be essential to fully realize the promise of 2D materials in decarbonizing the world-wide energy system.

5 Regulatory Frameworks and Policy Alignment

In addition to technological advancements, regulatory frameworks, scalability, and techno-economic viability play a significant role in determining whether electrochemical water splitting (EWS) technologies will be widely used, especially for sustainable hydrogen production. All these things are essential for making the global energy transition a reality and turning ideas that have only been tested in labs into solutions that can be sold.

Green technology, like as renewable hydrogen, is greatly influenced by government laws and regulations. To promote the use of hydrogen as a renewable fuel, the European Union has put in place a thorough framework that includes market mechanisms, regulations, and assistance measures to boost demand and supply [93]. Similarly, the goal of the Hydrogen Energy Earthshot effort, which is the brainchild of the United States Department of Energy, is to invest in infrastructure and innovate policy-driven so that clean hydrogen can be sold at $1/kg within 10 years [94]. Included in this larger plan to attain climate neutrality is the realization that a two-carrier method utilizing renewable power and hydrogen is necessary. An effective policy mix, including government R&D investment, is necessary to stimulate environmental advances and reduce emissions; but emission pricing is not enough on its own. All these programs show how important it is for the government to play a leading role in getting us to clean energy networks faster.

To generate a long-term market for hydrogen, green hydrogen certification systems must be developed. Efforts are ongoing to develop hydrogen guidelines and Guarantees of Origin (GO) systems, with programs such as CertifHy in Europe trying to achieve the same [95]. However, unified global standards remain a long way off, which may hold down international commerce and market adoption. Setting emission limitations, defining carbon accounting boundaries, and selecting whether clean hydrogen should be generated entirely using renewable energy sources are all important issues to handle [96]. Hassan et al. [97] also point out that regulatory frameworks, infrastructural development, and technological restrictions all impede wider implementation. To address these challenges, worldwide collaboration, innovative technology, and clear rules must be implemented to reassure investors and accelerate the adoption of green hydrogen across numerous sectors. Furthermore, establishments likewise the International Electrotechnical Commission (IEC) and the International Organization for Standardization (ISO) must develop harmonized frameworks for safety standards related to hydrogen storage, high-pressure operation, and the handling of electrolytes or toxic intermediates, especially in systems that use acidic or alkaline electrolytes. These frameworks are particularly critical in urban or decentralized hydrogen production to earn the confidence of the public and investors.

Furthermore, safety standards developed by entities such as the International Electrotechnical Commission (IEC) and the International Organization for Standardization (ISO) are required for safety standards concerning hydrogen storage, high-pressure operation, and the handling of electrolytes or toxic intermediates (particularly in systems using alkaline or acidic electrolytes). These frameworks are critical for maintaining public and investor trust, especially in urban or decentralized hydrogen generating situations.

6 Scalability of EWS Technologies

Scaling EWS from lab-scale demonstrations to industrial production requires many interconnected aspects, including materials scalability, system integration, operational dependability, and cost-effectiveness. Conventional water electrolyzers comprising of alkaline electrolyzers (AEL), PEM and AEM, each having unique scaling issues and benefits. For instance, PEM electrolyzers exhibit high current densities and compact designs, but they depend on rare and expensive noble metals (such as Ir and Pt) and fluorinated membranes, making scaling costly and resource intensive. AEL systems are more mature and less costly but have poorer efficiency and restricted dynamic operating capacity. Materials innovation in catalysts, notably the use of noble-metal-free 2D materials such as transition metal dichalcogenides (TMDs), transition metal trichalcogenides (TMTs), MXenes, MBenes, metal–organic frameworks (MOFs), graphdiyne and high-entropy materials, shows promise in overcoming these scaling constraints. However, obstacles persist in obtaining uniformity, repeatability, and performance stability on an industrial scale. Maintaining phase purity, defect distribution, and appropriate electrical structure in large-area MXene films or MOF-derived electrodes remains a difficult job, since synthesis techniques often lack the throughput and control necessary for mass manufacturing [98]. Engineering integration of EWS systems with renewable sources requires dynamic response capabilities, resilient power electronics, and intelligent control algorithms to deal with intermittency. Furthermore, large-scale deployment needs advances in water management systems, since feedwater quality influences system longevity. Desalination-integrated EWS is gaining popularity in coastal areas, despite introducing new cost and energy trade-offs. Modular designs and stacking electrolyzer units have shown potential for decentralized hydrogen production [99].

7 Techno-economic Analysis

A comprehensive techno-economic analysis is necessary to assess the commercial viability of EWS systems. The main statistic is the levelized cost of hydrogen (LCOH), which considers capital expenditure (CAPEX), operating expenditure (OPEX), system efficiency, plant lifespan, and electricity price. The current LCOH from renewable EWS varies from \$4 to \$6/kg H_2, compared to \$1–\$2/kg from fossil fuel-based techniques, such as steam methane reforming (IEA 2021) [100]. To achieve cost parity, significant CAPEX reductions are required via economies of scale, material innovation, and electrolyzer efficiency improvements. CAPEX is primarily driven by the electrolyzer stack (40–60%), BoP systems (e.g., gas purification, power conversion), and installation expenses. Expensive materials like as iridium oxide catalysts and perfluorinated membranes can increase the cost of PEM systems dramatically. Earth-abundant catalysts such as NiFe LDH, MoS_2, and graphene-based composites have showed promise in substituting precious metals without sacrificing performance or stability [44].

OPEX mostly comprises energy expenditures, which account for up to 60% of the overall hydrogen cost. Thus, co-locating EWS units with low-cost renewable energy sources—particularly solar PV and wind—is crucial. Australia, the Middle East, and portions of Africa have high solar irradiation and land availability, making them suitable locations for green hydrogen hubs (IRENA 2022) [101]. Innovative technologies such as load-following operation and hybrid renewable-EWS systems may minimize idle periods and maximize energy use, lowering LCOH even more.

Furthermore, new technologies including electrified seawater electrolysis, photoelectrochemical (PEC) water splitting, and thermally assisted EWS are being investigated to enhance energy efficiency and water supply, especially in dry locations. While these technologies are still in the early phases of research, they have the potential to challenge existing cost and infrastructure paradigms [102]. To entice private investment and hasten commercialization, governments and consortia have started to provide financial mechanisms like as tax credits (e.g., the US Inflation Reduction Act), carbon credits, green hydrogen subsidies, and contract-for-difference programs. According to the Hydrogen Council, green hydrogen may achieve a \$2/kg objective as early as 2030 with coordinated worldwide investment of more than \$300 billion.

Aside from technological and budgetary considerations, societal acceptability, infrastructural compatibility, and labour preparation also impact EWS adoption. Public opposition to hydrogen storage in residential areas, a shortage of skilled workers for electrolyzer maintenance, and inadequate pipeline infrastructure to securely transport hydrogen may all delay implementations. Policymakers must promote public interaction, set regulations and standards, and invest in training programs in addition to technology assistance. Digitalization, including AI-driven diagnostics, blockchain-based hydrogen certification, and IoT-based electrolyzer monitoring, may enable transparency, predictive maintenance, and traceability, bringing EWS in line with future smart grid and energy trading systems. In conclusion, the effective integration of EWS technologies into the global energy ecosystem is dependent not only on catalytic or electrochemical improvements, but also on enabling laws, scalable engineering designs, and comprehensive techno-economic planning. A collaborative structure combining business, academia, and governments is required to move EWS from laboratory curiosity to global energy backbone.

8 Conclusion

Hydrogen is considered a primary sustainable energy carrier, driven by the global demand for clean energy and the imperative to reduce greenhouse gas emissions. Electrochemical water splitting (EWS) is a key technique for its production. This chapter covered the basics of EWS, the mechanics of HER and OER, and the 2D materials of interest to improve these processes. The electrocatalytic activity of these materials may be improved by varying synthesis protocols, surface functionalization, doping, and hybrid composites. Long-term operational stability, reproducibility, and scalable production of high-performance materials under industrial conditions remain challenges. Beyond scientific and materials characteristics, the chapter has explored regulatory, scalability, and techno-economic factors that influence EWS technology acceptance. Laboratory research must be translated into commercial implementation through coordinated policy frameworks, worldwide certification standards, and public–private collaborations. To produce hydrogen at competitive costs, EWS must reduce capital and operational expenses by merging low-cost, earth-abundant catalysts with renewable energy sources. 2D materials have capability to transform water electrolysis and accelerate the global transition to a low-carbon, hydrogen-based energy future with interdisciplinary research, catalyst design innovation, supportive policy frameworks, and market-driven deployment methods.

References

1. Nnabuife, S.G., Oko, E., Kuang, B., Bello, A., Onwualu, A.P., Oyagha, S., Whidborne, J.: The prospects of hydrogen in achieving net zero emissions by 2050: a critical review. Sustain. Chem. Clim. Action **2**, 100024 (2023). https://doi.org/10.1016/j.scca.2023.100024
2. Sazali, N.: Emerging technologies by hydrogen: a review. Int. J. Hydrogen Energy **45**, 18753–18771 (2020). https://doi.org/10.1016/j.ijhydene.2020.05.021
3. Nazir, H., Muthuswamy, N., Louis, C., Jose, S., Prakash, J., Buan, M.E.M., Flox, C., Chavan, S., Shi, X., Kauranen, P., Kallio, T., Maia, G., Tammeveski, K., Lymperopoulos, N., Carcadea, E., Veziroglu, E., Iranzo, A., Kannan, A.M.: Is the H2 economy realizable in the foreseeable future? Part III: H2 usage technologies, applications, and challenges and opportunities. Int. J. Hydrogen Energy (2020). https://doi.org/10.1016/j.ijhydene.2020.07.256
4. El-Shafie, M., Kambara, S., Hayakawa, Y.: Hydrogen production technologies overview. J. Power Energy Eng. **07**, 107–154 (2019). https://doi.org/10.4236/jpee.2019.71007
5. Balat, M.: Possible methods for hydrogen production. Energy Sources Part A Recover. Util. Environ. Eff. **31**, 39–50 (2009). https://doi.org/10.1080/15567030701468068;WEBSITE:WEBSITE:TFOPB;PAGEGROUP:STRING:PUBLICATION

6. Babaa, S., Khzouz, M.: Hydrogen production methods: a literature review. Eur. J. Appl. Sci. **12**, 188–193 (2024). https://doi.org/10.14738/AIVP.125.17623
7. Raveendran, A., Chandran, M., Dhanusuraman, R.: A comprehensive review on the electrochemical parameters and recent material development of electrochemical water splitting electrocatalysts. RSC Adv. **13**, 3843–3876 (2023). https://doi.org/10.1039/d2ra07642j
8. Shiva Kumar, S., Lim, H.: An overview of water electrolysis technologies for green hydrogen production. Energy Rep. **8**, 13793–13813 (2022). https://doi.org/10.1016/j.egyr.2022.10.127
9. Zou, X., Zhang, Y.: Noble metal-free hydrogen evolution catalysts for water splitting. Chem. Soc. Rev. **44**, 5148–5180 (2015). https://doi.org/10.1039/c4cs00448e
10. Lehner, F., Hart, D.: The importance of water electrolysis for our future energy system. In: Electrochemical Power Sources: Fundamentals, Systems and Applications: Hydrogen Production by Water Electrolysis, pp. 1–36 (2021). https://doi.org/10.1016/B978-0-12-819424-9.00008-2
11. Ozcan, H., El-Emam, R.S., Celik, S., Amini Horri, B.: Recent advances, challenges, and prospects of electrochemical water-splitting technologies for net-zero transition. Clean. Chem. Eng. **8**, 100115 (2023). https://doi.org/10.1016/j.clce.2023.100115
12. Hayat, A., Sohail, M., Ali, H., Taha, T.A., Qazi, H.I.A., Ur Rahman, N., Ajmal, Z., Kalam, A., Al-Sehemi, A.G., Wageh, S., Amin, M.A., Palamanit, A., Nawawi, W.I., Newair, E.F., Orooji, Y.: Recent advances and future perspectives of metal-based electrocatalysts for overall electrochemical water splitting. Chem. Rec. **23**, e202200149 (2023). https://doi.org/10.1002/tcr.202200149
13. Li, X., Zhao, L., Yu, J., Liu, X., Zhang, X., Liu, H., Zhou, W.: Water splitting: from electrode to green energy system. Nano-Micro Lett. **12**, 131 (2020). https://doi.org/10.1007/s40820-020-00469-3
14. Tian, W., Zhang, X., Wang, Z., Cui, L., Li, M., Xu, Y., Li, X., Wang, L., Wang, H.: Amorphization activated RhPb nanflowers for energy-saving hydrogen production by hydrazine-assisted water electrolysis. Chem. Eng. J. **440**, 135848 (2022). https://doi.org/10.1016/j.cej.2022.135848
15. Zhao, Y., Sun, M., Wen, Q., Wang, S., Han, S., Huang, L., Cheng, G., Liu, Y., Yu, L.: Homologous NiCoP@NiFeP heterojunction array achieving high-current hydrogen evolution for alkaline anion exchange membrane electrolyzers. J. Mater. Chem. A (2022). https://doi.org/10.1039/d2ta01233b
16. Xu, H., Yuan, J., He, G., Chen, H.: Current and future trends for spinel-type electrocatalysts in electrocatalytic oxygen evolution reaction. Coord. Chem. Rev. **475**, 214869 (2023). https://doi.org/10.1016/j.ccr.2022.214869
17. Qiao, C., Zhang, Y., Zhu, Y., Cao, C., Bao, X., Xu, J.: One-step synthesis of zinc-cobalt layered double hydroxide (Zn-Co-LDH) nanosheets for high-efficiency oxygen evolution reaction. J. Mater. Chem. A **3**, 6878–6883 (2015). https://doi.org/10.1039/c4ta06634k
18. Liu, D., Xu, H., Wang, C., Ye, C., Yu, R., Du, Y.: In situ etch engineering of Ru doped NiFe(OH)x/NiFe-MOF nanocomposites for boosting the oxygen evolution reaction. J. Mater. Chem. A **9**, 24670–24676 (2021). https://doi.org/10.1039/D1TA06438J
19. Xu, H., Wang, K., He, G., Chen, H.: Concurrent alloying and vacancy engineering for intensifying hydrogen spillover towards alcohol-water co-electrolysis. J. Mater. Chem. A **11**, 17609–17615 (2023). https://doi.org/10.1039/d3ta01766d
20. Luo, X., Shao, Q., Pi, Y., Huang, X.: Trimetallic molybdate nanobelts as active and stable electrocatalysts for the oxygen evolution reaction. ACS Catal. **9**, 1013–1018 (2019). https://doi.org/10.1021/acscatal.8b04521
21. Gray, H.B.: Powering the planet with solar fuel. Nat. Chem. **1**, 7 (2009). https://doi.org/10.1038/nchem.141
22. Schroeder, D.V., Pribram, J.K.: An introduction to thermal physics. Am. J. Phys. **67**, 1284–1285 (1999). https://doi.org/10.1119/1.19116
23. Zhu, J., Hu, L., Zhao, P., Lee, L.Y.S., Wong, K.Y.: Recent advances in electrocatalytic hydrogen evolution using nanoparticles. Chem. Rev. **120**, 851–918 (2020). https://doi.org/10.1021/acs.chemrev.9b00248

24. Morales-Guio, C.G., Stern, L.A., Hu, X.: Nanostructured hydrotreating catalysts for electrochemical hydrogen evolution. Pubs.Rsc.Org **43**, 6555–6569 (2014). https://pubs.rsc.org/en/content/articlehtml/2014/cs/c3cs60468c
25. Shi, Y., Zhang, B.: Recent advances in transition metal phosphide nanomaterials: synthesis and applications in hydrogen evolution reaction. Chem. Soc. Rev. **45**, 1529–1541 (2016). https://doi.org/10.1039/c5cs00434a
26. Anantharaj, S., Karthik, P.E., Subramanian, B., Kundu, S.: Pt nanoparticle anchored molecular self-assemblies of DNA: an extremely stable and efficient HER electrocatalyst with ultralow Pt content. ACS Catal. **6**, 4660–4672 (2016). https://doi.org/10.1021/acscatal.6b00965
27. Wang, C., Dekrafft, K.E., Lin, W.: Pt nanoparticles@photoactive metal-organic frameworks: efficient hydrogen evolution via synergistic photoexcitation and electron injection. J. Am. Chem. Soc. **134**, 7211–7214 (2012). https://doi.org/10.1021/ja300539p
28. Zeradjanin, A.R., Masa, J., Spanos, I., Schlögl, R.: Activity and stability of oxides during oxygen evolution reaction—from mechanistic controversies toward relevant electrocatalytic descriptors. Front. Energy Res. **8**, 613092 (2021). https://doi.org/10.3389/fenrg.2020.613092
29. Rossmeisl, J., Qu, Z.W., Zhu, H., Kroes, G.J., Nørskov, J.K.: Electrolysis of water on oxide surfaces. J. Electroanal. Chem. **607**, 83–89 (2007). https://doi.org/10.1016/j.jelechem.2006.11.008
30. Alves, V.A., da Silva, L.A., Boodts, J.F.C., Trasatti, S.: Kinetics and mechanism of oxygen evolution on IrO2-based electrodes containing Ti and Ce acidic solutions. Electrochim. Acta **39**, 1585–1589 (1994). https://doi.org/10.1016/0013-4686(94)85139-5
31. McCrory, C.C.L., Jung, S., Peters, J.C., Jaramillo, T.F.: Benchmarking heterogeneous electrocatalysts for the oxygen evolution reaction. J. Am. Chem. Soc. **135**, 16977–16987 (2013). https://doi.org/10.1021/ja407115p
32. Zeng, K., Zhang, D.: Recent progress in alkaline water electrolysis for hydrogen production and applications. Prog. Energy Combust. Sci. **36**, 307–326 (2010). https://doi.org/10.1016/j.pecs.2009.11.002
33. Xia, X., Deng, N., Cui, G., Xie, J., Shi, X., Zhao, Y., Wang, Q., Wang, W., Tang, B.: NIR light induced H2 evolution by a metal-free photocatalyst. Chem. Commun. **51**, 10899–10902 (2015). https://doi.org/10.1039/c5cc02589c
34. Peng, J., Dong, W., Wang, Z., Meng, Y., Liu, W., Song, P., Liu, Z.: Recent advances in 2D transition metal compounds for electrocatalytic full water splitting in neutral media. Mater. Today Adv. **8**, 100081 (2020). https://doi.org/10.1016/j.mtadv.2020.100081
35. Physical Electrochemistry: Fundamentals, Techniques and Applications, Choice Rev. Online 49 (2011) 49-0279-49-0279. https://doi.org/10.5860/choice.49-0279
36. Paul, S.: Hydrogenation and dehydrogenation by catalysis. Ber. Dtsch. Chem. Ges. **44**, 17 (1911)
37. Skúlason, E., Tripkovic, V., Björketun, M.E., Gudmundsdóttir, S., Karlberg, G., Rossmeisl, J., Bligaard, T., Jónsson, H., Nørskov, J.K.: Modeling the electrochemical hydrogen oxidation and evolution reactions on the basis of density functional theory calculations. J. Phys. Chem. C **114**, 18182–18197 (2010). https://doi.org/10.1021/jp1048887
38. Nørskov, J.K., Bligaard, T., Logadottir, A., Kitchin, J.R., Chen, J.G., Pandelov, S., Stimming, U.: Trends in the exchange current for hydrogen evolution. J. Electrochem. Soc. **152**, J23 (2005). https://doi.org/10.1149/1.1856988
39. Conway, B.E., Tilak, B.V.: Interfacial processes involving electrocatalytic evolution and oxidation of H2, and the role of chemisorbed H. Electrochim. Acta **47**, 3571–3594 (2002). https://doi.org/10.1016/S0013-4686(02)00329-8
40. Trasatti, S.: Work function, electronegativity, and electrochemical behaviour of metals. III. Electrolytic hydrogen evolution in acid solutions. J. Electroanal. Chem. **39**, 163–184 (1972). https://doi.org/10.1016/S0022-0728(72)80485-6
41. Gao, Q., Zhang, W., Shi, Z., Yang, L., Tang, Y.: Structural design and electronic modulation of transition-metal-carbide electrocatalysts toward efficient hydrogen evolution. Adv. Mater. **31**, 1802880 (2019). https://doi.org/10.1002/adma.201802880

42. Jaramillo, T.F., Jørgensen, K.P., Bonde, J., Nielsen, J.H., Horch, S., Chorkendorff, I.: Identification of active edge sites for electrochemical H2 evolution from MoS2 nanocatalysts, Science (80-) **317**, 100–102 (2007). https://doi.org/10.1126/science.1141483
43. Greeley, J., Nørskov, J.K., Kibler, L.A., El-Aziz, A.M., Kolb, D.M.: Hydrogen evolution over bimetallic systems: understanding the trends. ChemPhysChem **7**, 1032–1035 (2006). https://doi.org/10.1002/cphc.200500663
44. She, Z.W., Kibsgaard, J., Dickens, C.F., Chorkendorff, I., Nørskov, J.K., Jaramillo, T.F.: Combining theory and experiment in electrocatalysis: insights into materials design. Science (80-) **355** (2017). https://doi.org/10.1126/science.aad4998
45. Stamenkovic, V.R., Mun, B.S., Arenz, M., Mayrhofer, K.J.J., Lucas, C.A., Wang, G., Ross, P.N., Markovic, N.M.: Trends in electrocatalysis on extended and nanoscale Pt-bimetallic alloy surfaces. Nat. Mater. **6**, 241–247 (2007). https://doi.org/10.1038/nmat1840
46. Kang, J., Hwang, J., Han, B.: First-principles computational screening of highly active pyrites catalysts for hydrogen evolution reaction through a universal relation with a thermodynamic variable. J. Phys. Chem. C **122**, 2107–2112 (2018). https://doi.org/10.1021/acs.jpcc.7b09294
47. Dau, H., Limberg, C., Reier, T., Risch, M., Roggan, S., Strasser, P.: The mechanism of water oxidation: from electrolysis via homogeneous to biological catalysis. ChemCatChem **2**, 724–761 (2010). https://doi.org/10.1002/cctc.201000126
48. Chen, Z., Fan, Q., Zhou, J., Wang, X., Huang, M., Jiang, H., Cölfen, H.: Toward understanding the formation mechanism and OER catalytic mechanism of hydroxides by in situ and operando techniques. Angew. Chemie Int. Ed. **62**, e202309293 (2023). https://doi.org/10.1002/anie.202309293
49. Zhu, K., Zhu, X., Yang, W.: Application of in situ techniques for the characterization of NiFe-based oxygen evolution reaction (OER) electrocatalysts. Angew. Chem. Int. Ed. **58**, 1252–1265 (2019). https://doi.org/10.1002/anie.201802923
50. Liu, Z., Wang, X., Xie, G., Ge, J.: Acid oxygen evolution reaction: mechanisms, design principles, and prospects for application in membrane electrodes. Chem. Eng. J. **499**, 155901 (2024). https://doi.org/10.1016/J.CEJ.2024.155901
51. Zhang, J., Bin Yang, H., Zhou, D., Liu, B.: Adsorption energy in oxygen electrocatalysis. Chem Rev. **122**, 17028–17072 (2022). https://doi.org/10.1021/acs.chemrev.1c01003
52. Koper, M.T.M.: Theory of multiple proton-electron transfer reactions and its implications for electrocatalysis. Chem. Sci. **4**, 2710–2723 (2013). https://doi.org/10.1039/c3sc50205h
53. Koper, M.T.M.: Thermodynamic theory of multi-electron transfer reactions: implications for electrocatalysis. J. Electroanal. Chem. **660**, 254–260 (2011). https://doi.org/10.1016/j.jelechem.2010.10.004
54. Rossmeisl, J., Dimitrievski, K., Siegbahn, P., Nørskov, J.K.: Comparing electrochemical and biological water splitting. J. Phys. Chem. C **111**, 18821–18823 (2007). https://doi.org/10.1021/jp077210j
55. Man, I.C., Su, H.Y., Calle-Vallejo, F., Hansen, H.A., Martínez, J.I., Inoglu, N.G., Kitchin, J., Jaramillo, T.F., Nørskov, J.K., Rossmeisl, J.: Universality in oxygen evolution electrocatalysis on oxide surfaces. ChemCatChem **3**, 1159–1165 (2011). https://doi.org/10.1002/CCTC.201000397
56. Wu, Z.P., Lu, X.F., Zang, S.Q., Lou, X.W.: Non-noble-metal-based electrocatalysts toward the oxygen evolution reaction. Adv. Funct. Mater. **30**, 1910274 (2020). https://doi.org/10.1002/adfm.201910274
57. Medford, A.J., Vojvodic, A., Hummelshøj, J.S., Voss, J., Abild-Pedersen, F., Studt, F., Bligaard, T., Nilsson, A., Nørskov, J.K.: From the Sabatier principle to a predictive theory of transition-metal heterogeneous catalysis. J. Catal. **328**, 36–42 (2015). https://doi.org/10.1016/j.jcat.2014.12.033
58. Cui, C., He, Z., Wu, Y., Cheng, X., Wu, H., Li, Y., Cao, Y., Wong, W.Y.: High-performance polymer solar cells based on a 2D-conjugated polymer with an alkylthio side-chain. Energy Environ. Sci. **9**, 885–891 (2016). https://doi.org/10.1039/c5ee03684d
59. Iglesias, A.I., Mishra, A., Vitart, V., Bykhovskaya, Y., Höhn, R., Springelkamp, H., Cuellar-Partida, G., Gharahkhani, P., Bailey, J.N.C., Willoughby, C.E., Li, X., Yazar, S., Nag, A.,

Khawaja, A.P., Polašek, O., Siscovick, D., Mitchell, P., Tham, Y.C., Haines, J.L., Kearns, L.S., Hayward, C., Shi, Y., Van Leeuwen, E.M., Taylor, K.D., Wang, J.J., Rochtchina, E., Attia, J., Scott, R., Holliday, E.G., Baird, P.N., Xie, J., Inouye, M., Viswanathan, A., Sim, X., Bonnemaijer, P., Rotter, J.I., Martin, N.G., Zeller, T., Mills, R.A., Staffieri, S.E., Jonas, J.B., Schmidtmann, I., Boutin, T., Kang, J.H., Lucas, S.E.M., Wong, T.Y., Beutel, M.E., Wilson, J.F., Allingham, R.R., Brilliant, M.H., Budenz, D.L., Christen, W.G., Fingert, J., Friedman, D.S., Gaasterland, D., Gaasterland, T., Hauser, M.A., Kraft, P., Lee, R.K., Lichter, P.R., Liu, Y., Loomis, S.J., Moroi, S.E., Pericak-Vance, M.A., Realini, A., Richards, J.E., Schuman, J.S., Scott, W.K., Singh, K., Sit, A.J., Vollrath, D., Weinreb, R.N., Wollstein, G., Zack, D.J., Zhang, K., Donnelly, P., Barroso, I., Blackwell, J.M., Bramon, E., Brown, M.A., Casas, J.P., Corvin, A., Deloukas, P., Duncanson, A., Jankowski, J., Markus, H.S., Mathew, C.G., Palmer, C.N.A., Plomin, R., Rautanen, A., Sawcer, S.J., Trembath, R.C., Wood, N.W., Spencer, C.C.A., Band, G., Bellenguez, C., Freeman, C., Hellenthal, G., Giannoulatou, E., Pirinen, M., Pearson, R., Strange, A., Su, Z., Vukcevic, D., Langford, C., Hunt, S.E., Edkins, S., Gwilliam, R., Blackburn, H., Bumpstead, S.J., Dronov, S., Gillman, M., Gray, E., Hammond, N., Jayakumar, A., McCann, O.T., Liddle, J., Potter, S.C., Ravindrarajah, R., Ricketts, M., Waller, M., Weston, P., Widaa, S., Whittaker, P., Uitterlinden, A.G., Vithana, E.N., Foster, P.J., Hysi, P.G., Hewitt, A.W., Khor, C.C., Pasquale, L.R., Montgomery, G.W., Klaver, C.C.W., Aung, T., Pfeiffer, N., MacKey, D.A., Hammond, C.J., Cheng, C.Y., Craig, J.E., Rabinowitz, Y.S., Wiggs, J.L., Burdon, K.P., Van Duijn, C.M., MacGregor, S.: Cross-ancestry genome-wide association analysis of corneal thickness strengthens link between complex and Mendelian eye diseases. Nat. Commun. **9**, 1864 (2018). https://doi.org/10.1038/s41467-018-03646-6

60. Wang, X., Song, S., Zhang, H.: A redox interaction-engaged strategy for multicomponent nanomaterials. Chem. Soc. Rev. **49**, 736–764 (2020). https://doi.org/10.1039/c9cs00379g
61. Grimaud, A., Diaz-Morales, O., Han, B., Hong, W.T., Lee, Y.L., Giordano, L., Stoerzinger, K.A., Koper, M.T.M., Shao-Horn, Y.: Activating lattice oxygen redox reactions in metal oxides to catalyse oxygen evolution. Nat. Chem. **9**, 457–465 (2017). https://doi.org/10.1038/nchem.2695
62. Bockris, J.O.M., Otagawa, T.: Mechanism of oxygen evolution on perovskites. J. Phys. Chem. **87**, 2960–2971 (1983). https://doi.org/10.1021/j100238a048
63. Rong, X., Parolin, J., Kolpak, A.M.: A fundamental relationship between reaction mechanism and stability in metal oxide catalysts for oxygen evolution. ACS Catal. **6**, 1153–1158 (2016). https://doi.org/10.1021/acscatal.5b02432
64. Yoo, J.S., Rong, X., Liu, Y., Kolpak, A.M.: Role of lattice oxygen participation in understanding trends in the oxygen evolution reaction on perovskites. ACS Catal. **8**, 4628–4636 (2018). https://doi.org/10.1021/acscatal.8b00612
65. Rong, C., Dastafkan, K., Wang, Y., Zhao, C.: Breaking the activity and stability bottlenecks of electrocatalysts for oxygen evolution reactions in acids. Adv. Mater. **35**, 2211884 (2023). https://doi.org/10.1002/adma.202211884
66. Sun, L., Feng, M., Peng, Y., Zhao, X., Shao, Y., Yue, X., Huang, S.: Constructing oxygen vacancies by doping Mo into spinel Co3O4 to trigger a fast oxide path mechanism for acidic oxygen evolution reaction. J. Mater. Chem. A **12**, 8796–8804 (2024). https://doi.org/10.1039/D4TA00655K
67. Lin, C., Li, J.L., Li, X., Yang, S., Luo, W., Zhang, Y., Kim, S.H., Kim, D.H., Shinde, S.S., Li, Y.F., Liu, Z.P., Jiang, Z., Lee, J.H.: In-situ reconstructed Ru atom array on α-MnO2 with enhanced performance for acidic water oxidation. Nat. Catal. **4**, 1012–1023 (2021). https://doi.org/10.1038/s41929-021-00703-0
68. Chen, F.Y., Wu, Z.Y., Adler, Z., Wang, H.: Stability challenges of electrocatalytic oxygen evolution reaction: from mechanistic understanding to reactor design. Joule **5**, 1704–1731 (2021). https://doi.org/10.1016/J.JOULE.2021.05.005/ASSET/1C18431D-A813-472F-AB2F-F9A0980B3812/MAIN.ASSETS/GR5.JPG
69. Rajendran, R.G.: Polymer electrolyte membrane technology for fuel cells. MRS Bull. **30**, 587–590 (2005). https://doi.org/10.1557/MRS2005.165/METRICS

70. Mauritz, K.A., Moore, R.B.: State of understanding of Nafion. Chem. Rev. **104**, 4535–4585 (2004). https://doi.org/10.1021/cr0207123
71. Xiao, F., Wang, Y.C., Wu, Z.P., Chen, G., Yang, F., Zhu, S., Siddharth, K., Kong, Z., Lu, A., Li, J.C., Zhong, C.J., Zhou, Z.Y., Shao, M.: Recent advances in electrocatalysts for proton exchange membrane fuel cells and alkaline membrane fuel cells. Adv. Mater. **33**, 2006292 (2021). https://doi.org/10.1002/adma.202006292
72. Ludueña, G.A., Kühne, T.D., Sebastiani, D.: Mixed Grotthuss and vehicle transport mechanism in proton conducting polymers from Ab initio molecular dynamics simulations. Chem. Mater. **23**, 1424–1429 (2011). https://doi.org/10.1021/CM102674U/ASSET/IMAGES/LARGE/CM-2010-02674U_0005.JPEG
73. Escorihuela, J., Narducci, R., Compañ, V., Costantino, F.: Proton conductivity of composite polyelectrolyte membranes with metal-organic frameworks for fuel cell applications. Adv. Mater. Interfaces **6** (2019). https://doi.org/10.1002/admi.201801146
74. Kraytsberg, A., Ein-Eli, Y.: Review of advanced materials for proton exchange membrane fuel cells. Energy Fuels **28**, 7303–7330 (2014). https://doi.org/10.1021/ef501977k
75. Sun, H., Xu, X., Kim, H., Jung, W.C., Zhou, W., Shao, Z.: Electrochemical water splitting: bridging the gaps between fundamental research and industrial applications. Energy Environ. Mater. **6** (2023). https://doi.org/10.1002/eem2.12441
76. Voiry, D., Shin, H.S., Loh, K.P., Chhowalla, M.: Low-dimensional catalysts for hydrogen evolution and CO2 reduction. Nat. Rev. Chem. **2** (2018). https://doi.org/10.1038/s41570-017-0105
77. Chia, X., Eng, A.Y.S., Ambrosi, A., Tan, S.M., Pumera, M.: Electrochemistry of nanostructured layered transition-metal dichalcogenides. Chem. Rev. **115**, 11941–11966 (2015). https://doi.org/10.1021/acs.chemrev.5b00287
78. Wu, X., Zhang, H., Zhang, J., Lou, X.W.: Recent advances on transition metal dichalcogenides for electrochemical energy conversion. Adv. Mater. **33** (2021). https://doi.org/10.1002/adma.202008376
79. Island, J.O., Molina-Mendoza, A.J., Barawi, M., Biele, R., Flores, E., Clamagirand, J.M., Ares, J.R., Sánchez, C., Van Der Zant, H.S.J., D'Agosta, R., Ferrer, I.J., Castellanos-Gomez, A.: Electronics and optoelectronics of quasi-1D layered transition metal trichalcogenides. 2D Mater. **4** (2017). https://doi.org/10.1088/2053-1583/aa6ca6
80. Naguib, M., Kurtoglu, M., Presser, V., Lu, J., Niu, J., Heon, M., Hultman, L., Gogotsi, Y., Barsoum, M.W.: Two-dimensional nanocrystals produced by exfoliation of Ti 3AlC 2. Adv. Mater. **23**, 4248–4253 (2011). https://doi.org/10.1002/adma.201102306
81. Luxa, J., Kupka, P., Lipilin, F., Šturala, J., Subramani, A., Lazar, P., Sofer, Z.: Hydrogen evolution reaction activity in Mo2TiC2Tx MXene derived from Mo2TiAlC2 MAX phase: insights from compositional transformations. ACS Catal. 15336–15347 (2024). https://doi.org/10.1021/ACSCATAL.4C04099/ASSET/IMAGES/LARGE/CS4C04099_0006.JPEG
82. Tang, Y., Yang, C., Xu, X., Kang, Y., Henzie, J., Que, W., Yamauchi, Y.: MXene nanoarchitectonics: defect-engineered 2D MXenes towards enhanced electrochemical water splitting. Adv. Energy Mater. **12** (2022). https://doi.org/10.1002/aenm.202103867
83. Rosenkranz, A., Zambrano, D., Przyborowski, A., Shah, R., Jastrzębska, A.M.: MAB-phases and beyond—a tribological success story? Adv. Mater. Interfaces (2022). https://doi.org/10.1002/admi.202200869
84. Hayat, A., Bashir, T., Ahmed, A.M., Ajmal, Z., Alghamdi, M.M., El-Zahhar, A.A., Sohail, M., Amin, M.A., Al-Hadeethi, Y., Ghasali, E., Raza, S., Orooji, Y.: Novel 2D MBenes-synthesis, structure, properties with excellent performance in energy conversion and storage: a review. Mater. Sci. Eng. R Reports **159**, 100796 (2024). https://doi.org/10.1016/J.MSER.2024.100796
85. Liu, X., Ge, X., Dong, Y., Fu, K., Meng, F., Si, R., Zhang, M., Xu, X.: First-principle calculations on the structure, electronic property and catalytic activity for hydrogen evolution reaction of 2D transition-metal borides. Mater. Chem. Phys. **253** (2020). https://doi.org/10.1016/j.matchemphys.2020.123334
86. Cheng, Y., Mo, J., Li, Y., Zhang, Y., Song, Y.: A systematic computational investigation of the water splitting and N2 reduction reaction performances of monolayer MBenes. Phys. Chem. Chem. Phys. **23**, 6613–6622 (2021). https://doi.org/10.1039/D0CP06405J

87. Shi, J., Yu, C., Kang, W., Xiao, X., Sun, X.: First-principles studies on transition metal doped Mo2B2 as anode material for Li-ion batteries. ChemistryOpen **13** (2024). https://doi.org/10.1002/open.202300313
88. Xiong, W., Dong, Z.: An emerging two-dimensional transition metal boride (MBene) material: recent experimental progress and prospects. Extrem. Mater. **1**, 11–26 (2025). https://doi.org/10.1016/J.EXM.2025.04.001
89. Liu, X., Gong, Y.: α-MoB2 nanosheets for hydrogen evolution in alkaline and acidic media. ACS Appl. Nano Mater. (2022). https://doi.org/10.1021/acsanm.2c00574
90. Ahmad, F., Rafiq, K., Najam, T., Hussain, E., Sohail, M., Abid, M.Z., Mahmood, A., Javed, M.S., Shah, S.S.A.: Metal-organic frameworks for electrocatalytic water-splitting: beyond the pyrolysis. Int. J. Hydrogen Energy **48**, 35075–35111 (2023). https://doi.org/10.1016/j.ijhydene.2023.05.247
91. Hui, L., Xue, Y., Huang, B., Yu, H., Zhang, C., Zhang, D., Jia, D., Zhao, Y., Li, Y., Liu, H., Li, Y.: Overall water splitting by graphdiyne-exfoliated and -sandwiched layered double-hydroxide nanosheet arrays. Nat. Commun. **9** (2018). https://doi.org/10.1038/s41467-018-07790-x
92. Zhao, Y., Wu, J., Cao, X., Li, D., P. Huang, H. Gao, Q. Gu, J. Zhang, G. Wang, H. Liu, High-entropy materials for water splitting: an atomic nanoengineering approach to sustainable hydrogen production. Adv. Mater. 2506117 (2025). https://doi.org/10.1002/ADMA.202506117
93. Graczyk, A., Brusiło, P., Graczyk, A.M.: Hydrogen as a renewable fuel of non-biological origins in the European Union—the emerging market and regulatory framework. Energies **18**, 617 (2025). https://doi.org/10.3390/EN18030617
94. Pivovar, B.S., Ruth, M.F., Myers, D.J., Dinh, H.N.: Hydrogen: targeting $1/kg in 1 decade. Electrochem. Soc. Interface **30**, 61–65 (2021). https://doi.org/10.1149/2.F15214IF
95. Velazquez Abad, A., Dodds, P.E.: Green hydrogen characterisation initiatives: definitions, standards, guarantees of origin, and challenges. Energy Policy **138** (2020). https://doi.org/10.1016/j.enpol.2020.111300
96. Gonçalves, I.M., Carvalho, T.V., Maria, G., Silva, M.: Challenges and prospects for hydrogen market in Brazil. Rev. Contemp. **5**, e7194 (2025). https://doi.org/10.56083/RCV5N1-016
97. Hassan, Q., Algburi, S., Sameen, A.Z., Salman, H.M., Jaszczur, M.: Green hydrogen: a pathway to a sustainable energy future. Int. J. Hydrogen Energy **50**, 310–333 (2024). https://doi.org/10.1016/j.ijhydene.2023.08.321
98. Anasori, B., Gogotsi, Y.: 2D Metal Carbides and Nitrides (MXenes): Structure, Properties and Applications. Springer International Publishing (2019). https://doi.org/10.1007/978-3-030-19026-2
99. Ginsberg, M., Zhang, Z., Atia, A.A., Venkatraman, M., Esposito, D.V., Fthenakis, V.M.: Integrating solar energy, desalination, and electrolysis. Sol. RRL **6** (2022). https://doi.org/10.1002/solr.202100732
100. Global Hydrogen Review 2021—Analysis—IEA, (n.d.). https://www.iea.org/reports/global-hydrogen-review-2021. Accessed 24 June 2025
101. The Geopolitics of the Energy Transformation: The Hydrogen Factor (n.d.)
102. Xu, S.W., Li, J., Zhang, N., Shen, W., Zheng, Y., Xi, P.: Recent advances in direct seawater splitting for producing hydrogen. Chem. Commun. **59**, 9792–9802 (2023). https://doi.org/10.1039/d3cc02074f

Unraveling Mechanistic Pathways in Electrochemical Water Splitting

Kassa Belay Ibrahim, Pratik Shinde, Elisa Moretti, and Alberto Vomiero

Abstract Electrochemical water splitting is a vital process for sustainable hydrogen production, yet its mechanistic pathways remain challenging to decipher due to the complexity of interfacial charge transfer and reaction intermediates. State-of-the-art electrocatalysts, primarily based on precious metals such as ruthenium, platinum, and iridium, exhibit excellent activity but suffer from high cost and scarcity, limiting their industrial scalability. Consequently, research has shifted toward alternative catalysts, including transition metal oxides, sulfides, phosphides, hydroxides, layered double hydroxides, and single-atom catalysts. These offer a promising balance of performance, cost, and abundance. Recent advancements in in-situ and operando characterization techniques, such as X-ray diffraction, X-ray absorption spectroscopy, Raman spectroscopy, and Fourier transform infrared spectroscopy, have significantly improved our understanding of catalytic dynamics under realistic reaction conditions. Complementary computational approaches, including density functional theory (DFT), provide atomic-level insights into reaction pathways and active site evolution. Despite these breakthroughs, challenges remain in capturing transient intermediate states and understanding structural transformations during long-term operation. This chapter integrates advanced experimental and computational methods to unravel the oxygen evolution reaction (OER) and hydrogen evolution reaction (HER) mechanistic pathways.

K. B. Ibrahim (✉) · P. Shinde · E. Moretti (✉) · A. Vomiero (✉)
Department of Molecular Sciences and Nanosystems, Ca' Foscari University of Venice, Venezia Mestre, Italy
e-mail: kassabelay.ibrahim@unive.it

E. Moretti
e-mail: elisa.moretti@unive.it

A. Vomiero
e-mail: alberto.vomiero@ltu.se

A. Vomiero
Division of Materials Science, Department of Engineering Sciences and Mathematics, Luleå University of Technology, Luleå, Sweden

O. P. Pandey et al. (eds.), *2D Materials for Electrochemical Water Splitting*, Engineering Materials, https://doi.org/10.1007/978-981-95-2003-9_2

Keywords Electrochemical water splitting · Hydrogen evolution reaction · Oxygen evolution reaction · In-situ and operando characterization · Density functional theory · Transition metal-based electrocatalysts

1 Introduction

1.1 Overview of Water Splitting as a Clean Energy Solution

One of the significant challenges we face, inevitable in the twenty-first century, is the intensifying energy demand, depletion of fossil fuels, and detrimental effects of the utilization of non-renewable resources on our environment. The main aim is to replace traditional coal, oil, and other fossil fuels with renewable energy, which includes clean energy and fuels, advanced energy storage, and efficient energy conversion technologies. However, this process is very difficult and full of challenges. Hydrogen is widely regarded as one of the most promising alternatives and potential energy carriers. Its advantages include earth abundance, high gravimetric energy density, and carbon-free emissions, making it a viable candidate to replace exhaustible fossil fuels. Hydrogen energy, heralded as a novel, clean, and highly efficient energy source, stands at the forefront of initiatives aimed at decarbonizing and fostering a sustainable, carbon-neutral economy. Due to this, it is termed as "fuel of the future," because of its clean, non-toxic, and emission-free fuel with an energy yield of 122 kJ g^{-1}, i.e., approximately 2.75 times higher than most hydrocarbon fuels [1]. It also offers excellent energy conversion efficiency, environmental compatibility, and produces only water as a byproduct, further reinforcing its role in sustainable energy systems. However, hydrogen does not exist in its pure form in nature. Instead, it is bound with other substances such as water, fossil fuels, biomass, natural gas, and other hydrocarbons, requiring extraction and processing as a clean energy source. Some of the various processes for producing hydrogen involve electrochemical [2, 3], thermochemical [4], and photochemical methods [4, 5]. Hydrogen can be produced from many different sources, like steam-methane reforming and *electrolysis* (splitting water with electricity). Hydrogen production currently occurs in large part via steam reforming hydrocarbons, which requires high temperatures and pressure, inevitably leading to the consumption of fossil fuels and carbon dioxide emissions. Among all, electrochemical water splitting is emerging as a pivotal clean energy solution, converting water into hydrogen and oxygen gases, thus providing a sustainable method for hydrogen production.

1.2 *Water Splitting as Future Energy Carrier (H_2 Economy)*

Green hydrogen production via water splitting is a cutting-edge technique for producing environmentally friendly H_2. Electrocatalytic water splitting offers a simple, safe, and high-purity alternative for hydrogen production, garnering significant attention in recent years. The production of hydrogen should be in the most spotless way to ensure its sustainability. During electrolysis, the complete water-splitting reaction occurs on two different electrode surfaces [6]. Under ideal conditions, the thermodynamic potential for driving the electrochemical water splitting process can be as low as 1.23 V, which corresponds to an energy input of $\Delta G = 237.1$ kJ mol^{-1} [7]. Thermodynamically, the production of hydrogen from water is an uphill process that is shown in the following formula:

$$H_2O(l) \rightarrow H_2(g) + \frac{1}{2}O_2(g): \Delta G = +237 \left(\frac{\text{kJ}}{\text{mol}}\right),$$

$$\Delta E = 1.23 \text{ V versus RHE}$$

Those reactions are kinetically sluggish and thermodynamically stable; thus, an additional voltage is required due to the built-in thermochemical barrier of this system, which is known as overpotential. Thus, the breakdown requires a high overpotential. To reduce the overpotential, a catalyst is needed that enables the dissociation of water at a faster rate. Depending on the type of external energy supply, water splitting (WS) can be catalyzed either by an external voltage (electrocatalysis) or by external light (photocatalysis). The electrocatalytic process of WS involves the decomposition of water into oxygen and hydrogen gas in response to an electric current passing through the water. Electrochemical water splitting is carried out in a cell with a cathode, an anode, and a separator. The whole reaction is made up of two half-reactions, which can be expressed in different ways depending on reaction conditions (Fig. 1).

2 Fundamentals of Water Electrolysis

2.1 *Oxygen Evolution Reaction (OER)*

The OER is a complex process involving multiple elementary reactions with intricate electron and proton transfer steps. One of the primary challenges in understanding the OER mechanism is the dynamic evolution of reactive sites on the catalyst surface during the reaction. This makes it difficult to precisely identify the reaction pathway. The generally accepted elemental steps that elucidate the overall reaction pathways for OER in alkaline and acidic media are illustrated in Fig. 2a. Taking the alkaline OER as an example, free OH^- ions in the solution first adsorb onto the catalyst's

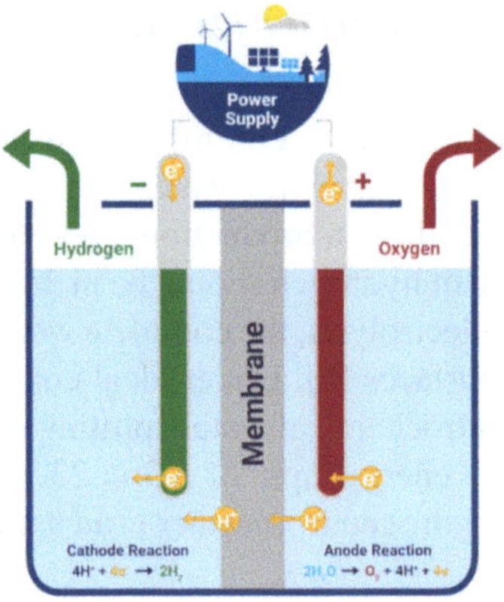

In acidic solution

Cathode : $4H^+ + 4e^- \rightarrow 2H_2$

Anode: $2H_2O \rightarrow O_2 + 4H^+ + 4e^-$

$$2H_2O \rightarrow 2O_2 + 2H_2$$

In alkaline solution

Cathode : $4H_2O + 4e^- \rightarrow 2H_2 + 4OH^-$

Anode: $4OH^- \rightarrow O_2 + 2H_2O + 4e^-$

$$2H_2O \rightarrow 2O_2 + 2H_2$$

Fig. 1 Water electrolyzers under different electrolytes

active site, forming an R–OH intermediate while transferring an electron from OH^- to the electrode. This intermediate is then oxidized to form M–O, which subsequently binds with another OH^- to generate M–OOH. Further oxidation of M–OOH leads to the release of O_2 molecules. Compared to the alkaline OER process, the acidic OER requires additional energy to promote water dissociation, making it more energy-intensive [8, 9].

Since the OER process involves the repeated adsorption and desorption of oxygen-containing intermediates, catalysts for OER are typically transition metal oxides, hydroxides, and high-valence metal salts with oxygen-functionalized groups. Additionally, OER catalysts often undergo strong oxidation during the reaction. Interestingly, some highly oxidizable materials, such as transition metal phosphides, selenides, and sulfides, have demonstrated excellent catalytic performance due to their ability to transform into active oxyhydroxide species under OER conditions [8]. A key factor governing OER activity is the bonding strength of the metal–oxygen (M–O) intermediates. If the M–O bond is too weak, the intermediate formation becomes difficult, hindering the reaction. Conversely, if the M–O bond is too strong,

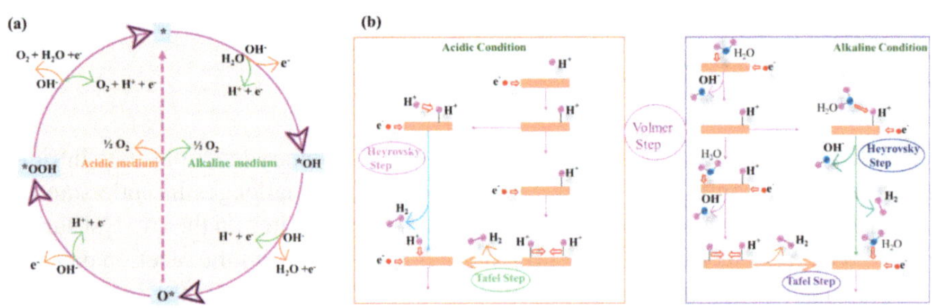

Fig. 2 Reaction mechanism under acidic and alkaline conditions. **a** Oxygen evolution reaction mechanism, wherein * represents the active sites of the catalyst. **b** Hydrogen evolution reaction

the intermediate becomes overly stable, preventing further reaction steps. An ideal OER catalyst should exhibit an optimal M–O bonding strength that facilitates both intermediate formation and desorption, ensuring an efficient catalytic cycle [10, 11].

OER in acidic electrolyte

$$H_2O + * \rightarrow HO^* + H^+ + e^-$$

$$HO^* \rightarrow O^* + H^+ + e^-$$

$$O^* + H_2O \rightarrow HOO^* + H^+ + e^-$$

$$HOO^* \rightarrow ^* + O_2 + H^+ + e^-$$

OER in basic electrolyte

$$OH + * \rightarrow HO^* + e^-$$

$$HO^* + OH^- \rightarrow O^* + H_2O + e^-$$

$$O^* + OH^- \rightarrow HOO^* + e^-$$

$$HOO^* + OH^- \rightarrow * + O_2 + H_2O + e^-$$

It occurs at the cathode, where protons (H^+) are reduced to form hydrogen gas (H_2).

2.2 *Hydrogen Evolution Reaction (HER)*

A complete electrolyzer for overall water splitting consists of a power source, an electrolyte, a cathode, and an anode. When a voltage is applied, water molecules decompose into hydrogen and oxygen. The different mechanisms of HER processes are illustrated in Fig. 2b, in which, step-by-step HER processes taking place under acidic and alkaline conditions are shown. The HER occurs at the cathode and follows the Volmer reaction, followed by either the Heyrovsky reaction or the Tafel reaction [12]. In the Volmer reaction, hydrogen adsorption takes place. In acidic electrolytes, free H^+ ions adsorb onto the catalyst's active sites and combine with electrons transferred from the electrode, forming hydrogen atoms bonded to metal sites (M–H). In alkaline electrolytes, due to the low proton concentration, H^+ is generated from the dissociation of water molecules, which requires a higher activation energy. The subsequent step in HER follows either the Heyrovsky or the Tafel reaction [13, 14]. In the Tafel reaction, two adsorbed hydrogen atoms (H_{ads}) recombine on the catalyst surface to form H_2, which then desorbs. Depending on the reaction conditions, HER proceeds via either the Volmer-Heyrovsky or Volmer-Tafel pathway [11]. Regardless of the mechanism, the adsorption and desorption of H_{ads} play a crucial role in determining the reaction rate [11].

The binding strength between H_{ads} and the catalyst surface significantly affects HER efficiency. Weak H_{ads} binding leads to poor hydrogen adsorption, hindering the reaction, whereas strong H_{ads} binding makes H_2 desorption difficult, limiting the reaction turnover [15]. The ideal HER catalyst should have adsorption-free energy (ΔG_H) close to zero, ensuring balanced adsorption and desorption [16]. Pt and Pt-based catalysts exhibit the most optimal ΔG_H and are currently considered the best HER catalysts. However, their high cost and scarcity restrict large-scale applications. To address this limitation, transition metals such as W, Co, and Ni, along with their derivatives like phosphides, sulfides, and selenides, have emerged as promising alternatives due to their significantly lower cost. Although their HER performance is still inferior to that of Pt-based catalysts, several advanced design strategies, including single-atom catalysts, atomic defect engineering, heteroatom doping, and heterostructure design, have proven effective in enhancing electrocatalytic activity [17, 18]. These approaches continue to improve catalyst performance, aiming to achieve HER activity comparable to that of Pt-based systems.

HER in acidic media (* stands for an active center on the electrocatalyst)

$$H^+ + e^- \rightarrow H^* \quad \text{(Volmer)}$$

$$H^+ + e^- + H^* \rightarrow H_2 \quad \text{(Heyrovsky)}$$

$$\text{(or) } 2H^* \rightarrow H_2 \quad \text{(Tafel)}$$

HER in Alkaline Electrolyte

$$H_2O + e^- \rightarrow H^* + OH^- \quad \text{(Volmer)}$$

$$H_2O + e^- \rightarrow H_2 + OH^- \quad \text{(Heyrovsky)}$$

$$\text{(or) } 2H^* \rightarrow H_2 \quad \text{(Tafel)}$$

2.3 OER and HER Mechanism on Different Catalyst Surfaces

During both the OER and the HER, the nature of active sites plays a crucial role in the adsorption and desorption of reaction intermediates, ultimately determining the catalytic activity. In most cases, highly active sites are dynamically generated through surface reconstruction. Therefore, understanding this dynamic reconstruction process is essential for uncovering the intrinsic catalytic mechanisms and identifying the true active species. Water electrolysis can occur in either acidic or alkaline environments, with a theoretical potential of 1.23 V. Both OER, and HER involve a transition where oxygen valence electrons are depleted, moving from OH^-/H_2O reactants (which have 8 valence electrons per oxygen atom, satisfying the octet rule) to O_2 products, where two unpaired electrons occupy π^* orbitals in the same spin direction [19]. The reaction proceeds through key intermediates, including *OH*, *O*, OOH*, and OO*, which bind to active sites at the electrolyte-electrocatalyst interface via chemisorption. The primary distinction between different OER mechanisms lies in the O–O bond formation step. The reaction sequence typically begins with the formation of metal hydroxide (MOH) intermediates, followed by their oxidation to metal oxide (MO) species. Dioxygen formation from MO can occur via two distinct pathways: direct splitting of two MO centers into O_2 or hydroperoxide formation, where the MO intermediate reacts with water (in acidic conditions) or OH^- (in alkaline/neutral conditions) to form a hydroperoxide species, which subsequently decomposes, releasing O_2 [20–22].

The specific OER reaction mechanism depends on the catalyst's properties and structure. Both the Eley–Rideal (ER) and Langmuir–Hinshelwood (LH) mechanisms involve four reaction steps, progressing from adsorbed *OH* to *O**, leading to metal-site oxidation. Despite variations in proposed mechanisms, most researchers agree that the key OER intermediates M–OH, M–O, and M–OOH (where M represents the catalyst's metal site) are common in both acidic and alkaline environments. A widely accepted OER mechanism at the anode illustrates that the reaction pathways share identical intermediates in acidic and alkaline conditions [23]. Two major O_2 formation routes exist: a direct pathway, where O_2 forms through the combination of two M–O species, and an indirect pathway, where M–OOH forms as an intermediate before decomposing into O_2. Although differences exist, M–O bonding interactions within OER intermediates (M–OH, M–O, M–OOH) are widely regarded as critical determinants of electrocatalytic performance. As a heterogeneous reaction, OER efficiency depends on the strength and stability of these bonds, which influence intermediate formation, reaction kinetics, and overall catalytic ability [24–26].

Under acidic conditions, the HER begins with the Volmer reaction, where hydrogen protons combine with active sites on the electrode surface to form adsorbed hydrogen intermediates (H*). Depending on the coverage of H*, the second step can proceed via either the Heyrovsky or Tafel reaction [27]. When the H* coverage is low, the H* species are sparsely distributed and do not readily combine [11]. In this case, H* reacts with protons from the electrolyte in the Heyrovsky reaction to produce hydrogen gas. In contrast, at high H* coverage, two H* species directly couple to form hydrogen gas without electron transfer, known as the Tafel reaction. Under alkaline conditions, the HER also starts with the Volmer reaction, but unlike acidic conditions where the proton source is H^+, water molecules supply the protons. Water molecules combine with electrons in this step to form OH^- and H*. The higher energy barrier associated with breaking the O–H bond in water restricts the efficiency of alkaline HER in industrial applications. Similarly, the Heyrovsky reaction in alkaline media involves H* reacting with water molecules to produce hydrogen gas. However, the Tafel reaction remains unaffected by pH and involves the direct coupling of two H* species to form hydrogen gas [18].

2.4 Electrochemical Water Splitting Testing Protocols

The evaluation of electrochemical performance, including activity and stability, is commonly conducted using a three-electrode setup (as shown in Fig. 3a) in laboratory settings. This configuration consists of a working electrode, a counter electrode, and a reference electrode, along with a gas inlet for O_2 or Ar/N_2 [28]. To minimize mass transport limitations, rotation is often applied, with a typical rotational speed of 1600 rpm. For assessing overall water splitting performance, a two-electrode setup is employed, where high-performance electrocatalysts for the OER and HER are attached to the anode and cathode, respectively, and immersed in an electrolyte. Significant research efforts are currently focused on designing bifunctional electrocatalysts (Fig. 3b) that can efficiently catalyze both the OER and HER, thereby simplifying the device and accelerating overall water electrolysis [29]. The equipment of a large-scale water electrolyzer involves open systems, in which reaction products and reactants are continuously circulated in and out of the electrochemical cells (Fig. 3c) [30]. The two electrodes are also immersed in the electrolyte but are separated by a diaphragm to prevent the cross-diffusion of generated oxygen and hydrogen gases. The membrane is the core that is independent of the water electrolyzers and ensures ion transportation toward the two electrodes. The membrane needs to be stable under operational conditions [31].

2.5 Electrochemical Water Splitting Testing Conditions

The electrolyte plays a pivotal role in water splitting by serving as an essential medium for ensuring efficient electrochemical reactions. One of its primary functions is to provide ionic conductivity, facilitating the movement of ions such as H^+, OH^-, or O^{2-} between the anode and cathode. This ionic flow maintains charge balance during the reactions, which is crucial for the continuity of the electrochemical process. Additionally, the electrolyte acts as a reactant source by supplying water molecules or hydroxide ions necessary for the OER and HER (Fig. 4a, b)

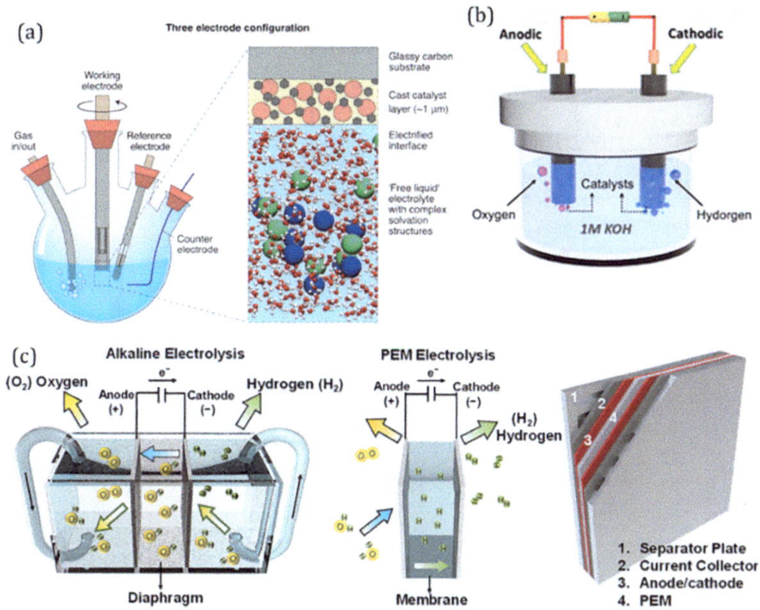

Fig. 3 **a** A schematic diagram of (**a**) a three-electrode configuration used for lab-scale evaluation of electrocatalytic performances. Adapted with permission from Ref. [32] (Copyright (2020) Springer Nature, Nature Materials). **b** The overall alkaline water electrolyzer at the lab scale. **c** Membrane electrode assembly for alkaline water electrolyzer, PEM water electrolyzer, and component overview for a typical PEM water electrolyzer. Adapted with permission from Ref. [7] (Copyright (2021) Wiley, Advanced Materials)

[33]. Beyond these basic roles, the electrolyte significantly influences the catalyst surface. Electrolyte ions interact with the active sites of catalysts, altering the adsorption energies of reaction intermediates, thereby impacting the catalytic activity and overall efficiency of the reactions. Electrolytes are broadly classified into acidic, alkaline, and neutral types, each offering distinct advantages and limitations. Acidic electrolytes, such as H_2SO_4 and $HClO_4$, are dominated by H^+ ions [34, 35]. In these media, HER proceeds through proton reduction, while OER involves the oxidation of water molecules to generate O_2. Acidic electrolytes are commonly used in proton exchange membrane (PEM) electrolyzers because of their high ionic conductivity, although their corrosive nature necessitates the use of noble metal catalysts like Pt and IrO_2. On the other hand, alkaline electrolytes like KOH and NaOH rely on OH^- ions. In this environment, HER involves water dissociation to produce H^+ and H_2, while OER oxidizes hydroxide ions to O_2. Alkaline media enable the use of non-precious metal catalysts, making them cost-effective and less prone to corrosion compared to acidic systems. Neutral or near-neutral electrolytes, such as Na_2SO_4 or phosphate buffers, present a trade-off with slower HER and OER kinetics due to limited ionic conductivity and higher reaction overpotentials [36, 37]. However, they are often employed in specialized applications where stability and environmental

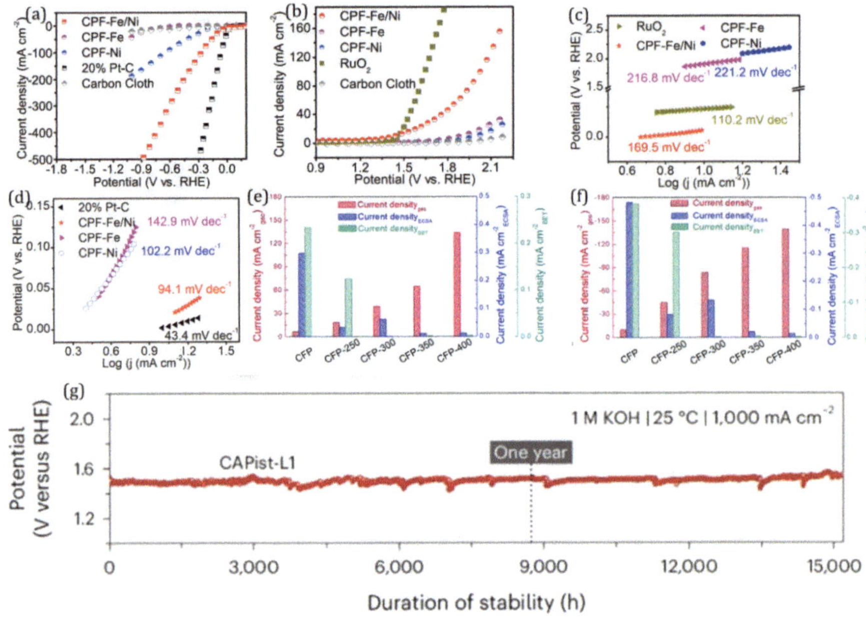

Fig. 4 Overall water-splitting performance of the CPF-Fe/Ni in 0.5 M H_2SO_4 and 1 M KOH. **a** HER. **b** OER. Tafel slopes. **c** HER. **d** OER. Adapted with permission from Ref. [43] (Copyright (2023) Springer Nature, Nature Communication). Current density normalized by ECSA, BET, and geometric areas. **e** OER. **f** HER. Adapted with permission from Ref. [44] (Copyright (2020) Royal Society of Chemistry. Journal of Material Chemistry A). **g** Materials stability and durability based on chronopotentiometry. Adapted with permission from Ref. [45] (Copyright (2024) Springer Nature. Nature Catalysis)

considerations take precedence. Each type of electrolyte offers unique benefits and challenges, underscoring the importance of tailoring the electrolyte composition to specific water-splitting technologies for optimal performance.

2.6 *Electrochemical Characterization and Performance Indicators*

A three-electrode cell setup is widely used for evaluating the electrochemical activity and stability of electrocatalysts. To mitigate contamination in alkaline electrolytes, plastic containers (e.g., PTFE, POM) are preferred over glass [38]. To maintain the equilibrium potential at the thermodynamic standard values, the electrolytes must be fully purged and saturated by high-purity H_2 and O_2 rather than inert Ar or N_2 throughout the electrochemical tests for HER and OER, respectively. Common working electrodes include defined geometric areas substrates including planar glassy carbon, rotating ring-disk electrodes, indium tin oxide glasses, porous metal or carbon meshes, and foams, with precise geometric areas defined for reproducibility. For electrode preparation, drop-casting methods are commonly used to deposit electrocatalyst powders. Counter electrodes should also be rationally selected. The projected area of a counter electrode needs to be larger than that of the working electrode to rapidly supply or sink electrons for the reactions at the working electrode. For the HER studies, high-purity graphite is recommended to replace noble metals such as Pt and Au, as the noble metals can be leached from the counter electrode and deposited on the working electrode under the cathodic reduction potential to enhance the electrocatalytic activity and electronic conductivity, leading to artifacts when evaluating the intrinsic activities of working electrodes. However, extra attention should be paid to the use of graphite as a counter electrode for long-term tests, as the oxidative by-products and inherent impurities from carbon may diffuse into the electrolytes. For the OER studies, noble metals can be used as the counter electrode as their leaching is minimized under the cathodic potential. It is also recommended to separate the working and counter-electrode compartments with an ion-selective membrane or diaphragm to reduce the impact of gas bubbles generated by the counter-electrode on the reactions occurring at the working electrode [39, 40]. The common reference electrodes include Hg/HgO, saturated calomel electrode (SCE) of Hg/Hg_2Cl_2, Ag/AgCl, Ag/Ag_2SO_4, and Hg/Hg_2SO_4 [41, 42]. For collecting the electrochemical data, the steady-state polarization curves are highly recommended instead of linear scanning voltammetry (LSV). The steady-state polarization curves can be obtained by either chronopotentiometry (CP) or chronoamperometry (CA) with defined step sizes of current or potential response and time intervals. This is important to ensure data reliability especially when the electrodes have large capacitive background currents arising from large electrochemical surface areas (ECSAs) or the electrodes themselves are prone to reduction and oxidation under the cathodic and anodic potentials convoluting the true HER and OER currents, respectively.

2.7 Key Electrochemical Performance Indicators

Many metrics and performance indicators are commonly used to evaluate the electrocatalytic performance of HER and OER electrocatalysts, including the RHE potential, specific activity, mass activity, overpotential, Tafel slope, catalytic stability, turnover number and frequency, faradaic efficiency and energy efficiency. The results provide a comparative analysis between the ideal and the actual catalyst by determining the overpotential, which serves as a key performance metric. The overpotential is calculated based on the difference between the thermodynamic potential required for the reaction and the actual potential at which the reaction occurs. The deviation of the real catalyst from the ideal scenario offers insight into its catalytic efficiency and potential limitations. Electrolytic water splitting is an inherently uphill reaction, as indicated by the positive value of Gibbs free energy (ΔG) and requires overcoming a substantial kinetic barrier. Catalysts play a critical role in reducing this kinetic barrier [17].

2.7.1 Key Electrochemical Metrics for Evaluating Water Splitting Catalysts

The thermodynamic potential for the electrochemical water splitting reaction is 1.23 V at 25 °C and 1 atm. However, due to the kinetic barrier for the reaction, water electrolysis requires a higher potential than the thermodynamic potential (1.23 V) for OER and 0 V for HER to overcome the kinetic barrier. The excess potential is also known as overpotential (η), mainly from the intrinsic activation barriers that exist on both the anode and cathode. Overpotential is a very important descriptor to evaluate the activity of the electrocatalysts. Usually, the overpotential value corresponding to the current density of 10 mA cm^{-2} is used to compare the activities among different catalysts. This current density corresponds to a 12.3% solar-to-hydrogen efficiency. In electrochemical water splitting, there are primarily three kinds of overpotentials: activation overpotential, concentration overpotential, and impedance overpotential [43]. The reversible potentials for HER and OER are 0 and 1.23 V versus standard hydrogen electrode (SHE), respectively. Hence, the equations $\eta_{OER} = E_{RHE} - 1.23$ V and $\eta_{HER} = E_{RHE} - 0$ V can be used for OER and HER to calculate their overpotentials at a desired current density without iR compensation.

$$\text{iR drop free } \eta_{OER} = E_{RHE} - 1.23\text{ V} - E_{iR}$$

$$\text{iR drop free } \eta_{HER} = E_{RHE} - 0\text{ V} - E_{iR}$$

Tafel slope: is another important notation, mostly used to assess the activity from the overpotential versus kinetic current. It can also be expressed by the equation: $\eta = a + b\log j$, where η is the overpotential, and j is the current density. The Tafel slope is an important way to measure the speed of certain processes in electrochemical water splitting, like the OER and the HER (Fig. 4c, d). It tells us a lot about how these reactions work by showing the slope of the straight line on a graph that connects the logarithm of the reaction rate (current density) to the electrode potential. If the Tafel slope is steeper, the response rate is more sensitive to changes in the electrode voltage. This parameter is crucial for knowing how electrochemical kinetics work when water splits [46]. In real life, a lower Tafel slope is better because it means the electrochemical response is faster and more efficient [47]. For example, a smaller Tafel slope means a significant current density increment as a function of the overpotential change, or in other words, faster electrocatalytic reaction kinetics. Current density plays a pivotal role in electrochemical reactions, particularly in water electrolysis for hydrogen and oxygen production [48]. Higher current densities result in faster reactions, but maintaining a balance among efficiency, energy consumption, and electrode stability is critical. Recent advancements in catalysts and materials have demonstrated the ability to achieve high current densities for both the HER and the OER, showcasing the potential for developing efficient and sustainable electrochemical systems. In addition to current density, the exchange current density is a key parameter that reflects the intrinsic charge transfer rate under equilibrium conditions. A higher exchange current density signifies a faster charge transfer rate and a lower reaction energy barrier. For an ideal electrocatalyst, a combination of a lower Tafel slope and a higher exchange current density is desirable, as these characteristics indicate improved catalytic performance and efficiency.

Faradaic efficiency (FE) is a quantitative parameter used to evaluate the efficiency of electron transfer in electrocatalytic processes, such as the HER and OER. It is defined as the ratio of the experimentally detected quantity of H_2 or O_2 gas to the theoretically calculated quantity. The theoretical gas production is typically determined through the integration of chronoamperometric or chronopotentiometric data, while the experimental values are obtained using techniques such as the water–gas displacement method or gas chromatography. In practice, the Faradaic efficiency is often less than 100% due to secondary reactions occurring at the electrode surface. This makes FE a critical parameter for assessing the efficiency of catalysts in electrochemical reactions. Additionally, low corrosion is an essential factor to consider when evaluating the long-term stability of a catalyst. Long-term measurements are required to determine whether a catalyst is suitable for practical applications [49]. As a result, many researchers calculate FE to assess the performance and efficiency of catalysts in real-world scenarios [50].

Turnover frequency (TOF): is a valuable parameter for describing the intrinsic catalytic activity of a catalyst by quantifying the reaction rate at its active sites. TOF measures how many reactants are converted into the desired product by the catalytic site per unit of time. However, determining the precise TOF value for most heterogeneous electrocatalysts can be challenging, as the exact number of active sites per electrode area is often estimated. Despite this imprecision, TOF remains a useful metric for comparing the catalytic activities of different catalysts, particularly within similar systems or under comparable conditions. In the context of electrochemical water splitting, which uses an electrolyzer and electric current to convert water into hydrogen and oxygen gases, TOF provides insight into the effectiveness of catalysts. Catalysts play a critical role in enhancing the reaction rate and improving efficiency [51]. By indicating how well a catalyst enables the transformation of water into hydrogen and oxygen, TOF helps researchers evaluate and compare the performance of various catalysts, ultimately aiding in the identification of more effective materials for the desired electrochemical processes [52].

Specific activity: The specific activity of an electrocatalyst is defined as the current density (j) normalized by the electrocatalyst area at a given potential. Three kinds of areas are always used to calculate the current density, including the geometric area (J_{geo}), the Brunauer–Emmett–Teller surface area (J_{BET}) (adsorption isotherm studies apply to determine the real surface area), and the electrochemical surface area (J_{ECSA}) [53, 54]. For an electrode with a smooth and flat surface, the current density can be normalized to the geometrical electrode surface. However, it is not appropriate to evaluate the activities of powder electrocatalysts and porous electrodes by normalizing their current densities with the geometrical electrode surfaces, as the roughness degree of the powders and porosity will affect the fair comparison of intrinsic activities. The inappropriate use of the geometrical electrode surface will lead to an artificial effect to boost the activity and cannot reflect the intrinsic activity of an electrocatalyst. Therefore, for powder electrocatalysts and porous electrodes, it is recommended to evaluate the specific activity by normalizing the current density with the ECSA or BET surface area [55]. Therefore, the accurate determination of real active sites and current density in OER and HER is influenced by the normalization of ECSA, BET, and geometric area. In recent work, the current densities are compared by normalizing them by the geometric area (J_{geo}), BET surface area (J_{BET}), and ECSA (J_{ECSA}). This comparison (Fig. 4e, f) reveals that J_{geo} (dimensions) indicates increasing current density following thermal treatment, but both J_{BET} (gas adsorption) and J_{ECSA} (electrolyte adsorption) show the converse variation.

Mass activity expressed in A g^{-1} or A mg^{-1}, is the current density normalized by the mass of the active electrocatalyst at a given potential and depends on factors such as particle size, morphology, and catalyst mass loading [55]. High mass activity is crucial for developing cost-effective and compact electrocatalysts for practical devices. Since most electrochemical metrics are mass-dependent, reporting the mass loading is essential. An ideal electrocatalyst should combine both high specific and mass activity. For practical applications, optimizing the mass loading ensures adequate coverage of the conductive substrate and the ideal thickness of the electrocatalyst layer, promoting efficient electron and mass transfer [54, 56].

Double layer capacitance: Two techniques are typically utilized to calculate the electrochemically active surface area (ECSA) of the electrocatalyst based on the capacitance of the electrochemical double layer. (1) Cyclic voltammogram (CVs) is used to identify a non-Faradaic capacitive current connected to double-layer charging [44], and (2) electrochemical impedance spectroscopy is used to quantify a frequency-dependent impedance (EIS) [57]. CV is used to define a potential range where no obvious Faradaic processes seem to take place to detect double-layer charge. The open-circuit potential (OCP) of the system serves as the center of this range, which is generally a 0.1 V potential window. It is considered that all measured current in this range of non-faradaic potential results from double-layer charging. Then, the charging current (i_c) is determined using CVs at various scan rates [58]. Then, the higher the ECSA will have the higher the activity. The ECSA value can be measured by double-layer capacitance (Cdl) in a non-faradaic potential window in either aqueous or aprotic electrolytes. Unfortunately, the specific capacitance used in this method to calculate ECSA depends on the surface-active compositions and other complex factors, and it is difficult to accurately obtain the true ECSA value, although a universal specific capacitance of 40 or 60 μF cm^{-2} regardless of materials has been widely used as an approximation. Alternatively, other methods can also be employed to measure the ECSAs of electrocatalysts including the hydrogen underpotential deposition, underpotential deposition of metal, CO stripping voltammetry, use of reversible redox probe molecules, and redox of surface metals. These methods are limited to some specific electrode compositions. Moreover, there are discrepancies in the ECSA values of the same material measured using different methods. Regardless of the adopted ECSA determination methods, it is challenging to determine the accurate true ECSA value. The BET measurement can be performed in the N_2 adsorption–desorption experiments at 77 K, which is a standard procedure to assess the porosity of heterogeneous powder catalysts and adsorbents. The BET surface area may not reflect the true ECSA value, especially for the materials with poor conductivity, since the sites where N_2 molecules are adsorbed may not be electrochemically active. Furthermore, the BET surface areas of monolithic foam electrodes are difficult to precisely measure [44].

Catalyst stability and durability: The stability of the electrocatalyst is a crucial factor in determining the efficiency and longevity of an electrochemical water-splitting system. Several factors, including the applied voltage and current density, can significantly impact the stability of the electrocatalyst during the electrochemical process [59]. Evaluating stability is essential to assess whether a catalyst is suitable for practical applications in water-splitting cells [60]. There are two typical methods for characterizing the stability of electrocatalysts. One method is chronoamperometry (I–t curve) or chronopotentiometry (E–t curve), which measures the current variation with time under a fixed potential or measures the potential change with time at a fixed current. For this measurement, the longer the tested current or potential remains constant, the better the stability of the catalyst. For the comparison with the different research groups, people usually set a current density larger than 10 mA cm^{-2} for at least a 10 h test. Current work by Li et al., reports long-term durability of more than 15,000 h (Fig. 4g) based on NiFe anode catalysts for anion exchange membrane water electrolysis [45]. Another method is cyclic voltammetry (CV), which measures the current by cycling the potential, usually requiring more than 5000 cycles of run at a specific scan rate (e.g. 50 mV s^{-1}). Linear sweep voltammetry (LSV) is typically applied to examine the overpotential shift before and after CV cycling at a specific current density [61]. The smaller the change of overpotential is, the better the electrocatalyst's stability [62].

3 Catalysts of Interest

For practical electrocatalyst design, key factors include activity, stability, and cost, with equal emphasis on electrochemical performance and affordability. In commercial alkaline water electrolysis, electrodes must exhibit corrosion resistance, abundant active sites, high conductivity, and excellent structural integrity. Commonly, cathodes are made of Ni, but alloying Ni with metals (e.g., W, Mo, Cr, Co, Fe, Zn) or metallic oxides (e.g., MoO_2, MoO_3, CeO_2) mitigates issues like increased overpotential due to Ni-hydride formation. Anodes are also typically made of Ni to reduce costs. These materials perform well in alkaline conditions and are cost-effective, making alkaline water electrolyzers more affordable than other types. However, these electrode materials are unsuitable for acidic environments like proton exchange membrane (PEM) electrolysis, which requires precious metal-based materials. Pt and Pt–Pd are used for

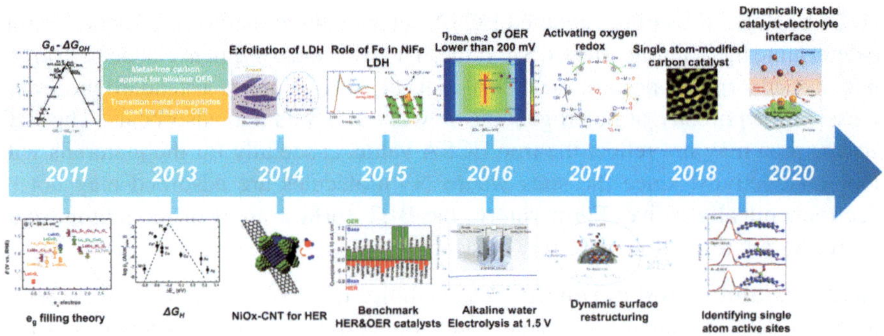

Fig. 5 Timeline of recent developments of non-precious catalysts. Adapted with permission from Ref. [70] (Copyright (2020) Royal Society of Chemistry. Chemical Society review)

cathodic HER, while IrO_2 and RuO_2 are used for anodic OER, reflecting the higher cost of PEM electrolyzers. Ideal electrodes should possess several fundamental properties: (1) high intrinsic activity for OER and HER; (2) long-term stability; (3) high electronic conductivity; (4) large surface area; (5) favorable hydrophilicity; and (6) high gas permeability. The optimization of surface properties, such as the coordination environment, oxidation state, and electronic structure of active centers, allows for fine-tuning of adsorption energies for key intermediates. Consequently, research efforts target the development of efficient and stable non-noble metal catalysts for alkaline water electrolyzers (AWEs) shown in Fig. 5 [28]. Transition metal oxides containing Fe, Co, and Ni are promising candidates for alkaline OECs due to their better performance, relative abundance, tunable 3d electron configuration, and versatile structures [63, 64]. Branches of water-splitting catalyst include layered double hydroxide (LDH) [65], metal chalcogenides/phosphides, and perovskite family [66, 67]. Still, most were restricted to the alkaline electrolyte [64, 68–70].

Here, we present a summary of some catalytic materials for applications for fundamental research.

Perovskite family: Perovskites, with the formula $A_{n+1}B_nO_{3n+1}$ (A is alkali metal/alkaline earth metal, B is transition metal), are constructed by periodically packed rock-salt monolayers (AO) and perovskite (ABO_3) multilayers (Fig. 6a) [71]. The B transition metal is at the center of the octahedron, forming a backbone by sharing corners, while A fills the open spaces to complete the perovskite structure [72]. Transition metals in perovskites can have oxidation states of 3+, 4+, or a mixture of both, depending on the oxidation state of A. Perovskites like $(A_1^{2+})(B^{4+})O_3$, $(A_2^{3+})(B^{3+})O_3$, or $(A_1^{2+})_{0.5}(A_2^{3+})_{0.5}(B^{3.5+})O_3$ can have desired transition metal oxidation states by doping alkaline-earth metal (A_1) and rare-earth metal (A_2) [73]. Different research based on $Ba_{0.5}Sr_{0.5}Co_{0.8}Fe_{0.2}O_{3-\delta}$, $BaNi_{0.83}O_{2.5}$, $LaMO_3$ (M = Mn, Co, Ni), $SrFeO_3$ and $CaFeO_3$ on OER and HER were reported [74, 75].

Spinel family: Spinel compounds have diverse compositions, with over 100 reported so far. Their general formula is AB_2O_4 (Fig. 6b), where A and B atoms typically

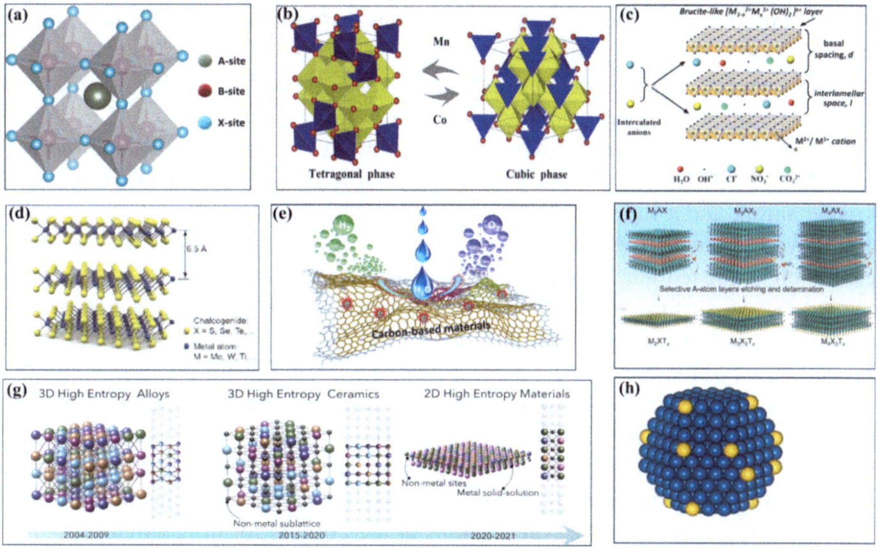

Fig. 6 General crystal structure of different OER/HER catalysts. **a** Perovskites. Adapted with permission from Ref. [90] (Copyright (2022) Wiley). **b** Spinels oxides. Adapted with permission from Ref. [91] (Copyright (2019) Cell Press). **c** Layered double hydroxides. Adapted with permission from Ref. [92] (Copyright (2016) Wiley). **d** Transition metal chalcogenides. Adapted with permission from Ref. [93] (Copyright (2015) Springer Nature). **e** Carbon-based materials. Adapted with permission from Ref. [94] (Copyright (2016) Wiley). **f** MXenes. Adapted with permission from Ref. [95] (Copyright (2020) Springer Nature). **g** High Entropy Materials. Adapted with permission from Ref. [96] (Copyright (2023) Springer Nature). **h** Single-atom catalysts. Adapted with permission from Ref. [97] (Copyright (2013) American Chemical Society)

consist of group 2, group 13, and first-row transition metals. The structure of spinel is more complicated than perovskite. It has octahedral (O_h) and tetrahedral (T_d) sites, leading to two types: normal spinel with A^{2+} in Td and B^{3+} in O_h sites and inverse spinel with A^{2+} in O_h and B^{3+} in T_d and O_h sites. Transition metals in spinels can have tetrahedral and octahedral coordination, leading to distinct d-band splitting (T_d: e_gt_{2g}; O_h: $t_{2g}e_g$) based on crystal field theory. Spinel compounds are stable and good conductors in alkaline solutions, making them suitable electrocatalysts for OER. Fe-based (ferrite) and Co-based spinel oxides are common, often doped with transition/alkaline metals such as Mn, Ni, Cu, Zn, and Li [76]. For ferrite spinels, $CoFe_2O_4$ > $CuFe_2O_4$ > $NiFe_2O_4$ > $MnFe_2O_4$ was found to be the trend for OER activity [77]. Incorporating Ni, Cu, and Li in Co-based spinels improved OER activity except for Mn, whose inclusion reduced activity. This may be due to the effect on Jahn–Teller distortion, as observed in the $Mn_{3-x}Co_xO_4$ system, which has better activity as Mn content decreases [78].

Layered double hydroxides (LDHs) family: represent an emerging class of 2D layered materials belonging to the group of hydrotalcite-like (HT) compounds represented by the formula: $[M_{1-x}{}^{2+}M_x{}^{3+}(OH)_2]^{x+}\cdot[A_{x/n}]^{n-}\cdot mH_2O$ (Fig. 6c), where M^{2+}

and M^{3+} are the divalent and trivalent metal ions, respectively, A^{n-} are inorganic or organic anions, *m* is the number of interlayer water, and $x = M^{3+}/(M^{2+} + M^{3+})$ is the layer charge density or molar ratio [79–81]. The M(II) and M(III) indicate divalent and trivalent metal cations. The divalent metal ion M^{2+} (e.g., Ca^{2+}, Zn^{2+}, Mg^{2+}, Ni^{2+}) is octahedrally coordinated to six OH^- hydroxyl groups and trivalent ions M^{3+} (e.g., Al^{3+}, Fe^{3+}, Cr^{3+}, In^{3+}). Additionally, A indicates the interlayer inorganic or organic anion, and x indicates the molar ratio. Recently, LDHs have been excellent alternatives in this field. Among them, Ni-based LDH materials have attracted significant interest due to their effective charge transport, appreciable stability, immensely exposed active surface sites, composition based on earth-abundant materials, and impressive catalytic activity [82, 83]. Despite the great progress, low conductivity and self-aggregation are still the current primary issues for NiFe-LDHs, inhibiting the optimization of electrocatalytic performances [80]. These compounds, which include metal hydroxides ($M(OH)_2$; M = transition metals) and metal oxyhydroxides (MOOH), exhibit strong electrocatalytic activity for OER [80]. The crystal structure of LDHs features a layered stacking arrangement, with transition metals at the center of octahedra and oxygen anions at the corners forming [MO_6] subunits. LDHs can play different roles in a catalytic system. Specifically, they can be:

- **Supporting surfaces**—The incorporation of new species introduces reactivity that pristine LDHs inherently lack. In this role, LDHs provide chemical confinement and enhance catalyst selectivity but do not actively participate in the reaction steps.
- **Source of catalytically active species**—LDHs transform a new material that exhibits the desired catalytic properties.
- **Absorbers**—Catalytic species are adsorbed within the interlayer galleries and can later be exchanged with the external environment. LDHs must first be exposed to an excess of the target species, which are subsequently released. In this scenario, LDHs function as reservoirs or create a catalytic environment, but the active role is attributed to another atomic or molecular entity.
- **Intrinsic catalysts**—The catalytic properties originate from the LDH lattice itself, without requiring the incorporation of additional species.

Transition metal chalcogenides (TMCs): include two subsets: the well-known TMDs with the form of MX_2 (M = Mo, W) (Fig. 6d) and the less explored transition metal trichalcogenides (TMTs) with the form of MX_3 (M = Ti, Zr, Hf), where X represents chalcogens (S, Se, Te). TMCs form a diverse class of 2D materials that inherently address the gapless nature of graphene while exhibiting superior electroconductivity compared to metal oxides, making them highly desirable for electrochemical applications. TMCs make a cost-effective electrode material with high activity toward HER, OER, and ORR. Multiple studies confirm the value of the usage of transition metal chalcogenides in areas of catalysis and electrochemistry [84, 85]. Compounds based on Co, Fe, Mn, and Ni are widely studied for the HER and OER due to their excellent conductivity and resistance to electrolyte corrosion [86]. These materials exhibit OER/HER activities comparable to metal oxides, with overpotentials typically in the range of 200–300 mV at 10 mA/cm^2. Incorporating foreign metal ions, such as in ternary $FeNiS_2$ ultrathin nanosheets, further enhances OER performance. For example, Ni-doping $Co_{0.85}Se$ improves its conductivity and OER activity, reducing overpotentials from 324 to 255 mV. However, phase transformations from sulfides to oxides or oxyhydroxides during OER can influence the interpretation of overpotential data. Metal dichalcogenides like $CoSe_2$ and $NiSe_2$ are particularly promising for OER. $CoSe_2$ nanobelts on nitrogen-doped graphene achieved an overpotential of 366 mV and a low Tafel slope of 40 mV/dec, while Fe-doped $NiSe_2$ oxide catalysts exhibited an impressively low overpotential of 195 mV and a Tafel slope of 28 mV/dec. Hybrid materials combining metal chalcogenides with oxides, such as Fe_3O_4, Mn_3O_4, or CeO_2, show enhanced OER activity due to synergistic effects. Phase transformation is not universal among metal chalcogenides. For instance, Ni_3Se_2 demonstrates intrinsic OER activity, comparable to NiFe LDHs, with an overpotential of 310 mV at 10 mA/cm^2 but a higher Tafel slope (97.2 mV/dec), indicating a distinct rate-determining step. Ni_3Se_2 avoids significant phase changes, while its high-index facet Ni_3S_2 nanosheets deliver exceptional HER and OER performance in alkaline conditions, maintaining stability for over 150 h.

Carbon-based materials

Carbon-based materials (Fig. 6e) have been extensively investigated as water-splitting electrocatalysts for their high electrical conductivity. They are promising and cost-effective alternatives to noble-metal electrocatalysts [87]. Although great progress has been made in designing high-performance carbon-based electrocatalysts, several issues are challenging and await being solved before carbon-based materials can be widely used in electrocatalysis. Remarkably, the OER activity of some carbon-based materials is comparable to that of noble-metal materials [88]. However, the HER activity of the former is yet to be improved greatly to compete with the latter. Moreover, theoretical calculations have revealed that the electrocatalytic activity of carbon-based materials is still inferior to that of metal-based materials in most cases. Therefore, effective designing strategies are highly desired to break the scaling relationships that limit the improvement in the activities of carbon-based electrocatalysts. Moreover, carbon-based materials are thermodynamically unstable under acidic OER conditions, which leads to poor stability among mechanical and chemical reactions in the process of electrocatalysis [89].

Emerging Materials: Aimed at enhancing overall efficiency and reducing the fabrication cost of water electrolysis, considerable attempts have been made to exploit more promising material systems in fundamental research. The field of heterogeneous catalysis has seen remarkable advances with the emergence of new materials that offer enhanced catalytic properties. These materials, ranging from 2D materials (graphene) and TMDs to MXenes, high-entropy alloys, and single-atom catalysts (SACs), are reshaping the landscape of catalytic reactions by enabling more efficient, selective, and durable processes. These new materials bring novel physical, chemical, and electronic properties that significantly improve catalytic activity and performance in various reactions, from energy conversion to chemical synthesis.

MXenes: one of the latest, and relatively large, family of 2D materials is transition-metal carbides and nitrides, called MXenes. MXenes are demonstrated by the chemical formula $M_{n+1}X_nT_x$, (where M represents Ti, V, Mo), X is C and/or N (n = 1, 2 or 3), and T_x is the surface-terminating functionality (e.g., –O, –OH, and/or –F). $Ti_3C_2T_x$ the first MXenes family was discovered in 2011 and has greatly expanded and made an impact in the fields of energy. Thanks to Gogotsi's group efforts MXenes has become a promising area of research. MXenes are an interesting class of materials for OER/HER electrocatalysis since they are chemically stable in acidic media, resistant to poisoning, have excellent mechanical durability, and possess high electronic conductivity. MXenes, which are single-layered counterparts of non-layered TMC, were proposed to be the promising anode materials for metal ion batteries because of their good conductivity. Different types of MXenes phase structures like M_2AX, M_3AX_2, and M_4AX_3, are shown in Fig. 6f. As a result, the MXenes family is quite diverse with multiple forms: single metal structures ($Ti_3C_2T_x$, Ti_2CT_x, V_2CT_x, etc.), ordered double TM MXenes (i.e., $Mo_2Ti_2C_3T_x$, $Mo_2TiC_2T_x$, $Cr_2TiC_2T_x$, etc.), solid solution MXenes (i.e., $Ti_{2-y}V_yCT_x$, $Mo_{4-y}V_yC_3T_x$, $Ti_{2-y}Nb_yCT_x$), and ordered divacancy MXenes ($Mo_{1.33}CT_x$, $W_{1.33}CT_x$, etc.) [86].

High entropy materials (HEMs): a relatively new class of catalysts represented with distinct structures as shown in Fig. 6g, have gained attention for their complex, multi-component composition that enables them to exhibit extraordinary catalytic properties. The concept of high entropy at the nanoscale was introduced in 2017, further expanded to layered two-dimensional (2D) materials in 2020–2022 with entropy-stabilized multiple principal elements in oxide, chalcogenide, carbide/carbonitride (MXenes), and hydroxide forms with promising practical applications [98]. These materials typically consist of five or more metals with molar concentrations of 5–35 at% in near-equal proportions, creating a "high-entropy" phase that leads to unique physical and chemical characteristics. HEMs are considered promising catalysts for catalysis because of their adjustable constituents, abundant active sites, unsaturated coordination, and stable structures. The typical definition of HEMs is described as materials containing five or more metal cations with different radii and electronegativity in equimolar or near equimolar ratios [99]. While high entropy 2D materials are still in their infancy, the combination of their high surface area and functionalities of 2D materials with entropy stabilization brings the promise to accelerate materials discovery for applications such as energy conversion or storage. In 2015, Rost et al., reported an entropy-stabilized oxide with a rock salt structure $MgNiCuZnCoO_x$ [100]. Entropy has been perceived as an important driving force in catalytic reactions. Recent advances in HEMs offered a rich source for investigating the functions of entropy in catalysis. Depending on entropy contributions, the multi-atomic species, which are integrated into the same lattice, can drastically expand multi-catalytic active sites to further boost catalytic activity. Furthermore, the derived disordered structures and randomized location of metal elements in HEMs not only can lower the energy barriers during catalytic reactions but also induce more surface defects and unsaturated sites as well as stabilized structures, which are positive for adsorption–desorption of reactants and charge transfer [101]. In addition, the multi-component high entropy catalysts also make a significant contribution to overcoming the challenges in conventional enthalpy-dominated systems. Due to the rapid development of HEMs in various catalytic reactions, there is a pressing need to conduct a comprehensive review to emphasize their prospects and challenges. It has been found that conventional and advantageous properties are observed in HEAs, such as increased hardness and wear resistance, overturning of the strength–ductility trade-off, and exceptional fracture resistance under cryogenic conditions, accompanied by the four core effects high-entropy, severe lattice distortion, sluggish diffusion kinetics and cocktail effects as first proposed and recently re-examined by Yeh and colleagues. Moreover, as a class of materials, HEMs contain oxides, alloys, carbides, fluorides, and sulfides [102].

Single-atom catalysts (SACs) are defined as ensembles that consist of an atom of any element (metal, metalloid, non-metal, halogen) that is spatially isolated from atoms of the same chemical identity (e.g., adsorbed, embedded, immobilized, grafted, or confined, Fig. 6h) and directly bound to a solid carrier. Depending on the specific interaction and associated degree of orbital overlap, the electronic structure of the minority element can be significantly hybridized with the host (e.g., heteroatom-doped carbons). In the case of single-atom alloys (SAA), the isolated atom may change from a metallic to a free-atom-like state [103]. SACs have emerged as one of the most recent innovative developments in heterogeneous catalysis. SACs are a class of catalysts in which individual, catalytically active metal atoms are isolated and anchored onto the support. These catalysts are notable for their ability to achieve optimal metal utilization, where every metal atom is exposed to reactants and available for catalytic reactions. SACs have garnered significant attention for their unique properties, distinguishing them from metal nanoparticles (NPs) and mononuclear metal compounds. They exhibit extraordinary catalytic activity, selectivity, and stability in oxidation, water–gas shift, and hydrogenation reactions. These enhanced properties are largely attributed to the low-coordination environment of the metal atoms, along with high atom utilization and strong metal-support interactions [104]. Size reduction of metal particles also benefits the performance of catalysts in other aspects: (1) low-coordination environment of metal centers [105], which is ascribed to the unsaturated metal atoms exposed to the small-sized particles [105, 106], (2) quantum size effects, where confinement of electrons leads to discrete energy level distributions and a distinctive HOMO–LUMO gap [107]; (3) Metal–support interactions, originating from the chemical bonding effect between metal and supports and the associated interface, as well as the charge transfer between metal species [108] and supports [108, 109]. Different types of SACs exist, according to the chemical interactions between the mononuclear metal atom and supports, including single metal atoms anchored to metal oxides, metal surfaces, graphene, etc. It is important not to confuse SACs with single-site heterogeneous catalysts (SSHCs). By definition, a SSHC may consist of one or more atoms, including clusters of atoms, as long as all clusters exhibit the same catalytic behavior [110].

4 Catalyst Surface Characterization

There is a rapid establishment and development of various characterization techniques to elucidate the electronic and geometric configuration of materials. These techniques are prerequisites for revealing the relationship between the structure and catalytic performance, which allows a more in-depth and precise understanding of the catalytic reaction mechanism.

Raman spectroscopy: it is based on Raman scattering, where light interacts with molecular bonds, leading to elastic and inelastic scattering. In the spontaneous Raman effect, incident photons excite molecules to a virtual energy state, and when the molecules relax, they emit photons with different frequencies due to changes in vibrational or rotational energy levels. This frequency shift, known as the Raman shift, provides information about molecular vibrations. In situ Raman spectroscopy is particularly useful for studying catalyst behavior during electrochemical reactions, as it reveals bond formation and conversion. The technique is enhanced by inelastic light scattering on roughened metal surfaces like silver, gold, or copper (surface-enhanced Raman spectroscopy, SERS) [111]. The mechanism of signal enhancement in SERS is attributed to the excitation of localized surface plasmons, in which the oscillating electric field associated with the incident light interacts with the electron cloud in metal nanoparticles and amplifies the inelastically scattered light. For example, Deng et al. utilized in situ electrochemical Raman (EC-Raman) spectroscopy to monitor the surface structure of Cu, Cu_2O, CuO, and $Cu(OH)_2$ catalysts during water oxidation in 0.1 M KOH. Their in-situ Raman spectra, recorded at 1.7 V versus RHE, revealed the formation of Cu^{3+} oxide (peak at 603 cm^{-1}), a metastable species that emerged exclusively on CuO and $Cu(OH)_2$ electrodes. This species facilitated water oxidation with an efficiency ten times higher than that of the other two electrodes, indicating that Cu^{3+} species generated under applied bias serve as the catalytically active sites for the oxygen evolution reaction, an insight enabled by in situ EC-Raman spectroscopy [112].

Molecular Vibration Spectroscopy: Fourier transform infrared spectroscopy (FT-IR) utilizes infrared light to induce molecular vibration and thereby identify chemical compounds by their unique IR absorption spectrum [113]. It operates on the principle that when infrared radiation interacts with a sample, specific chemical bonds within the sample absorb frequencies of infrared light. This absorption causes the bonds to vibrate at characteristic frequencies, which are indicative of the sample's molecular structure. FTIR spectroscopy is a powerful tool for identifying molecular vibrations, functional groups, and chemical bonding within a material. FTIR spectroscopy offers distinct advantages, such as real-time monitoring of chemical changes, non-invasive analysis, molecular-level insights, and applicability across a wide range of samples and disciplines. However, these techniques also have limitations, including the need for specialized equipment and technical expertise, sensitivity constraints, the generation of average data from bulk materials, and challenges in data interpretation, particularly when dealing with complex spectra that require advanced modeling and analysis. Such diverse analytical approaches facilitate a comprehensive understanding of both solution-phase and surface-bonded species, therefore advancing the comprehension of electrochemical reactions.

Scanning Transmission Electron Microscopy (STEM): emerges as a powerful tool used to characterize electrocatalyst morphological and compositional features at atomic resolution [114]. This characterization technique enables the study of structures at sub-angstrom resolution, making it invaluable for the precise identification of atomic configurations [115]. Notably, STEM has played a crucial role in materials characterization and has been extensively utilized in experimental research focused on the construction and analysis of catalysts. It is primarily employed to determine the position and distribution of atomic pair structures anchored on supports, leveraging atomic number (Z-value) contrast to differentiate between dual-atom species in active sites and the supporting substrate [116, 117].

X-ray powder diffraction (XRD): This provides valuable crystallographic insights, including lattice parameters and phase purity. XRD can be integrated with electrochemical techniques and is a powerful tool for studying dynamic changes in electrode materials during electrochemical reactions. Operando XRD enables real-time monitoring of crystal structure evolution, composition changes, and crystallite size variations, providing insights into reaction pathways, active sites, and degradation mechanisms [118]. In recent years, advancements in in-situ X-ray diffraction (XRD) techniques, particularly Crystal Truncation Rods (CTR) diffraction and Grazing Incidence Diffraction (GID), have significantly improved the ability to characterize material surfaces in real-time. CTR diffraction is employed to investigate the surface and interface structures of single crystals. When a crystal is truncated, it produces a distinct diffraction pattern along rods that extend perpendicular to the surface, known as crystal truncation rods. The intensity distribution along these rods provides detailed information about surface atomic positions and reconstructions.

GID, on the other hand, is an X-ray diffraction technique where the incident beam interacts with the sample surface at a very shallow angle typically less than a few degrees. This approach enhances surface sensitivity, enabling the study of thin films, surfaces, and interfaces while minimizing interference from the bulk material [119, 120]. The continuous development of these in situ XRD techniques allows for dynamic monitoring of catalysts, particularly crystalline materials, during reactions. By offering real-time insights into reaction kinetics, phase stability, and catalyst performance under operando conditions, these methods play a crucial role in optimizing electrocatalytic systems for various applications.

X-ray photoelectron spectroscopy (XPS): XPS is one of the most powerful techniques for studying the elemental composition and chemical states of the surface of materials [121]. The principle of XPS is to radiate the sample with X-rays of known energy so that the inner electrons or valence electrons of atoms or molecules are excited and emitted. Electrons excited by photons are called photoelectrons, and the energy of photoelectrons can be measured, resulting in the knowledge of the binding energy within the material. We can obtain the surface information of materials according to it. In the past, in-situ XPS was combined with synchrotron radiation photoelectron spectroscopy [122]. However, more convenient ambient pressure X-ray photoelectron spectroscopy (AP-XPS) is now compatible with in-situ experiments. When we apply potential to a sample, the dynamic changes of the existing form of an element and charge distribution for a material surface can be studied by AP-XPS. There are two principal methods to obtain depth-dependent elemental and chemical information using XPS: (1) depth-profiling by etching (for instance with argon ions), where surface layers are gradually removed by sputtering, while XPS spectra are obtained at various stages during the removal of surface layers, and (2) non-destructive depth profiling through the variation of the escape-depth of the electrons, either by varying the detection angle relative to the sample surface or by measuring photoelectrons with different kinetic energies [123].

X-ray absorption spectroscopy is a technique used to analyze the energy-dependent fine structure of the X-ray absorption coefficient used to reveal the changes in an electrocatalyst's oxidation states, coordination geometry, and electronic transitions of atoms during electrochemical reactions [124]. Depending on the analytical application, this technique has two parts: (i) X-ray absorption near-edge structure (XANES): highly sensitive to the oxidation state and electronic configuration of the detected elements, as the energy of core electrons is influenced by the distribution of electrons in the valence state. (ii) Extended X-ray absorption fine structure (EXAFS): provides information about the local environment of an element regarding the average bond length of neighboring atoms, coordination number, type of the neighboring atoms, mean-square disorder of neighboring atoms, bond length, and the chemical identity of the nearest atom [125]. In detail, the evolution of the coordinated environment of an absorbing element can be further extracted by in situ EXAFS spectra, studying the dynamic changes of the scattering path with increasing potential to obtain the coordination form of atoms. Also, we can develop the changes of chemicals under applied potential according to XANES [126].

In-situ/operando techniques: the in situ/operando characterization techniques employed in electrochemical catalysis mainly involve molecular vibration spectroscopy, X-ray-related spectroscopy, and electron microscopy [127]. By definition, "in situ" describes the collection of catalytic behavior in the same conditions relevant to practical catalytic operation, whereas "operando" combines in situ characterization of a working catalyst during genuine reaction conditions with simultaneous measurement of catalytic performance [128]. We can analyze the changes in specific absorption peaks to determine the reaction environment, reaction path, catalyst state, and active site through theoretical calculation, electrochemical methods (kinetics study), and multispectral resolution spectroscopy, including in situ Raman spectroscopy, Fourier transform infrared spectroscopy, X-ray powder diffraction, and fluorescence spectroscopy, etc. [129, 130]. As shown in Fig. 7, combining in-situ/operando characterization techniques with theoretical studies enables a deeper understanding of the reaction mechanism and the identification of the real active site before and after surface reconstruction during electrochemical measurements.

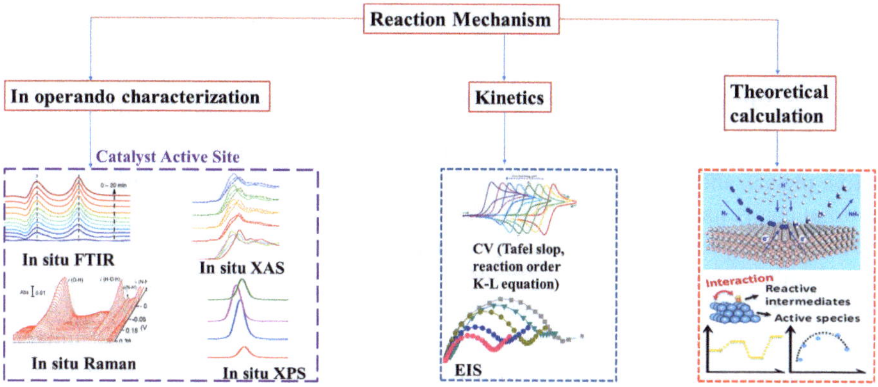

Fig. 7 Illustration of in-situ/operando characterization techniques, kinetics, and theoretical calculation for determining reaction mechanisms in water splitting

5 Summary and Future Perspectives

5.1 Summary

Hydrogen is a promising substitute for fossil fuels due to its high gravimetric energy density and zero pollution emissions, making it a cornerstone for clean and renewable energy systems. Among the various hydrogen production methods, water splitting emerges as a key technology for energy conversion and storage. However, its global adoption is limited, contributing to only 4% of the world's hydrogen production, primarily due to the sluggish kinetics of the OER and HER, which require high overpotentials and hinder energy efficiency.

5.2 Future Directions

Advances in Catalyst Development for Water Splitting

Hydrogen production via electrochemical water splitting is a promising alternative to conventional energy sources. However, the sluggish kinetics of both the cathodic HER and the anodic OER significantly hinder its industrial application. Developing highly active and stable catalysts is crucial to overcoming these limitations.

State-of-the-art catalysts for HER predominantly operate in acidic electrolytes, while OER catalysts are more effective in alkaline media. The development of a single catalyst that performs efficiently in both electrolytes remains an open challenge.

Recently, catalysts based on transition metal compounds, including layered double hydroxides, chalcogenides, spinels, perovskites, single-atom catalysts, and high-entropy materials, have emerged as potential solutions. However, the underlying reaction mechanisms on these catalytic surfaces are still not fully understood, necessitating further investigation.

Understanding Mechanisms and Structural Transformations

A detailed understanding of structural transformations and the identification of real active phases and sites during catalytic reactions are vital for the rational design of optimal catalysts. Integration of in situ characterization techniques with theoretical modeling has proven to be an advanced approach to unravel reaction mechanisms, intermediates, and pathways. Advanced characterization techniques such as XAS, XRD, XPS, FTIR, and Raman spectroscopy, combined with theoretical studies, are playing a crucial role in unraveling these mechanisms. These insights are instrumental in guiding the development of electrocatalysts with enhanced activity and stability.

Challenges in Standardizing Catalyst Evaluation

Despite significant progress, directly comparing the performance of various nanostructured catalysts remains challenging. Differences in catalyst mass loadings, substrate materials, and electrochemical measurement methods can significantly affect electron transfer rates and performance descriptors. Establishing standardized evaluation protocols is essential for effective benchmarking and electrocatalyst screening across research groups.

Recent innovations, including nanostructure and lattice oxygen engineering, offer promising pathways for designing active, stable, and cost-effective OER and HER electrocatalysts. These advances are expected to accelerate the mass commercialization of water-splitting technologies, driving the hydrogen economy toward broader industrial adoption.

References

1. Kapdan, I.K., Kargi, F.: Bio-hydrogen production from waste materials. Enzyme Microb. Technol. **38**, 569–582 (2006)
2. Keçebaş, A., Kayfeci, M., Bayat, M.: Electrochemical hydrogen generation. In: Solar Hydrogen Production, pp. 299–317. Elsevier (2019)
3. Chi, J., Yu, H.: Water electrolysis based on renewable energy for hydrogen production. Chin. J. Catal. **39**, 390–394 (2018)
4. Abbassi-Mohadjel, N.: Sustainable water-splitting methods to produce hydrogen. McGill Green Chem. J. **72**.
5. Park, H.G., Holt, J.K.: Recent advances in nanoelectrode architecture for photochemical hydrogen production. Energy Environ. Sci. **3**, 1028–1036 (2010)
6. Ibrahim, K.B., Tsai, M.-C., Chala, S.A., Berihun, M.K., Kahsay, A.W., Berhe, T.A., Su, W.-N., Hwang, B.-J.: A review of transition metal-based bifunctional oxygen electrocatalysts. J. Chin. Chem. Soc. **66**, 829–865 (2019)
7. An, L., Wei, C., Lu, M., Liu, H., Chen, Y., Scherer, G.G., Fisher, A.C., Xi, P., Xu, Z.J., Yan, C.H.: Recent development of oxygen evolution electrocatalysts in acidic environment. Adv. Mater. **33**, 2006328 (2021)

8. Wong, M.-K., Foo, J.J., Loh, J.Y., Ong, W.-J.: Leveraging dual-atom catalysts for electrocatalysis revitalization: exploring the structure-performance correlation. Adv. Energy Mater. **14**, 2303281 (2024)
9. Li, X., Cheng, Z., Wang, X.: Understanding the mechanism of the oxygen evolution reaction with consideration of spin. Electrochem. Energy Rev. **4**, 136–145 (2021)
10. Ha, M.-A., Larsen, R.E.: Multiple reaction pathways for the oxygen evolution reaction may contribute to IrO_2 (110)'s high activity. J. Electrochem. Soc. **168**, 024506 (2021)
11. Wei, J., Zhou, M., Long, A., Xue, Y., Liao, H., Wei, C., Xu, Z.J.: Heterostructured electrocatalysts for hydrogen evolution reaction under alkaline conditions. Nano-Micro Lett. **10**, 1–15 (2018)
12. Durst, J., Siebel, A., Simon, C., Hasché, F., Herranz, J., Gasteiger, H.A.: New insights into the electrochemical hydrogen oxidation and evolution reaction mechanism. Energy Environ. Sci. **7**, 2255–2260 (2014)
13. Jiang, J., Xu, G., Gong, B., Zhu, J., Wang, W., Zhao, T., Feng, Y., Wu, Q., Liu, S., Zhang, L.: Hydrogen spillover mechanism of superaerophobic $NiSe_2$-Ni_5P_4 electrocatalyst to promote hydrogen evolution in saline water. Adv. Funct. Mater. **35**, 2412685 (2025)
14. Zahra, R., Drissi, N., Kumar, A.: Enhanced HER activity through $FeAlO_3$ perovskite oxide incorporated with conducting polymer PANI. Inorg. Chem. Commun. **172**, 113609 (2025)
15. Chatenet, M., et al.: Water electrolysis: from textbook knowledge to the latest scientific strategies and industrial developments. Chem. Soc. Rev. **51**, 4583–4762 (2022)
16. Lasia, A.: Mechanism and kinetics of the hydrogen evolution reaction. Int. J. Hydrogen Energy **44**, 19484–19518 (2019)
17. Wang, S., Lu, A., Zhong, C.-J.: Hydrogen production from water electrolysis: role of catalysts. Nano Converg. **8**, 4 (2021)
18. Lv, C. et al.: One-step construction of $Fe_{0.5}Mo_{0.5}Se_2$ with P doping on carbon cloth for efficient hydrogen evolution in acidic and alkaline media. Int. J. Hydrogen Energy **51**, 413–422 (2024)
19. Amin, H.M.A., Apfel, U.-P.: Metal-rich chalcogenides as sustainable electrocatalysts for oxygen evolution and reduction: state of the art and future perspectives. Eur. J. Inorg. Chem. **2020**, 2679–2690 (2020)
20. Jiang, Q., et al.: A highly efficient high-entropy metal hydroxymethylate electrocatalyst for oxygen evolution reaction. Chem. Eng. J. **453**, 139510 (2023)
21. Chen, S., Wang, Y., Wang, Z., Zhang, K.: Advances in nonprecious metal catalysts for efficient water oxidation in alkaline media. Ionics **29**, 9–32 (2023)
22. Kim, Y., Choi, E., Kim, S., Byon, H.R.: Layered transition metal oxides (LTMO) for oxygen evolution reactions and aqueous Li-ion batteries. Chem. Sci. **14**, 10644–10663 (2023)
23. Sha, S., Ge, R., Li, Y., Cairney, J.M., Zheng, R., Li, S., Liu, B., Zhang, J., Li, W.: High-entropy catalysts for electrochemical water-electrolysis of hydrogen evolution and oxygen evolution reactions. Front. Energy **18**, 265–290 (2024)
24. Wang, Z., Liu, S., Duan, W., Xing, Y., Hu, Y., Ma, Y.: Transition metal selenides as catalysts for electrochemical water splitting. Int. J. Hydrogen Energy **60**, 1414–1432 (2024)
25. Zeng, Y., Zhao, M., Huang, Z., Zhu, W., Zheng, J., Jiang, Q., Wang, Z., Liang, H.: Surface reconstruction of water splitting electrocatalysts. Adv. Energy Mater. **12**, 2201713 (2022)
26. Yan, Z., Liu, H., Hao, Z., Yu, M., Chen, X., Chen, J.: Electrodeposition of (hydro)oxides for an oxygen evolution electrode. Chem. Sci. **11**, 10614–10625 (2020)
27. Yang, G., Zhang, C., Chen, Z., Wang, J., Gao, G., Li, Z., Huang, R., Cui, Y.: Molybdenum single-atom solid-acid catalyst for the hydrogen evolution reaction in the alkaline electrolyte. ACS Catal. **15**, 2270–2281 (2025)
28. McCrory, C.C., Jung, S., Ferrer, I.M., Chatman, S.M., Peters, J.C., Jaramillo, T.F.: Benchmarking hydrogen evolving reaction and oxygen evolving reaction electrocatalysts for solar water splitting devices. J. Am. Chem. Soc. **137**, 4347–4357 (2015)
29. Liu, Z., Tan, H., Liu, D., Liu, X., Xin, J., Xie, J., Zhao, M., Song, L., Dai, L., Liu, H.: Promotion of overall water splitting activity over a wide pH range by interfacial electrical effects of metallic NiCo-nitrides nanoparticle/$NiCo_2O_4$ nanoflake/graphite fibers. Adv. Sci. **6**, 1801829 (2019)

30. Wei, C., Rao, R.R., Peng, J., Huang, B., Stephens, I.E., Risch, M., Xu, Z.J., Shao-Horn, Y.: Recommended practices and benchmark activity for hydrogen and oxygen electrocatalysis in water splitting and fuel cells. Adv. Mater. **31**, 1806296 (2019)
31. Sun, H., Xu, X., Kim, H., Jung, W., Zhou, W., Shao, Z.: Electrochemical water splitting: bridging the gaps between fundamental research and industrial applications. Energy Environ. Mater. **6**, e12441 (2023)
32. Lagadec, M.F., Grimaud, A.: Water electrolysers with closed and open electrochemical systems. Nat. Mater. **19**, 1140–1150 (2020)
33. Li, R.-J., Niu, W.-J., Zhao, W.-W., Yu, B.-X., Cai, C.-Y., Xu, L.-Y., Wang, F.-M.: Achievements and challenges in surfactants-assisted synthesis of MOFs-derived transition metal–nitrogen–carbon as a highly efficient electrocatalyst for ORR, OER, and HER. Small **21**, 2408227 (2025)
34. Kitiphatpiboon, N., Chen, M., Feng, C., Li, S., Abudula, A., Guan, G.: Highly efficient electrocatalysts for seawater electrolysis under high current density: a critical review. MetalMat **1**, e11 (2024)
35. Chen, W., Yu, M., Liu, S., Zhang, C., Jiang, S., Duan, G.: Recent progress of Ru single-atom catalyst: synthesis, modification, and energetic applications. Adv. Funct. Mater. **34**, 2313307 (2024)
36. Jeong, U., Lee, G., Kim, H.J., Kim, K., Han, J.-I.: Electrochemically-mediated treatment of sulfur dioxide and purification of carbon dioxide coupled with hydrogen production. Sep. Purif. Technol. **330**, 125438 (2024)
37. Siva, P., Vasu, K.: Efficient hydrogen evolution from degenerate-doped P-type WS_2 electrocatalysts over a wide pH range. J. Mater. Chem. A **12**, 11101–11108 (2024)
38. Li, X., Zhao, L., Yu, J., Liu, X., Zhang, X., Liu, H., Zhou, W.: Water splitting: from electrode to green energy system. Nano-Micro Lett. **12**, 131 (2020)
39. Wang, H., Wang, M., Wang, J.: How to choose suitable reference electrode and aqueous electrolyte to avoid error in electrochemical measurements? Curr. Chin. Sci. **3**, 204–212 (2023)
40. Troudt, B.K., Rousseau, C.R., Dong, X.I., Anderson, E.L., Bühlmann, P.: Recent progress in the development of improved reference electrodes for electrochemistry. Anal. Sci. **38**, 71–83 (2022)
41. Yang, X.-H., Cuesta, A., Cheng, J.: Computational Ag/AgCl reference electrode from density functional theory-based molecular dynamics. J. Phys. Chem. B **123**, 10224–10232 (2019)
42. Kawashima, K., Marquez-Montes, R.A., Son, Y.J., Milton, A.E.F., Le, T.V., Guo, C., Smith, L.A., Chukwuneke, C.E., Vaidyula, R.R., Mullins, C.B.: In: Considerations of Reference Electrodes and Liquid Junction Potentials for Accurate Electrocatalysis Studies, Electrochemical Society Meeting Abstracts, vol. 244, pp. 2575–2575. The Electrochemical Society, Inc. (2023)
43. Zang, Y., Lu, D.-Q., Wang, K., Li, B., Peng, P., Lan, Y.-Q., Zang, S.-Q.: A pyrolysis-free Ni/Fe bimetallic electrocatalyst for overall water splitting. Nat. Commun. **14**, 1792 (2023)
44. Ren, H., Pan, Y., Sorrell, C.C., Du, H.: Assessment of electrocatalytic activity through the lens of three surface area normalization techniques. J. Mater. Chem. A **8**, 3154–3159 (2020)
45. Li, Z., et al.: Seed-assisted formation of NiFe anode catalysts for anion exchange membrane water electrolysis at industrial-scale current density. Nat. Catal. **7**, 944–952 (2024)
46. Geng, Q., Li, Y., Xie, H.: Interface engineering of CoS/SnS_2 nanosheets for boosting electrocatalytic overall water splitting in alkaline medium. Vacuum **203**, 111308 (2022)
47. Prakash, C., Sahoo, P., Yadav, R., Pandey, A., Singh, V.K., Dixit, A.: Nanoengineered Zn-modified nickel sulfide (NiS) as a bifunctional electrocatalyst for overall water splitting. Int. J. Hydrogen Energy **48**, 21969–21980 (2023)
48. Wu, S., Liu, J., Cui, B., Li, Y., Liu, Y., Hu, B., He, L., Wang, M., Zhang, Z., Tian, K.: Fluorine-doped nickel cobalt oxide spinel as efficiently bifunctional catalyst for overall water splitting. Electrochim. Acta **299**, 231–244 (2019)
49. Yang, F., Luo, Y., Yu, Q., Zhang, Z., Zhang, S., Liu, Z., Ren, W., Cheng, H.M., Li, J., Liu, B.: A durable and efficient electrocatalyst for saline water splitting with current density exceeding 2000 mA cm^{-2}. Adv. Funct. Mater. **31**, 2010367 (2021)

50. Shen, F., Wang, Y., Qian, G., Chen, W., Jiang, W., Luo, L., Yin, S.: Bimetallic iron-iridium alloy nanoparticles supported on nickel foam as highly efficient and stable catalyst for overall water splitting at large current density. Appl. Catal. B **278**, 119327 (2020)
51. Nisar, L., Sadaqat, M., Hassan, A., Shah, A., Najam-Ul-Haq, M., Ashiq, M.N., Ehsan, M.F., Joya, K.S.: Ultrathin CoTe nanoflakes electrode demonstrating low overpotential for overall water splitting. Fuel **280**, 118666 (2020)
52. Wang, Y., Jiao, Y., Yan, H., Yang, G., Tian, C., Wu, A., Liu, Y., Fu, H.: Vanadium-incorporated CoP_2 with lattice expansion for highly efficient acidic overall water splitting. Angew. Chem. **134**, e202116233 (2022)
53. Tsai, F.-T., Deng, Y.-T., Pao, C.-W., Chen, J.-L., Lee, J.-F., Lai, K.-T., Liaw, W.-F.: The HER/OER mechanistic study of an FeCoNi-based electrocatalyst for alkaline water splitting. J. Mater. Chem. A **8**, 9939–9950 (2020)
54. Parra-Puerto, A., Ng, K.L., Fahy, K., Goode, A.E., Ryan, M.P., Kucernak, A.: Supported transition metal phosphides: activity survey for HER, ORR, OER, and corrosion resistance in acid and alkaline electrolytes. ACS Catal. **9**, 11515–11529 (2019)
55. Qi, Z., Xiao, C., Liu, C., Goh, T.W., Zhou, L., Maligal-Ganesh, R., Pei, Y., Li, X., Curtiss, L.A., Huang, W.: Sub-4 nm PtZn intermetallic nanoparticles for enhanced mass and specific activities in catalytic electrooxidation reaction. J. Am. Chem. Soc. **139**, 4762–4768 (2017)
56. Shanmugapriya, S., Ganeshbabu, M., Surendran, S., Lee, Y.S., Selvan, R.K.: The trifunctional electrocatalytic activity of hierarchically structured porous carbon derived from environmentally malignant Prosopis juliflora. Carbon Trends **17**, 100419 (2024)
57. Nazari, M., Ghaemmaghami, M.: Approach to evaluation of electrocatalytic water splitting parameters, reflecting intrinsic activity: toward the right pathway. ChemSusChem **16**, e202202126 (2023)
58. McCrory, C.C.L., Jung, S., Peters, J.C., Jaramillo, T.F.: Benchmarking heterogeneous electrocatalysts for the oxygen evolution reaction. J. Am. Chem. Soc. **135**, 16977–16987 (2013)
59. Yang, J., Mohmad, A.R., Wang, Y., Fullon, R., Song, X., Zhao, F., Bozkurt, I., Augustin, M., Santos, E.J., Shin, H.S.: Ultrahigh-current-density niobium disulfide catalysts for hydrogen evolution. Nat. Mater. **18**, 1309–1314 (2019)
60. Zhang, Z., et al.: Electrochemical deposition as a universal route for fabricating single-atom catalysts. Nat. Commun. **11**, 1215 (2020)
61. Anantharaj, S., Ede, S.R., Karthick, K., Sam Sankar, S., Sangeetha, K., Karthik, P.E., Kundu, S.: Precision and correctness in the evaluation of electrocatalytic water splitting: revisiting activity parameters with a critical assessment. Energy Environ. Sci. **11**, 744–771 (2018)
62. Wang, Z., Hou, Q., Zhang, X., Liu, H., Lin, X., Chen, H., Ding, R., Wei, S., Lu, X., Liu, S.: Surface reconfiguration of silver nanoparticle doped cobalt phosphate for enhancing overall water electrocatalysis in alkali. Appl. Surf. Sci. **639**, 158165 (2023)
63. Roger, I., Shipman, M.A., Symes, M.D.: Earth-abundant catalysts for electrochemical and photoelectrochemical water splitting. Nat. Rev. Chem. **1**, 0003 (2017)
64. Kang, Q., Lai, D., Tang, W., Lu, Q., Gao, F.: Intrinsic activity modulation and structural design of NiFe alloy catalysts for an efficient oxygen evolution reaction. Chem. Sci. **12**, 3818–3835 (2021)
65. Septiani, N.L.W., et al.: Self-assembly of two-dimensional bimetallic nickel-cobalt phosphate nanoplates into one-dimensional porous chainlike architecture for efficient oxygen evolution reaction. Chem. Mater. **32**, 7005–7018 (2020)
66. Craig, M.J., Coulter, G., Dolan, E., Soriano-López, J., Mates-Torres, E., Schmitt, W., García-Melchor, M.: Universal scaling relations for the rational design of molecular water oxidation catalysts with near-zero overpotential. Nat. Commun. **10**, 4993 (2019)
67. Chen, Y., Rui, K., Zhu, J., Dou, S.X., Sun, W.: Recent progress on nickel-based oxide/(oxy)hydroxide electrocatalysts for the oxygen evolution reaction. Chem. - A Eur. J. **25**, 703–713 (2019)
68. Chen, Y., Rui, K., Zhu, J., Dou, S.X., Sun, W.: Recent progress on nickel-based oxide/(oxy) hydroxide electrocatalysts for the oxygen evolution reaction. Chem. - A Eur. J. **25**, 703–713 (2019)

69. Wang, G.B., Lin, L.L., Wang, H.T., Chen, H.M.: Novel insight from in-/ex-situ investigation toward intercalated water in high-performance oxygen evolution electrocatalysts and their mechanisms. J. Chin. Chem. Soc. **70**, 2064–2088 (2023)
70. Wang, J., Gao, Y., Kong, H., Kim, J., Choi, S., Ciucci, F., Hao, Y., Yang, S., Shao, Z., Lim, J.: Non-precious-metal catalysts for alkaline water electrolysis: operando characterizations, theoretical calculations, and recent advances. Chem. Soc. Rev. **49**, 9154–9196 (2020)
71. Vojvodic, A., Nørskov, J.K.: Optimizing perovskites for the water-splitting reaction. Science **334**, 1355–1356 (2011)
72. Suntivich, J., May, K.J., Gasteiger, H.A., Goodenough, J.B., Shao-Horn, Y.: A perovskite oxide optimized for oxygen evolution catalysis from molecular orbital principles. Science **334**, 1383–1385 (2011)
73. Yagi, S., et al.: Covalency-reinforced oxygen evolution reaction catalyst. Nat. Commun. **6**, 8249 (2015)
74. Lee, J.G., Hwang, J., Hwang, H.J., Jeon, O.S., Jang, J., Kwon, O., Lee, Y., Han, B., Shul, Y.-G.: A new family of perovskite catalysts for oxygen-evolution reaction in alkaline media: $BaNiO_3$ and $BaNi_{0.83}O_{2.5}$. J. Am. Chem. Soc. **138**, 3541–3547 (2016)
75. Anusha, H.S., Jijoe, S.P., Tenzin, T., Divya, V., Anilkumar, K.M., Yashas, S.R., Shivaraju, H.P.: $MgFeO_3$ perovskite nanostructures: a new frontier in photoelectrochemical water splitting. Int. J. Hydrogen Energy **88**, 617–625 (2024)
76. Nikolov, I., Darkaoui, R., Zhecheva, E., Stoyanova, R., Dimitrov, N., Vitanov, T.: Electrocatalytic activity of spinel related cobalties $MxCo_3{}^{-}XO_4$ (M = Li, Ni, Cu) in the oxygen evolution reaction. J. Electroanal. Chem. **429**, 157–168 (1997)
77. Shetty, K., Renuka, L., Nagaswarupa, H.P., Nagabhushana, H., Anantharaju, K.S., Rangappa, D., Prashantha, S.C., Ashwini, K.: A comparative study on $CuFe_2O_4$, $ZnFe_2O_4$ and $NiFe_2O_4$: morphology, impedance and photocatalytic studies. Mater. Today: Proc. **4**, 11806–11815 (2017)
78. Li, M., Xiong, Y., Liu, X., Bo, X., Zhang, Y., Han, C., Guo, L.: Facile synthesis of electrospun MFe_2O_4 (M = Co, Ni, Cu, Mn) spinel nanofibers with excellent electrocatalytic properties for oxygen evolution and hydrogen peroxide reduction. Nanoscale **7**, 8920–8930 (2015)
79. Arrabito, G., Bonasera, A., Prestopino, G., Orsini, A., Mattoccia, A., Martinelli, E., Pignataro, B., Medaglia, P.G.: Layered double hydroxides: a toolbox for chemistry and biology. Crystals **9**, 361 (2019)
80. Ibrahim, K.B., Su, W.N., Tsai, M.C., Kahsay, A.W., Chala, S.A., Birhanu, M.K., Lee, J.F., Hwang, B.J.: Heterostructured composite of NiFe-LDH nanosheets with Ti_4O_7 for oxygen evolution reaction. Mater. Today Chem. **24**, 100824 (2022)
81. Ibrahim, K.B., Su, W.-N., Tsai, M.-C., Chala, S.A., Kahsay, A.W., Yeh, M.-H., Chen, H.-M., Duma, A.D., Dai, H., Hwang, B.-J.: Robust and conductive Magnéli phase Ti_4O_7 decorated on 3D-nanoflower NiRu-LDH as high-performance oxygen reduction electrocatalyst. Nano Energy **47**, 309–315 (2018)
82. Wang, Y., et al.: Atomically targeting NiFe LDH to create multivacancies for OER catalysis with a small organic anchor. Nano Energy **81**, 105606 (2021)
83. Alobaid, A., Wang, C., Adomaitis, R.A.: Mechanism and kinetics of HER and OER on NiFe LDH films in an alkaline electrolyte. J. Electrochem. Soc. **165**, J3395 (2018)
84. Palchoudhury, S., Ramasamy, K., Han, J., Chen, P., Gupta, A.: Transition metal chalcogenides for next-generation energy storage. Nanoscale Adv. **5**, 2724–2742 (2023)
85. Wang, F., Shifa, T.A., Zhan, X., Huang, Y., Liu, K., Cheng, Z., Jiang, C., He, J.: Recent advances in transition-metal dichalcogenide based nanomaterials for water splitting. Nanoscale **7**, 19764–19788 (2015)
86. Belay Ibrahim, K., Ahmed Shifa, T., Zorzi, S., Getaye Sendeku, M., Moretti, E., Vomiero, A.: Emerging 2D materials beyond Mxenes and TMDS: transition metal carbo-chalcogenides. Prog. Mater. Sci. **144**, 101287 (2024)
87. Wang, X., Vasileff, A., Jiao, Y., Zheng, Y., Qiao, S.Z.: Electronic and structural engineering of carbon-based metal-free electrocatalysts for water splitting. Adv. Mater. **31**, 1803625 (2019)

88. Hu, C., Paul, R., Dai, Q., Dai, L.: Carbon-based metal-free electrocatalysts: from oxygen reduction to multifunctional electrocatalysis. Chem. Soc. Rev. **50**, 11785–11843 (2021)
89. Wang, T., Cao, X., Jiao, L.: MOFs-derived carbon-based metal catalysts for energy-related electrocatalysis. Small **17**, 2004398 (2021)
90. Wu, H.-S., Murti, B.T., Singh, J., Yang, P.-K., Tsai, M.-L.: Prospects of metal-free perovskites for piezoelectric applications. Adv. Sci. **9**, 2104703 (2022)
91. Xu, Z., Zhang, Y., Li, X., Qin, L., Meng, Q., Zhang, G., Fan, Z., Xue, Z., Guo, X., Liu, Q.: Template-free synthesis of stable cobalt manganese spinel hollow nanostructured catalysts for highly water-resistant Co oxidation. iScience **21**, 19–30 (2019)
92. Richetta, M., Digiamberardino, L., Mattoccia, A., Medaglia, P., Montanari, R., Pizzoferrato, R., Scarpellini, D., Varone, A., Kaciulis, S., Mezzi, A.: Surface spectroscopy and structural analysis of nanostructured multifunctional (Zn, Al) layered double hydroxides. Surf. Interface Anal. **48**, 514–518 (2016)
93. Kuc, A., Heine, T., Kis, A.: Electronic properties of transition-metal dichalcogenides. MRS Bull. **40**, 577–584 (2015)
94. Hu, C., Dai, L.: Carbon-based metal-free catalysts for electrocatalysis beyond the ORR. Angew. Chem. Int. Ed. **55**, 11736–11758 (2016)
95. Hong, W., Wyatt, B.C., Nemani, S.K., Anasori, B.: Double transition-metal Mxenes: atomistic design of two-dimensional carbides and nitrides. MRS Bull. **45**, 850–861 (2020)
96. Nemani, S.K., Torkamanzadeh, M., Wyatt, B.C., Presser, V., Anasori, B.: Functional two-dimensional high-entropy materials. Commun. Mater. **4**, 16 (2023)
97. Yang, X.-F., Wang, A., Qiao, B., Li, J., Liu, J., Zhang, T.: Single-atom catalysts: a new frontier in heterogeneous catalysis. Acc. Chem. Res. **46**, 1740–1748 (2013)
98. Niu, B., Zhang, F., Ping, H., Li, N., Zhou, J., Lei, L., Xie, J., Zhang, J., Wang, W., Fu, Z.: Sol-gel autocombustion synthesis of nanocrystalline high-entropy alloys. Sci. Rep. **7**, 3421 (2017)
99. Yang, B., Zhang, Y., Pan, H., Si, W., Zhang, Q., Shen, Z., Yu, Y., Lan, S., Meng, F., Liu, Y.: High-entropy enhanced capacitive energy storage. Nat. Mater. **21**, 1074–1080 (2022)
100. Rost, C.M., Sachet, E., Borman, T., Moballegh, A., Dickey, E.C., Hou, D., Jones, J.L., Curtarolo, S., Maria, J.-P.: Entropy-stabilized oxides. Nat. Commun. **6**, 8485 (2015)
101. Zhan, C., et al.: Subnanometer high-entropy alloy nanowires enable remarkable hydrogen oxidation catalysis. Nat. Commun. **12**, 6261 (2021)
102. Wang, Y., Mi, J., Wu, Z.-S.: Recent status and challenging perspective of high entropy oxides for chemical catalysis. Chem. Catal. **2**, 1624–1656 (2022)
103. Greiner, M.T., Jones, T.E., Beeg, S., Zwiener, L., Scherzer, M., Girgsdies, F., Piccinin, S., Armbrüster, M., Knop-Gericke, A., Schlögl, R.: Free-atom-like D states in single-atom alloy catalysts. Nat. Chem. **10**, 1008–1015 (2018)
104. Liu, J.: Catalysis by supported single metal atoms. ACS Catal. **7**, 34–59 (2017)
105. Lopez, N., Janssens, T., Clausen, B., Xu, Y., Mavrikakis, M., Bligaard, T., Nørskov, J.K.: On the origin of the catalytic activity of gold nanoparticles for low-temperature Co oxidation. J. Catal. **223**, 232–235 (2004)
106. Van Hardeveld, R., Hartog, F.: The statistics of surface atoms and surface sites on metal crystals. Surf. Sci. **15**, 189–230 (1969)
107. Valden, M., Lai, X., Goodman, D.: Onset of catalytic activity of gold clusters on Titania with the appearance of nonmetallic properties. Science **281**, 1647–1650 (1998)
108. Campbell, C.T.: Electronic perturbations. Nat. Chem. **4**, 597–598 (2012)
109. Yoon, B., Hakkinen, H., Landman, U., Worz, A.S., Antonietti, J.-M., Abbet, S., Judai, K., Heiz, U.: Charging effects on bonding and catalyzed oxidation of Co on Au_8 clusters on MgO. Science **307**, 403–407 (2005)
110. Thomas, J.M., Raja, R., Lewis, D.W.: Single-site heterogeneous catalysts. Angew. Chem. Int. Ed. **44**, 6456–6482 (2005)
111. Langer, J., et al.: Present and future of surface-enhanced Raman scattering. ACS Nano **14**, 28–117 (2020)

112. Deng, Y., Handoko, A.D., Du, Y., Xi, S., Yeo, B.S.: In situ Raman spectroscopy of copper and copper oxide surfaces during electrochemical oxygen evolution reaction: identification of CuIII oxides as catalytically active species. ACS Catal. **6**, 2473–2481 (2016)
113. Kurouski, D., Dazzi, A., Zenobi, R., Centrone, A.: Infrared and Raman chemical imaging and spectroscopy at the nanoscale. Chem. Soc. Rev. **49**, 3315–3347 (2020)
114. Moshtaghpour, A., Velazco-Torrejon, A., Nicholls, D., Robinson, A.W., Kirkland, A.I., Browning, N.D.: Diffusion distribution model for damage mitigation in scanning transmission electron microscopy. J. Microsc. **297**, 57–77 (2025)
115. Wu, Y., Wu, P., Tang, Y., Fu, R., Cui, Y., Chen, J., Kübel, C., Xu, F.: Deciphering unexplored reversible potassium storage and small volume change in a CaV_4O_9 anode with in situ transmission electron microscopy. Adv. Funct. Mater. **34**, 2314344 (2024)
116. Zhang, J., et al.: Unveiling charge utilization mechanisms in ferroelectric for water splitting. Nat. Commun. **16**, 1515 (2025)
117. Chu, T., Zhou, L., Zhang, B., Xuan, F.-Z.: Accurate atomic scanning transmission electron microscopy analysis enabled by deep learning. Nano Res. **17**, 2971–2980 (2024)
118. Schlüter, N., Novák, P., Schröder, D.: Nonlinear electrochemical analysis: worth the effort to reveal new insights into energy materials. Adv. Energy Mater. **12**, 2200708 (2022)
119. Amirbeigiarab, R., Tian, J., Herzog, A., Qiu, C., Bergmann, A., Roldan Cuenya, B., Magnussen, O.M.: Atomic-scale surface restructuring of copper electrodes under Co_2 electroreduction conditions. Nat. Catal. **6**, 837–846 (2023)
120. Duquesne, E., et al.: Operando XRD and electrogravimetry coupling to analyze species transfers during redox processes in Ni/Fe-layered double hydroxide. J. Mater. Chem. A **10**, 24783–24792 (2022)
121. Stevie, F.A., Donley, C.L.: Introduction to X-ray photoelectron spectroscopy. J. Vac. Sci. Technol. A **38** (2020)
122. Fadley, C.S.: X-ray photoelectron spectroscopy: progress and perspectives. J. Electron. Spectrosc. Relat. Phenom. **178–179**, 2–32 (2010)
123. Greczynski, G., Haasch, R.T., Hellgren, N., Lewin, E., Hultman, L.: X-ray photoelectron spectroscopy of thin films. Nat. Rev. Methods Prim. **3**, 40 (2023)
124. Meilak, J.A., et al.: An X-ray absorption spectroscopy investigation into the fundamental structure of liquid metal alloys. Small Sci. **4**, 2400317 (2024)
125. Xu, Y.N., Mei, B., Xu, Q., Fu, H.Q., Zhang, X.Y., Liu, P.F., Jiang, Z., Yang, H.G.: In situ/operando synchrotron radiation analytical techniques for Co_2/Co reduction reaction: from atomic scales to mesoscales. Angew. Chem. Int. Ed. e202404213 (2024)
126. Chantler, C.T., Bunker, G., D'Angelo, P., Diaz-Moreno, S.: X-ray absorption spectroscopy. Nat. Rev. Methods Prim. **4**, 89 (2024)
127. Chen, M., Liu, D., Qiao, L., Zhou, P., Feng, J., Ng, K.W., Liu, Q., Wang, S., Pan, H.: In-situ/operando Raman techniques for in-depth understanding on electrocatalysis. Chem. Eng. J. **461**, 141939 (2023)
128. Chala, S.A., Tsai, M.-C., Olbasa, B.W., Lakshmanan, K., Huang, W.-H., Su, W.-N., Liao, Y.-F., Lee, J.-F., Dai, H., Hwang, B.J.: Tuning dynamically formed active phases and catalytic mechanisms of in situ electrochemically activated layered double hydroxide for oxygen evolution reaction. ACS Nano **15**, 14996–15006 (2021)
129. Chen, Z., Fan, Q., Zhou, J., Wang, X., Huang, M., Jiang, H., Cölfen, H.: Toward understanding the formation mechanism and OER catalytic mechanism of hydroxides by in situ and operando techniques. Angew. Chem. Int. Ed. **62**, e202309293 (2023)
130. Song, J., Qian, Z.-X., Yang, J., Lin, X.-M., Xu, Q., Li, J.-F.: In situ/operando investigation for heterogeneous electro-catalysts: from model catalysts to state-of-the-art catalysts. ACS Energy Lett. **9**, 4414–4440 (2024)

Electrochemical Diagnostics in Water Splitting: Methods and Best Practices

Suraj Kumar, Priyakshi Bora, and Dipak Sinha

Abstract Electrochemical water splitting offers an efficient pathway for hydrogen generation, playing a key role in advancing green energy solutions. While developing efficient electrocatalysts is crucial, ensuring accurate evaluation of their performance is equally important. However, the rapid expansion of materials science and the growing interdisciplinary interest in electrochemistry have led to widespread misinterpretation of data, particularly among beginners to the field. This chapter lays out essential guidelines for assessing water-splitting electrocatalysts. It covers critical factors like experimental setup, electrode selection (reference, counter and working), and electrolytes. Additionally, it outlines key metrics for evaluating catalytic activity, selectivity, and stability, along with recommendations and best practices for standardized data reporting to enhance clarity and reproducibility in the field.

1 Introduction

Electrochemical water splitting, when driven by green energy, is a highly efficient technique for the large-scale production of green hydrogen. The procedure encompasses two principal half-reactions: the hydrogen evolution reaction (HER) at the cathode and oxygen evolution reaction (OER) at the anode (Fig. 1a) [1]. The HER is a relatively simple process including a series of adsorption, discharge, and desorption reactions involving protons or water molecules to generate hydrogen gas. Conversely, the OER is a multi-step, complex reaction that entails the formation of several intermediates including the reduction and self-oxidation of electrocatalysts [2]. Despite its promise, water splitting faces significant kinetic limitations, particularly for the OER, which requires considerable overpotentials (η) to proceed efficiently. Addressing these issues requires the design of efficient, cost-effective, earth-abundant and durable electrocatalysts that can achieve high current densities while operating at low overpotentials [2, 3]. Some researchers employ systematic computational and experimental

S. Kumar · P. Bora · D. Sinha (✉)
Department of Chemistry, School of Sciences, Nagaland University, Lumami, Nagaland, India
e-mail: dipaksinha@gmail.com

O. P. Pandey et al. (eds.), *2D Materials for Electrochemical Water Splitting*, Engineering Materials, https://doi.org/10.1007/978-981-95-2003-9_3

studies to develop more efficient catalysts, whereas others focus on designing and engineering new electrode structures/catalysts [2, 4, 5]. Such efforts have advanced the understanding of critical electrochemical reactions, including HER and OER. Despite these advances, comparing the catalytic properties of different nanomaterials still poses considerable challenge. To ensure continued progress in the field, it is imperative to adopt a standardized methodology for evaluating and benchmarking the catalytic performance of nanomaterials [2, 3, 6–8]. This chapter outlines recommended experimental practices, guidelines for effectively reporting electrocatalytic results, and a clear overview of the dos and don'ts when evaluating different types of electrocatalysts for water splitting.

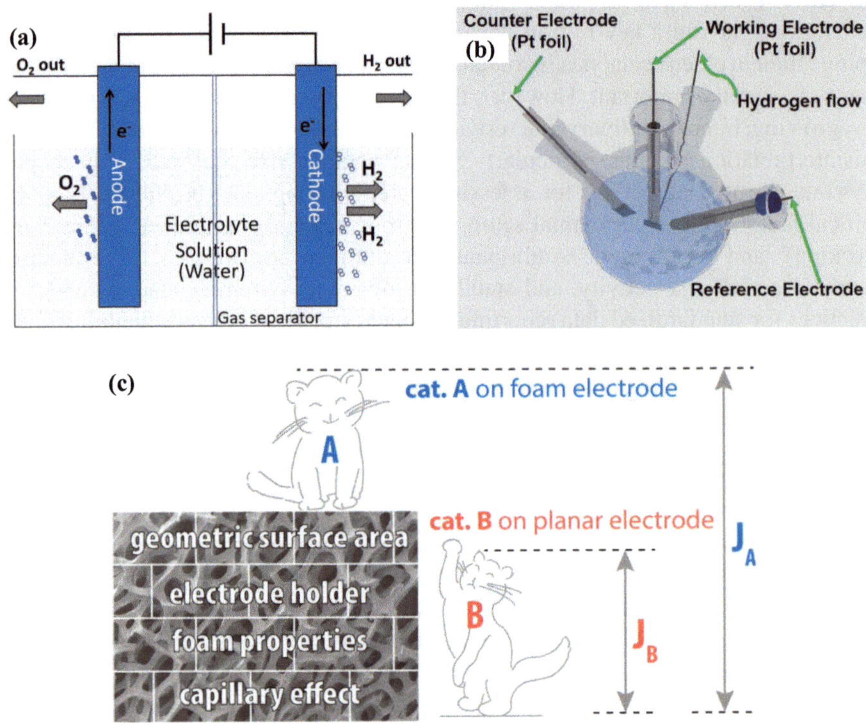

Fig. 1 **a** Diagram of a two-electrode water electrolyzer or fuel cell. Adapted with permission from [12]. © 2021, Yue et al.; exclusive licensee Elsevier Ltd. Distributed under a Creative Commons Attribution License 4.0 (CC BY) https://creativecommons.org/licenses/by/4.0/. **b** Illustration of a three-electrode cell configuration, including a working electrode, reference electrode, counter electrode, and a gas inlet. Adapted with permission from [13]. © 2020 American Chemical Society. **c** When comparing catalyst performance, those supported by foam-type electrodes (Cat A) seem "superior" compared to those on planar electrodes (Cat B). Adapted with permission from [14]. © 2020 American Chemical Society

2 Experimental Setup

2.1 *Electrochemical Cell*

A good experimental setting is crucial for comprehending the chemistry and practical uses of the catalyst [9]. Numerous studies suggest that a three-electrode configuration, comprising a working electrode, a reference electrode, a counter electrode and a gas inlet is preferred over a two-electrode system (Fig. 1b). The three-electrode system enables the assessment of the individual overpotential of the working electrode, allowing for accurate evaluation of the catalyst in water-splitting operations. In contrast, a two-electrode system, which is used in devices like electrolyzers and fuel cells, only measures the cumulative overpotential of the entire system, making them unsuitable for studying behaviour of individual electrodes [2].

Further, it is essential to conduct the electrochemical test with a clean electrolyte. This can be achieved by carefully selecting suitable cell materials, as the cell itself can influence the activity results. For neutral and acidic electrolytes, glass cells are typically preferred, whereas polymer-based materials like polytetrafluoroethylene (PTFE) or polyoxymethylene (POM) are ideal for alkaline environments due to the potential corrosion of glass in such conditions. Moreover, the purity of precursor chemicals utilized for the preparation of the electrolyte is crucial. Impurities, particularly in alkaline solutions, may deposit on the working electrode and compromise results. Therefore, high-purity reagents should be employed to ensure consistent and reliable outcomes [10]. Additionally, to eliminate potential contaminants that might be present in the cell, thorough cleaning of the electrochemical cell and associated components is essential. For example, the cell can be soaked in a concentrated NaOH solution/piranha solution for a minimum of 24 h before use.

Another essential component of the electrochemical cell design is the gas inlet within the three-electrode system. It supplies the required reactants, such as H_2 for HER and O_2 for OER. Some studies have examined the impact of gas saturation on electrochemical reactions and found that in Ar-enriched $HClO_4$, the onset potential for HER of Pt-based catalysts occurs before 0 V versus RHE. Likewise, for OER, studies on perovskite catalysts have shown higher currents under Ar saturation compared to O_2 saturation, highlighting the importance of proper gas management [11]. Therefore, continuous gas purging is indispensable for ensuring consistent and reliable activity measurements for HER and OER, regardless of whether the reaction entails gas evolution or consumption.

2.2 *Reference and Counter Electrode Selection*

The selection of proper reference electrode and counter electrode is crucial for accurate and reliable potential measurements in electrochemical studies. A reference electrode must maintain consistency and precision in its potential, as this forms the foundation for controlling and monitoring the potential applied to catalysts. Among the frequently employed reference electrodes, the saturated calomel electrode (SCE) and silver/silver chloride electrode (Ag/AgCl) are widely preferred. However, their usage requires careful consideration of factors such as concentration of KCl solution, chloride ion (Cl^-) leaching and compatibility with the electrolyte [2]. Studies have also shown that the KCl concentration in the reference electrode affects its potential. For instance, the potential of an Ag/AgCl reference electrode is 0.197 V relative to the standard hydrogen electrode (SHE) with KCl (saturated) but shifts to 0.288 V when using 0.1 M KCl [2].

Alkaline electrolytes also present challenges for Cl^- containing reference electrodes such as Ag/AgCl or SCE, because they may exhibit potential shifts due to interactions with hydroxide ions (OH^-), which can alter the pH of KCl filling solutions or lead to oxide formation. For these environments, Hg/HgO electrodes are more stable and reliable. Modifying the electrode design, for instance, replacing the frit material with alkaline-stable components, can further improve performance [15, 16]. To ensure consistency across different studies and electrolytes, all potentials should be referenced to the reversible hydrogen electrode (RHE) scale. This conversion accounts for pH variations and simplifies comparisons of data under different conditions [17]. Thus, selecting the appropriate reference electrode and accurately calibrating it are essential steps for achieving reliable and reproducible results in electrocatalytic research.

Similarly, choosing an appropriate counter electrode is crucial. A suitable counter electrode must facilitate electron transfer without restricting the reaction occurring at the working electrode. Materials like Pt are frequently used, due to their ability to sustain high currents and balance charge during reactions such as hydrogen and oxygen evolution. However, while Pt is considered robust, it is not without drawbacks. Recent studies have shown substantial Pt dissolution during electrochemical cycling in both acidic and alkaline conditions [18, 19]. This dissolution can lead to Pt being deposited onto the working electrode, especially in experiments with Pt group metal-free catalysts. Such contamination alters the electrochemical performance, often resulting in misleadingly high activity measurements and erroneous conclusions [9]. To mitigate these issues, researchers are actively investigating alternatives to Pt as counter electrode [20]. Carbon-based materials like graphite rods and glassy carbon, are promising options. Additionally, separating the counter and working electrode compartments with a proton-exchange membrane like Nafion helps minimize cross-contamination. However, even carbon electrodes have potential issues, as they may release impurities or gases like CO and CO_2 under oxidizing conditions [21]. Thus, a reliable counter electrode should meet several critical criteria, including a higher current than the working electrode, providing a uniform electric field, and preventing contamination of the working electrode. Addressing these aspects ensures enhanced accuracy and reliability in electrocatalytic measurements, which is especially important when investigating Pt group metal-free catalytic systems.

2.3 Working Electrode

Producing dependable electrochemical data requires careful attention to the design, preparation, and cleanliness of the working electrode. The selection of substrate is a key step in accurately evaluating an electrocatalyst's efficacy. A substrate acts as a stable, electrically conductive platform for depositing the catalyst material [3]. Glassy carbon is a widely used substrate for assessing electrocatalytic activity owing to its electrochemical inertness, high conductivity, and stability within the relevant potential window for reactions like HER and OER. However, glassy carbon has limitations, particularly in OER studies, where carbon corrosion becomes significant [3]. When the potential surpasses its equilibrium value of 0.207 V versus RHE, carbon becomes thermodynamically unstable. Its electrochemical oxidation produces soluble and insoluble organic and inorganic compounds, such as CO and CO_2, which can interfere with measurements [22].

Alternative substrates, such as carbon cloth and nickel foam, are commonly used. However, these substrates lack well-defined geometric areas, which hinders precise evaluation of the specific and mass activity of catalyst materials (Fig. 1c). For comprehensive guidance on using foam-based electrodes, the authors encourage readers to consult additional reports [7, 14].

Additionally, using a rotating disk electrode (RDE) is recommended for evaluating the kinetics of HER and OER. The RDE configuration provides well-defined mass transport of hydrogen or oxygen, effectively isolating kinetic currents from mass-transport-limited reactions (e.g., the oxygen reduction reaction (ORR) and hydrogen oxidation reaction (HOR)). The Teflon-shrouded RDE, originally introduced by Frumkin and later optimized by Schmidt and colleagues for use with ink-casted electrodes, continues to be the most widely adopted configuration [2]. Another commonly used option is the rotating ring-disk electrode (RRDE), which includes a concentric second working electrode, typically a Au or Pt ring, alongside the single RDE.

Researchers must also focus on the precise preparation of ink-casted working electrodes, paying close attention to factors such as catalyst loading, the incorporation of conductive additives, and Nafion content [2]. Firstly, the catalyst loading must be optimized; too little leads to incomplete coating and poor mass transport, while too much causes aggregation and higher mass transport losses. A typical ink-casted layer is around 1 μm thick. Secondly, when preparing electrodes with freestanding catalyst particles, a conductive, catalytically inert additive should be included to preserve the catalyst's structure. Common choices like porous carbon materials (e.g., Tanaka Kikinzoku Kogyo, Japan (TKK), Ketjen Black, Vulcan) are effective but may corrode before OER begins, limiting their stability for long-term tests [23]. However, they remain suitable for short-term OER activity screening and for reactions like HER. Finally, Nafion loading should be minimized to avoid thick films that increase mass transport resistance and affect the precision of kinetic current measurements [2].

2.4 Electrolyte Preparation

The selection of electrolyte greatly affects the performance of electrochemical systems, with variations in pH (acidic, neutral, or alkaline) and the type of solvent used (such as organic solvents or ionic liquids) playing key roles [24]. Commonly employed acidic electrolytes for HER studies include sulfuric acid (H_2SO_4) [25], hydrochloric acid (HCl) [26], and perchloric acid ($HClO_4$) [26]. On the other hand, alkaline electrolytes like potassium hydroxide (KOH) [27] and sodium hydroxide (NaOH) [28] are widely used, with lithium hydroxide (LiOH) occasionally utilized for specific applications [26]. Beyond traditional liquid electrolytes, solid-state electrolytes and ionic liquids have also garnered significant attention for their advantageous characteristics, such as high electrical conductivity, low vapour pressure and functional versatility [29].

The preparation process of electrolytes, is highly sensitive to impurities, which may stem from a range of sources, including commercial reagents, environmental exposure, and materials used in electrochemical setups. Impurities in alkaline electrolytes, for instance, can result from CO_2 absorption from the air, forming dissolved carbonates that alter the effective molarity and lead to inaccurate pH estimations. Hygroscopic alkali solids, which readily absorb moisture, further contribute to inaccuracies in concentration if not handled in controlled environments. To mitigate these issues, specific measures are necessary. For preventing CO_2 absorption, deionized water can be pre-boiled prior to use and inert gas purging can be applied during electrolyte preparation [30]. Handling hygroscopic solids in dry environments ensures accurate weighing, while checking alkali purity using certificates of analysis from manufacturers helps maintain the correct concentration. Contamination from glassware is another significant concern, particularly when using strong alkaline solutions. Borosilicate glass can undergo etching, releasing metal contaminants into the electrolyte. Replacing glassware with plasticware, especially for concentrated alkaline solutions used over extended periods, is an effective solution [31].

Another source of impurities stems from the degradation of materials, such as cation exchange membranes, and the leaching of metals from electrodes, cleaning tools, and other components. Metal impurities, including iron (Fe), titanium (Ti), manganese (Mn), and aluminium (Al), can significantly impact electrochemical performance. For example, while trace iron can enhance OER activity, it can also poison electrocatalysts over time. Such effects are particularly problematic during prolonged testing, emphasizing the need for high-quality materials and consistent monitoring of electrolyte purity [30].

Cations and anions in electrolytes also play a significant role in electrochemical water-splitting processes. Cations, including alkali metal ions (Li^+, K^+, Na^+, Cs^+), interact with water molecules, surface adsorbates, and the electrode surface through noncovalent interactions [32]. These interactions can alter the solvation environment and change how reactive intermediates behave at the electrified interface. For example, in high-pH electrolytes, larger cations like Cs^+ improve catalytic activity more effectively than smaller cations like Li^+ [33]. Anions also have a significant impact. Certain anions, such as sulfate (SO_4^{2-}), bisulfate (HSO_4^-), and chloride (Cl^-), can adsorb onto electrode surfaces and interfere with reaction processes and diminish catalytic performance [34]. These adsorbed anions can poison the catalyst, slowing down reactions. In contrast, perchlorate (ClO_4^-) ions are considered less reactive and are often used in HER studies because they do not negatively affect the catalyst or the reaction kinetics [35]. To minimize the negative effects of cations and anions, careful control of electrolyte composition is essential. Selecting cations and anions that do not interfere with the catalyst is crucial. For example, larger cations like K^+ or Cs^+ may be preferred for their weaker hydration and better stabilization of reaction intermediates. Anions that are less likely to adsorb onto the electrode surface, such as ClO_4^-, should be used to avoid poisoning the catalyst. Thus, electrolyte optimization represents a pragmatic and impactful approach for enhancing the overall efficiency of water electrolysis systems.

3 Evaluation of Performance of Electrocatalysts

To ascertain if a catalyst is suitable as an electrocatalyst for water splitting, it is indispensable to evaluate its activity, selectivity, and stability with the highest level of accuracy (Scheme 1). An optimal electrocatalyst is characterised by its high activity, selectivity and stability [3, 36]. These parameters, along with the methods and best practices for evaluating water-splitting electrocatalysts, are discussed in detail in the next sections [6].

3.1 Activity

There are multiple ways to express activity. Activity markers are generally classified into apparent and intrinsic types. Apparent activity depends on parameters like catalyst loading and surface area, whereas intrinsic activity remains constant regardless of these variables. Intrinsic markers are crucial for designing efficient electrocatalytic systems with high apparent activity. However, most recent studies on water splitting focus solely on apparent markers, neglecting intrinsic ones [6]. Among these, the most commonly used activity marker is overpotential (η). Overpotential is defined as the additional electromotive force required by a catalyst to initiate a reaction beyond its equilibrium potential. In other words, η represents the difference between the potential needed to attain a specific current density and the thermodynamic potential for OER (1.23 V) or HER (0 V), as illustrated in Fig. 2a [37]. The overpotential at a given current density, excluding iR compensation, is expressed by the equations, $\eta_{HER} = E_{RHE} - 0$ V and, $\eta_{OER} = E_{RHE} - 1.23$ V for HER and OER respectively. A higher η will result in greater electrical energy consumption, which reduces energy conversion efficiency. A highly efficient electrocatalyst can deliver a higher current density while consuming less energy [36, 37].

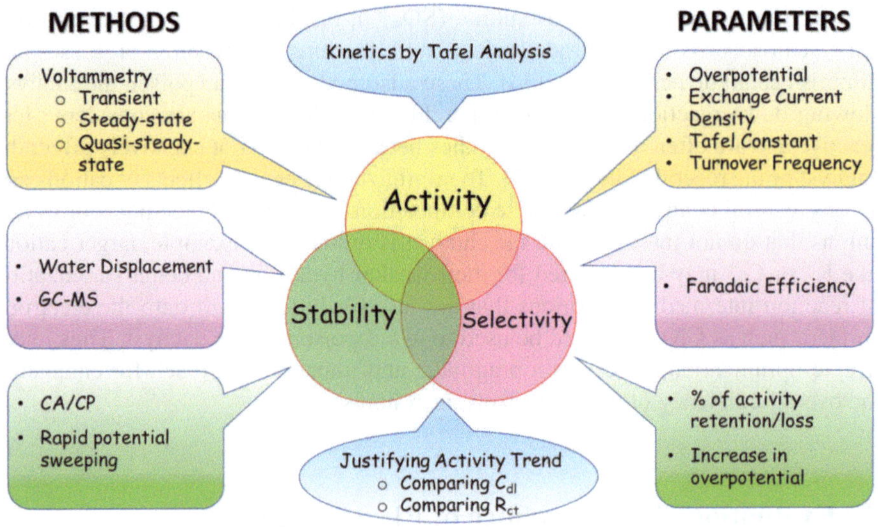

Scheme 1 Parameters and techniques for assessing water-splitting electrocatalysts. Reproduced with permission from [6]. © 2022, Anantharaj et al.; exclusive licensee Royal Society of Chemistry. Distributed under a Creative Commons Attribution License 3.0 (CC BY-NC) https://creativecommons.org/licenses/by-nc/3.0/

The standard reference for current density, denoted as η_{10} is 10 mA cm^{-2} and serves as the standard for comparing HER and OER electrocatalysts. This value represents the current density required to achieve a 10% solar-to-fuel efficiency under 1 sun illumination [2]. η_{10} is now widely recognized as the most significant activity parameter for ranking different HER/OER electrocatalysts. However, it is crucial to recognize that η_{10} does not entirely represent the intrinsic activity of an electrocatalyst, as the current responses are normalized based on the electrode's geometrical surface area, which may not accurately reflect the actual surface area of the active interface [2, 37]. Additionally, in industrial electrolyzers, a much higher current density of approximately 200–400 mA cm^{-2} is necessary to enable continuous H_2 production [37]. Furthermore, η is typically reported relative to the RHE. In most published studies, the transformation from the practically applied reference electrode to the RHE is determined using the Nernst equation. However, this correction method can introduce errors in assessing electrocatalytic activity as the potential of a specific reference electrode calculated using the Nernst equation and calibrated using cyclic voltammetry (CV) can differ by up to 30 mV [13].

Niu and colleagues [13] conducted experimental calibrations of Hg/Hg_2Cl_2, Hg/HgO and Ag/AgCl reference electrodes in a three-electrode system, revealing significant discrepancies between the theoretically calculated values from the Nernst equation and the experimentally calibrated RHE values (Fig. 2c–e). Therefore, it is recommended to calibrate the reference electrode during electrocatalyst activity evaluation. Additionally, it is a good practice to report experimental conditions like pH and

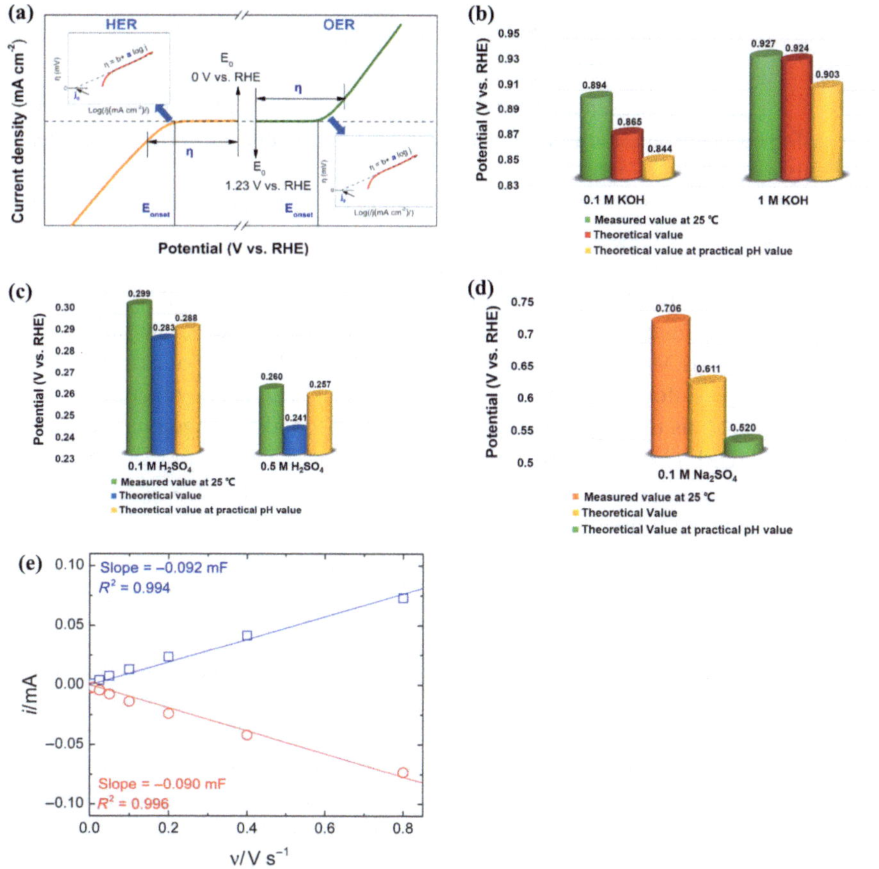

Fig. 2 **a** Representative polarization curves for HER and OER. The Tafel plots obtained from the corresponding polarization curves are shown in the inset. Adapted with permission from [38]. © 2019 American Chemical Society. RHE-calibrated potential determined through experimental measurement at 25 °C and theoretical calculation using the Nernst equation of the **b** Hg/HgO electrode, **c** Hg/Hg_2Cl_2 electrode, **d** Ag/AgCl electrode. Adapted with permission from [13]. © 2019 American Chemical Society. **e** Double-layer capacitance measurements determine the ECSA of electrodeposited NiO_X in 1 M NaOH by plotting anodic and cathodic charging currents at 0.05 V (vs. SCE) against the scan rate. Adapted with permission from [39]. © 2013 American Chemical Society

temperature, as these factors influence reference electrode calibration and should be clearly documented.

One important intrinsic activity marker commonly reported in HER/OER studies is turnover frequency (TOF). TOF measures the quantity of reactant molecules a catalyst converts into hydrogen or oxygen per each active site of a catalyst per second at a specified operating potential [37]. It is represented as,

$$\mathrm{TOF} = \frac{\mathrm{j} \times \mathrm{N_A}}{\mathrm{n} \times \mathrm{F} \times \Gamma}$$

where, j, N_A, n, F, Γ represent current density (A cm^{-2}), Avogadro's number, number of transferred electrons, Faraday constant and total concentration of active sites of the catalysts.

To date, no reliable method has been developed to precisely measure the true number of active sites (S_{Real}), largely due to the complex surface characteristics such as roughness and porosity of the substrates [2, 3, 7]. However, the electrochemical active surface area (ECSA) is commonly considered as a practical approximation for S_{Real}. Three primary methods are commonly employed to estimate the ECSA of electrocatalysts: (i) Measuring the non-Faradaic double layer capacitance (C_{dl}); (ii) Calculating the coulombic charge associated with surface redox reactions of electrocatalysts, such as underpotential deposition (UPD) and hydrogen underpotential deposition (HUPD) and (iii) Deriving adsorption capacitance from EIS (Ca) [7].

The C_{dl} measurement method uses scan rate-dependent CV graphs and is calculated using the following equation [2, 7]:

$$\mathrm{C_{dl}} = \frac{\mathrm{i}}{\nu}$$

where ν and i represent the scan rate and current density. When current (i) is plotted against scan rate (v), a linear relationship is observed, where the slope represents the double-layer capacitance (C_{dl}) (Fig. 2e). To estimate the ECSA, C_{dl} is converted using a standard specific capacitance value, for instance, 60 mF cm^{-2} for CO, which contributes to the inaccuracy in the ECSA calculated from C_{dl} measurements [2]. This conversion factor can vary considerably depending on the surface structure, even for Pt. For instance, the specific C_{dl} values for Pt (100), Pt (110), Pt (111), and Pt (533) surfaces are 116, 72, 88 and 64 μF cm_{Pt}^{-2}, respectively [2]. Thus, highlighting the need for more detailed investigations.

Alternative methods for estimating ECSA include UPD and HUPD, which provide surface area approximations for various metal electrocatalysts [7]. The HUPD method estimates ECSA by measuring the adsorption and desorption of a submonolayer of hydrogen atoms on the electrocatalytic surface. This process is represented by the reaction: $H^+ + e^- + * \rightarrow H^*$, where * represents an active site on the catalyst surface. To calculate ECSA using HUPD, CV is typically performed at ν of 50 mV s^{-1} within a potential window ranging from 0.03 to 1.1 V (vs. RHE). Throughout the experiment, a continuous flow of argon (Ar) is maintained to remove dissolved oxygen (O_2), which could induce the oxygen reduction reaction (ORR) and interfere with ECSA estimation [2].

After subtracting the double-layer currents, the area under HUPD region (0.05–0.4 V vs. RHE) is integrated and averaged for both the positive and negative going scan directions, and the resulting Coulombic charge is then used to determine the ECSA of Pt by considering a specific charge of 210 μC cm_{Pt}^{-2} [2, 16]. However, HUPD is not without limitations and presents certain ambiguities, even when applied to pure metal electrocatalysts. For instance, in the case of pure Pt metals, the total Coulombic charge linked to hydrogen adsorption differs based on the particular Pt surface structure [7]. Further, the precision decreases even more when employing multimetallic systems, such as alloys [2].

In UPD, the electrocatalyst's surface is coated with a single layer of atoms (for instance, Pb, Cu or Ag). The Coulombic charge involved in stripping these metal atoms (e.g., $*Cu \rightarrow Cu^{2+} + 2e^- + *$) is subsequently converted to surface area by applying a specific charge value (e.g., 420 μC $cm_{Cu}{}^{-2}$ for Cu_{upd} stripping at Pt) [7]. The ECSA determined through UPD depends on the extent of surface coverage. Any alterations in the chemical state of metal ions following the development of a three-dimensional structure on the electrocatalyst surface can introduce errors in accurately estimating the surface area [2, 7].

The adsorption capacitance (Ca) derived from EIS can also serve as a method for estimating the electroactive surface area of oxide electrodes. However, converting Ca to ECSA has only been established for NiO_X films so far, and reported specific capacitance values required for this conversion remain scarce [40]. The Brunauer–Emmett–Teller (BET) method, which is based on N_2 adsorption, is another technique for estimating the surface area of electrocatalysts [2]. Currently, the BET method is considered the best option, to determine the surface area particularly for metal oxides [41]. BET assesses the electrocatalytically active sites on which species like H_2, H_2O or O_2 are adsorbed and provides a fairly reliable representation of the ECSA of drop-cast electrodes. This reliability is supported by the observed correlation between surface area derived from the BET approximation and the Coulombic charge under the redox peak and C_{dl}, both of which are quantitative indicators of ECSA [2]. Therefore, it is recommended to report both BET surface area and ECSA, as this provides a complete assessment of the potential surface area [3].

Additionally, the Tafel slope serves a valuable parameter for assessing the reaction kinetics and mechanisms occurring at the electrode surface. Tafel plots, which are generated by plotting overpotential against the logarithm of current density, are commonly employed to determine both the Tafel slope (Fig. 3a, b) and the exchange current (i_O) [3, 6, 36, 37]. In simple terms, the Tafel slope represents the number of millivolts (mV) needed to amplify the current by a factor of 10, and is expressed in mV/decade (mV dec^{-1}) [42]. Catalysts with lower Tafel slopes are more efficient, since they need smaller increase in overpotential to reach high current densities. One can derive Tafel plots for both HER and OER using various electrochemical techniques, including voltammetry, chronoamperometry, and EIS [36]. However, studies have shown that obtaining Tafel plots from potentiodynamic polarization curves, like CV and linear sweep voltammetry (LSV) can lead to significant inaccuracies [3]. This is due to the fact that Tafel analysis, by definition, should be conducted using steady-state responses [36]. For instance, S. Anantharaj and colleagues [43] constructed Tafel curves for Pt/C (20 wt%) using LSV, CA and EIS methods and found that in 0.5 M H_2SO_4 ($v = 5$ mV s^{-1}), the Tafel slope obtained from LSV was 43 mV dec^{-1}, while EIS and CA yielded values of 28 mV dec^{-1} and 33 mV dec^{-1} respectively. Given the expected Tafel slope for Pt/C is 30 mV dec^{-1}, the curve obtained through EIS was the most precise among the three methods.

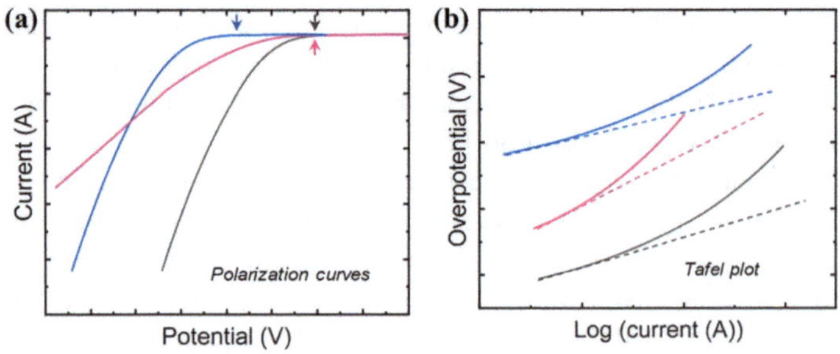

Fig. 3 **a** Polarization curves and **b** associated Tafel plots. Adapted with permission from [44]. © 2018 American Chemical Society

Further, analyzing Tafel slope may be complicated by several experimental factors, including bubble formation, mass transport limitations, difficulties in measuring ohmic resistance, and variations in ohmic resistance with applied potential. To minimize errors, these issues must be carefully addressed [42]. For accurate Tafel slope determination, it is advisable to use thin catalyst layers with high reactant concentrations under conditions that enhance mass transport. A summary of these practical considerations is illustrated in Scheme 2.

The following are some best practices for determining the Tafel slope [6, 42, 45]:

1. The catalyst should be activated before starting the Tafel study, either by CV cycling or using CA method, to achieve a constant current at a specific potential.
2. For accurate Tafel slope determination, it is advisable that measurements should cover a range of at least two decades. Tafel lines derived from fractional decades may lead to inaccurate values.
3. If the Tafel lines are not perfectly linear over the selected potential range, it is recommended to fit them with multiple slopes as needed. It is not recommended to extrapolate a non-linear Tafel line to determine the slope.
4. It is advisable to avoid running CA for extended periods, as this may lead to excessive bubble formation, potentially blocking active sites and altering reactivity.
5. It is recommended that the potential be iR-corrected and plotted against log i to accurately obtain the Tafel slope.
6. To determine the exchange current density, the Tafel slope should be extrapolated to the log i axis at zero overpotential. Taking the antilog of this intercept will provide the exchange current density.

3.2 Selectivity

Selectivity is generally not a major challenge for electrocatalysts, as potential competing reactions are largely absent. Standard reduction potentials, whether experimentally determined or derived from thermochemical data, enable researchers to

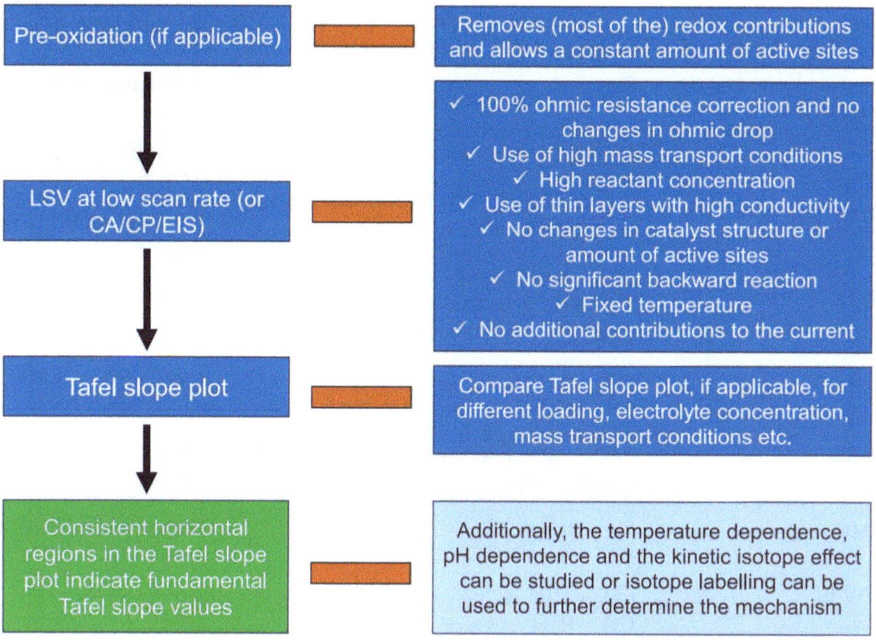

Scheme 2. Practical factors to consider for conducting a kinetically relevant Tafel slope analysis. Reproduced with permission from [42]. © 2024, Heijden et al.; exclusive licensee American Chemical Society. Distributed under a Creative Commons Attribution License 4.0 (CC BY) https://creativecommons.org/licenses/by/4.0/

anticipate which reactions might occur under specific electrochemical potentials. Table 1 displays the standard reduction potentials associated with several pertinent processes in the field of electrocatalysis [8].

For gaseous products such as those from HER and OER, selectivity is typically assessed by measuring the Faradaic efficiency (FE). FE represents the proportion of moles of product actually obtained to the theoretical quantity predicted from the total charge passed, expressed as either a percentage or a fraction. In other words, it is the ratio of the experimentally measured gas volume (hydrogen or oxygen) produced to the theoretically calculated volume [8, 36]. It is expressed as:

$$\eta_{\text{Faraday}} = \frac{V_{\text{gas}}\ (\text{Produced})}{V_{\text{gas}}\ (\text{Calculated})}$$

Accurate measurements of FE are crucial, not only to describe the reaction selectivity but also to validate claims regarding the stability and activity of the electrocatalyst.

Several techniques can be used to measure FE as illustrated in Fig. 4. For HER and OER, an effective approach involves using in-line gas chromatography-mass

Table 1 Standard potentials versus normal hydrogen electrode (NHE), associated reactions, and techniques for measuring FE. Adapted with permission from [8]

Standard potential (E°) (vs. NHE) (V)	Reaction	Abbreviation	Techniques for FE
1.23	$O_2 + 4H^+ + 4e^- \rightleftharpoons 2H_2O$	ORR/OER	Gas collection, sensors, GC
0.70	$O_2 + 2H^+ + 2e^- \rightleftharpoons H_2O_{(aq)}$	ORR	UV-Vis, RRDE, titration
0.27	$N_2 + 8H^+ + 6e^- \rightleftharpoons 2NH_4^+{}_{(aq)}$	N_2RR	UV-Vis, NMR, MS
0.17	$CO_2 + 8H^+ + 8e^- \rightleftharpoons CH_4 + 2H_2O$	$CO_2RR(CH_4)$	GC, GC-MS, DEMS
0.02	$CO_2 + 6H^+ + 6e^- \rightleftharpoons CH_3OH + H_2O$	$CO_2RR(CH_3OH)$	GC-MS, DEMS, NMR
0.00	$2H + 2e^- \rightleftharpoons H_2$	HER/HOR	Gas collection, sensors, GC
−0.10	$CO_2 + 2H^+ + 2e^- \rightleftharpoons CO + H_2O$	$CO_2RR(CO)$	RRDE, GC, GC-MS, DEMS
−0.12	$CO_2 + 2H^+ + 2e^- \rightleftharpoons HCOOH_{(aq)}$	$CO_2RR(HCOOH)$	NMR, LC-MS, RRDE

Abbreviations GC: Gas chromatography; RRDE: Rotating ring-disk electrode; NMR: Nuclear magnetic resonance; MS: Mass spectrometry; GC-MS: Gas chromatography-mass spectrometry; DEMS: Differential electrochemical mass spectrometry; LC-MS: Liquid chromatography-mass spectrometry

spectrometry (GC-MS), electrochemical mass spectrometry (EC-MS) or differential electrochemical mass spectrometry (DEMS) [6, 8]. These techniques allow for real-time, precise quantification of the products, enabling nearly instantaneous calculation of the selectivity. Additionally, by comparing the experimentally measured gas evolution rate with the theoretically calculated gas amounts over specific time intervals, one can gain insights into how the reaction rate is influenced by variations in the applied potential [36].

Beyond these methods, RRDE setup offers a real-time approach for determining FE during the OER process [6, 36]. This technique operates by capturing reaction products generated at the disk electrode (generator electrode), which is coated with the electrocatalyst, onto the ring electrode (collector electrode), where they are subsequently reduced or oxidized back [8]. FE can be determined on the basis of Faraday's laws of electrolysis by evaluating the charge transfer during redox reactions at both the ring and disk electrodes, assuming the collection efficiency (N_{CL}) of the RRDE setup is accurately known. It is recommended to carefully calibrate the collection efficiency of the ring with a one-electron redox system, such as $Fe\ (CN)_6^{4-}/Fe\ (CN)_6^{3-}$ in the relevant supporting electrolyte. Additionally, it is important to note that the

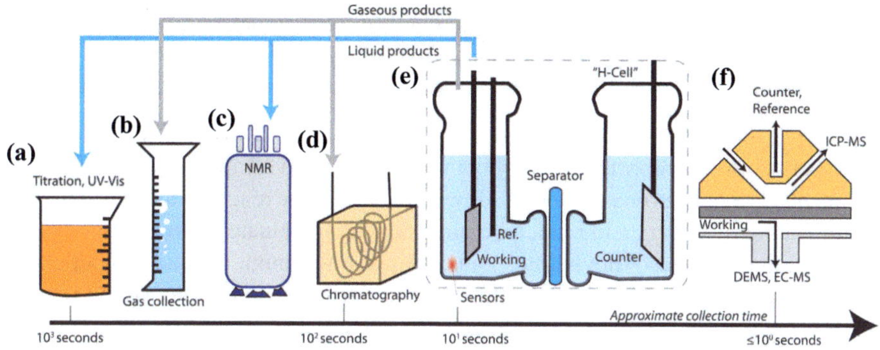

Fig. 4 Instruments and techniques employed to measure FE include various batch techniques such as **a** titration, **b** collection of gases, **c** NMR, **d** GC. **e** H-cells are employed to segregate the working electrode and counter electrode, enabling separate collection of generated products for product collection. **f** Electrochemical flow cells allow for near-real-time detection of product using techniques like DEMS, ICP-MS and EC-MS. Reproduced with permission from [8]. © 2023, Kempler et al.; exclusive licensee Springer Nature. Distributed under a Creative Commons Attribution License 4.0 (CC BY) https://creativecommons.org/licenses/by/4.0/

ORR can disrupt almost all other reduction reactions, making it essential to remove O_2 from both the electrolyte and the reaction headspace [8]. In this case, FE is calculated using the following equation:

$$\mathrm{FE}\ (\%) = \frac{j_r \times n_d}{j_d \times n_r \times N_{CL}}$$

Here, j_r and j_d denote the currents measured at the ring and disk electrodes, respectively, while n_d and n_r represent the number of transferred electrons per molecule of product at the disk electrode and per molecule of product consumed at the ring electrode. For HER, both n_d and n_r are 2, whereas for OER, they are 4 [6].

While the mentioned techniques are efficient, fast, and convenient for determining FE, these advanced setups may not be accessible to everyone. Consequently, one of the most widely used methods is the water-displacement setup, where the volume of gas generated at the working electrode is collected and measured using an inverted burette or graduated cylinder. Since the gas is captured in an inverted graduated cylinder containing water (often coloured for better visibility), conducting pre-electrolysis before starting the measurements is essential to ensure accurate results [6, 8]. One limitation of this method is the assumption that all reaction products are completely transformed into the gas phase. However, if the electrolyte becomes supersaturated, it can lead to an underestimated FE value. To obtain more accurate FE measurements, it is important to test at current densities greater than 10 mA cm^{-2} to induce gas nucleation on the surface of the electrode. Furthermore, it is advisable to provide sufficient electrolysis duration to ensure that the headspace gas volume significantly exceeds any dissolved gases, while minimizing the electrolyte volume to mitigate gas entrapment [8]. An electrocatalyst must achieve a FE of at least 90% to qualify as commercially viable for water splitting applications [36].

The following are some of the best practices for determining Faradaic efficiency [6, 8, 36]:

1. It is essential to report volume of both the reactor and sample for batch measurements. For online measurements, flow rates exiting the reactor should be provided to enable reviewers and readers to independently evaluate the results.
2. Including standard deviations values from minimum of three independent measurements is recommended to enhance clarity and reliability of the results.
3. It is recommended to use DEMS, GC-MS, RRDE and EC-MS for time-sensitive and real-time measurements.
4. Before utilizing the RRDE for selectivity analysis, it is essential to first establish the collection efficiency (N_{CL}).
5. It is advisable to use chemically inert dyes in the water displacement method.

3.3 Stability

The stability of an electrocatalyst is a critical factor in determining its suitability for large-scale commercial applications, as it directly impacts its performance over extended operational periods. Even if an electrocatalyst exhibits exceptional activity and selectivity, it cannot be considered ideal without sufficient stability [6]. The stability of electrocatalysts is typically evaluated using CV or LSV techniques. In these assessments, the electrocatalyst is exposed to extreme conditions, such as high overpotentials (500–600 mV) or elevated current densities (1000–2000 mA cm^{-2}) for extended durations, often lasting several days. The resulting changes in performance metrics, such as overpotential or current density, are monitored [6]. These harsh conditions can induce severe corrosion, detachment, deactivation, and leaching of the catalyst material. Any catalyst that endures these rigorous conditions over several days is generally considered suitable for scaling up and bulk electrolysis operations [36].

The required number of cycles for assessing the stability of electrocatalysts varies between HER and OER. For HER, it is often possible to conduct thousands of cycles since the polarization curve begins at 0 V versus NHE. In contrast, for OER, it is rare to find catalysts that remain stable for more than 250–1000 cycles. A change in either the onset potential or the overpotential at a specified current density (e.g., 10, 50, or 100 mA cm^{-2} for highly efficient catalysts with prominent redox peaks) serves as a key indicator of an electrocatalyst's performance. A catalyst is deemed acceptable if the overpotential increase remains within 30 mV and the overall activity loss after the stability test remains below 5%. Factors such as the choice of substrate, electrode type, and catalyst fixation method also significantly affects the stability of the catalyst [36]. Further, it is advisable to conduct stability experiments at slightly increased temperatures, as cell temperature typically rises during bulk electrolysis. For bifunctional electrocatalysts capable of performing both cathodic HER and anodic OER, a reverse polarity test is sometimes conducted [6, 46]. In this test, the HER and OER activities are assessed both before and after the system's polarity is reversed. This polarity reversal is typically achieved using chronoamperometry or chronopotentiometry, and the changes in activity are assessed through transient techniques.

The following are some recommended best practices for conducting stability assessments [6, 36]:

1. It is recommended to use electrodes with a surface area of at least 1 cm^2 to ensure accurate and reliable results.
2. Complete wettability of the electrode surface should be ensured and care must be taken to mitigate capillary action in three-dimensional electrodes.
3. It is advised to minimize the presence of multiple heterojunctions in the circuit.
4. It is suggested to facilitate the quick elimination of gas bubbles from the working electrode to enhance mass transfer. This can be achieved by conducting stability studies under high current density/overpotentials or by stirring the solution.
5. It is advisable to avoid assessing the stability of foam-type electrodes at low overpotential or current, as this causes gas bubbles to form slowly, potentially blocking most of the active sites for a prolonged period.

References

1. Bora, P., Kumar, S., Sinha, D.: 2D transition metal dichalcogenides for efficient hydrogen generation. Mater. Today Sustain. **27**, 100914 (2024). https://doi.org/10.1016/j.mtsust.2024.100914
2. Wei, C., Rao, R.R., Peng, J., Huang, B., Stephens, I.E.L., Risch, M., Xu, Z.J., Shao-Horn, Y.: Recommended practices and benchmark activity for hydrogen and oxygen electrocatalysis in water splitting and fuel cells. Adv. Mater. **31**, 1806296 (2019). https://doi.org/10.1002/adma.201806296
3. An, H., Park, W., Shin, H., Chung, D.Y.: Recommended practice for measurement and evaluation of oxygen evolution reaction electrocatalysis. EcoMat **6**, e12486 (2024). https://doi.org/10.1002/eom2.12486
4. Hwang, J., Rao, R.R., Giordano, L., Katayama, Y., Yu, Y., Shao-Horn, Y.: Perovskites in catalysis and electrocatalysis. Science **358**, 751–756 (2017). https://doi.org/10.1126/science.aam7092
5. Kuo, D.-Y., Kawasaki, J.K., Nelson, J.N., Kloppenburg, J., Hautier, G., Shen, K.M., Schlom, D.G., Suntivich, J.: Influence of surface adsorption on the oxygen evolution reaction on IrO_2 (110). J. Am. Chem. Soc. **139**, 3473–3479 (2017). https://doi.org/10.1021/jacs.6b11932
6. Anantharaj, S., Noda, S.: Dos and don'ts in screening water splitting electrocatalysts. Energy Adv. **1**, 511–523 (2022). https://doi.org/10.1039/D2YA00076H
7. Huang, C., Chu, P.K.: Recommended practices and benchmarking of foam electrodes in water splitting. Trends Chem. **4**, 1065–1077 (2022). https://doi.org/10.1016/j.trechm.2022.09.008
8. Kempler, P.A., Nielander, A.C.: Reliable reporting of Faradaic efficiencies for electrocatalysis research. Nat. Commun. **14**, 1158 (2023). https://doi.org/10.1038/s41467-023-36880-8
9. Chen, J.G., Jones, C.W., Linic, S., Stamenkovic, V.R.: Best practices in pursuit of topics in heterogeneous electrocatalysis. ACS Catal. **7**, 6392–6393 (2017). https://doi.org/10.1021/acscatal.7b02839

10. Subbaraman, R., Danilovic, N., Lopes, P.P., Tripkovic, D., Strmcnik, D., Stamenkovic, V.R., Markovic, N.M.: Origin of anomalous activities for electrocatalysts in alkaline electrolytes. J. Phys. Chem. C **116**, 22231–22237 (2012). https://doi.org/10.1021/jp3075783
11. Rincón, R.A., Ventosa, E., Tietz, F., Masa, J., Seisel, S., Kuznetsov, V., Schuhmann, W.: Evaluation of perovskites as electrocatalysts for the oxygen evolution reaction. ChemPhysChem **15**, 2810–2816 (2014). https://doi.org/10.1002/cphc.201402137
12. Yue, M., Lambert, H., Pahon, E., Roche, R., Jemei, S., Hissel, D.: Hydrogen energy systems: a critical review of technologies, applications, trends and challenges. Renew. Sustain. Energy Rev. **146**, 111180 (2021). https://doi.org/10.1016/j.rser.2021.111180
13. Niu, S., Li, S., Du, Y., Han, X., Xu, P.: How to reliably report the overpotential of an electrocatalyst. ACS Energy Lett. **5**, 1083–1087 (2020). https://doi.org/10.1021/acsenergylett.0c00321
14. Zheng, W., Liu, M., Lee, L.Y.S.: Best practices in using foam-type electrodes for electrocatalytic performance benchmark. ACS Energy Lett. **5**, 3260–3264 (2020). https://doi.org/10.1021/acsenergylett.0c01958
15. Stevens, M.B., Enman, L.J., Batchellor, A.S., Cosby, M.R., Vise, A.E., Trang, C.D.M., Boettcher, S.W.: Measurement techniques for the study of thin film heterogeneous water oxidation electrocatalysts. Chem. Mater. **29**, 120–140 (2017). https://doi.org/10.1021/acs.chemmater.6b02796
16. Sheng, W., Gasteiger, H.A., Shao-Horn, Y.: Hydrogen oxidation and evolution reaction kinetics on platinum: acid versus alkaline electrolytes. J. Electrochem. Soc. **157**, B1529 (2010). https://doi.org/10.1149/1.3483106
17. Pourbaix, M.: Applications of electrochemistry in corrosion science and in practice. Corros. Sci. **14**, 25–82 (1974). https://doi.org/10.1016/S0010-938X(74)80006-5
18. Lopes, P.P., Strmcnik, D., Tripkovic, D., Connell, J.G., Stamenkovic, V., Markovic, N.M.: Relationships between atomic level surface structure and stability/activity of platinum surface atoms in aqueous environments. ACS Catal. **6**, 2536–2544 (2016). https://doi.org/10.1021/acscatal.5b02920
19. Chen, R., Yang, C., Cai, W., Wang, H.-Y., Miao, J., Zhang, L., Chen, S., Liu, B.: Use of platinum as the counter electrode to study the activity of nonprecious metal catalysts for the hydrogen evolution reaction. ACS Energy Lett. **2**, 1070–1075 (2017). https://doi.org/10.1021/acsenergylett.7b00219
20. Wu, G., More, K.L., Johnston, C.M., Zelenay, P.: High-performance electrocatalysts for oxygen reduction derived from polyaniline. Iron Cobalt Sci. **332**, 443–447 (2011). https://doi.org/10.1126/science.1200832
21. Ashton, S.J., Arenz, M.: Comparative DEMS study on the electrochemical oxidation of carbon blacks. J. Power Sources **217**, 392–399 (2012). https://doi.org/10.1016/j.jpowsour.2012.06.015
22. Edgington, J., Deberghes, A., Seitz, L.C.: Glassy carbon substrate oxidation effects on electrode stability for oxygen evolution reaction catalysis stability benchmarking. ACS Appl. Energy Mater. **5**, 12206–12218 (2022). https://doi.org/10.1021/acsaem.2c01690
23. Han, B., Risch, M., Belden, S., Lee, S., Bayer, D., Mutoro, E., Shao-Horn, Y.: Screening oxide support materials for OER catalysts in acid. J. Electrochem. Soc. **165**, F813 (2018). https://doi.org/10.1149/2.0921810jes
24. Trasatti, S.: Work function, electronegativity, and electrochemical behaviour of metals: III. Electrolytic hydrogen evolution in acid solutions. J. Electroanal. Chem. Interfacial Electrochem. **39**, 163–184 (1972). https://doi.org/10.1016/S0022-0728(72)80485-6
25. Nørskov, J.K., Bligaard, T., Logadottir, A., Kitchin, J.R., Chen, J.G., Pandelov, S., Stimming, U.: Trends in the exchange current for hydrogen evolution. J. Electrochem. Soc. **152**, J23 (2005). https://doi.org/10.1149/1.1856988
26. Gebremariam, G.K., Jovanović, A.Z., Dobrota, A.S., Skorodumova, N.V., Pašti, I.A.: Hydrogen evolution volcano(es)—from acidic to neutral and alkaline solutions. Catalysts **12** (2022). https://doi.org/10.3390/catal12121541
27. Sheng, W., Myint, M., Chen, J.G., Yan, Y.: Correlating the hydrogen evolution reaction activity in alkaline electrolytes with the hydrogen binding energy on monometallic surfaces. Energy Environ. Sci. **6**, 1509–1512 (2013). https://doi.org/10.1039/C3EE00045A

28. Badawy, W.A., Feky, H.E., Helal, N.H., Mohammed, H.H.: Cathodic hydrogen evolution on molybdenum in NaOH solutions. Int. J. Hydrogen Energy **38**, 9625–9632 (2013). https://doi.org/10.1016/j.ijhydene.2013.05.098
29. Chen, K., Xu, B., Shen, L., Shen, D., Li, M., Guo, L.-H.: Functions and performance of ionic liquids in enhancing electrocatalytic hydrogen evolution reactions: a comprehensive review. RSC Adv. **12**, 19452–19469 (2022). https://doi.org/10.1039/D2RA02547G
30. Márquez, R.A., Kawashima, K., Son, Y.J., Castelino, G., Miller, N., Smith, L.A., Chukwuneke, C.E., Mullins, C.B.: Getting the basics right: preparing alkaline electrolytes for electrochemical applications. ACS Energy Lett. **8**, 1141–1146 (2023). https://doi.org/10.1021/acsenergylett.2c02847
31. Hausmann, J.N., Traynor, B., Myers, R.J., Driess, M., Menezes, P.W.: The pH of aqueous NaOH/KOH solutions: a critical and non-trivial parameter for electrocatalysis. ACS Energy Lett. **6**, 3567–3571 (2021). https://doi.org/10.1021/acsenergylett.1c01693
32. Zhang, R., Pearce, P.E., Duan, Y., Dubouis, N., Marchandier, T., Grimaud, A.: Importance of water structure and catalyst-electrolyte interface on the design of water splitting catalysts. Chem. Mater. **31**, 8248–8259 (2019). https://doi.org/10.1021/acs.chemmater.9b02318
33. Strmcnik, D., Kodama, K., van der Vliet, D., Greeley, J., Stamenkovic, V.R., Marković, N.M.: The role of non-covalent interactions in electrocatalytic fuel-cell reactions on platinum. Nat. Chem. **1**, 466–472 (2009). https://doi.org/10.1038/nchem.330
34. Goyal, A., Koper, M.T.M.: The interrelated effect of cations and electrolyte pH on the hydrogen evolution reaction on gold electrodes in alkaline media. Angew. Chem. Int. Ed. **60**, 13452–13462 (2021). https://doi.org/10.1002/anie.202102803
35. Sheng, W., Zhuang, Z., Gao, M., Zheng, J., Chen, J.G., Yan, Y.: Correlating hydrogen oxidation and evolution activity on platinum at different pH with measured hydrogen binding energy. Nat. Commun. **6**, 5848 (2015). https://doi.org/10.1038/ncomms6848
36. Raveendran, A., Chandran, M., Dhanusuraman, R.: A comprehensive review on the electrochemical parameters and recent material development of electrochemical water splitting electrocatalysts. RSC Adv. **13**, 3843–3876 (2023). https://doi.org/10.1039/D2RA07642J
37. Yu, Z.-Y., Duan, Y., Feng, X.-Y., Yu, X., Gao, M.-R., Yu, S.-H.: Clean and affordable hydrogen fuel from alkaline water splitting: past, recent progress, and future prospects. Adv. Mater. **33**, 2007100 (2021). https://doi.org/10.1002/adma.202007100
38. Yu, J., He, Q., Yang, G., Zhou, W., Shao, Z., Ni, M.: Recent advances and prospective in ruthenium-based materials for electrochemical water splitting. ACS Catal. **9**, 9973–10011 (2019). https://doi.org/10.1021/acscatal.9b02457
39. McCrory, C.C.L., Jung, S., Peters, J.C., Jaramillo, T.F.: Benchmarking heterogeneous electrocatalysts for the oxygen evolution reaction. J. Am. Chem. Soc. **135**, 16977–16987 (2013). https://doi.org/10.1021/ja407115p
40. Watzele, S., Bandarenka, A.S.: Quick determination of electroactive surface area of some oxide electrode materials. Electroanalysis **28**, 2394–2399 (2016). https://doi.org/10.1002/elan.201600178
41. Jung, S., McCrory, C.C.L., Ferrer, I.M., Peters, J.C., Jaramillo, T.F.: Benchmarking nanoparticulate metal oxide electrocatalysts for the alkaline water oxidation reaction. J. Mater. Chem. A Mater. **4**, 3068–3076 (2016). https://doi.org/10.1039/C5TA07586F
42. van der Heijden, O., Park, S., Vos, R.E., Eggebeen, J.J.J., Koper, M.T.M.: Tafel slope plot as a tool to analyze electrocatalytic reactions. ACS Energy Lett. **9**, 1871–1879 (2024). https://doi.org/10.1021/acsenergylett.4c00266
43. Anantharaj, S., Ede, S.R., Karthick, K., Sam Sankar, S., Sangeetha, K., Karthik, P.E., Kundu, S.: Precision and correctness in the evaluation of electrocatalytic water splitting: revisiting activity parameters with a critical assessment. Energy Environ. Sci. **11**, 744–771 (2018). https://doi.org/10.1039/C7EE03457A
44. Voiry, D., Chhowalla, M., Gogotsi, Y., Kotov, N.A., Li, Y., Penner, R.M., Schaak, R.E., Weiss, P.S.: Best practices for reporting electrocatalytic performance of nanomaterials. ACS Nano **12**, 9635–9638 (2018). https://doi.org/10.1021/acsnano.8b07700

45. Singha Roy, S., Madhu, R., Karmakar, A., Kundu, S.: From theory to practice: a critical and comparative assessment of Tafel slope analysis techniques in electrocatalytic water splitting. ACS Mater. Lett. **6**, 3112–3123 (2024). https://doi.org/10.1021/acsmaterialslett.4c00831
46. Krstajić, N.V., Jović, V.D., Gajić-Krstajić, Lj., Jović, B.M., Antozzi, A.L., Martelli, G.N.: Electrodeposition of Ni–Mo alloy coatings and their characterization as cathodes for hydrogen evolution in sodium hydroxide solution. Int. J. Hydrogen Energy **33** 3676–3687 (2008). https://doi.org/10.1016/j.ijhydene.2008.04.039

Cutting-Edge Approaches to Self-supported 2D Electrocatalyst Synthesis

M. I. Chebanenko, D. S. Dmitriev, A. A. Lobinsky, and V. I. Popkov

Abstract The growing need for environmentally friendly energy solutions has spurred the creation of self-supporting two-dimensional electrocatalysts, which provide improved durability, a high surface area, and adjustable electronic characteristics. This chapter presents a comprehensive review of their synthesis methods, structural diversity, and performance optimization strategies. It begins by categorizing key 2D electrocatalysts, including transition metal dichalcogenides (TMDs), MXenes, layered double hydroxides (LDHs), hybrid structures, and metal-free catalysts, highlighting their distinct physicochemical properties and catalytic activity. The discussion then shifts to the role of different substrate types—metal, carbon-based, transparent conductive, and non-conductive—in influencing electrocatalytic performance and long-term stability. A systematic analysis of advanced fabrication techniques follows, covering electrodeposition, solution-based synthesis (SILAR, gas–liquid interface methods, and spin coating), solvothermal approaches, and vapor deposition techniques (CVD and PVD), with an emphasis on their advantages, limitations, and applications in hydrogen evolution and electrochemical energy conversion. By integrating recent advances in material design and synthesis, this chapter underscores the growing significance of self-supported 2D electrocatalysts in next-generation energy technologies and lays the groundwork for developing scalable and high-performance electrocatalysts.

Keywords Nanostructures · Thin films · Electrochemistry · Surface reactions · Coating · Interfaces

1 Introduction

The global transition toward sustainable energy sources necessitates the development of efficient and scalable technologies for clean energy conversion and storage. Hydrogen (H_2) has become a topic of great interest as a potential energy source

M. I. Chebanenko (✉) · D. S. Dmitriev · A. A. Lobinsky · V. I. Popkov
Hydrogen Energy Laboratory, Ioffe Institute, St. Petersburg, Russia
e-mail: m_chebanenko@list.ru

O. P. Pandey et al. (eds.), *2D Materials for Electrochemical Water Splitting*, Engineering Materials, https://doi.org/10.1007/978-981-95-2003-9_4

due to its high energy content and its environmentally friendly combustion process, which results in the production of water [1]. Among the most viable methods for producing "green" H_2, water electrolysis offers a carbon-free approach. Nevertheless, its widespread implementation is constrained by its high energy consumption and slow reaction kinetics, which significantly affect the efficiency of the process and its economic viability. by high energy consumption and sluggish reaction kinetics, which significantly impact process efficiency and economic feasibility [2]. One of the key challenges in improving electrolysis efficiency is the development of advanced electrocatalysts that facilitate hydrogen evolution at lower overpotentials, thereby reducing energy input and enhancing system performance.

To address these challenges, self-supported 2D electrocatalysts have emerged as a promising substitute for traditional catalytic systems [3]. These materials offer several distinct advantages, including high surface area, which increases the density of active catalytic sites, and superior electrical conductivity, enabling rapid charge transfer. Additionally, their structural tunability and chemical stability allow for optimization under specific operational conditions, leading to improved performance in hydrogen evolution reactions [4]. Given these advantages, self-supported 2D electrocatalysts have the potential to outperform traditional catalyst architectures, particularly in electrocatalytic and electrochemical energy conversion applications. However, realizing their full potential demands meticulous management of their synthesis and structural engineering, necessitating advanced fabrication strategies.

This chapter provides a comprehensive review of cutting-edge synthesis techniques for self-supported 2D electrocatalysts, emphasizing their role in optimizing catalytic activity, stability, and large-scale applicability. The discussion begins with an overview of the key material classes, including transition metal dichalcogenides (TMDs), recognized for their exceptional catalytic efficiency and durability; MXenes, distinguished by their high conductivity and tunability; and hybrid structures, which leverage synergistic interactions to enhance catalytic activity. Other promising 2D materials are also considered [5].

Following this, the chapter explores the critical role of substrate selection in catalyst synthesis. The mechanical support, electronic properties, and chemical stability of the electrocatalytic system depend significantly on the substrate type, which may include metal, carbon-based, transparent conductive, or non-conductive substrates. Key selection criteria such as chemical inertness, mechanical integrity, and interfacial compatibility are discussed, highlighting their influence on the overall performance of self-supported electrocatalysts.

The final and central focus of this chapter is an in-depth examination of modern synthesis approaches, including electrodeposition, wet-chemical techniques (such as the SILAR method and gas–liquid interface methods), solvothermal strategies, and vapor-phase deposition techniques (CVD and PVD). Each method is analyzed in terms of its advantages, limitations, and practical applications in hydrogen evolution and electrochemical energy conversion. By integrating recent advancements in nanostructure engineering and material optimization, this review aims to provide a foundational understanding of how synthesis strategies can be tailored to enhance catalytic performance and long-term stability.

Ultimately, this chapter serves as a framework for future research and innovation in the field of self-supported 2D electrocatalysts, offering valuable insights for their application in hydrogen energy systems, electrochemical devices, and beyond.

2 Key Classes of 2D Electrocatalysts

2.1 Transition Metal Dichalcogenides (TMDs)

Transition metal dichalcogenides (TMDs) represent a significant class of two-dimensional materials that have garnered considerable research interest in recent decades. These materials feature a layered 2D structure composed of transition metal atoms bonded to chalcogens, typically following the general formula MX_2, where M denotes a transition metal atom, and X represents a chalcogen (S, Se, or Te).

The structural properties of TMDs were first described by Linus Pauling in 1923, sparking extensive research into this class of compounds. By the late 1960s, approximately 40 TMDs with layered structures had been identified, laying the foundation for intensive studies of their unique physicochemical properties [6].

Due to their unique optical properties and high specific surface area, TMDs—such as MoS_2—exhibit promising potential as electrocatalysts for the hydrogen evolution reaction (HER). These materials not only demonstrate competitive performance compared to conventional catalysts but also offer new opportunities for optimization and efficiency enhancement in electrochemical processes.

In a recent study [7], a porous NiMo@MoS_2 material was synthesized using microwave sintering powder metallurgy followed by hydrothermal treatment. The electrocatalytic properties of the resulting electrode material were evaluated in the electrochemical reforming of a 1.0 M KOH solution. It was shown that the MoS_2 coating significantly enhances proton adsorption and dissociation from water molecules while increasing the electrochemically active surface area (ECSA). The optimized NiMo@MoS_2 composite electrode achieves hydrogen evolution at a low overpotential of 89 mV at a current density of 10 mA/cm^2, with a Tafel slope of 84.0 mV/dec. Moreover, this material exhibits excellent long-term stability, maintaining continuous operation for 25 h without significant performance degradation.

Beyond their outstanding electrocatalytic performance, TMD-based electrocatalysts are also highly promising for photocatalytic water splitting, enabling efficient solar energy conversion into hydrogen. Additionally, these materials participate in redox reactions, expanding their applications in fuel cells and rechargeable batteries. Furthermore, some TMDs have demonstrated catalytic activity for ammonia synthesis and carbon fixation, broadening their potential in sustainable energy and chemical industries.

2.2 MXenes

In 2011, a research team from Drexel University (USA) introduced a novel class of two-dimensional materials called MXenes. These materials were derived from transition metal carbides and exhibited distinctive structural and electrical characteristics, making them a significant addition to the family of 2D materials.

MXenes possess a hexagonal layered structure (space group $P6_3/mmc$) and are described by the general formula $M_nX_mT_x$ where M stands for a transition metal, X denotes either carbon or nitrogen, and T corresponds to surface functional groups such as –OH, –O, or –F. Due to their high electrical conductivity, mechanical strength, and chemical stability, MXenes are promising candidates for a variety of potential uses.

One of the most promising uses of MXenes is in electrocatalysis, particularly for hydrogen evolution reactions (HER) and CO_2 reduction reactions (CO_2RR). Studies indicate that MXenes can significantly enhance the efficiency of these processes, paving the way for the creation of environmentally friendly energy sources and carbon capture technologies.

In a groundbreaking study [8], the electrocatalytic performance of 2D titanium carbonitride MXene in the hydrogen evolution reaction was evaluated for the first time. The findings revealed that this 2D titanium carbonitride exhibits significantly higher catalytic performance than titanium carbide. The recorded HER overpotential and Tafel slope were 53 mV and 86 mV/dec, respectively, indicating a substantial advantage of this material over conventional transition metal carbide-based electrocatalysts. Moreover, these results established that the obtained overpotential was the lowest among all known undoped MXenes derived from transition metal carbides. Additionally, the study demonstrated that 2D titanium carbonitride exhibits excellent long-term stability, further enhancing its prospects for practical use in the fields of electrocatalysis and sustainable energy technologies.

2.3 Layered Double Hydroxides (LDHs)

The development of layered double hydroxides (LDHs) dates back to 1842, when Hochstetter first discovered natural minerals of this class. In 1942, Feitknecht synthesized [Mg–Al–CO_3]-LDH and introduced the concept of a layered structure. Since the 1970s, LDHs have been actively employed as catalysts and catalyst supports in various catalytic processes [9]. These materials consist of alternating metal and hydroxide layers, forming a highly tunable and functional class of materials.

The general chemical formula of LDHs is given by $[M_{1-x}{}^{2+}\ M_x{}^{3+}(OH)_2]^{x+}[A_{x/n}]^{n-}\cdot mH_2O$, where M^{2+} and M^{3+} represent divalent and trivalent metal cations, respectively, while A^{n-} denotes an exchangeable anion that can be replaced through ion-exchange processes.

The unique layered structure of LDHs allows them to absorb light and facilitate photochemical reactions, making them attractive for applications in hydrogen energy and fuel cells. Their high surface area and ion-exchange capabilities enhance catalytic efficiency, particularly in hydrogen evolution reactions (HER) and methanol oxidation processes. Furthermore, LDHs serve as effective catalyst supports, enhancing the catalytic activity and stability of other active materials.

A recent study [10] demonstrated the electrocatalytic potential of LDHs not only for hydrogen production but also for the decomposition of acetonitrile. In this work, Co LDH and Ni–Co LDH were synthesized via solution precipitation onto nickel foam substrates, using Co-based zeolitic imidazolate frameworks as precursors. The results showed that the charging and discharging time for Ni–Co LDHs was more than twice as long as for Co LDHs, suggesting a higher concentration of active catalytic sites. Additionally, the Tafel slope for Ni–Co LDH was recorded at 68 mV dec^{-1}, significantly lower than 146 mV dec^{-1} for Co LDH, demonstrating a higher electron transfer rate. These electrochemical experiments confirm that Ni–Co LDH exhibits superior charge transfer properties, enhancing electrooxidation reactions and making it a promising candidate for electrocatalysis and other electrochemical applications.

2.4 Hybrid Structures

Hybrid structures, also referred to as heterostructures, are unique materials composed of multiple layers of different substances, each with its own unique physical and chemical characteristics. The concept of heterostructures was first explored in semiconductor research, particularly in the 1970s, when studies focused on gallium arsenide (GaAs) and other III–V semiconductors.

The functional properties of heterostructures are highly dependent on their composition and architecture. For example, graphene-based heterostructures exhibit exceptional electrical conductivity and mechanical strength, while transition metal oxide heterostructures display unique optical and electronic characteristics, making them promising candidates for photonics and catalysis applications.

Heterostructures have also gained significant attention as electrocatalysts. In a recent study [11], a CoFe@NiFe heterostructure supported on nickel foam (NF) was synthesized by electrodepositing NiFe-LDH onto CoFe-LDH nanosheets. The hydrogen evolution reaction (HER) activity of the CoFe@NiFe/NF sample was evaluated in a 1.0 M KOH solution, revealing an overpotential of 240 mV and a Tafel slope of 88 mV/dec. The core–shell CoFe@NiFe/NF architecture exhibited the largest electrochemically active surface area (ECSA), nearly 2.5 times greater than that of CoFe-LDH. As a result, the study demonstrated that this heterostructure achieves performance comparable to state-of-the-art electrocatalysts, maintaining long-term stability for over 30 h of continuous operation.

2.5 Metal-Free Electrocatalysts

Metal-free two-dimensional (2D) materials represent a unique class of materials that have only recently gained significant research attention in the twenty-first century. The first references to such materials are associated with graphene, which was successfully isolated as a single-atom-thick layer of graphite by Andre Geim and Konstantin Novoselov in 2004. However, interest in metal-free 2D materials grew substantially after the discovery of black phosphorus (BP) and graphitic carbon nitride (g-C_3N_4) opening new research directions in quantum electronics, photonics, and catalytic materials.

Metal-free 2D electrocatalysts exhibit high catalytic activity and selectivity, making them promising candidates for various applications, including hydrogen production, ammonia synthesis, and organic oxidation reactions. The functional properties of these materials vary depending on their atomic structure and chemical composition, allowing for tailored electrocatalytic performance.

In a recent study [12], Iram et al. synthesized a $SmAlS_3$/g-C_3N_4 nanocomposite via a hydrothermal method and evaluated its performance in electrocatalytic reforming of an alkaline aqueous solution. The results showed that the nanocomposite exhibited enhanced electrocatalytic efficiency compared to its individual components, $SmAlS_3$ and g-C_3N_4. The recorded overpotentials for the oxygen evolution reaction (OER) and hydrogen evolution reaction (HER) were 278 mV and 305 mV, respectively, with corresponding Tafel slopes of 49 mV dec^{-1} for OER and 63 mV dec^{-1} for HER.

Another study [13] demonstrated the feasibility of BP-based electrocatalysts incorporating PtRu nanoclusters. The synthesized PtRu/BP nanocomposite exhibited high electrocatalytic activity in hydrogen evolution from alkaline aqueous solutions, achieving an HER current density of 88.5 mA cm^{-2}. Theoretical modeling further confirmed that electronic interactions between BP and PtRu accelerate water dissociation, leading to a significant enhancement in HER activity.

Overall, metal-free 2D materials represent a promising research direction with significant potential for scientific and technological applications, making them an important focus for future advancements in catalysis, energy conversion, and beyond.

The analysis of key self-supported 2D electrocatalysts highlights their diverse structural and electronic properties, enabling advancements in hydrogen and oxygen evolution reactions and other energy conversion processes. Materials such as TMDs, MXenes, LDHs, hybrid structures, and metal-free electrocatalysts demonstrate unique advantages, with performance largely influenced by their morphology, active site distribution, and conductivity. Nevertheless, beyond the selection of materials, the choice of the substrate is of utmost importance for ensuring mechanical stability, efficient electron transport, and long-term reliability. The next section explores various substrates, including metal foams, carbon-based supports, and transparent conductive oxides, assessing their impact on electrocatalytic efficiency and scalability.

3 Substrates for Self-supported 2D Electrocatalysts

A critical factor influencing the efficiency of 2D electrocatalysts is the choice of substrate used for their synthesis. This section explores several commonly employed substrate types in the fabrication of self-supported electrocatalysts, including conductive metal substrates, carbon-based substrates, fluorine-doped tin oxide (FTO) and indium tin oxide (ITO)-coated glass, and non-conductive substrates.

3.1 Metal Substrates

Metallic substrates, such as foam, foil, and mesh, play a crucial role in the synthesis of 2D electrocatalysts due to their high electrical conductivity and mechanical strength [14]. Metal foams offer exceptional support for active substances and possess a highly interconnected network, which expands the surface area available for electrochemical reactions and improves the accessibility of electrolytes to catalytic sites, making them perfect for use in electrodes. Metal foils, commonly made of stainless steel or copper, are also widely used as substrates. They offer high conductivity and can be easily processed into various shapes and structures, allowing for optimization in specific electrocatalytic reactions. Additionally, metal meshes facilitate efficient ion transport and provide direct exposure to active materials, contributing to improved catalyst performance.

3.2 Carbon-Based Substrates

Carbon-based substrates, including carbon cloth, carbon paper, graphene paper, and carbon felt, are gaining increasing popularity as supports for 2D electrocatalysts due to their high conductivity, flexibility, and corrosion resistance [15]. Carbon cloth offers excellent mechanical strength and flexibility, making it suitable for flexible electrode fabrication. Additionally, it provides good electrical conductivity and a large surface area for the adsorption of active materials. Carbon paper, a lightweight and porous substrate, enables a high specific surface area and improved access to active sites, thereby enhancing electrocatalytic efficiency.

Graphene paper, composed of stacked graphene layers, exhibits exceptional electrical and mechanical properties, making it an excellent substrate for 2D electrocatalysts by ensuring high conductivity and structural stability. Carbon felt, with its highly porous structure, allows for efficient electrolyte penetration and ion transport, thereby facilitating improved catalytic performance. These characteristics make carbon-based substrates highly promising for the development of efficient electrocatalytic systems.

3.3 Transparent Conductive Substrates

Glass-based substrates, such as fluorine-doped tin oxide (FTO) and indium tin oxide (ITO), are commonly employed as conductive supports for 2D electrocatalysts. These substrates offer high electrical conductivity and are suitable for both electrochemical and photoelectrochemical processes [16].

FTO substrates provide high transparency and conductivity, making them ideal for photovoltaic applications. Additionally, they exhibit excellent corrosion resistance, which is essential for ensuring long-term electrode stability. ITO substrates also feature high conductivity and transparency, enabling their use in various electrochemical and optoelectronic applications. Their integration as substrates for 2D electrocatalysts enhances both catalytic efficiency and operational stability, making them valuable components in advanced energy conversion systems.

3.4 Non-conductive Substrate

Recent research has also explored non-traditional non-conductive substrates, including paper, fabrics, and sponges, for the synthesis of 2D electrocatalysts [17]. These materials offer unique advantages in electrocatalyst fabrication and present new opportunities for designing low-cost, flexible, and efficient electrode supports.

Paper-based substrates, due to their lightweight and porous nature, can be utilized to create cost-effective and efficient electrodes. They provide excellent adsorption of active materials and can be easily processed into desired shapes and structures. Additionally, paper is an environmentally friendly and widely available material, making it an attractive option for sustainable electrode development.

Textile-based substrates, including fabrics made from polymeric or plant-derived fibers, offer flexibility and mechanical durability, essential for wearable and flexible electrode applications. These substrates enable the fabrication of lightweight and robust electrode architectures, suitable for various electrochemical systems.

Sponge-based materials, characterized by their highly porous structures, provide a large surface area that enhances electron transport. This makes them particularly suitable for electrode applications where maximizing surface availability for 2D material deposition and improving access to active catalytic sites are critical.

Thus, the choice of an appropriate substrate for 2D electrocatalyst synthesis is crucial for achieving high efficiency and stability. The explored substrate types offer diverse advantages and open new avenues for research aimed at optimizing electrocatalyst performance across different applications.

The selection of appropriate substrates for self-supported 2D electrocatalysts plays a crucial role in determining their mechanical stability, electron transport efficiency, and long-term performance. The analysis of metal, carbon-based, transparent conductive, and non-conductive substrates has demonstrated that each type offers unique advantages, including flexibility, chemical resistance, high conductivity and scalability, which directly impact electrocatalytic activity and practical applicability. However, beyond the choice of material and substrate, the synthesis techniques used to fabricate self-supported 2D electrocatalysts are equally critical in achieving optimal structural and functional properties. The next section explores cutting-edge synthesis methods, including electrodeposition, solution-based techniques, solvothermal strategies, and vapor deposition, highlighting their advantages, challenges, and potential for large-scale implementation in hydrogen evolution and electrochemical energy conversion technologies.

4 Advanced Synthesis of Self-supported 2D Electrocatalysts

Currently, a wide range of synthesis techniques exist for fabricating self-supported electrode materials using various types of substrates. For most of these methods, chemical stability and high surface area of the substrate are key parameters.

However, for certain approaches, such as electrochemical synthesis, the electrical conductivity of the substrate is also a critical factor. In these methods, substrates are generally classified into metallic substrates (e.g., nickel foam, copper foam) and carbon-based substrates (e.g., carbon fibers, carbon felt). A similar classification applies to substrates used in solvothermal synthesis techniques, where the choice of substrate is of utmost importance in shaping the structural and functional characteristics of self-supported electrode materials.

4.1 Electrodeposition Methods

Electrodeposition as a synthesis method for 2D electrocatalysts is of considerable interest due to its ability to precisely control process parameters and accommodate various operational modes. The primary electrochemical synthesis modes include galvanostatic and potentiostatic deposition, as well as pulse and potentiodynamic deposition, the latter being implemented via cyclic voltammetry (Fig. 1).

The most commonly used metallic substrates in these deposition modes are copper and nickel foams, characterized by their highly developed morphology and large specific surface area. However, to reduce production costs and enhance resource efficiency in the fabrication of self-supported electrodes, there has been a growing trend toward the use of carbon-based nonwoven substrates, such as carbon cloth, carbon fiber, and carbon felt. These materials can be utilized for electrodeposition either in their pristine form or after surface galvanization with nickel or copper. This technological approach enables metal resource conservation while preserving the substrate's well-developed morphology and high specific surface area.

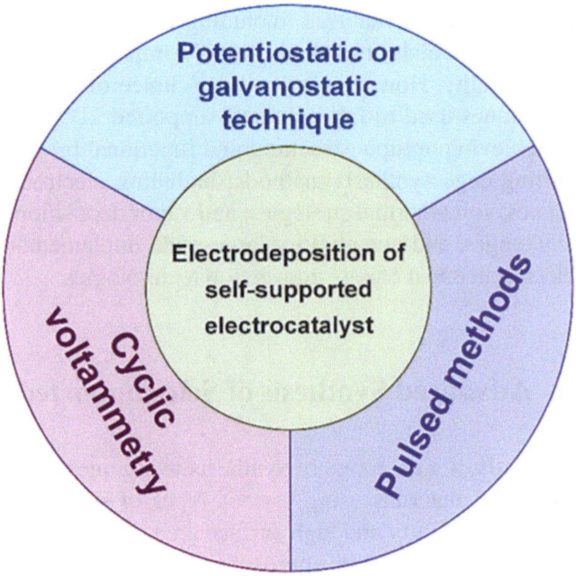

Fig. 1 Electrochemical deposition techniques for self-supported electrocatalysts

4.1.1 Metal Substrates

Electrodeposited Electrocatalytic Systems on Metallic Substrates

The primary electrocatalytic systems deposited on metallic substrates via electrodeposition primarily consist of polymetallic systems based on iron-group elements (Fe, Ni, Co), doped with either transition metals (Mo, W, Sn) or nonmetals such as sulfur or phosphorus.

In studies [18, 19], copper plates and copper foams were used as substrates, respectively. The Ni–Fe system was synthesized on a copper plate using a chloride-based electrolyte composed of $NiCl_2 \times 6H_2O$ (1 M), $FeCl_2 \times 4H_2O$ (0.15 M), and $CaCl_2 \times 2H_2O$ (1.2 M), with the pH adjusted to 2.5. Electrodeposition was carried out under galvanostatic conditions at a current density of 40 mA cm^{-2} and a temperature of 60 °C for 10 min, using nickel plates as anodes. A key parameter in this synthesis was the optimized Fe^{2+} concentration in the electrolyte, ensuring a well-developed deposit morphology and low overpotentials for water-splitting reactions (HER and OER). The deposition process resulted in the formation of urchin-like nanostructures on the copper substrate surface, which not only increased the electrochemically active surface area of the electrocatalyst but also facilitated the desorption of gas bubbles from the electrode surface. Electrochemical testing revealed overpotentials of −124 mV@10 mA cm^{-2} and −243 mV@100 mA cm^{-2} for HER, and 292 mV@10 mA cm^{-2} and 374 mV@100 mA cm^{-2} for OER. The Tafel slopes were determined to be −114 mV dec^{-1} for HER and 86 mV dec^{-1} for OER. Long-term testing at a current density of 100 mA cm^{-2} demonstrated HER stability, with an overpotential of −250 mV maintained over 10 h of continuous operation.

The Fe–Co–Ni polymetallic system was synthesized via cyclic voltammetry, utilizing an optimized number of cycles (N = 35) and a scan rate of 10 mV s^{-1}. Electrodeposition was performed within a potential range of −440 to −840 mV versus RHE at 45 °C in a three-electrode cell, with Pt foil serving as the counter electrode and $Hg|Hg_2Cl_2$ as the reference electrode. The electrolyte composition included $NiCl_2 \times 6H_2O$ (0.11 M), $FeCl_2 \times 4H_2O$ (0.18 M), and $CoCl_2 \times 6H_2O$ (0.04 M), with the pH adjusted to 2.0 using 1 M HCl. The deposition process on copper foam resulted in the formation of a metallic coating with an Fe:Co:Ni atomic ratio of 27:42:31. The synthesized electrocatalyst exhibited an HER overpotential of −65 mV@10 mA cm^{-2} and −214 mV@100 mA cm^{-2}, with a Tafel slope of −117 mV dec^{-1}. The improved performance was attributed to both the transition from a copper plate to copper foam as a substrate and the use of cyclic voltammetry as the synthesis method, which allowed better control over the morphology of the electrode material. The stability of the Fe–Co–Ni system was comparable to that of Ni–Fe, with a recorded HER overpotential of −255 mV@100 mA cm^{-2} during 3 h of continuous operation.

Impact of Electrodeposition Mode on the Catalytic Performance of 2D Electrocatalysts

The influence of electrodeposition mode on the catalytic properties of 2D electrocatalysts was investigated in studies [20–22], focusing on the synthesis of the Ni–Fe–Sn system on nickel foam. Yihui Wu et al. demonstrated that the electrode material's applicability depended on the electrodeposition mode, with cyclic voltammetry favoring cathodic applications (HER) and galvanostatic deposition being more suitable for anodic applications (OER). Both synthesis methods used an electrolyte containing $NiSO_4 \times 6H_2O$ (100 g L^{-1}), $FeSO_4 \times 7H_2O$ (30 g L^{-1}), $SnCl_2 \times 2H_2O$ (3 g L^{-1}), H_3BO_3 (20 g L^{-1}), NaCl (30 g L^{-1}), and sodium citrate ($C_6H_5Na_3O_7 \times 2H_2O$, 70 g L^{-1}), with the pH adjusted to 4.0 using 1 M HCl and 1 M KOH. Galvanostatic deposition was carried out at a current density of 20 mA cm^{-2} and 40 °C for 1 h, using nickel plates as anodes. In contrast, the cyclic voltammetry approach employed a three-electrode cell with a graphite counter electrode and an $Hg|Hg_2Cl_2$ reference electrode, with electrodeposition performed in the potential range of 150 to −850 mV versus $Hg|Hg_2Cl_2$ at a scan rate of 6 mV s^{-1} for seven cycles.

A major advantage of cyclic voltammetry as an electrodeposition method is the formation of coatings with stronger adhesion to the substrate, improving electrical contact and reducing charge transfer resistance. According to Yihui Wu et al., the Ni–Fe–Sn system synthesized via galvanostatic deposition exhibited excellent OER performance, with an overpotential of 253 mV@10 mA cm^{-2} and a Tafel slope of 61.5 mV dec^{-1}. Conversely, the same system synthesized via voltammetry demonstrated an HER overpotential of −103 mV@10 mA cm^{-2} and a Tafel slope of 97.4 mV dec^{-1}.

One of the most effective approaches is low-frequency pulsed electrodeposition, as demonstrated in [22]. Researchers synthesized a bifunctional 2D electrocatalyst using an electrolyte similar to that in [20, 21], applying pulsed deposition at frequencies of 60, 100, and 140 Hz with a current density of 30 mA cm^{-2}. The optimal catalytic performance for both HER and OER was achieved at 100 Hz with a duty cycle of 80%. The recorded overpotentials were −27 mV@10 mA cm^{-2} and −367 mV@200 mA cm^{-2} for HER, and 201 mV@10 mA cm^{-2} and 495 mV@200 mA cm^{-2} for OER. The Tafel slopes for the electrode material obtained via pulsed deposition were −98.6 mV dec^{-1} for HER and 32.4 mV dec^{-1} for OER. Stability tests of the Ni–Fe–Sn system synthesized under these conditions demonstrated continuous operation for 55 h at a cell voltage of 1.55 V.

As evident from these studies, pulsed and potentiodynamic electrodeposition methods yield superior results in electrocatalyst synthesis. Pulsed electrodeposition generates a series of low-frequency rectangular pulses, forming a dense array of catalytically active centers in the form of fine-grained deposits on the substrate surface. Cyclic voltammetry, which applies a triangular waveform with extended cycle periods and polarity reversal, enables the formation of more compact coatings with lower charge transfer resistance. Both methods effectively enhance the catalytic properties of electrodeposited electrocatalysts by either creating a well-developed morphology (a dense array of active sites) or reducing the energy barriers for electrochemical reactions by improving the electrical conductivity of the coating.

Nonmetal-Doped Electrocatalysts

Examples of nonmetal doping in electrocatalysts can be found in studies [23, 24]. In [23], a potentiostatic approach was used to synthesize an Ni-S system on nickel foam as a cathodic material for water electrolysis. Electrodeposition was conducted from an electrolyte containing nickel sulfate ($NiSO_4 \times 6H_2O$, 100 g L^{-1}) as the nickel source and thiourea (CH_4N_2S, 100 g L^{-1}) as the sulfur source. The electrolyte also included H_3BO_3 (40 g L^{-1}), NaCl (20 g L^{-1}), sodium citrate (70 g L^{-1}), and 5-sulfosalicylic acid (5 g L^{-1}), with a pH of 4.0 and a temperature of 40 °C. The electrodeposition was carried out in a three-electrode cell with a graphite counter electrode and an $Hg|Hg_2Cl_2$ reference electrode at −900 mV versus $Hg|Hg_2Cl_2$ for 20 min without stirring. Electrochemical testing revealed an HER overpotential of −58 mV@10 mA cm^{-2}, with a Tafel slope of −81.6 mV dec^{-1}. Long-term stability tests demonstrated stable HER performance over 12 h, with overpotentials of −70 mV@10 mA cm^{-2} and −175 mV@50 mA cm^{-2}.

In study [24], Toghraei et al. synthesized an Ni–Mo–P system on nickel foam using galvanostatic electrodeposition. The process was conducted in a three-electrode cell with a platinum counter electrode and an $Hg|Hg_2Cl_2$ reference electrode. Electrodeposition was performed at a current density of 60 mA cm^{-2} for 1 h at room temperature (18–25 °C). The electrolyte consisted of sodium citrate ($C_6H_5Na_3O_7 \times 2H_2O$, 0.3 M), nickel chloride ($NiCl_2 \times 6H_2O$, 0.1 M) as the nickel source, sodium hypophosphite ($NaH_2PO_2 \times H_2O$, 0.1 M) as the phosphorus source, and sodium molybdate ($NaMoO_4 \times 4H_2O$, 0.02 M) as the molybdenum source. The incorporation of phosphorus into the coating facilitated HER initiation by enhancing proton adsorption at negatively charged catalytic sites. Meanwhile, the Ni–Mo pair promoted the formation of Mo–H hydride bonds, which improved HER kinetics and reduced the thermodynamic energy barrier for charge transfer. The combined use of Ni–Mo–P resulted in an accelerated electrochemical reaction by lowering charge transfer resistance. The HER overpotentials for this material were recorded at -63 mV@10 mA cm^{-2} and -140 mV@100 mA cm^{-2}, with a Tafel slope of $-$87.3 mV dec^{-1}. Long-term stability tests (10 h) showed only a slight increase in HER overpotential, reaching -150 mV@100 mA cm^{-2}.

4.1.2 Carbon Substrates

The application of nonwoven carbon materials as substrates for electrodeposition is largely similar to the use of metallic foams. Both polymetallic systems and metal catalysts doped with sulfur or phosphorus can be deposited onto these substrates. However, key differences arise due to the intrinsic properties of carbon materials. First, their hydrophobic surface requires either the addition of a surfactant to the electrolyte or substrate pretreatment. Second, carbon materials undergo thermal decomposition at relatively low temperatures (400–500 °C in air), which enables the transformation of fiber networks into tubular structures, thereby increasing the catalytic surface area.

Study [25] demonstrated the potentiostatic electrodeposition of a bifunctional Ni–Co–P catalytic system on carbon cloth. The carbon substrate was pretreated by sequential ultrasonic cleaning in acetone, ethanol, and bidistilled water for 20 min, followed by drying at 60 °C. Electrodeposition was carried out using an electrolyte composed of $NiCl_2 \times 6H_2O$ (0.1 M), $CoCl_2 \times 6H_2O$ (0.1 M), and $NaH_2PO_2 \times H_2O$ (0.5 M) in a three-electrode cell with a graphite counter electrode and an $Hg|Hg_2Cl_2$ reference electrode. The process was conducted at a potential of -1 V versus $Hg|Hg_2Cl_2$ for 1 h. The resulting electrocatalyst exhibited excellent performance for both the cathodic and anodic reactions of water splitting. The HER overpotential for the synthesized system was -120 mV@10 mA cm^{-2} with a Tafel slope of -118.4 mV dec^{-1}. For the OER process, the Ni–Co–P system demonstrated an overpotential of 276 mV@10 mA cm^{-2} with a Tafel slope of -111.4 mV dec^{-1}. When used as both the cathode and anode, the material maintained stable operation for 36 h at an applied voltage of 1.69 V@10 mA cm^{-2}.

The synthesis of tubular Ni-Cu structures using carbon felt was proposed in study [26] through an electrochemical template-assisted approach. This method involved the formation of copper microtubes, followed by nickel electrodeposition. The overall synthesis process is illustrated in Fig. 2.

In the first step, copper was electrodeposited onto carbon felt from an electrolyte containing $CuSO_4 \times 5H_2O$ (35 g L^{-1}), potassium-sodium tartrate ($KNaC_4H_4O_6 \times 4H_2O$, 180 g L^{-1}), and NaOH (50 g L^{-1}). The carbon felt was pretreated via ultrasonic cleaning in ethanol for 5 min. Electrodeposition was carried out under galvanostatic

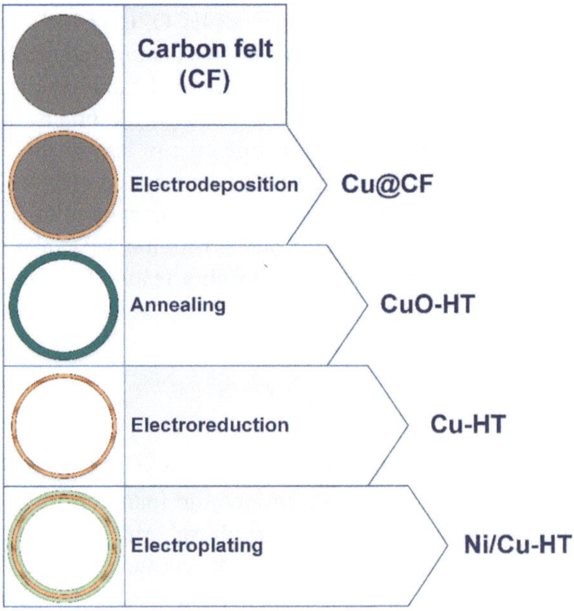

Fig. 2 Schematic representation of electrochemical template synthesis of Ni–Cu microtubes

conditions at a current density of 10 mA cm^{-2} for 100 min, using copper plates as anodes. The obtained intermediate product was dried at 100 °C and subsequently annealed in a muffle furnace at 450 °C for 4 h. During annealing, the carbon felt was burned off, and the copper structure oxidized to CuO, forming copper oxide microtubes. The microtubes were then reduced via electrolysis in 1 M Na_2SO_4 at a current of 50 mA for 45 min. Finally, nickel was electroplated onto the copper microtube array from an electrolyte containing $NiSO_4 \times 7H_2O$ (75 g L^{-1}), Na_2SO_4 (50 g L^{-1}), H_3BO_3 (25 g L^{-1}), and KCl (10 g L^{-1}) at a current density of 25 mA cm^{-2} for 15 min.

The synthesized Ni–Cu microtubes exhibited excellent HER activity, with overpotentials of −61 mV@10 mA cm^{-2} and −346 mV@100 mA cm^{-2}, along with a Tafel slope of −108.9 mV dec^{-1}. Long-term stability tests demonstrated sustained operation for 12 h, maintaining an overpotential of −160 mV@10 mA cm^{-2}.

4.2 *Solution-Based Synthesis*

Solution-based or "wet chemical" synthesis methods are typically employed for the fabrication of multilayered nanostructured materials and materials with 2D and 3D morphologies. Among these methods, those that utilize "soft chemistry" conditions—i.e., processes conducted under ambient or near-ambient conditions—are particularly advantageous. These approaches minimize energy consumption and enhance cost-effectiveness in the production of electrocatalytic materials. Notable examples of such synthesis techniques include the Successive Ionic Layer Adsorption and Reaction (SILAR) method, the Solid–Liquid–Gas Interface (SLGI) and Gas–Solid Interface Transition (GSIT) techniques, and the spin coating technique, which will be discussed in detail in the following sections.

4.2.1 Successive Ionic Layer Adsorption and Reaction (SILAR) Method

The Successive Ionic Layer Adsorption and Reaction (SILAR) method is a bottom-up synthesis technique in which chemical reactions occur sequentially at the solid–liquid interface. Numerous two-dimensional (2D) inorganic materials, such as oxides, oxyhydroxides, sulfides, fluorides, and noble metal complexes, are frequently fabricated using this technique. The advantages and applications of SILAR for the synthesis of electroactive materials—such as electrocatalysts, photocatalysts, electrochromic coatings, and electrode materials for supercapacitors and metal-ion batteries—have been extensively reviewed in studies [27–29].

Fundamentals of the SILAR Process

The SILAR synthesis process involves the sequential adsorption of cationic and anionic precursor species from aqueous solutions onto a substrate, followed by the removal of unreacted excess ions. This process ultimately leads to the formation of an insoluble product, which constitutes the resulting thin-film coating. To achieve the desired film thickness, the deposition cycle can be repeated multiple times.

The deposition process consists of several key steps to ensure uniform growth and film integrity:

1. Adsorption of the primary ionic precursor: The formation of an adsorbed monolayer of the initial precursor on the substrate surface is critical for obtaining a homogeneous and well-ordered coating.
2. Removal of excess ions: To prevent non-uniform deposition, the substrate is rinsed in a washing solution to eliminate loosely adsorbed ions.
3. Reaction with the secondary ionic precursor: The substrate is immersed in a solution containing the secondary precursor, triggering either an ion-exchange or redox reaction, which leads to the formation of the desired compound.
4. Final rinsing: The substrate undergoes another washing step before returning to the initial solution for the next deposition cycle.

This cyclic process (Fig. 3) continues until the desired film thickness and composition are achieved.

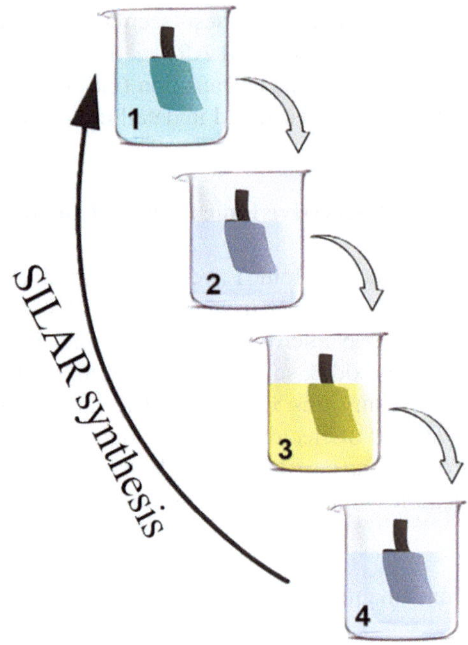

Fig. 3 Stages of the SILAR method for thin film deposition

Advantages of the SILAR Method

SILAR has established itself as a simple and energy-efficient method for depositing thin films of oxides, oxyhydroxides, hydroxides, sulfides, and more complex heterostructures on various substrates. Several unique features and benefits of this technique include:

1. Substrate versatility: The process is applicable to substrates with complex geometries and high surface areas, including both conductive materials (e.g., metals, metal foams, carbon-based materials) and non-conductive materials (e.g., glass, plastics, textiles).
2. Use of cost-effective precursors: The method employs readily available and inexpensive inorganic salts, making it economically viable.
3. Precise control over deposition parameters: The film growth rate, thickness, and stoichiometry can be fine-tuned by adjusting precursor concentration, composition, and immersion time.
4. Adjustable film thickness: By varying the number of deposition cycles, thin films can be fabricated within a broad thickness range—from nanometer-scale coatings to several microns.
5. Ambient processing conditions: Deposition typically occurs under standard atmospheric conditions, reducing the need for expensive equipment and minimizing energy consumption.

6. Post-deposition treatment flexibility: Films can undergo post-synthesis annealing to enhance crystallinity, promote grain growth, or refine structural properties.
7. Tailored crystallographic orientation: The ability to regulate ionic reactions at the substrate-solution interface allows for the fabrication of coatings with controlled grain structure and preferred crystallographic orientation.
8. Nanocrystal growth control: The extensive tunability of synthesis parameters enables precise regulation of nucleation, crystal growth, morphology, and stoichiometry, thereby optimizing the material's electrocatalytic properties.

The SILAR method stands out as a versatile, cost-effective, and energy-efficient approach for depositing thin films with precise control over morphology, composition, and thickness. Its compatibility with a wide range of substrates, along with the ability to fine-tune deposition conditions and post-synthesis modifications, makes it an attractive choice for fabricating high-performance electrocatalytic materials. These advantages collectively position SILAR as a highly adaptable technique for the development of advanced functional coatings in various electrochemical and energy applications.

Electrocatalytic Materials Synthesized via SILAR

Among the primary 2D electrocatalytic materials synthesized using the SILAR method are chalcogenides, layered oxides, and transition metal hydroxides. These materials exhibit high electrochemically active surface areas, which provide numerous catalytic sites. Their ultrathin crystal structure facilitates rapid charge transfer within the electroactive layer, enhances diffusion processes, and promotes efficient gas bubble detachment, thus improving electrocatalytic performance.

Metal Chalcogenides

Thin film metal chalcogenides are one of the most frequently synthesized materials by SILAR due to the simplicity of the process and the availability of anionic precursors which are widely available. Traditional SILAR-based fabrication of metal chalcogenide thin films relies on the use of dissolved metal salts as cationic precursors and water-soluble chalcogenides as anionic precursors. The driving force for chalcogenide formation on the substrate surface is an ion-exchange reaction, facilitated by the low solubility of the resulting metal chalcogenide compound in aqueous media.

Metal Oxides and Hydroxides

SILAR-based deposition of metal oxides and hydroxides has gained significant attention due to the high electrochemical activity and chemical stability of these materials. The general approach for oxide and hydroxide synthesis differs from that of chalcogenides, as it lacks an anionic precursor capable of directly delivering O^{2-} ions to form an oxide structure. Instead, metal salts (such as acetates, chlorides, nitrates, and sulfates) are typically used as cationic precursors. The formation of oxides and hydroxides is induced by immersing the substrate in either heated distilled water, alkaline solutions, or oxidizing agents (e.g., H_2O_2, $KMnO_4$, NaClO), triggering a redox reaction that drives oxide precipitation.

A recent study [30] demonstrated the potential of SILAR for the synthesis of a highly efficient CuO-based photoelectrocatalyst. The authors established a correlation between precursor pH levels and the resulting CuO nanolayer morphology, highlighting its significant impact on photoelectrocatalytic performance.

Layered Double Hydroxides (LDHs)

Layered double hydroxides (LDHs) of transition metals can also be readily synthesized via SILAR. LDHs are characterised by their large surface area, layered structure and efficient anionic exchange capacity, which make them highly attractive for use in electrolytic and photocatalytic water splitting in alkaline environments.

For instance, studies [31, 32] were the first to demonstrate the SILAR-based synthesis of CoCr-LDH and NiCr-LDH nanocrystals. In this process, aqueous solutions of cobalt and nickel salts were used as cationic precursors, while hexahydroxychromium complexes ($[Cr(OH)_6]^{3-}$) served as anionic precursors. The resulting deposition yielded ultrathin (3–5 nm) 2D LDH nanosheets on the substrate surface. These nanostructures exhibited exceptional electrocatalytic activity and stability for water-splitting reactions, specifically oxygen evolution reaction (OER) and hydrogen evolution reaction (HER).

The electrocatalytic performance of these materials was remarkable:

- CoCr-LDH achieved an HER overpotential of 175 mV at 10 mA cm^{-2};
- NiCr-LDH exhibited HER and OER overpotentials of 212 mV and 325 mV at 10 mA cm^{-2}, respectively.

These findings highlight the effectiveness of the SILAR method for synthesizing high-performance 2D electrocatalytic materials with controlled morphology, composition, and electrochemical properties.

4.2.2 Gas–Liquid Interface Techniques

Recent advancements in electrocatalyst synthesis have led to the development of Solid–Liquid–Gas Interface (SLGI) techniques, where chemical reactions occur at gas–liquid and gas–liquid–solid phase boundaries. These methods belong to the template-assisted synthesis category, where templates such as gas bubbles, liquid droplets, or the surface of an aqueous solution serve as structural guides, enabling the fabrication of materials with unique morphologies. Examples of such structures include core–shell architectures, microspheres, microtubes, micro-rolls, and nanospirals. A comprehensive review [33] discusses various SLGI-based approaches for synthesizing electrocatalysts for HER, OER, and ORR (Fig. 4). However, further research is required to deepen the understanding of physicochemical mechanisms occurring at gas–liquid (G-L) and gas–liquid–solid (G-L-S) interfaces to optimize these synthesis methods.

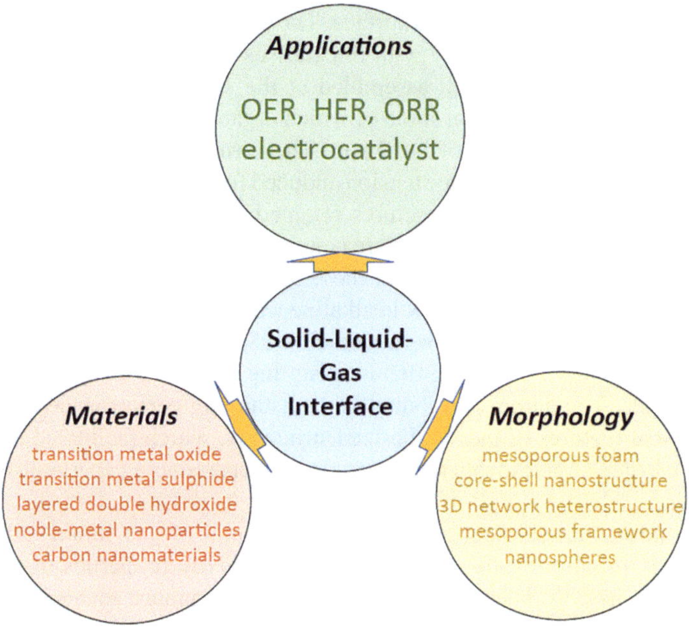

Fig. 4 Examples of electrocatalyst structures obtained via the solid–liquid–gas interface technique

The Gas-Solution Interface Technique (GSIT) is a specific case of template-assisted synthesis at phase boundaries, where a chemical reaction occurs between two reactants—one in the gas phase and the other dissolved in liquid [34]. This process results in a thin gradient film due to the spatial separation and diffusion of the reactants along the interface. Gradients in thin films may occur due to differences in morphology, density or chemical composition, which under certain conditions cause interfacial forces to cause the initially planar layers to curl into microtubes or microlayers, especially during the process of lyophilic drying.

In a series of studies by L. Gulina et al., novel methods for producing microrolls and nanospirals of transition metal-based inorganic compounds via interfacial reactions were presented. One particularly interesting example is the synthesis of iron-based nanospirals, as described in [35]. In this process, a thin $Fe(OH)_3$ film was formed by reacting an aqueous iron salt solution with gaseous ammonia. Upon drying, the film transformed into $Fe(OH)_3$ nanospirals with diameters of up to 10 μm. These spirals were subsequently converted into Fe_2O_3 nanospirals through annealing in air, and then further reduced in a hydrogen atmosphere to yield metallic iron nanospirals. Importantly, both annealing and reduction steps preserved the original morphology of the nanospirals. Electrochemical testing revealed that Fe_2O_3 nanospirals exhibited exceptional performance as HER electrocatalysts, highlighting the potential of this interfacial synthesis approach for advanced catalytic applications.

Study [36] investigated a chemical reaction between gaseous ozone and a manganese acetate aqueous solution droplet, resulting in the in-situ formation of

MnO_2 nanocrystals on the droplet's surface. The obtained MnO_2 nanocrystals exhibited a 2D nanosheet morphology with thicknesses ranging from 5 to 8 nm. Spontaneously, these nanosheets self-assembled at the solution-gas interface, forming distinct bands 1–2 μm in width, some of which organized into ordered honeycomb-like structures with cell sizes of 5–20 μm. The formation of these self-assembled structures was attributed to surface tension-induced forces arising from droplet curvature. After drying, the MnO_2 structures retained their morphology and exhibited fractal-like patterns with a geometric area equivalent to that of the original droplet.

Electrochemical characterization of these MnO_2-based electrocatalysts revealed that they were highly active for OER in alkaline water electrolysis. The most efficient catalysts demonstrated an OER overpotential of 284 mV at 10 mA cm^{-2} and a Tafel slope of 37.7 mV dec^{-1}, ranking them among the best pure manganese oxides for electrocatalysis. The authors attributed this outstanding performance to the unique morphological features of the synthesized structures, which facilitated the efficient release of oxygen bubbles from the electrode surface, reducing energy losses during electrolysis.

These findings emphasize the significant potential of gas–liquid–solid interfacial synthesis in tailoring the morphology and catalytic properties of electrocatalysts, opening new avenues for designing high-performance materials for energy conversion applications.

4.2.3 Spin Coating

Another promising approach in wet chemistry synthesis is the spin coating technique [37]. This technique is frequently used for depositing uniform thin films on a variety of flat surfaces. The spin coating system consists of a rotating disk onto which the substrate is secured, while the reagent solution is dispensed onto its surface drop by drop. The film formation process occurs in four stages: liquid deposition, high-speed rotation, controlled liquid outflow, and solvent evaporation. Centrifugal force causes the solution to spread uniformly as the substrate spins quickly, and the solvent evaporates, leaving behind a homogeneous thin coating of the produced material. The rotation speed and solution volume significantly influence the properties of the resulting thin films, allowing precise control over synthesis parameters.

Currently, spin coating is considered one of the fastest, most cost-effective, and simplest methods for obtaining uniform thin-film samples. One of its key advantages is the ability to produce highly homogeneous layers on both conductive and non-conductive substrates. The final film thickness primarily depends on the substrate's rotation speed and the viscosity of the precursor solution. The method's benefits include rapid synthesis, minimal equipment requirements, precise control over film thickness (down to tens of nanometers), and compatibility with a wide range of substrates.

The successful application of spin coating for the synthesis of photoelectrocatalytic materials is demonstrated in study [38], where α-Fe_2O_3-based thin films with thicknesses of up to 30 nm were fabricated. The researchers established a correlation between film thickness and photocurrent density values (34.6, 7.8, and 13.5 $\mu A\ cm^{-2}$), demonstrating the ability to fine-tune synthesis parameters. Additionally, study [39] utilized spin coating to fabricate a photoelectrocatalyst based on $SrCl_2$-IrO_x, deposited onto a polycrystalline silicon substrate. The material exhibited high photoanodic activity, generating a primary photocurrent at 1.06 V versus RHE. This work highlights the effectiveness of spin coating for producing photoelectrocatalytic coatings on silicon surfaces, avoiding potential etching effects that may occur in other deposition techniques used for photoactive layer formation.

4.3 Solvothermal Synthesis

The solvothermal (or hydrothermal) approach has become a widely adopted method for synthesizing 2D electrode materials, particularly for the fabrication of heteroatomic structures based on MXenes, layered double hydroxides (LDHs), and two-dimensional arrays (2D arrays). This technique enables precise control over material composition and morphology by utilizing high-temperature and high-pressure conditions in a sealed reaction environment.

Metallic foams are commonly used as substrates due to their widespread availability and ease of use. However, carbon-based materials, such as carbon fiber, carbon cloth, and carbon paper, are also employed, though less frequently. Their lower adoption rate is partially attributed to the need for additional surface pretreatment, similar to electrochemical synthesis methods. This pretreatment step is essential for removing trapped gases (degassing) and enhancing the substrate's wettability, which facilitates uniform deposition and adhesion of the synthesized material.

4.3.1 Metal Substrates

A representative example of the solvothermal synthesis of self-supported electrode materials utilizing MXenes is described in study [40]. In this work, a $Ti_3C_2T_x$ layered structure, obtained through the etching of Ti_3AlC_2, was combined with cetyltrimethylammonium bromide (CTAB) and thioacetamide (TAA) in a Teflon-lined autoclave. A nickel foam substrate was also introduced into the reactor, serving both as the substrate and the active surface for the solvothermal reaction. The synthesis was carried out at 180 °C for 10 h, resulting in the formation of a 2D nanowall structure of $Ti_3C_2T_x$ coupled with Ni_3S_2 on the nickel foam surface.

The fabricated electrode material exhibited a HER overpotential of −72 mV at 10 mA cm^{-2} with a Tafel slope of −45 mV dec^{-1}. Furthermore, long-term electrochemical stability tests demonstrated continuous operation at a current density of 20 mA cm^{-2} for 12 h under a constant overpotential of −169 mV.

Another widely studied 2D material for electrocatalytic water splitting is the nanorod-based structure. In study [41], an array of Cu/CuO@Co/Co_3O_4 nanorods was grown directly on the surface of copper foam. The synthesis consisted of three stages:

1. The copper foam was immersed in a 2.5 M NaOH + 0.125 M $(NH_4)_2S_2O_8$ solution, leading to the formation of $Cu(OH)_2$ nanorod arrays on its surface.
2. The hydrothermal process was carried out in a solution of 0.075 M $CoSO_4 \times 7H_2O$ and 0.35 M urea (CH_4N_2O) at 85 °C for 2 h, resulting in the deposition of $Cu(OH)_2@Co_2CO_3(OH)_2$ onto the nanorods.
3. The obtained precursor was annealed in a reducing atmosphere (H_2/Ar) at 300 °C for 3 h, yielding the final Cu/CuO@Co/Co_3O_4 structure.

The synthesized electrocatalyst demonstrated HER overpotentials of −82 mV at 10 mA cm^{-2} and −120 mV at 50 mA cm^{-2}, along with a Tafel slope of − 48 mV dec^{-1}. Long-term stability tests showed that the material maintained stable performance for 36 h, sustaining an overpotential of −100 mV at 10 mA cm^{-2}.

4.3.2 Carbon Substrates

A representative approach to modern hydrothermal synthesis using a carbon-based substrate is described in study [42]. In this work, a 30 mL Teflon-lined autoclave was loaded with $Co(NO_3)_2 \times 6H_2O$ (0.05 M), NH_4F (0.09 M), CH_4N_2O (0.17 M), and a 2 × 4 cm^2 piece of carbon cloth (CC). The synthesis was carried out at 120 °C for 8 h, followed by ethanol and distilled water washing and subsequent drying at 40 °C for 6 h.

The resulting electrode material, $Co(CO_3)_{0.5}OH$@CC, exhibited a nanoneedle array structure on the surface of the carbon fibers. This morphology significantly enhances the electrocatalytically active surface area (by up to 40 times) and facilitates more efficient gas bubble detachment. The oxygen evolution reaction (OER) overpotential for $Co(CO_3)_{0.5}OH$@CC was measured at 317 mV at 10 mA cm^{-2}, with a Tafel slope of 146 mV dec^{-1}.

Long-term electrochemical testing (27 h) demonstrated that applying 1.5 V in an electrolyzer with a graphite cathode and the synthesized anode sustained a current density of 10 mA cm^{-2}, confirming the electrode's stability and efficiency in catalytic applications.

4.4 Vapor Deposition Techniques

4.4.1 Chemical Vapor Deposition (CVD)

Chemical Vapor Deposition (CVD) is one of the most widely used methods for depositing thin films through chemical reactions between gaseous precursors. During the CVD process, gas-phase molecules react on a heated substrate, forming a solid deposit. This technique enables precise control over thickness, structure, and morphology of the resulting 2D materials while also allowing the synthesis of high-purity films, which is crucial for achieving high catalytic activity and stability.

CVD is extensively employed in various fields, including fuel cell materials, photovoltaic devices, and electrocatalysts for hydrogen evolution reactions (HER).

The CVD synthesis process (Fig. 5) can be described as follows:

1. The gas-phase precursors enter the reaction chamber and move toward the deposition zone.
2. Inside the reactor, chemical reactions occur, transforming the gas-phase precursors into film-forming compounds and by-products.
3. The reacted precursors diffuse toward the substrate surface.
4. The precursors adsorb onto the substrate, initiating the deposition process.
5. The adsorbed precursors diffuse across the surface to specific growth sites.
6. Unwanted by-products desorb from the surface, ensuring an uninterrupted deposition process.
7. The desorbed by-products are carried away by the gas flow and removed from the reactor, maintaining process purity and preventing the accumulation of contaminants.

There are several types of CVD processes, each distinguished by the method used to initiate chemical reactions. The following are the five most commonly applied CVD techniques for synthesizing 2D materials.

Thermal CVD

Thermal CVD operates by heating both the substrate and gaseous precursors to high temperatures (typically 500–1000 °C or higher), causing decomposition or reaction of the precursors. The thermal energy activates molecular interactions, leading to the formation of the target material.

There are two primary types of reaction chambers:

– Hot-Wall CVD—The entire reaction chamber, including its walls, is uniformly heated, ensuring a high deposition rate and film uniformity.

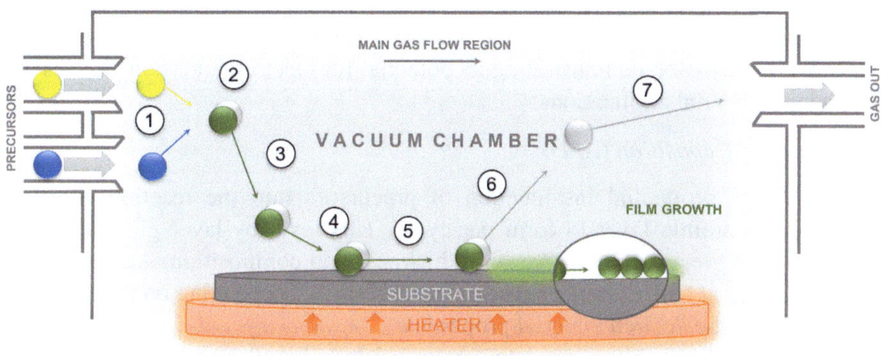

Fig. 5 Schematic illustration of a typical thermal CVD process

– Cold-Wall CVD—Only the substrate is heated via microwave or radiofrequency radiation, keeping the chamber walls cool. This approach allows for deposition of temperature-sensitive materials while improving film purity.

Low-Pressure CVD (LPCVD)

LPCVD is conducted under reduced pressure conditions, minimizing contamination risks and improving film uniformity. This approach is particularly successful for semiconductors and 2D materials like graphene and molybdenum disulfide (MoS_2).

The reaction is initiated by heating the substrate and precursors to a temperature at which they start reacting. Lower pressure reduces molecular collisions, leading to higher film purity and homogeneity. The typical operating temperatures range from 300 to 800 °C, depending on the target material.

Plasma-Enhanced CVD (PECVD)

PECVD allows for the creation of thin films with adjustable electrical and mechanical properties, including conductivity, high dielectric permittivity, hardness, and flexibility. These properties make PECVD an ideal choice for applications in microelectronics, optoelectronics, and flexible electronics.

The chemical reactions in PECVD are driven by plasma generation, which activates the gas molecules, producing reactive radicals that facilitate film formation on the substrate. Unlike conventional CVD, PECVD allows deposition at lower temperatures, making it ideal for substrates and materials that are sensitive to heat.

Inductively Coupled Plasma CVD (ICPCVD)

ICPCVD is frequently used for depositing ultra-pure and highly uniform thin films. Like PECVD, this method utilizes plasma to initiate chemical reactions, but in this case, the plasma is generated through inductive heating. This ensures a more uniform energy distribution within the reaction chamber and precise control over plasma parameters.

In study [43], ICPCVD was successfully employed to synthesize hexagonal boron nitride (h-BN) using borazine as the precursor. The study examined the influence of reaction temperature and gas flow rates (H_2 and N_2) on the quality of the deposited films. The optical properties of the synthesized h-BN exceeded those of commercially available h-BN, demonstrating its potential for electron emission devices and expanded industrial applications.

Atomic Layer Deposition (ALD)

ALD involves sequential introduction of precursors into the reaction chamber, allowing one atomic layer to form per cycle. This layer-by-layer growth mechanism provides precise control over film thickness and composition. ALD reactions typically occur at temperatures between 50 and 300 °C, depending on the precursor chemistry and the desired material properties.

In study [44], ALD was used to fabricate MoS_2-based electrode materials on 3D-printed titanium electrodes. The deposition of 2D MoS_2 layers significantly reduced the HER overpotential, with the largest reduction (−0.4 V) observed after 30 ALD cycles. The material's stability was confirmed through chronopotentiometric testing over 250 cycles, highlighting ALD as an effective method for improving electrocatalyst performance.

CVD is a powerful technique for synthesizing 2D electrocatalysts, offering precise structural control and high material purity. Despite its high cost and process complexity, its ability to produce high-performance electrocatalysts for hydrogen evolution, oxygen reduction, and other electrochemical applications makes it an essential tool in electrocatalysis research and development.

4.4.2 Physical Vapor Deposition (PVD)

Physical Vapor Deposition (PVD) is a method of depositing materials that involves the process of vaporizing materials and then condensing them onto a surface. During the PVD process, a solid target material is evaporated in a vacuum or controlled atmosphere, after which the vapor condenses onto the substrate surface, forming thin films. This method enables precise control over material properties by altering the parameters of deposition, including the deposition rate, chamber pressure, and the distance between the target and the substrate.

The key stages of the PVD process (Fig. 6) can be outlined as follows:

1. The target material is heated to high temperatures, initiating evaporation. This can occur in vacuum or an inert atmosphere to prevent oxidation.
2. The vaporized material travels upward through the vacuum or gas medium toward the substrate.
3. Upon reaching the substrate, the material condenses due to the temperature gradient, forming the initial film layers.
4. Once the desired film thickness is achieved, the deposition process is terminated.

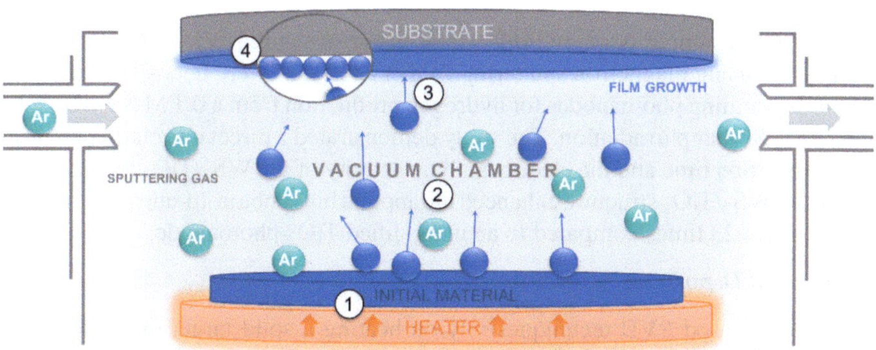

Fig. 6 Fundamental steps of the physical vapor deposition (PVD) process

PVD techniques can generally be categorized into two major groups: sputtering-based and evaporation-based deposition methods.

Sputtering Deposition

The process of sputtering involves the application of physical or magnetic force to a solid target, resulting in the release of atoms or molecules from the target surface and their deposition onto a substrate. This method produces films with high adhesion and uniform thickness, making it particularly valuable for microelectronics and optoelectronics.

1. Magnetron Sputtering
   - Utilizes a magnetic field to enhance the sputtering efficiency.
   - Produces denser and higher-quality coatings due to the high degree of ionization and particle energy.
2. Ion-Beam Sputtering
   - Accelerated ions bombard the target, knocking out atoms, which are then deposited onto the substrate.
   - Enables precise control over film thickness and composition.

Advantages of sputtering deposition:

- high deposition rate;
- excellent adhesion and uniformity;
- capability to deposit complex multilayer structures.

Limitations of sputtering deposition:

- challenging thickness control, particularly over large deposition areas;
- high-energy particle bombardment may induce structural defects, negatively impacting electrical and optical properties;
- difficulties in multilayer fabrication, as each new layer may interact with the previous one, requiring additional process optimization.

A notable example is study [45], where heterostructured WS_2/TiO_2 films were synthesized using magnetron sputtering. WS_2 nanosheets were deposited onto TiO_2 nanorods, forming photoanodes for hydrogen production from a 0.1 M Na_2SO_4 solution under Xe-lamp irradiation. The study demonstrated a direct correlation between WS_2 deposition time and the electrocatalytic activity of the WS_2/TiO_2 material. The optimized WS_2/TiO_2 structure enhanced the applied bias photon-to-current efficiency (ABPE) by 3.23 times compared to an unmodified TiO_2 photoanode.

Evaporation Deposition

Evaporation-based PVD techniques rely on heating a solid target material until it transitions into the vapor phase, which then condenses onto the substrate, forming a thin film.

1. Thermal Evaporation

The substance to be processed is subjected to heat until it transforms into a gaseous state, after which it is cooled and deposited onto a colder surface, resulting in the creation of a thin coating.

2. Electron-Beam Evaporation

A focused electron beam is employed to heat and vaporize the target material at high temperatures, enabling rapid and precise deposition.

3. Pulsed Laser Deposition (PLD)

A high-energy laser is used to remove material from a solid surface, creating a cloud of plasma that is then deposited onto a substrate. Enables high control over film composition and structure.

4. Cathodic Arc Deposition

A high-voltage arc discharge is applied between a cathode (target material) and an anode, leading to vaporization and ionization of the cathode material. The ionized atoms condense onto the substrate, forming a film with high adhesion and tunable properties.

In study [46], graphitic carbon nitride (g-C_3N_4) thin films were synthesized using thermal evaporation. This approach allowed precise control over the material's optical properties, optimizing it for applications such as solar-to-chemical energy conversion and environmental pollutant removal.

Challenges and Future Prospects

While PVD techniques offer numerous advantages, they also present certain challenges:

- high equipment costs;
- limited efficiency for complex geometries;
- potential structural defects due to high-energy particle interactions.

However, advancements in process optimization and emerging innovations continue to expand PVD's applications across aerospace, medical, and automotive industries. These improvements enable the development of next-generation functional coatings, further solidifying PVD as a critical deposition technique in advanced materials engineering.

5 Conclusion

The development of self-supported 2D electrocatalysts represents a transformative step in advancing hydrogen evolution reactions (HER) and electrochemical energy conversion technologies. This chapter has explored the key classes of 2D electrocatalysts, including transition metal dichalcogenides (TMDs), MXenes, layered double hydroxides (LDHs), hybrid structures, and metal-free catalysts, emphasizing their distinct physicochemical properties and catalytic performance. Additionally, it has critically examined the role of substrates, such as metal foams, carbon-based materials, and transparent conductive supports, in optimizing electrocatalytic efficiency and long-term stability. Finally, an extensive review of state-of-the-art synthesis techniques—including electrodeposition, solution-based synthesis, solvothermal methods, and vapor deposition techniques (CVD and PVD)—was provided, highlighting their advantages, limitations, and scalability for practical applications.

The characteristics of the structure and electronic features of two-dimensional electrocatalysts are crucial in determining their electrochemical performance, durability, and ability to transfer charge. TMDs, such as MoS_2 and WS_2, demonstrate high catalytic efficiency due to their layered structure and ability to facilitate proton adsorption and dissociation. Similarly, MXenes exhibit exceptional conductivity and tunability, making them promising candidates for HER and CO_2 reduction reactions. The inclusion of LDHs has further expanded the scope of self-supported electrocatalysts, as their high surface area and ion-exchange properties enhance catalytic performance in water electrolysis and methanol oxidation processes. Additionally, hybrid structures, which integrate multiple materials with complementary functionalities, offer synergistic effects that improve catalytic activity. The growing interest in metal-free electrocatalysts, such as graphitic carbon nitride (g-C_3N_4) and black phosphorus, underscores the potential for developing cost-effective and sustainable alternatives to traditional metal-based catalysts.

Beyond material selection, substrate choice is a critical factor in ensuring mechanical stability, electronic conductivity, and electrochemical durability. Metal foams (e.g., nickel, copper) provide high conductivity and porosity, facilitating efficient charge transport and electrolyte penetration. However, carbon-based substrates, such as carbon paper and carbon fabric, offer high flexibility, chemical inertness, and corrosion resistance, making them suitable for long-term electrocatalytic applications. Transparent conductive substrates, such as fluorine-doped tin oxide (FTO) and indium tin oxide (ITO)-coated glass, have demonstrated potential for photoelectrochemical applications by integrating light-harvesting capabilities with efficient electrocatalysis.

The review of advanced synthesis techniques underscores the importance of precise control over material morphology, crystallinity, and electronic structure. Electrodeposition methods, such as galvanostatic and cyclic voltammetry deposition, enable highly tunable and scalable fabrication of polymetallic and doped electrocatalysts. Solution-based approaches, including Successive Ionic Layer Adsorption and Reaction (SILAR) and Gas–Liquid Interface techniques, offer cost-effective and scalable routes for fabricating 2D electrocatalysts with well-defined nanostructures. Solvothermal methods, particularly for MXenes and LDH-based materials, enable heteroatom doping and structural modulation, further enhancing catalytic activity. Finally, vapor deposition methods, such as Chemical Vapor Deposition (CVD) and Physical Vapor Deposition (PVD), provide an unparalleled ability to synthesize ultra-thin, high-purity films with precisely controlled thickness and composition.

Despite these advancements, several key challenges remain in the field of self-supported 2D electrocatalysts. First, the scalability of synthesis techniques remains a critical bottleneck, as many high-performance electrocatalysts require complex and time-consuming fabrication methods that may not be economically viable for large-scale industrial applications. Second, long-term stability under real-world operating conditions is essential, as material degradation, catalyst poisoning, and mechanical instability can significantly impact performance. Finally, it is essential to continue improving heterostructures and hybrid materials to fully exploit their collaborative effects and discover novel catalytic capabilities.

Future investigations should concentrate on developing cost-effective and scalable methods for synthesis that preserve the high catalytic activity of 2D materials while ensuring mechanical robustness and chemical stability. The integration of artificial intelligence (AI)-driven material design, high-throughput screening, and machine learning algorithms can accelerate the discovery of new 2D electrocatalyst compositions with optimized catalytic properties. Additionally, the exploration of novel substrate-engineered systems, such as 3D-printed architectures and flexible electrodes, holds promise for advancing next-generation hydrogen evolution technologies.

In conclusion, self-supported 2D electrocatalysts offer an exciting frontier for sustainable energy conversion and electrochemical applications. Their high surface area, tunable electronic properties, and structural versatility provide a strong foundation for the development of efficient and durable hydrogen evolution catalysts. By addressing scalability, long-term stability, and hybrid material integration, future advancements in this field will pave the way for the widespread adoption of hydrogen-based energy solutions, contributing to a more sustainable and carbon–neutral future.

References

1. Pashchenko, D.: Green hydrogen as a power plant fuel: What is energy efficiency from production to utilization? Renew. Energy **223**, 120033 (2024). https://doi.org/10.1016/j.renene.2024.120033
2. Zeng, K., Zhang, D.: Recent progress in alkaline water electrolysis for hydrogen production and applications. Prog. Energy Combust. Sci. **36**, 307–326 (2010). https://doi.org/10.1016/j.pecs.2009.11.002
3. Ma, T.Y., Dai, S., Qiao, S.Z.: Self-supported electrocatalysts for advanced energy conversion processes. Mater. Today **19**, 265–273 (2016). https://doi.org/10.1016/j.mattod.2015.10.012
4. Fu, K., Yuan, D., Yu, T., Lei, C., Kou, Z., Huang, B., Lyu, S., Zhang, F., Wan, T.: Recent advances on two-dimensional nanomaterials supported single-atom for hydrogen evolution electrocatalysts. Molecules **29**(18), 4304 (2024). https://doi.org/10.3390/molecules29184304
5. Sobhani Bazghale, F., Gilak, M.R., Zamani Pedram, M., Torabi, F., Naikoo, G.A.: 2D nanocomposite materials for HER electrocatalysts—a review. Heliyon **10**, e23450 (2024). https://doi.org/10.1016/j.heliyon.2023.e23450
6. Askari, M.B., Salarizadeh, P., Veisi, P., Samiei, E., Saeidfirozeh, H., Tourchi Moghadam, M.T., Di Bartolomeo, A.: Transition-metal dichalcogenides in electrochemical batteries and solar cells. Micromachines **14**, 691 (2023). https://doi.org/10.3390/mi14030691

7. Zhang, L.W., Xu, J.L., Tang, J., Li, L.L., Luo, J.M.: Synthesis of self-supporting bulk porous NiMo@MoS_2 electrocatalyst to enhance hydrogen evolution in alkaline conditions. J. Mater. Res. Technol. **17**, 1167–1175 (2022). https://doi.org/10.1016/j.jmrt.2022.01.068
8. Liang, K., Tabassum, A., Kothakonda, M., Zhang, X., Zhang, R., Kenney, B., Koplitz, B.D., Sun, J., Naguib, M.: Two-dimensional titanium carbonitride MXene as a highly efficient electrocatalyst for hydrogen evolution reaction. Mater. Rep. Energy **2**, 100075 (2022). https://doi.org/10.1016/j.matre.2021.100075
9. Fan, K., Xu, P., Li, Z., Shao, M., Duan, X.: Layered double hydroxides: next promising materials for energy storage and conversion. Next Mater. **1**, 100040 (2023). https://doi.org/10.1016/j.nxmate.2023.100040
10. Xiang, M., Zhang, T., Tan, F., Cai, S., Xu, Z.: Self-supported nickel-cobalt layered double hydroxide for efficient primary amine electrooxidation and decoupled water electrolysis. Catal. Commun. **179**, 106698 (2023). https://doi.org/10.1016/j.catcom.2023.106698
11. Yang, R., Zhou, Y., Xing, Y., Li, D., Jiang, D., Chen, M., Shi, W., Yuan, S.: Synergistic coupling of CoFe-LDH arrays with NiFe-LDH nanosheets for highly efficient overall water splitting in alkaline media. Appl. Catal. B Environ. **253**, 131–139 (2019). https://doi.org/10.1016/j.apcatb.2019.04.054
12. Iram, F., Mallah, A.M., Safdar, M., Khitouni, M., Hammami, B., Alsehaibani, H.A.H., Alkhaibari, I.S.: Fabrication and OER/HER electrochemical activity of Sm, Al-assisted g-C_3N_4 nanocomposite for alkaline water splitting. J. Alloys Compd. **1014**, 178603 (2025). https://doi.org/10.1016/j.jallcom.2025.178603
13. Li, Y., Pei, W., He, J., Liu, K., Qi, W., Gao, X., Zhou, S., Xie, H., Yin, K., Gao, Y., He, J., Zhao, J., Hu, J., Chan, T.S., Li, Z., Zhang, G., Liu, M.: Hybrids of PtRu nanoclusters and black phosphorus nanosheets for highly efficient alkaline hydrogen evolution reaction. ACS Catal. **9**, 10870–10875 (2019). https://doi.org/10.1021/acscatal.9b03506
14. Cheliotis, I., Zergioti, I.: A review on transfer methods of two-dimensional materials. 2D Mater. **11**, 22004 (2024). https://doi.org/10.1088/2053-1583/ad2f43
15. Ramesh, M., Rajeshkumar, L., Bhoopathi, R.: Carbon substrates: a review on fabrication, properties, and applications. Carbon Lett. **31**, 557–580 (2021). https://doi.org/10.1007/s42823-021-00264-z
16. Marinova, V., Petrov, S., Petrova, D., Napoleonov, B., Chau, N.H.M., Lan, Y.P., Strijkova, V., Hsu, K.Y., Dimitrov, D., Lin, S.H.: Effect of transparent conductive layers on the functionality of liquid crystal devices: comparison of AZO, FTO, and ITO. Opt. Mater. X **22**, 100330 (2024). https://doi.org/10.1016/j.omx.2024.100330
17. Yang, H., Driess, M., Menezes, P.W.: Self-supported electrocatalysts for practical water electrolysis. Adv. Energy Mater. **11**(39), 2102074 (2021). https://doi.org/10.1002/aenm.202102074
18. Hatami, E., Toghraei, A., Barati Darband, G.: Electrodeposition of Ni–Fe micro/nano urchin-like structure as an efficient electrocatalyst for overall water splitting. Int. J. Hydrogen Energy **46**, 9394–9405 (2021). https://doi.org/10.1016/j.ijhydene.2020.12.110
19. Kaboli, A., Esfandiari, N., Darband, G.B., Sharifi, R., Aliofkhazraei, M., Rouhaghdam, A.S.: Electrodeposition of Fe-Co-Ni coating by cyclic voltammetry for efficient hydrogen production. J. Electroanal. Chem. **958**, 118151 (2024). https://doi.org/10.1016/j.jelechem.2024.118151
20. Wu, Y., Gao, Y., He, H., Zhang, P.: Electrodeposition of self-supported Ni–Fe–Sn film on Ni foam: an efficient electrocatalyst for oxygen evolution reaction. Electrochim. Acta **301**, 39–46 (2019). https://doi.org/10.1016/j.electacta.2019.01.151
21. Hassan, M.H., Kamel, S., Safaraliev, M., Kokin, S.: Improved techno-economic optimization of hybrid solar/wind/fuel cell/diesel systems with hydrogen energy storage. Int. J. Hydrogen Energy **68**, 998–1018 (2024). https://doi.org/10.1016/j.ijhydene.2024.04.124
22. You, S., Wu, Y., Wang, Y., He, Z., Yin, L., Zhang, Y., Sun, Z., Zhang, Z.: Pulse-electrodeposited Ni–Fe–Sn films supported on Ni foam as an excellent bifunctional electrocatalyst for overall water splitting. Int. J. Hydrogen Energy **47**, 29315–29326 (2022). https://doi.org/10.1016/j.ijhydene.2022.06.265

23. Wu, Y., Lian, J., Wang, Y., Sun, J., He, Z., Gu, Z.: Potentiostatic electrodeposition of self-supported Ni–S electrocatalyst supported on Ni foam for efficient hydrogen evolution. Mater. Des. **198**, 109316 (2021). https://doi.org/10.1016/j.matdes.2020.109316
24. Toghraei, A., Shahrabi, T., Barati Darband, G.: Electrodeposition of self-supported Ni-Mo-P film on Ni foam as an affordable and high-performance electrocatalyst toward hydrogen evolution reaction. Electrochim. Acta **335**, 135643 (2020). https://doi.org/10.1016/j.electacta.2020.135643
25. Sheng, M., Yang, Y., Bin, X., Que, W.: One-step electrochemical synthesis and surface reconstruction of NiCoP as an electrocatalyst for bifunctional water splitting. Materials (Basel) **16**(4), 1529 (2023). https://doi.org/10.3390/ma16041529
26. Dmitriev, D.S., Tenevich, M.I., Popkov, V.I.: Electrochemical template synthesis of Ni–Cu bilayer hollow microtubes for green hydrogen production through electrocatalytic reforming of ethanol. Int. J. Hydrogen Energy **47**, 17971–17981 (2022). https://doi.org/10.1016/j.ijhydene.2022.03.277
27. Soonmin, H.: Recent advances in the growth and characterizations of SILAR-deposited thin films. Appl. Sci. **12**(16), 8184 (2022). https://doi.org/10.3390/app12168184
28. Pratiwi, R., Chusniyah, D.A., Yanti, W.: Fadliah, SILAR as an effective photocatalyst synthesis method for performance improvement in hydrogen recovery on water splitting process: a brief review. IOP Conf. Ser.: Earth Environ. Sci. **1451**, 012047 (2025). https://doi.org/10.1088/1755-1315/1451/1/012047
29. Ratnayake, S.P., Ren, J., Colusso, E., Guglielmi, M., Martucci, A., Della Gaspera, E.: SILAR deposition of metal oxide nanostructured films. Small **17**(49), 2101666 (2021). https://doi.org/10.1002/smll.202101666
30. Arulkumar, E., Thanikaikarasan, S., Rajkumar, S., Wondimu, W.: Influence of solution pH dependency on structure, optical, and photoelectrochemical characteristics of SILAR-deposited copper oxide thin films. Heliyon **10**, e33579 (2024). https://doi.org/10.1016/j.heliyon.2024.e33579
31. Lobinsky, A.A., Popkov, V.I.: Direct SILD synthesis of efficient electroactive materials based on ultrathin nanosheets of amorphous CoCr-LDH. Mater. Lett. **322**, 132472 (2022). https://doi.org/10.1016/j.matlet.2022.132472
32. Lobinsky, A.A., Dmitriev, D.S., Popkov, V.I.: Novel route of NiCr-layered double hydroxide nanosheets synthesis on nickel foam as a bifunctional electrocatalyst for water splitting process. Int. J. Hydrogen Energy **48**, 22495–22501 (2023). https://doi.org/10.1016/j.ijhydene.2023.04.174
33. Jiang, H., Luo, R., Li, Y., Chen, W.: Recent advances in solid–liquid–gas three-phase interfaces in electrocatalysis for energy conversion and storage. EcoMat **4**, 1–38 (2022). https://doi.org/10.1002/eom2.12199
34. Gulina, L.B., Tolstoy, V.P., Solovev, A.A., Gurenko, V.E., Huang, G., Mei, Y.: Gas-solution interface technique as a simple method to produce inorganic microtubes with scroll morphology. Prog. Nat. Sci. Mater. Int. **30**, 279–288 (2020). https://doi.org/10.1016/j.pnsc.2020.05.001
35. Gulina, L.B., Tolstoy, V.P., Lobinsky, A.A., Petrov, Y.V.: Formation of Fe and Fe_2O_3 microspirals via interfacial synthesis. Part. Part. Syst. Char. **35**, 1–7 (2018). https://doi.org/10.1002/ppsc.201800186
36. Tolstoy, V.P., Vladimirova, N.I., Gulina, L.B.: Formation of ordered honeycomb-like structures of manganese oxide 2D nanocrystals with the birnessite-like structure and their electrocatalytic properties during oxygen evolution reaction upon water splitting in an alkaline medium. ACS Omega **4**, 22203–22208 (2019). https://doi.org/10.1021/acsomega.9b03499
37. Tyona, M.D.: A theoretical study on spin coating technique. Adv. Mater. Res. **2**, 195–208 (2013). https://doi.org/10.12989/amr.2013.2.4.195
38. Kyesmen, P.I., Nombona, N., Diale, M.: Effects of film thickness and coating techniques on the photoelectrochemical behaviour of hematite thin films. Front. Energy Res. **9**, 1–13 (2021). https://doi.org/10.3389/fenrg.2021.683293
39. Ben-Naim, M., Palm, D.W., Strickler, A.L., Nielander, A.C., Sanchez, J., King, L.A., Higgins, D.C., Jaramillo, T.F.: A spin coating method to deposit iridium-based catalysts onto silicon for

water oxidation photoanodes. ACS Appl. Mater. Interfaces **12**, 5901–5908 (2020). https://doi.org/10.1021/acsami.9b20099

40. Tie, L., Li, N., Yu, C., Liu, Y., Yang, S., Chen, H., Dong, S., Sun, J., Dou, S., Sun, J.: Self-supported nonprecious MXene/Ni_3S_2 electrocatalysts for efficient hydrogen generation in alkaline media. ACS Appl. Energy Mater. **2**, 6931–6938 (2019). https://doi.org/10.1021/acsaem.9b01529
41. Cai, Z., Li, A., Zhang, W., Zhang, Y., Cui, L., Liu, J.: Hierarchical Cu@Co-decorated CuO@Co_3O_4 nanostructure on Cu foam as efficient self-supported catalyst for hydrogen evolution reaction. J. Alloys Compd. **882**, 160749 (2021). https://doi.org/10.1016/j.jallcom.2021.160749
42. Yan, Y.: Facile synthesis of carbon cloth supported cobalt carbonate hydroxide hydrate nanoarrays for highly efficient oxygen evolution reaction. Front. Chem. **9**, 1–7 (2021). https://doi.org/10.3389/fchem.2021.754357
43. Yamamoto, M., Murata, H., Miyata, N., Takashima, H., Nagao, M., Mimura, H., Neo, Y., Murakami, K.: Low-temperature direct synthesis of multilayered h-BN without catalysts by inductively coupled plasma-enhanced chemical vapor deposition. ACS Omega **8**, 5497–5505 (2023). https://doi.org/10.1021/acsomega.2c06757
44. Urbanová, V., Plutnar, J., Pumera, M.: Atomic layer deposition of electrocatalytic layer of MoS_2 onto metal-based 3D-printed electrode toward tailoring hydrogen evolution efficiency. Appl. Mater. Today **24**, 101131 (2021). https://doi.org/10.1016/j.apmt.2021.101131
45. Ladhane, S., Shah, S., Doiphode, V., Shinde, P., Punde, A., Kale, D., Rahane, S., Thombare, J., Hase, Y., Waghmare, A., Bade, B., Prasad, M., Patole, S.P., Jadkar, S.: Direct growth of WS_2 nanosheets using RF-magnetron sputtering on hydrothermally grown TiO_2 nanorods for enhancing photoelectrochemical water splitting. Mater. Chem. Phys. **334**, 130440 (2025). https://doi.org/10.1016/j.matchemphys.2025.130440
46. Chebanenko, M.I., Lebedev, L.A., Ugolkov, V.L., Prasolov, N.D., Nevedomskiy, V.N., Popkov, V.I.: Chemical and structural changes of g-C_3N_4 through oxidative physical vapor deposition. Appl. Surf. Sci. **600**, 154079 (2022). https://doi.org/10.1016/j.apsusc.2022.154079

2D Transition Metal Dichalcogenides for Water Splitting

Zi-Yu Chen, Kuo-Tai Hsu, and Chuan-Pei Lee

Abstract In recent years, the global development of sustainable energy technologies has garnered increasing attention. Among these, water splitting has emerged as a promising method for renewable energy conversion, offering significant potential for hydrogen production. Two-dimensional transition metal dichalcogenides (2D-TMDs) have been recognized as ideal catalytic materials attributable to their unique layered structure, abundant active sites, and tunable electronic properties. Compared to conventional noble metal catalysts, TMDs offer advantages such as excellent stability, low cost, and high activity. Furthermore, optimizing material structures, enhancing electrical conductivity, and elucidating catalytic mechanisms, through heterostructure design and advanced analytical techniques, could further improve the catalytic efficiency of 2D-TMDs. These developments are expected to drive the improvement of high-performance water-splitting catalysts and contribute to the progress of sustainable energy technologies. This chapter summarizes recent advances in the application of 2D-TMDs for the oxygen evolution reaction (OER) and hydrogen evolution reaction (HER), with a particular focus on their synthesis methods and catalytic performance, providing insights on their practical implementation in industrial water-splitting systems.

Keywords Hydrogen evolution reaction · Oxygen evolution reaction · Transition metal dichalcogenides · Water splitting

1 Introduction

Under the growing global energy demand, the search for renewable and clean energy sources has become an urgent priority. Water-splitting technology, which converts water into oxygen and hydrogen, is considered a key approach to achieving a hydrogen economy. The water-splitting process primarily involves two half-reactions: the HER and the OER [1, 2]. HER occurs at the cathode, where protons

Z.-Y. Chen · K.-T. Hsu · C.-P. Lee (✉)
Department of Applied Physics and Chemistry, University of Taipei, Taipei, Taiwan
e-mail: CPLee@utaipei.edu.tw

O. P. Pandey et al. (eds.), *2D Materials for Electrochemical Water Splitting*, Engineering Materials, https://doi.org/10.1007/978-981-95-2003-9_5

gain electrons to generate hydrogen, while OER is conducted at the anode, involving the oxidation of water molecules to produce oxygen, but these reactions suffer from kinetic limitations, necessitating additional energy input [3, 4]. Therefore, finding efficient catalysts to reduce the enhance conversion efficiency and overpotential is crucial issues.

1.1 The Principles and Mechanisms of the HER and the OER

HER primarily involves the transfer of electrons from water molecules or protons on the catalyst surface, bringing about the formation of hydrogen. Its mechanism can be divided into two main stages [5–7]. The first is the Volmer step, in which hydrogen ions (H^+) adsorb onto the catalyst surface and undergo electrochemical reduction:

- Volmer step (electrochemical reduction of protons):

$$H_3O^+ + e^- \rightarrow H* + H_2O \text{ (acidic media)}$$

$$H_2O + e^- \rightarrow H* + OH^- \text{ (alkaline media)}$$

In an acidic environment, hydrogen ions gain electrons on the electrode surface, and they are converted into adsorbed hydrogen atoms. In an alkaline environment, water molecules release protons and accept electrons, forming adsorbed hydrogen and hydroxide ions. This process determines the catalyst's hydrogen adsorption capability, which in turn the hydrogen generation rate [8, 9]. Subsequently, the adsorbed hydrogen atoms undergo further conversion into hydrogen through either the Heyrovsky or Tafel step.

- Heyrovsky reaction (proton-coupled electron transfer):

$$H* + H_3O^+ + e^- \rightarrow H_2 + H_2O \text{ (acidicmedia)}$$

$$H* + H_2O + e^- \rightarrow H_2 + OH^- \text{ (alkaline media)}$$

- Tafel reaction (hydrogen recombination):

$$2H* \rightarrow H_2$$

In the Heyrovsky reaction, the adsorbed hydrogen atom undergoes an electrochemical reaction with a nearby proton source (water molecules in an alkaline environment or hydrogen ions in an acidic environment) while simultaneously gaining an electron, ultimately forming hydrogen. In the Tafel reaction, two adsorbed hydrogen atoms directly undergo chemical recombination to form hydrogen. This mechanism indicates that an ideal catalyst should possess an appropriate hydrogen adsorption capability, allowing hydrogen atoms to adsorb onto the catalyst surface efficiently while also desorbing at the right moment to form hydrogen, thereby enhancing the efficiency of HER [10].

OER primarily involves the oxidation process of water molecules, where electron transfer occurs on the electrode surface through the catalyst, ultimately leading to the formation of oxygen. Its mechanism can be divided into several key steps [11–13], with the first being the oxidation of water to form hydroxyl radicals:

$$H_2O \rightarrow OH * + H^+ + e^- \text{ (acidic media)}$$

$$OH^- \rightarrow OH * + e^- \text{ (alkaline media)}$$

In an acidic environment, water molecules lose electrons and decompose into hydroxyl radicals and hydrogen ions, whereas in an alkaline environment, hydroxide ions release electrons to form hydroxyl radicals. The efficiency of this process is affected by the catalyst, as it can lower the reaction energy barrier, facilitate electron transfer, and accelerate oxygen generation. Subsequently, the hydroxyl radicals undergo further oxidation to form peroxides or oxygen atoms:

$$OH* \rightarrow O * + H^+ + e^- \text{ (acidic media)}$$

$$OH* \rightarrow O * + H_2O + e^- \text{ (alkaline media)}$$

In this step, hydroxyl radicals undergo further electron loss, converting into reactive oxygen species. These oxygen species interact with each other, facilitating the formation of oxygen and ultimately completing the OER:

$$O * + H_2O \rightarrow OOH * + e^-, \ OOH* \rightarrow O_2 + H^+ + e^- \text{ (acidic media)}$$

$$O * + OH^- \rightarrow OOH * + e^-, \ OOH* \rightarrow O_2 + H_2O + e^- \text{ (alkaline media)}$$

The entire OER process requires multiple electron transfer steps, which involves a higher activation energy making its kinetics slower than that of HER [14, 15].

In traditional catalysts, platinum (Pt) is most commonly used for HER [5, 10], while ruthenium oxide (RuO_2) and iridium oxide (IrO_2) are typically employed for OER [16]. Although these noble metal catalysts exhibit resource scarcity, their high cost, excellent catalytic activity, and stability issues limit their large-scale applications. Consequently, the search for low-cost, highly efficient alternative catalysts has become a major research focus. Specifically, 2D-TMDs have demonstrated great potential in water-splitting catalysis attributable to their unique layered structure, outstanding electronic properties, and abundant active sites [17].

1.2 Overview of 2D Materials

Two-dimensional materials possess atomic-scale thickness and can be controllably grown and transferred onto target substrates. One prominent example is graphene [18], a single-layer honeycomb structure composed of carbon atoms, which exhibits exceptional mechanical, thermal, and electrical properties. However, graphene inherently lacks a band gap, limiting its tunability in certain electronic and optoelectronic applications [19, 20]. To overcome this limitation, scientists have turned their attention to other 2D materials, such as 2D-TMDs [21, 22]. These materials consist of transition metals (e.g., tungsten, molybdenum) and chalcogen elements (e.g., selenium, sulfur), offering tunable band gaps and abundant active sites, making them highly suitable for applications in catalysis and beyond.

1.3 Properties and Advantages of 2D-TMDs in Water Splitting

2D-TMDs exhibit significant advantages in water splitting reactions due to their electronic properties and unique structural. First, the 2D structure endows TMDs with a higher specific surface area, providing more active sites and thereby enhancing catalytic performance [23, 24]. Unlike bulk materials, 2D-TMDs possess monolayer or few-layer structures, where edge sites and defect sites play a crucial role in HER and OER. For instance, WS_2 and MoS_2 exhibit excellent HER activity at sulfur edge sites (S-edge sites) [25, 26], and the number of these active sites can be further increased through layer thickness modulation, doping modifications, and other strategies [27–29]. Second, the layered structure of 2D-TMDs facilitates efficient electron and proton transport [30], and this makes rapid electron transfer critical in water electrolysis reactions. The tunable electronic structure of 2D-TMDs allows researchers to optimize their catalytic performance through doping, strain engineering, or heterostructure design. For example, doping with non-metal elements (e.g., N, P, S) or transition metals (e.g., Co, Ni) can effectively modulate the d-orbital electron distribution of TMDs, improving hydrogen adsorption free energy (ΔG_H) and thereby enhancing HER activity. Additionally, compared to metal oxides (e.g., RuO_2 and IrO_2), TMDs exhibit superior chemical stability in OER, preventing structural degradation of noble metal catalysts during prolonged reaction cycles [21].

1.4 Synthesis Methods of 2D-TMDs

In the fabrication of TMDs, scientists continuously explore more efficient and controllable synthesis strategies to ensure that these materials achieve their full potential in catalysis, optoelectronic devices, and energy storage applications. The three common methods for synthesizing TMDs are hydrothermal synthesis, chemical vapor deposition (CVD), and exfoliation methods [31–33].

In hydrothermal synthesis [34], researchers typically select metal salts as precursors and use an appropriate sulfur source for the reaction. These precursors are dissolved in a solvent and then transferred into a Teflon-lined stainless steel autoclave, where the reaction takes place under moderate temperature conditions (e.g., 140–180 °C) for several hours to promote crystal growth. By adjusting the reaction time and precursor the morphology, concentration and the crystallinity of the final product can be controlled, leading to the formation of high-quality 2D-TMDs (Fig. 1a).

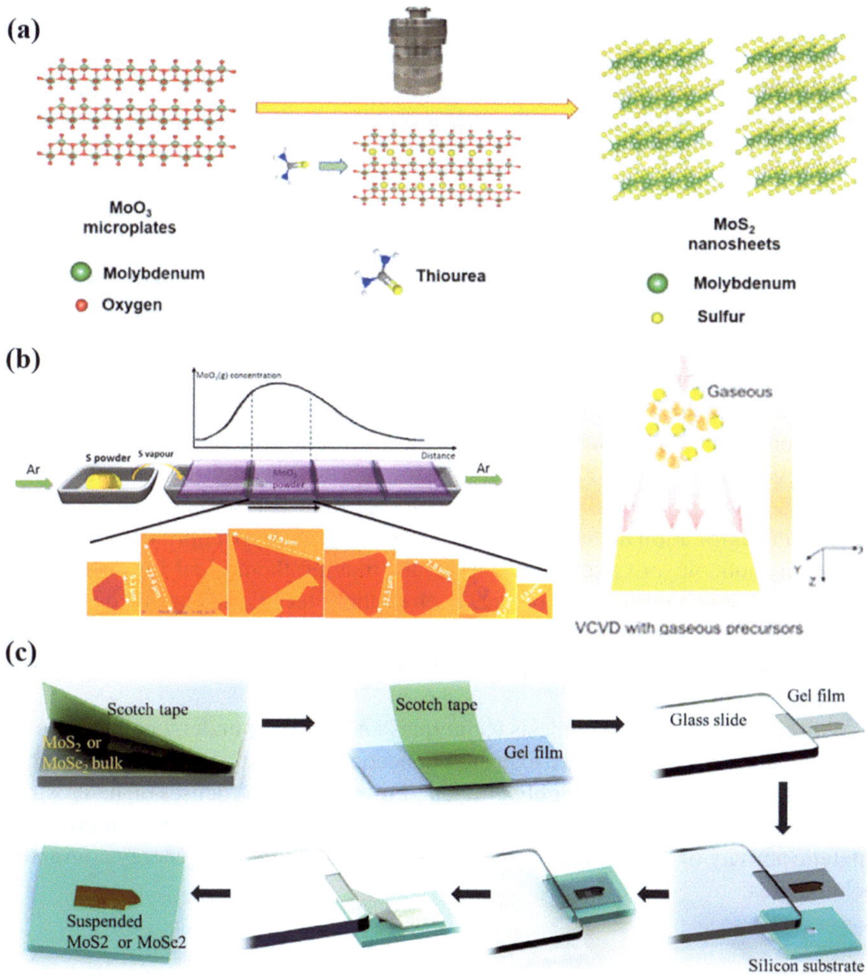

Fig. 1 **a** Hydrothermal method, **b** Chemical vapor deposition (CVD), and **c** Mechanical exfoliation method [41]

Compared to the solution-based hydrothermal method, CVD is a widely used technique for synthesizing TMDs [35–37], enabling the growth of high-quality thin-film structures through vapor-phase reactions (Fig. 1b). During the CVD process, metal oxides and sulfur sources are used as precursors and transported to the substrate surface by an inert gas flow. Under high-temperature conditions, the metal oxide precursors are gradually reduced and react with sulfur, ultimately forming monolayer or few-layer TMDs films on the substrate. Additionally, carrier gas pressure and temperature control play a crucial role in determining the growth mechanism and morphology of the film, thereby regulating the layer number and the crystallinity of TMDs. These factors are critical in defining the performance of TMDs, making CVD technology an essential approach for the controlled synthesis of 2D materials and the development of electronic devices.

However, when researchers need to obtain high-quality monolayer or few-layer TMDs directly from bulk materials, exfoliation methods become the simplest and most straightforward option (Fig. 1c). Mechanical exfoliation is the most traditional and commonly used approach [38, 39], where researchers repeatedly peel MoS_2 or WS_2 crystals using adhesive tape and then transfer the thin layers onto a SiO_2/Si substrate. Although this method can produce high-quality monolayer TMDs, it suffers from extremely low yield and is not suitable for large-scale applications. In contrast, liquid-phase exfoliation offers better scalability technique [40]. In this method, TMDs powders are first dispersed in an suitable solvent and then subjected to ultrasonic treatment (sonication) to break large particles into monolayer or few-layer nanosheets. Afterward, centrifugation is used to remove unexfoliated particles, resulting in a uniformly dispersed TMDs nanosheet solution.

Briefly, the hydrothermal method is well-suited for catalysis and energy storage applications, as it allows for doping-induced electronic structure modulation. CVD has emerged as the dominant technique for fabricating high-quality monolayer TMDs, making it particularly suitable for optoelectronic and semiconductor devices. Meanwhile, exfoliation methods provide a simple and effective approach to obtaining high-purity monolayer or few-layer materials, making them ideal for fundamental research and application development. Each of these techniques has its own advantages, which determine the broad applicability of TMDs across various fields. Together, they provide critical support for the advancement of electronic technologies and future energy.

With advancements in synthesis techniques and a deeper understanding of material mechanisms. 2D-TMDs are expected to become the next generation of low-cost water-splitting catalysts and highly efficient. Future research focuses on heterostructure design, doping modifications, and defect engineering to further improve the catalytic activity of TMDs and promote their application in clean energy conversion technologies.

2 2D-TMDs for Water Splitting Reactions

2.1 2D-TMDs in Hydrogen Evolution Reaction

In recent years, 2D-TMDs have demonstrated significant potential in the field of water electrolysis. Among them, their application in the HER has garnered particular attention. HER requires highly efficient catalysts to reduce the reaction overpotential and enhance reaction kinetics. Due to their exposed edge sites, tunable electronic structure, and excellent conductivity [42].

Ravikumar et al. synthesized nanoflower-like structures of MoS_2 and $MoSe_2$ using the hydrothermal method and evaluated their electrocatalytic activity in the HER [43]. The study found that at an overpotential of 300 mV versus RHE, the $MoSe_2$-modified glassy carbon electrode ($MoSe_2$/GC) achieved a current density of 10.96 mA/cm^2, whereas MoS_2/GC required an overpotential of 357 mV to reach the same current density (Fig. 2a). These results show that both $MoSe_2$ and MoS_2 exhibit good HER activity. However, compared to MoS_2/GC, $MoSe_2$/GC possesses a larger interlayer spacing (Fig. 2b) and superior electronic conductivity, which allows for faster electron transfer and a lower overpotential (110 mV vs. RHE) in the HER process. Additionally, MoS_2/GC demonstrated excellent structural stability, maintaining over 90% of its initial catalytic activity in long-term stability tests. Its smaller interlayer spacing (0.62 nm) helps stabilize the structure and reduce potential catalytic site loss. This study confirms that both $MoSe_2$ and MoS_2 are excellent HER electrocatalysts, with $MoSe_2$ exhibiting superior performance. These findings show a critical foundation for the subsequent design of $MoSe_2$–MoS_2 heterostructures.

Furthermore, Das et al. investigated the electrocatalytic performance of bulk WS_2 as a catalyst for the HER and found that its catalytic activity was limited by a large overpotential and poor electron transfer capability [44]. The study results showed that bulk WS_2 exhibited an onset potential of 288 mV versus RHE in the HER, with a Tafel slope of 138 mV/dec., indicating sluggish HER kinetics. Electrochemical testing revealed that WS_2 exhibited a high charge transfer resistance (R_{ct}), indicative of poor electron transport efficiency, which negatively affected its HER activity (Fig. 2c). Due to these limitations inherent to pure TMDs, researchers have explored various engineering strategies, such as doping modulation, heterostructure design, and defect engineering, to enhance their catalytic activity and stability. For instance, Das et al. employed Zn doping WS_2 (Zn-WS_2) to improve the HER performance of WS_2 (Fig. 2d), with the primary objective of enhancing its electron transfer capability and increasing the utilization of active sites. Their findings revealed that WS_2-30 (30 mg Zn doping) exhibited an onset potential of 180 mV versus RHE in the HER, with the Tafel slope reduced to 70 mV/dec., demonstrating a significant improvement in HER activity. Electrochemical testing further confirmed that Zn doping effectively reduced the charge transfer resistance (R_{ct}) of WS_2, indicating enhanced electron transport capability. Additionally, HER stability testing demonstrated that Zn-WS_2 maintained stable current density after 1000 cycles of cyclic voltammetry (CV), highlighting its excellent long-term catalytic stability (Fig. 2c). This study provides

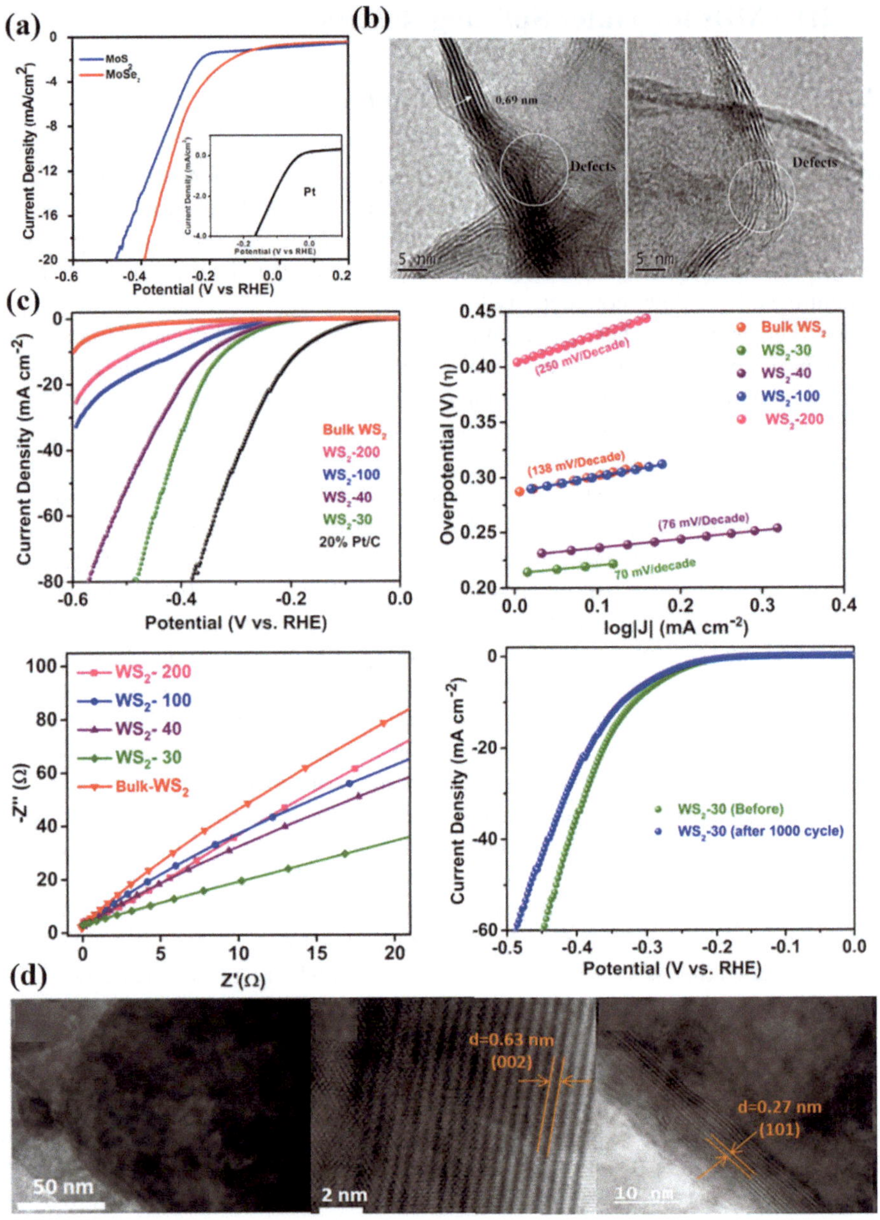

Fig. 2 **a** Electrochemical performance and **b** electron microscopy characterization of $MoSe_2$ and MoS_2 [43]; **c** electrochemical performance and **d** electron microscopy characterization of WS_2 and Zn-WS_2 [44]

strong evidence that Zn doping can effectively modulate the active site distribution and electronic structure of WS_2, significantly enhancing its HER electrocatalytic activity.

On the other hand, Yang et al. employed a tungsten (W) doping strategy to improve the HER activity of $MoSe_2$. Through vacancy engineering, they successfully optimized the electronic structure of $MoSe_2$, enhancing hydrogen adsorption capability and charge transfer rate [45]. Specifically, the introduction of W atoms induced the formation of selenium vacancies (Se vacancies), which further facilitated the 2H $\rightarrow$ 1T phase transition, thereby improving the intrinsic conductivity and catalytic activity of $MoSe_2$. Furthermore, Raman spectroscopy analysis and X-ray photoelectron spectroscopy (XPS) (Fig. 3a) confirmed that W doping effectively increased the number of active sites and optimized the electronic structure, further enhancing the kinetics of the HER. Electrochemical testing revealed that W-doped $MoSe_2$ (V-$WMoSe_2$) exhibited an overpotential of only 247 mV at a current density of 10 mA/cm^2, with a reduced Tafel slope and an increased turnover frequency (Fig. 3b). This study demonstrates that tuning the electronic structure of $MoSe_2$ by W doping and introducing Se vacancies can significantly enhance HER activity, providing a low-cost and efficient HER catalytic strategy. The incorporation of W atoms promotes the 2H-1T phase transition of $MoSe_2$ and induces electronic structure optimization, improving hydrogen adsorption capability and charge transfer rate.

In addition to optimization through doping modulation, the design of heterostructures is also a crucial strategy for improving the photocatalytic performance of 2D-TMDs. Satheesh et al. designed and synthesized a WS_2/Ag_3PO_4 heterostructure catalyst, which integrates the highly active edge sites of WS_2 with the excellent photocatalytic and charge transfer capabilities of Ag_3PO_4 to enhance HER activity [40]. The study found that when WS_2 and Ag_3PO_4 were combined in an optimal ratio of 3 wt% (A3W), the HER activity was significantly improved, with catalytic performance surpassing that of individual WS_2 or Ag_3PO_4. Electrochemical testing (Fig. 3c) revealed that WS_2/Ag_3PO_4 exhibited an overpotential of only 206 mV at a current density of 10 mA/cm^2 during the HER, coupled with a significantly reduced Tafel slope. Furthermore, a 50-h long-term stability test demonstrated that the catalyst maintained over 90% of its initial catalytic activity. Additionally, electrochemical impedance spectroscopy (EIS) analysis confirmed that the WS_2/Ag_3PO_4 heterostructure exhibited a lower charge transfer resistance (R_{ct}), indicating excellent charge transport capability, which further contributed to enhancing HER activity. This study demonstrates that the synergistic effect of the WS_2/Ag_3PO_4 heterostructure can significantly improve HER activity, providing insights into the development of efficient HER photocatalysts.

Furthermore, the combination of heterostructure design and doping modulation has also been demonstrated to further enhance HER electrocatalytic performance. Chandrakantha et al. developed an LrGO-ZnO/MoS_2 heterostructure catalyst, where laser-reduced graphene oxide (LrGO) and ZnO-doped MoS_2 formed a mixed-phase structure (Fig. 4a) to improve HER activity [46]. The study revealed that the high conductivity of LrGO facilitated electron transfer, ZnO acted as an electron acceptor to enhance charge separation efficiency, and the 1T-2H phase structure of

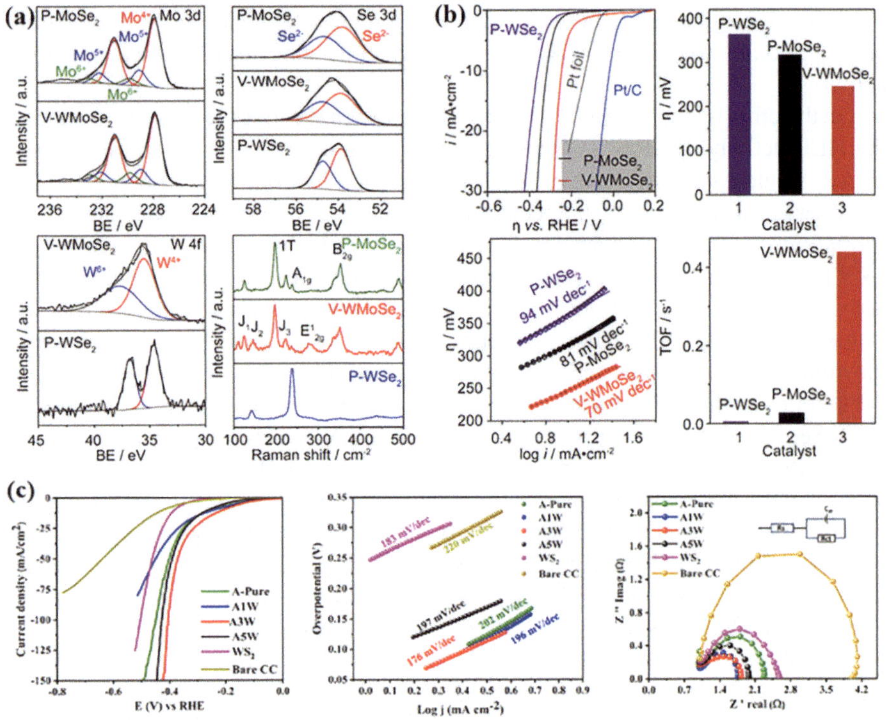

Fig. 3 **a** XPS and Raman spectroscopy of the P-$MoSe_2$ and V-$WMoSe_2$; **b** Electrochemical performance of P-$MoSe_2$, P-WSe_2, and V-$WMoS_2$ [45]; **c** The electrochemical performance of WS_2/Ag_3PO_4 in the HER [40]

MoS_2 provided additional catalytic active sites. Electrochemical testing (Fig. 4b, c) demonstrated that the HER catalytic activity of LrGO-ZnO/MoS_2 was significantly enhanced. In acidic conditions (pH 0), the catalyst required only 94 mV of overpotential to reach a current density of 10 mA/cm^2, while under alkaline conditions (pH 14), the overpotential remained as low as 191.5 mV, maintaining high catalytic efficiency. Tafel slope analysis further indicated that the catalyst exhibited low reaction resistance, highlighting its excellent kinetic properties. Moreover, after 4000 CV cycles and 24 h of continuous operation, the catalyst retained stable catalytic activity, demonstrating its potential for practical applications. This study confirms that the synergistic effect of the LrGO-ZnO/MoS_2 heterostructure effectively enhances HER activity, enabling outstanding electrocatalytic performance in both acidic and alkaline solutions.

Overall, these studies have confirmed the potential of 2D-TMDs in HERs and demonstrated how various modification strategies can enhance their catalytic performance, making them strong candidates to replace noble metal catalysts. Future research will further optimize these strategies to advance the development of the high-efficiency HER catalysts, providing critical support of green energy technologies.

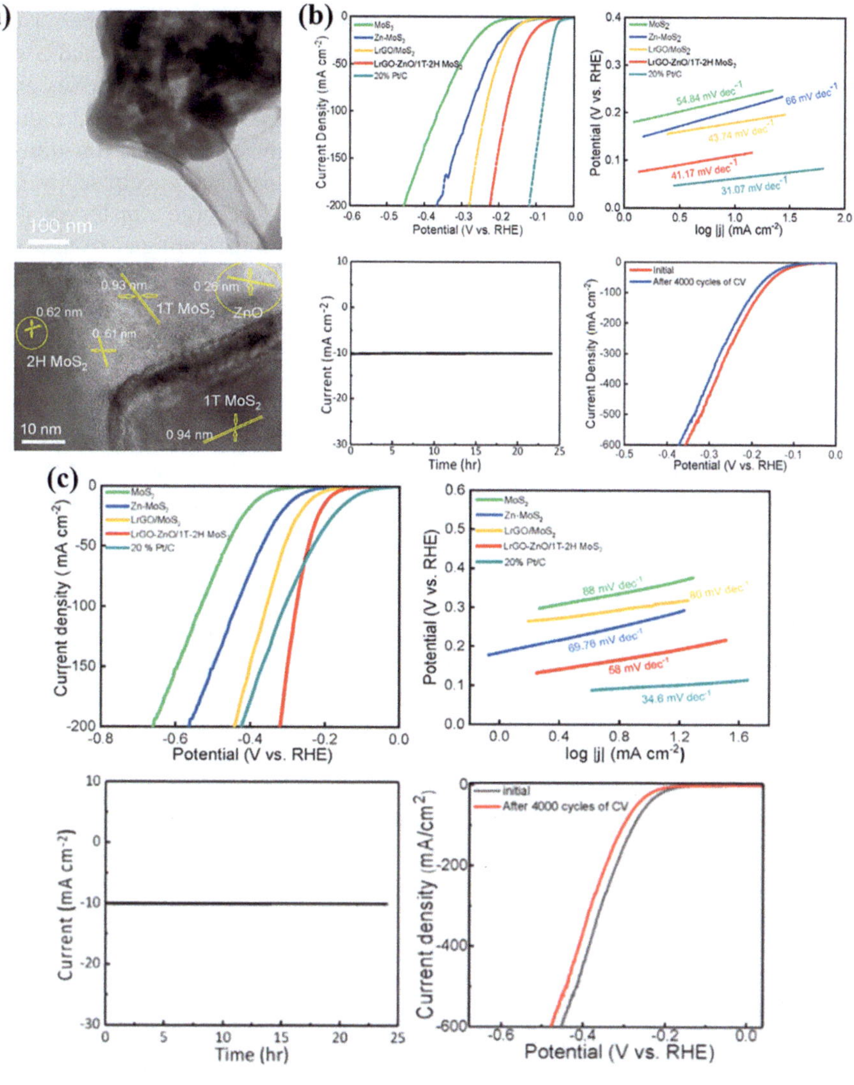

Fig. 4 **a** The morphologies of synthesized LrGO-ZnO/1T-2H MoS_2 and electrochemical performance in **b** 0.5 M H_2SO_4 solutions and **c** 1 M KOH solutions [46]

2.2 2D-TMDs in Oxygen Evolution Reaction

After exploring the application of 2D-TMDs on HER, we further focus on their potential in the OER.

Kankan Liu et al. investigated $MoSe_2$ and WSe_2 as transition metal diselenides (TMDs) for their electrocatalytic potential in the OER [47]. The study results indicated that $MoSe_2$ show an overpotential of 270 mV at a current density of 10 mA/cm^2, while WSe_2 required 300 mV, suggesting that $MoSe_2$ demonstrated superior OER activity. Additionally, Tafel slope measurements showed that $MoSe_2$ had a slope of 88 mV/dec., compared to 97 mV/dec. for WSe_2, indicating that $MoSe_2$ exhibited better OER kinetics, facilitating more efficient electron transfer and reducing the reaction energy barrier (Fig. 5a). To further analyze their electron transport capability, researchers conducted electrochemical impedance spectroscopy (EIS) measurements (Fig. 5b), which revealed that $MoSe_2$ had a charge transfer resistance (R_{ct}) of 5.57 Ω, significantly lower than WSe_2 (9.05 Ω), indicating a stronger electron transport capability that enhances OER efficiency. Although $MoSe_2$ and WSe_2 demonstrated catalytic activity in the OER, they were still constrained by high overpotentials and slow electron transfer rates, making them less competitive compared to highly efficient OER catalysts. Therefore, researchers designed a heterostructure catalyst, $MoSe_2$–$NiWSe_2$, which exhibited superior performance compared to individual $MoSe_2$, WSe_2, or $NiWSe_2$ catalysts. This catalyst was synthesized using the hydrothermal method, forming a nanoflower-like structure where $MoSe_2$ and $NiWSe_2$ layers were interwoven (Fig. 5c), reinforcing interfacial coupling interactions and further improving electron transport capability. Electrochemical testing showed that $MoSe_2$–$NiWSe_2$ required only 200 mV overpotential at a current density of 10 mA/cm^2, and a Tafel slope of 81 mV/dec., outperforming all individual materials (Fig. 5a). This result demonstrated faster OER kinetics, enhanced electron transport, and reduced reaction energy barriers, confirming a significant improvement in catalytic activity. Furthermore, EIS measurements revealed that $MoSe_2$–$NiWSe_2$ exhibited an R_{ct} of only 1.27 Ω, indicating greatly improved electron transport, which further accelerated the OER rate (Fig. 5b). The electrochemical active surface area of $MoSe_2$–$NiWSe_2$ was measured at 2.25 mF/cm^2, confirming that its heterostructure significantly increased the availability of catalytic active sites. The outstanding OER catalytic performance of $MoSe_2$–$NiWSe_2$ can be ascribed to multiple synergistic factors. In conclusion, this study demonstrates that the $MoSe_2$–$NiWSe_2$ heterostructure catalyst exhibits significantly enhanced OER catalytic performance, highlighting its potential as an efficient OER electrocatalyst.

In employing the doping modulation strategy, Liu et al. doped oxophilic metals into MoS_2 to enhance its OER activity through surface reorganization engineering (Fig. 6a) [48]. Through dual doping of Al and Co, they successfully tuned the electronic structure of MoS_2 and facilitated the formation of M–O (M = Co, Mo, Al) bonds on the catalytic surface. These oxide sites effectively reduced the adsorption strength of reaction intermediates (OH, OOH**), thereby improving OER kinetics. This approach not only provided more stable reaction sites but also minimized energy loss during the OER process, resulting in a notably lower overpotential of just 300 mV at 10 mA/cm^2 for $AlCo_3$-MoS_2, which was significantly lower than that of unmodified MoS_2 (437.5 mV) and other MoS_2-based catalysts. Furthermore, the electrocatalyst showed a Tafel slope of only 46.7 mV/dec., demonstrating its superior OER kinetics (Fig. 6b).

More importantly, this bimetallic doping strategy not only influences the intrinsic electronic structure of MoS_2 but also induces surface reconstruction, further leading to the formation of oxyhydroxide phases, which enhance the catalyst's activity in

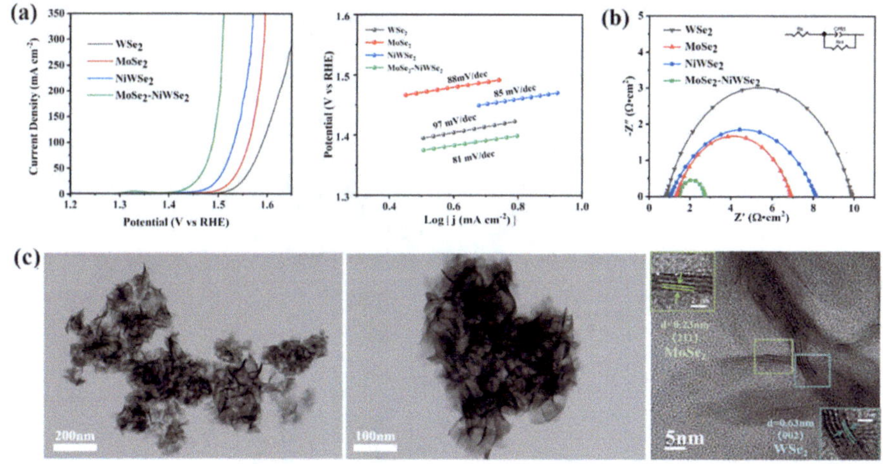

Fig. 5 **a** Electrochemical performance, **b** electrochemical impedance, and **d** C_{dl} curve of $MoSe_2$, WSe_2, and $MoSe_2$–$NiWSe_2$; **c** electron microscopy characterization of $MoSe_2$–$NiWSe_2$ [47]

alkaline environments. Additionally, by tuning the Co^{3+}/Co^{2+} ratio, the study revealed a synergistic effect between the doped metals, which can improved electron transfer rates also significantly lowered the reaction energy barrier of OER (Fig. 6c). This study demonstrates that disrupting Mo–S bonds through metal doping and introducing oxygen vacancies can substantially improve the OER activity of MoS_2, making it a highly efficient and cost-effective OER electrocatalyst.

Additionally, Alahmadi et al. investigated the application of aluminum (Al)-doped tungsten disulfide (Al-WS_2) as an OER electrocatalyst and evaluated its catalytic performance [49]. The researchers synthesized Al-WS_2 via a one-step hydrothermal method and found that it formed a skeleton-like nanostructure, which contributed to enhanced catalytic activity. Electrochemical testing revealed that 0.04% Al-WS_2 exhibited an overpotential of only 650 mV at a current density of 6 mA/cm^2 and a Tafel slope of 227 mV/dec., demonstrating superior catalytic performance compared to pristine WS_2 and indicating a significant improvement in OER kinetics. Furthermore, the study examined the influence of catalyst loading on OER activity, showing that when the loading of Al-WS_2 was 2 μL, it achieved optimal OER activity, corresponding to an overpotential of 440 mV. However, as the electrocatalyst loading increased (4 μL, 6 μL, 8 μL), the overpotential gradually increased (673 mV, 683 mV, and 692 mV, respectively), suggesting that excessive catalyst accumulation might hinder electron transport and negatively impact catalytic performance. In summary, Al-WS_2 demonstrated superior OER activity, characterized by a lower overpotential, faster electron transfer rate, and improved reaction kinetics, confirming that Al doping significantly enhances the electrocatalytic performance of WS_2 in the OER.

On the other hand, Mahmood et al. explored the role of defects in OER activity using first-principles calculations [50]. Their study revealed that introducing sulfur vacancies (S vacancies) and molybdenum vacancies (Mo vacancies) into the layered

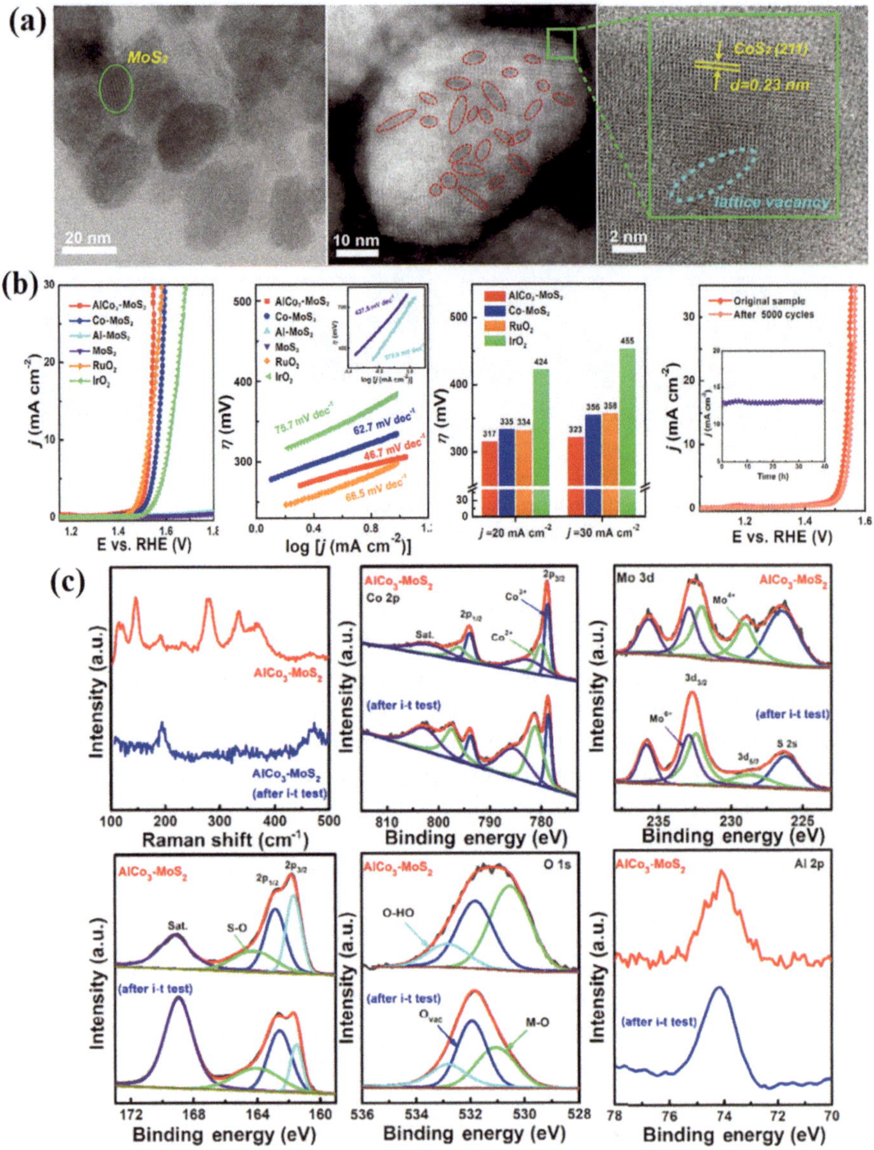

Fig. 6 **a** Electron microscopy characterization of $AlCo_3$–MoS_2 nanosheets, **b** electrochemical performance and stability in the OER, and **c** XPS and Raman spectra before and after the i-t test [48]

WS_2/MoS_2 structure could significantly modify its electronic structure and enhance catalytic performance. When Mo vacancies were present in the Mo layer, the OER overpotential decreased from 2.15 to 1.48 V, demonstrating that defects effectively improved charge transfer and reduced the reaction energy barrier. Furthermore, the study showed that S vacancies were more stable than Mo vacancies, but Mo vacancies had a greater impact on promoting the OER. This finding suggests that in TMDs heterostructures, precisely controlling the type and location of defects can significantly optimize OER activity. The key insight from this research is that defects play a crucial role in enhancing charge separation and catalytic activity in TMDs. By tuning the concentration and distribution of Mo and S vacancies, the study demonstrated that it is possible to modify the band structure of the material, thereby improving the adsorption energy of oxygen intermediates. Additionally, band alignment within the heterostructure was identified as a key factor influencing OER efficiency. Mahmood et al. used first-principles calculations to illustrate how defects impact band alignment, facilitating more efficient charge transfer to the catalytic sites. These findings show valuable insights into the development of efficient TMD-based OER catalysts.

Beyond defect engineering, this study also highlights the significant impact of interlayer interactions on electron transport within heterostructures. In particular, when defects are introduced into the vertical WS_2/MoS_2 structure, the interlayer coupling effect influences the overall catalytic performance. This finding suggests that by designing heterostructures with optimized defect concentrations and interlayer spacing, OER catalytic activity can be further enhanced. Therefore, this study proposes a novel design strategy, combining defect engineering and heterostructure modulation to optimize the catalytic performance of TMDs in the OER.

In summary, the optimization of oxophilic metal doping, heterostructure design, and defect engineering not only overcomes the intrinsic limitations of TMDs in the OER but also opens new pathways for the future design of 2D-TMD-based OER catalysts. With the continuous advancement of these technologies, 2D-TMDs, as highly efficient and cost-effective OER catalysts, are expected to play an increasingly critical role in renewable energy conversion. Future research directions may include further improving the stability of catalytic sites, optimizing doping concentrations, and exploring the synergistic effects in 2D-TMDs composites to maximize the catalytic efficiency of the OER.

2.3 *Bifunctional 2D-TMDs Electrocatalysts*

In both HER and OER, 2D-TMDs have demonstrated excellent catalytic performance. However, to improve the overall water-splitting efficiency, researchers have been dedicated to developing bifunctional catalysts that can efficiently catalyze both OER and HER simultaneously. These catalysts must not only exhibit superior hydrogen adsorption and desorption capabilities in HER but also provide stable oxygen intermediate binding sites in OER [51]. Recently, several studies have successfully enhanced the bifunctional catalytic performance of 2D-TMDs in HER and OER through heterostructure design, electronic structure modulation, and surface engineering, showcasing remarkable catalytic activity.

Firstly, Mishra et al. investigated the bifunctional electrocatalytic performance of MoS_2 nanolayers in HER and OER, focusing on their electron transport capability and catalytic kinetics [52]. MoS_2 was synthesized via the solvothermal method, and its layered structure (Fig. 7a) provided abundant active sites, facilitating electron transfer and improving reaction efficiency. However, single MoS_2 still exhibited certain limitations in catalytic performance, particularly in reaction kinetics and electron transport capability. In the HER, MoS_2 exhibited an overpotential of 136 mV at 10 mA/cm^2, with a Tafel slope of 120 mV/dec., indicating a relatively low energy demand. However, its reaction kinetics still had room for optimization, primarily constrained by the electron transfer step. EIS measurements showed that its charge transfer resistance (R_{ct}) was 9.72 Ω, demonstrating good electron transport capability, while double-layer capacitance (C_{dl}) measurements demonstrated an electrochemical active surface area of 89.02 cm^2, suggesting a high density of active sites that contribute to HER efficiency (Fig. 7b). In contrast, the electrocatalytic performance of MoS_2 in OER was more constrained. At 10 mA/cm^2, its overpotential reached 340 mV, and its Tafel slope was 189 mV/dec., reflecting a slower electron transfer rate and suboptimal reaction kinetics. EIS measurements showed that its R_{ct} was 11.87 Ω indicating that its catalytic activity was restricted by relatively high charge transfer resistance (Fig. 7c). Overall, while MoS_2 nanolayers provide certain HER and OER activity, their catalytic performance in both reactions remains limited.

To overcome these limitations, researchers have explored heterostructure design as a means to enhance its electrocatalytic activity. Among these approaches, Tian et al. carried out first-principles calculations to systematically explore the application of graphene/MoS_2 heterostructures in electrocatalytic water splitting [19]. MoS_2 and graphene are two of the most representative 2D materials, each possessing catalytic properties and possessing unique electronic. Monolayer MoS_2 features a stable 2D structure, excellent catalytic activity, and tunable band gap, but its low electron mobility and high photogenerated carrier recombination rate limit its catalytic efficiency. In contrast, graphene exhibits high electrical conductivity and high carrier mobility, but its zero-bandgap nature restricts its applications in photoelectrocatalysis. To address these limitations, researchers proposed combining MoS_2 and graphene in a heterostructure to optimize their electrocatalytic performance. Their findings revealed that the formation of this heterostructure significantly influenced surface charge distribution, thereby modulating catalytic activity for both OER and HER. Computational results indicated that the MoS_2 layer surface exhibited strong hydrogen adsorption capability, enabling excellent HER performance, while the graphene layer, due to its high conductivity and low overpotential, facilitated the OER. Furthermore, they investigated the influence of an external electric field on the band alignment of the heterostructure and found that applying an appropriate electric field could further reduce HER/OER overpotentials, thereby enhancing catalytic efficiency. These results suggest that combining graphene and MoS_2 not only optimizes active sites for different reaction mechanisms but also improves the catalyst's conductivity, enabling more efficient bifunctional catalytic applications.

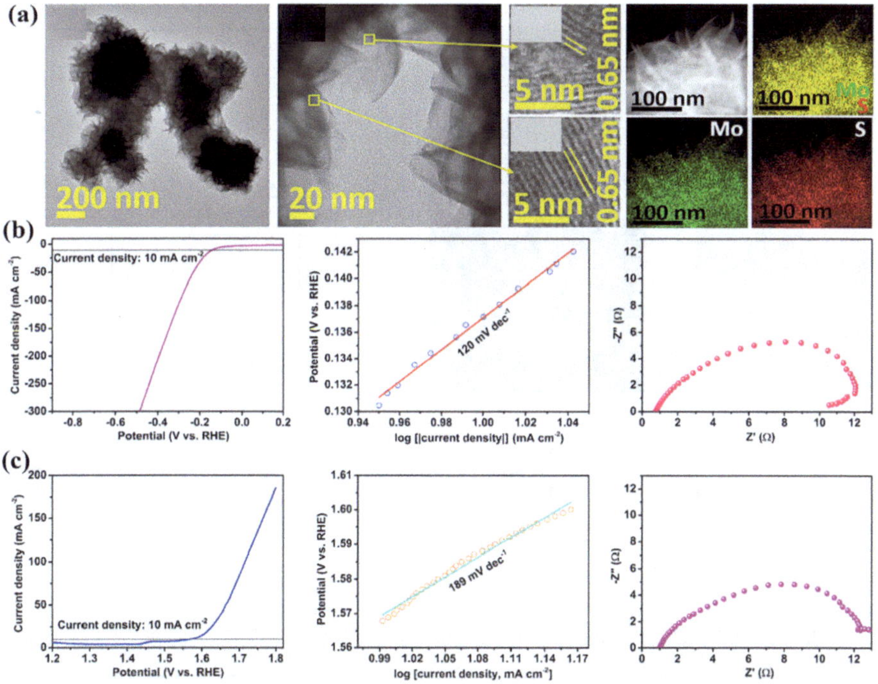

Fig. 7 **a** Electron microscopy characterization of MoS_2 and its electrochemical performance in **b** HER and **c** OER [52]

Building upon this concept, Shivakumar et al. designed and synthesized a NiFe-LDH (layered double hydroxide)/$MoSe_2$ heterostructure, utilizing the formation of a p–n junction to enhance HER and OER activity (Fig. 8a) [53]. The heterostructure was fabricated using a two-step hydrothermal process, and Mott-Schottky measurements confirmed that electrons transferred from NiFe-LDH (n-type) to $MoSe_2$ (p-type), forming a built-in electric field (Fig. 8b). This effectively promoted exciton dissociation and improved the reaction kinetics of the electrocatalytic process. Electrochemical testing revealed that in the HER, the heterostructure exhibited an overpotential of only 87 mV at a current density of 10 mA/cm^2, with a Tafel slope of 57 mV/dec. (Fig. 8c), indicating extremely low kinetic resistance and excellent hydrogen evolution capability. In the OER, NiFe-LDH/$MoSe_2$ required an overpotential of 316 mV at 10 mA/cm^2, with a Tafel slope of 70 mV/dec. (Fig. 8d), demonstrating stable oxygen evolution performance. Moreover, the catalyst exhibited long-term stability under both alkaline and acidic solution, maintaining high catalytic activity even after 40 h of continuous operation. This study confirms that heterostructure design incorporating a p–n junction can significantly enhance charge transfer efficiency, reduce charge recombination, and consequently improve HER/OER electrocatalytic performance.

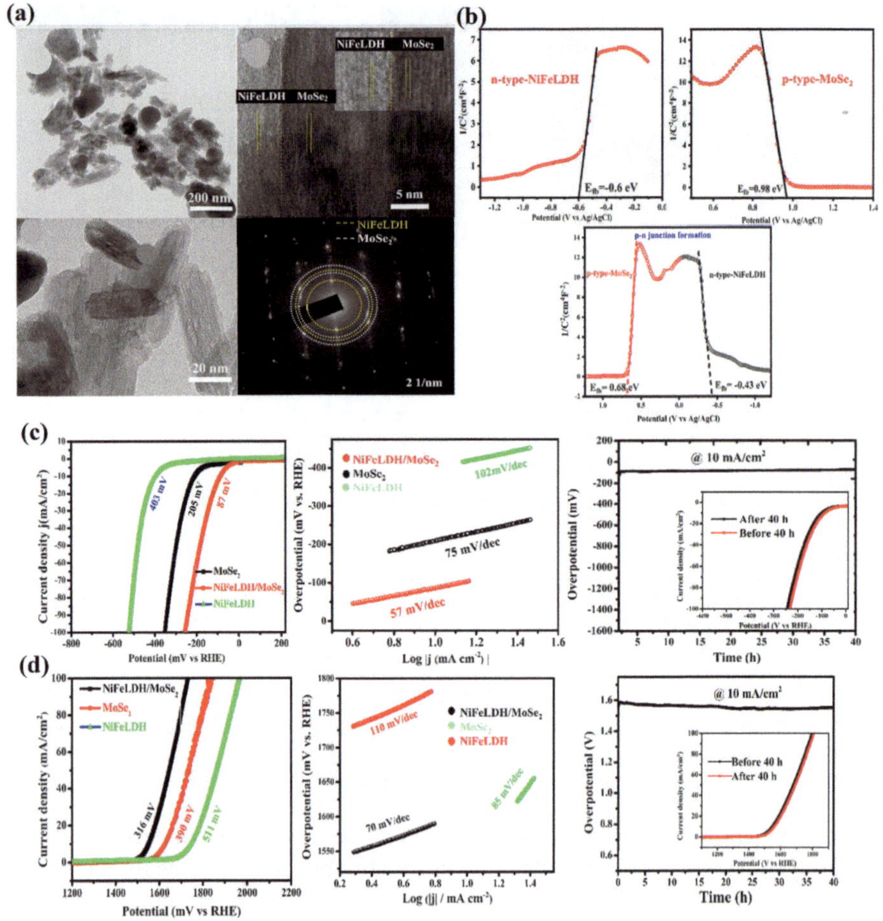

Fig. 8 **a** Electron microscopy characterization of NiFe-LDH/$MoSe_2$, **b** Mott-Schottky plots, and electrochemical performance and stability in **c** HER and **d** OER [53]

Additionally, Kotesh Kumar Mandari et al. developed a $NiO/MoS_2/BiVO_4$ ternary heterostructure to enhance HER and OER electrocatalytic performance [54]. In this study, MoS_2 nanosheets and $BiVO_4$ were synthesized using the hydrothermal method, while NiO was deposited onto the $MoS_2/BiVO_4$ surface through an ultrasonication process. The formation of this heterostructure facilitated charge transfer from Ni^{2+} to Mo^{6+}/Mo^{4+}, thereby enhancing active sites and improving charge and mass transfer capabilities (Fig. 9a). Electrochemical testing of HER electrocatalytic performance showed that the $NiO/MoS_2/BiVO_4$ heterostructure exhibited an overpotential of only 95 mV at a current density of 10 mA/cm^2 (Fig. 9b). The Tafel slope was measured at 72 mV/dec., and R_{ct} was significantly reduced, indicating improved charge transport efficiency. For OER performance in a 1.0 M KOH alkaline solution, the $NiO/MoS_2/BiVO_4$ heterostructure shows an overpotential of only

300 mV at 10 mA/cm^2, with a Tafel slope of 90 mV/dec. (Fig. 9c). Furthermore, long-term stability tests demonstrated that the catalyst maintained stable catalytic activity even after 24 h, with no significant voltage drift. These findings indicate that by leveraging heterostructure design, electronic structure modulation, and built-in electric field regulation, the bifunctional catalytic performance of 2D-TMDs in HER and OER can be significantly enhanced.

Sreekumar et al. investigated the bifunctional electrocatalytic activity of Ni-WS_2 coated stainless steel mesh (Ni-WS_2@SSM) in HER and OER [55]. The research team synthesized the catalyst using an electrodeposition method and optimized the deposition conditions to obtain an active surface with a nano-micro hierarchical structure (Fig. 10a). In HER testing, Ni-WS_2@SSM exhibited an overpotential of 89 mV at a current density of 10 mA/cm^2, significantly lower than that of pure Ni-coated SSM (189 mV). Additionally, the Tafel slope was measured at 102 mV/dec. (Fig. 10b), indicating faster reaction kinetics. EIS measurements showed a relatively

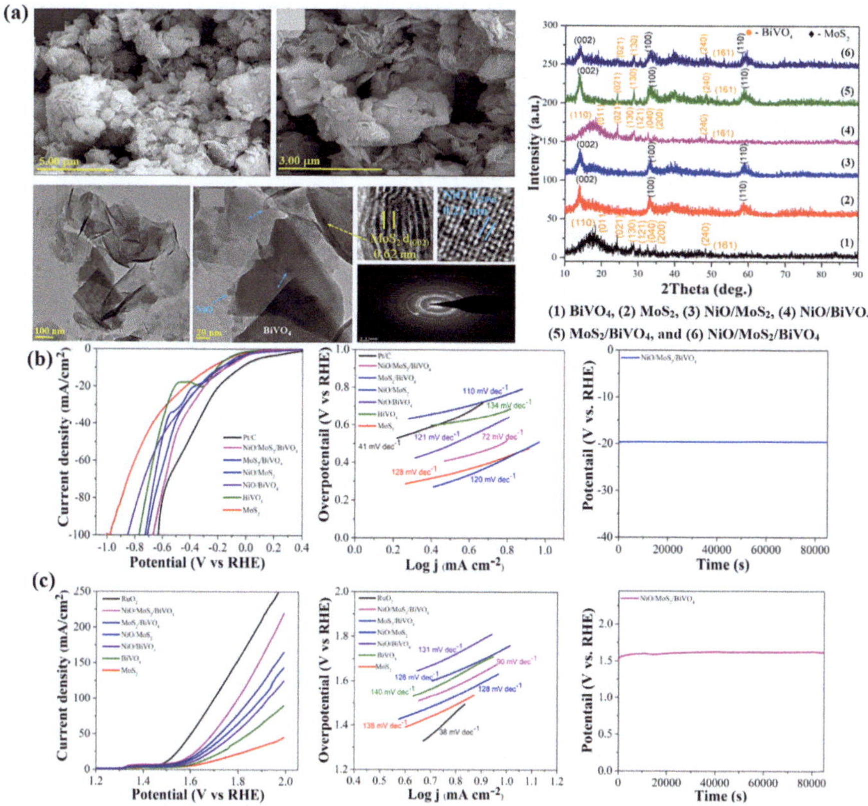

Fig. 9 **a** Electron microscopy characterization and crystal structure identification of NiO/MoS_2/$BiVO_4$, and electrochemical performance in **b** HER and **c** OER [54]

low R_{ct} value, suggesting excellent electron transport capability. For OER performance, Ni-WS_2@SSM displayed an overpotential of 230 mV, which was smaller than that of pure Ni-coated SSM (320 mV). The Tafel slope was measured at 99 mV/dec. (Fig. 10c), demonstrating higher electron transfer efficiency. After a 50-h stability test, Ni-WS_2@SSM maintained stable catalytic activity in both OER and HER, confirming its durability. The high electrocatalytic performance of Ni-WS_2@SSM is attributed to its nano-micro hierarchical structure, which provides a high specific surface area, promoting for the full exposure of catalytic active sites. Additionally, the synergistic effect between WS_2 and Ni lowers the reaction energy barrier, thereby enhancing reaction rates. Furthermore, the binder-free growth method eliminates additional electronic resistance, further improving electron transport efficiency. In summary, Ni-WS_2@SSM exhibits excellent HER and OER electrocatalytic activity, characterized by faster electron transport rates and high stability, demonstrating its potential application in water-splitting technologies.

Overall, the performance of the TMD-based catalysts in HER and OER is influenced by material composition, heterostructure design, electronic structure modulation, and interfacial engineering. Through doping, defect engineering, and heterostructure design, their catalytic performance can be further optimized, making them highly efficient and stable catalysts for water splitting. As shown in Table 1, various 2D-TMDs heterostructures have been evaluated as catalysts for HER and

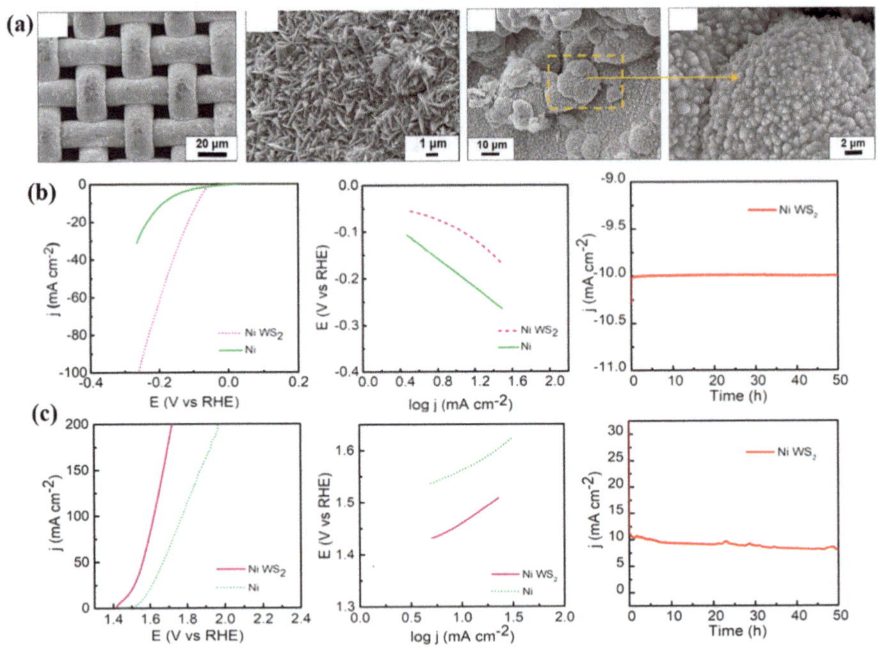

Fig. 10 **a** Electron microscopy characterization of Ni-WS_2@SSM, and electrochemical performance in **b** HER and **c** OER [55]

OER, with a focus on comparing their overpotential, Tafel slope, and stability to assess their catalytic efficiency. Future research should aim to further optimize the synthesis methods, interfacial interactions, and electronic structure modulation of these materials, facilitating their advancement in renewable energy conversion applications.

Table 1 Comparison list of catalytic activity of 2D-TMDs composite materials in HER and OER

Materials	Electrolyte	Overpotential mV@10 mA/cm^2		Tafel slope (mV/dec.)	Stability	Refs.
		HER	OER			
$MoSe_2$	0.5 M	300	–	–	–	[43]
MoS_2	0.5 M	357	–	–	–	[43]
Zn-WS_2	0.5 M H_2SO_4	180	–	70	1000 cycles	[44]
WS_2	0.5 M H_2SO_4	231	–	138	1000 cycles	[44]
V-$WMoSe_2$	0.5 M H_2SO_4	247	–	70	14 h	[45]
WS_2/Ag_3PO_4	0.5 M H_2SO_4	206	–	176	50 h	[40]
LrGO-ZnO/1T-2H MoS_2	0.5 M H_2SO_4/1 M KOH	94 (acidic)/191.5 (alkaline)	–	41.17 (acidic)/58 (alkaline)	4000 cycles	[46]
$MoSe_2$	1 M KOH	–	270	88	–	[47]
WSe_2	1 M KOH	–	300	97	–	[47]
$MoSe_2$ + $NiWSe_2$	1 M KOH	–	200	81	40 h	[47]
$AlCo_3$–MoS_2	0.1 M KOH	–	300	46.7	40 h/5000 cycles	[48]
$AlWS_2$	1 M KOH	–	440	227	1000 cycles	[49]
MoS_2	1 M OH	136	340	120 (HER)/189 (OER)	–	[52]
NiFe-LDH/$MoSe_2$	0.5 M H_2SO_4 (HER)/1 M KOH (OER)	87	316	57 (HER)/70 (OER)	40 h	[53]
NiO/MoS_2/$BiVO_4$	1 M KOH	95	300	72 (HER)/90 (OER)	24 h	[54]
Ni-WS_2	1 M KOH	89	230	102 (HER)/99 (OER)	50 h	[55]

3 Conclusion

HER and OER catalysts play a crucial role in sustainable energy conversion and storage. Herein, we systematically reviewed the progress of 2D-TMDs as catalysts for HER and OER, examining the correlation between their structural design, electronic modulation, and catalytic performance. Through literature analysis, it has been found that an efficient HER/OER catalyst should possess high conductivity, optimal adsorption energy for reactants and intermediates, and abundant active/defect sites to enhance catalytic performance and stability. technologies, enabling the transformation of clean energy into storable chemical fuels and driving the development of a hydrogen economy. However, TMDs still face several challenges in HER/OER research, including long-term catalyst stability, controllability of active sites, and feasibility of large-scale production. Future research can focus on the following directions: Mechanistic studies and material optimization, development of in-situ characterization techniques, interdisciplinary applications and technological integration. With continuous advancements in material design, electronic structure modulation, and surface engineering techniques, 2D-TMDs are expected to become highly efficient, stable, and cost-competitive alternatives to noble metal catalysts, providing significant support for the development of sustainable hydrogen production and green energy technologies.

Acknowledgements Thanks for the financial support provided by the National Science and Technology Council (NSTC) of Taiwan (grant number: 113-2221-E-845-001).

References

1. Yin, Z., Zhang, B., Zhang, Q., Zhou, Y., Li, J., Wang, Y.P., Wan, L., Ma, Z.: Graphene quantum dots/Fe_2 $(MoO_4)_3$ nanocomposites on Ni form as an efficient (pre)catalyst for OER and HER. J. Mol. Struct. **1329**, 141402 (2025). https://doi.org/10.1016/j.molstruc.2025.141402
2. Sajid, M., Qamar, M.A., Farhan, A., Qayyum, W., Khalid, A., Nawaz, A., Lee, S.L., Nawaz, H.: Emerging paradigms in two-dimensional materials: classification, synthesis, and the role of defects in electrocatalysis for water splitting and oxygen reduction reaction. J. Environ. Chem. Eng. **12**(5), 113784 (2024). https://doi.org/10.1016/j.jece.2024.113784
3. Sadiq, M., Kumar, A., Chandra, S., Makasana, J., Ballal, S., Gouadria, S., Pathak, P.K., Saini, S., Chaudhary, R.R., Mishra, V.: Boosting the catalytic performance of spinel sulphides by incorporating g-C_3N_4 for enhanced oxygen evolution reaction (OER). Diam. Relat. Mater. **153**, 111980 (2025). https://doi.org/10.1016/j.diamond.2025.111980
4. Zhou, H., Hou, J., Zhang, L., Kong, D., Wang, H., Zeng, L., Ren, Z., Xu, T., Wang, Y.: In-situ fabrication of vertical heterogeneous nickel diselenide-molybdenum diselenide architectures as bifunctional overall water-splitting electrocatalyst. J. Mater. Sci. Technol. **177**, 256–263 (2024). https://doi.org/10.1016/j.jmst.2023.09.010
5. Cao, D., Tie, M., Zhang, G., Huang, X.: Research progress of Ce-based electrocatalysts in hydrogen evolution reaction. J. Rare Earths 1936 (2025). https://doi.org/10.1016/j.jre.2025.01.015

6. Mathew, S., Sunajadevi, K.R.P., Pinheiro, D., Selvaraj, M.: Revealing the synergistic potential of Ti_3C_2 MXene/vanadium sulfide composite for enhancing electrochemical water splitting. **112**, 189–197 (2025). https://doi.org/10.1016/j.ijhydene.2025.02.383
7. Wert, S., Iffelsberger, C., Novčić, K.A., Pumera, M.: Corrosion of catalyst in high resolution: layered transition metal dichalcogenides electrocatalyse water splitting and corrode during the process. J. Catal. **416**, 85–91 (2022). https://doi.org/10.1016/j.jcat.2022.10.023
8. Ahmed, K., Hameed, S., Patchigolla, K., Dawood, N., Ghouri, Z.K.: Carbon-based electrocatalysts for hydrogen evolution reaction. Energy Convers. Manage. **26**, 100892 (2025). https://doi.org/10.1016/j.ecmx.2025.100892
9. Alfa, I., Hafeez, H.Y., Mohammed, J., Abdu, S., Suleiman, A.B., Ndikilar, C.E.: A recent progress and advancement on MoS_2-based photocatalysts for efficient solar fuel (hydrogen) generation via photocatalytic water splitting. Int. J. Hydrogen Energy **71**, 1006–1025 (2024). https://doi.org/10.1016/j.ijhydene.2024.05.203
10. Zhao, S., Pan, Y.: Coordination structures design of single-atom catalysts for enhanced hydrogen evolution reaction. Sustain. Chem. Energy Mater. **1**, 100003 (2025). https://doi.org/10.1016/j.scenem.2025.100003
11. Sarfraz, B., Bashir, I., Rauf, A.: CuS/NiFe-LDH/NF as a bifunctional electrocatalyst for hydrogen evolution (HER) and oxygen evolution reactions (OER). Fuel **337**, 127253 (2023). https://doi.org/10.1016/j.fuel.2022.127253
12. Garcia, V.G., Inacio, G.J., Filho, L.F., Pacheco, L.T., Pansini, F.N.N., Menezes, M.G., Paz, W.S.: Exploring the potential of α-Ge(111) monolayer in photocatalytic water splitting for hydrogen production. FlatChem. **48**, 100753 (2024). https://doi.org/10.1016/j.flatc.2024.100753
13. Zhao, W., Yang, X., Gu, Y., Li, S., Hu, Y., Tang, Y., Wang, M.: Surface reconstruction engineering of amorphous Ni-Fe phosphate electrocatalyst for highly stable alkaline water splitting. J. Power Sources **637**, 236560 (2025). https://doi.org/10.1016/j.jpowsour.2025.236560
14. Munawar, T., Bashir, A., Iqbal, F., Rafaqat, M., Guo, C., Zhiani, M., Zhao, C., Tong, Y., Khan, S.A., Koc, M., Yan, C.-F.: G-C_3N_4-modulated CdS-V_2O_5/g-C_3N_4 nanosheets as efficient alkaline OER and HER electrocatalyst for water-splitting. Fuel **385**, 134100 (2025). https://doi.org/10.1016/j.fuel.2024.134100
15. Cogal, S., Cogal, G.C., Mičušík, M., Michalcova, A., Šlouf, M., Omastová, M.: Conducting polymer-templated and nonnoble metal doped $MoSe_2$ hybrids as bifunctional electrocatalysts for overall water splitting. J. Electroanal. Chem. **946**, 117728 (2023). https://doi.org/10.1016/j.jelechem.2023.117728
16. Shi, H., Zhu, Y., Chen, J., Wang, Z., Zhan, J., Li, S., Han, P., Guo, X.: Reveal the effects of iron content in NiFex-LDH on its morphology, crystallinity, and electronic structure, and their relationship with OER activity. Int. J. Hydrogen Energy **97**, 874–881 (2025). https://doi.org/10.1016/j.ijhydene.2024.11.481
17. Aftab, S., Goud, B.S., Mukhtar, M., Haider, Z., Kabir, F., Nazir, G., Aslam, M.J., Aslam, M., Hegazy, H.H., Kim, J.H., Rajpar, A.H.: Perovskite photovoltaics with cutting-edge strategies in 2D-TMDs-based interfacial layer optimization. Mater. Today Sustain. **28**, 100982 (2024). https://doi.org/10.1016/j.mtsust.2024.100982
18. Liu, Y., Ma, Y., Yang, J., Zhang, S., Wu, N., Wang, P., Wang, L.: Overcoming chemical stability challenges of nanoporous graphene separation membranes in harsh environments. Chem. Eng. J. **486**, 160721 (2025). https://doi.org/10.1016/j.cej.2025.160721
19. Shi, L.-N., Cui, L.-T., Ji, Y.-R., Xie, Y., Zhu, Y.-R., Yi, T.-F.: Towards high-performance electrocatalysts: activity optimization strategy of 2D MXenes-based nanomaterials for water-splitting. Coord. Chem. Rev. **469**, 214668 (2022). https://doi.org/10.1016/j.ccr.2022.214668
20. Tian, L., He, C., Hu, J., Yang, J., Li, X., Chen, Z.: Electronic and optical properties and electrocatalytic water splitting in a graphene/MoS_2 heterojunction. Phys. B **685**, 416062 (2024). https://doi.org/10.1016/j.physb.2024.416062
21. Yang, Y.-S., Zhang, Y., Luo, Y.-F., Sun, S.-Z., Li, J.-H., Duan, L., Xie, J.: Direct Z-scheme ZrS_2/InP heterostructure as an efficient photocatalyst for overall water-splitting under acidic, alkaline and neutral environments. Mater. Sci. Semicond. Process. **185**, 108857 (2025). https://doi.org/10.1016/j.mssp.2024.108857

22. Wang, M., Feng, Z., Shi, P., Zhang, C., Wang, P., Li, L., Shen, Y., Lin, L.: High-throughput screening of promising bifunctional catalysts for OER/ORR in disulfides. Surf. Interfaces. **53**, 105069 (2024). https://doi.org/10.1016/j.surfin.2024.105069
23. Ge, D., Luo, R., Wang, X., Yang, L., Xiong, W., Wang, F.: Internal and external electric field tunable electronic structures for photocatalytic water splitting: Janus transition-metal chalcogenides/C_3N_4 van der Waals heterojunctions. Appl. Surf. Sci. **566**, 150639 (2021). https://doi.org/10.1016/j.apsusc.2021.150639
24. Sohn, I., Wi, S., Kim, Y., Shin, D., Kim, M., Lee, S., Yoon, H., Yoo, J., Chung, S.-M., Kim, H.: Selective passivation of 2D-TMD surface defects by atomic layer deposited Al_2O_3 to enhance recovery properties of gas sensor. Appl. Surf. Sci. **646**, 158906 (2024). https://doi.org/10.1016/j.apsusc.2023.158906
25. Islam, A., Khan, S., Mim, J.J., Akter, M.F., Islam, M.S., Oliullah, S., Hossain, N.: Advances in 2D heterostructures for quantum computing applications: a review. Inorg. Chem. Commun. **174**, 113980 (2025). https://doi.org/10.1016/j.inoche.2025.113980
26. He, C., Tian, L., Hu, J., Peng, M., Li, X.: Hydrogen production from photocatalytic water splitting in the hBNC/MoS_2 heterojunction: first-principles calculations. Phys. B **694**, 416438 (2024). https://doi.org/10.1016/j.physb.2024.416438
27. Zhang, W., Liu, H., Zhao, X., Chen, J., Li, C.: Cobalt-doped WSe_2@conducting polymer hybrid for enhanced electrocatalytic performance. Electrochim. Acta **435**, 142857 (2024). https://doi.org/10.1016/j.electacta.2024.142857
28. Kim, Y.S., Park, T.H., Lee, M.J.: Experimental investigation on the electrodeposited nickel-based catalysts for alkaline water electrolysis. Renew. Energy **196**, 123456 (2024). https://doi.org/10.1016/j.renene.2024.123456
29. Ricardo Marroquin, J.F., Ghiglieno, F., Archanjo, B.S., Roncaratti, L.F., Felix, J.F.: Defect engineering via gamma irradiation in scalable mechanical exfoliation of 2D materials for enhanced electrocatalysis. Appl. Surf. Sci. **630**, 157952 (2024). https://doi.org/10.1016/j.apsusc.2024.157952
30. Alfalasi, W., Feng, Y.P., Tit, N.: Designing a functionalized 2D-TMD (MoX_2, X = S, Se) hosting half-metallicity for selective gas-sensing applications: atomic-scale study. Acta Mater. **246**, 118655 (2023). https://doi.org/10.1016/j.actamat.2022.118655
31. Long, L.N., Bich, T.T.N.: Hydrothermal-based controllable synthesis and structural-optical characterization of 2D-MoS_2/graphene nanocomposites for optoelectronic applications. Opt. Mater. **156**, 115968 (2024). https://doi.org/10.1016/j.optmat.2024.115968
32. Romanov, R.I., Zabrosaev, I.V., Chouprik, A.A., Zarubin, S.S., Yakubovsky, D.I., Zavidovskiy, I.A., Bolshakov, A.D., Markeev, A.M.: Impact of water vapor on the 2D MoS_2 growth in metal-organic chemical vapor deposition. Vacuum **230**, 113739 (2024). https://doi.org/10.1016/j.vacuum.2024.113739
33. Jarolímková, J., Neubertová, V., Severa, K., Daniš, S., Cais, J., Lyutakov, O., Kolská, Z.: Hydrogen evolution reaction at different pH of 2D MoS_2 prepared via liquid-phase exfoliation. Surf. Interfaces **28**, 106086 (2025). https://doi.org/10.1016/j.surfin.2025.106086
34. Jaberi, S.Y.S., Ghaffarinejad, A., Khajehsaeidi, Z., Sadeghi, A.: The synthesis, properties, and potential applications of CoS_2 as a transition metal dichalcogenide (TMD). Int. J. Hydrogen Energy **48**, 15831–15878 (2023). https://doi.org/10.1016/j.ijhydene.2023.01.05
35. Rai, H., Ponomarev, I., Thakur, D., Balakrishnan, V., Polcar, T., Gosvami, N.N.: Impact of wrinkle orientation on the nanotribological properties of chemical vapor deposited TMD monolayers using AFM. Appl. Surf. Sci. **672**, 160884 (2024). https://doi.org/10.1016/j.apsusc.2024.160884
36. Liang, Z., Wang, M., Liu, S., Hassan, M., Zhang, X., Lei, S., Qiao, G., Liu, G.: One-pot hydrothermal synthesis of self-assembled MoS_2/WS_2 nanoflowers for chemiresistive room-temperature NO_2 sensors. Sens. Actuat. B Chem. **403**, 135215 (2024). https://doi.org/10.1016/j.snb.2023.135215
37. Dang, Z., Tang, Z., Wu, J., Chang, Y., Wang, Y.: Unraveling the reaction networks and key pathways during the gas phase stage in CVD synthesis of MoS_2. Chem. Eng. J. **503**, 157957 (2025). https://doi.org/10.1016/j.cej.2024.157957

38. Selamneni, V., Sahatiya, P.: Mixed dimensional transition metal dichalcogenides (TMDs) vdW heterostructure-based photodetectors: a review. Microelectron. Eng. **269**, 111926 (2023). https://doi.org/10.1016/j.mee.2022.111926
39. Kamila, S., Kandasamy, M., Chakraborty, B., Jena, B.K.: Recent development on engineered TMDs for charge storage performance: experimental and theoretical investigations. J. Energy Storage **89**, 111614 (2024). https://doi.org/10.1016/j.est.2024.111614
40. Satheesh, D., Baskar, L., Jayavelu, Y., Dekshinamoorthy, A., Sakthinathan, V.R., Vijayaraghavan, S., Krishnan, K., Rajendran, R., Pachaiappan, R., Daniel, P.J., Manavalan, K.: Exfoliated WS_2 nanosheet—Ag_3PO_4 heterogeneous composite catalyst for effective electrochemical water-splitting. Mater. Lett. **366**, 136478 (2024). https://doi.org/10.1016/j.matlet.2024.136478
41. Thakur, D., Porwal, C., Chauhan, V.S., Balakrishnan, V., Vaish, R.: 2D transition metal dichalcogenides: synthesis methods and their pivotal role in photo, piezo, and photo-piezocatalytic processes. Sep. Purif. Technol. **337**, 126462 (2024). https://doi.org/10.1016/j.seppur.2024.126462
42. Raj, A., Kumar, M., Singh, D.V., Anshul, A.: A critical review on transition metal dichalcogenides (TMDs): an efficiency booster for perovskite solar cells. FlatChem. **44**, 100629 (2024). https://doi.org/10.1016/j.flatc.2024.100629
43. Ravikumar, C.H., Nair, G.V., Muralikrishna, S., Nagaraju, D.H., Balakrishna, R.G.: Nanoflower-like structures of $MoSe_2$ and MoS_2 as efficient catalysts for hydrogen evolution. Mater. Lett. **220**, 133–135 (2018). https://doi.org/10.1016/j.matlet.2018.03.012
44. Das, A., Roy, D., Das, B.K., Ansari, M.I., Chattopadhyay, K.K., Sarkar, S.: Zinc doping induced WS_2 accelerating the HER and ORR kinetics: a theoretical and experimental validation. Catal. Today **423**, 113921 (2023). https://doi.org/10.1016/j.cattod.2022.10.003
45. Yang, H., Zhang, P., Zheng, Q., Ali, T., Raza, S.: Vacancy engineering in tungsten induced electronic structure optimization of $MoSe_2$ for enhanced electrocatalytic hydrogen evolution reaction. Int. J. Hydrogen Energy **48**, 35433–35442 (2023). https://doi.org/10.1016/j.ijhydene.2023.05.299
46. Chandrakantha, K.S., Zuo, Z., Chandrashekar, B.N., Amini, A., Rangappa, K.S., Srikantaswamy, S., Cheng, C.: Synergistic effects of 1T–2H MoS_2 and laser-reduced graphene oxide–ZnO scaffold composite catalyst for efficient hydrogen evolution reaction. Mater. Today Energy **45**, 101683 (2024). https://doi.org/10.1016/j.mtener.2024.101683
47. Liu, K., Yu, Y., Cheng, J., Wang, Y., Guo, F., Lei, S., Li, M., Yun, Y., Shi, R., Song, W.: Coupling interface constructions of flower-like structure $MoSe_2$–$NiWSe_2$ for efficient oxygen evolution reaction and urea oxidation reaction. Mater. Today Chem. **38**, 102071 (2024). https://doi.org/10.1016/j.mtchem.2024.102071
48. Mu, X., Zhu, Y., Gu, X., Dai, S., Mao, Q., Bao, L., Li, W., Liu, S., Bao, J., Mu, S.: Awakening the oxygen evolution activity of MoS_2 by oxophilic-metal induced surface reorganization engineering. J. Energy Chem. **62**, 546–551 (2021). https://doi.org/10.1016/j.jechem.2021.04.019
49. Alahmadi, M., BenAoun, S.: A single-step synthesis of aluminum-doped WS_2 for oxygen evolution reaction with loading influence on catalytic performance. J. Saudi Chem. Soc. **28**, 101845 (2024). https://doi.org/10.1016/j.jscs.2024.101845
50. Mahmood, A., Lu, G., Wang, X., Wang, Y., Xie, X., Sun, J.: Investigating the stability and role of defects in vertically aligned WS_2/MoS_2 heterojunctions on OER activity using first principles study. J. Power Sources **551**, 232208 (2022). https://doi.org/10.1016/j.jpowsour.2022.232208
51. Chai, H., Luo, Q., Chen, W., Li, Y., Zhao, M., Shi, J., Tang, Y., Dai, X.: The first-principles calculations of photocatalytic water splitting and photoelectric properties of two-dimensional $Mo_xW_{1-x}S_2$ and $MoS_{2x}Se_{2(1-x)}$ alloys. Sol. Energy Mater. Sol. Cells **262**, 112552 (2023). https://doi.org/10.1016/j.solmat.2023.112552
52. Mishra, R.K., Kumar, V., Trung, L.G., Choi, G.J., Ryu, J.W., Kumar, P., Bhardwaj, R., Lee, S.H., Gwag, J.S.: Bifunctionality of MoS_2 nanolayer catalyst for water-splitting reactions of hydrogen and oxygen. Mater. Lett. **338**, 134026 (2023). https://doi.org/10.1016/j.matlet.2023.134026

53. Shivakumar, P., Monika, M.N., Deepu, M., Manjunatha kumara, K.S., Budagumpi, S., Nagaraju, D.H.: Amorphous and crystalline heterostructure $MoSe_2$/NiFe-LDH through p-n junction formation for electrochemical water splitting-synergistic back bonding effect. Int. J. Hydrogen Energy **65**, 74–82 (2024). https://doi.org/10.1016/j.ijhydene.2024.03.263
54. Mandari, K.K., Pandey, S., Kang, M.: Highly efficient ternary NiO/MoS_2/$BiVO_4$ heterostructure for electrocatalytic HER/OER applications. Int. J. Hydrogen Energy **52**, 275–287 (2024). https://doi.org/10.1016/j.ijhydene.2023.07.058
55. Sreekumar, R., Pillai, S.R., Baby, R.R., Suseelamma, S.S.: Morphology tuned Ni-WS_2 coated stainless steel mesh as an efficient bi-functional electrocatalyst for water splitting. Mater. Today Commun. **44**, 111860 (2025). https://doi.org/10.1016/j.mtcomm.2025.111860

2D Trichalcogenides for HER and OER

Md. Rakib Khan, Paricha Jebin, Jannatul Shahrin Ananna, and A. J. Saleh Ahammad

Abstract Considering the rapid depletion of non-renewable energy (e.g., fossil fuels), the development of new renewable energy sources has garnered significant attention from the world's scientific community. Water splitting can serve as an alternative clean energy source to address this energy issue. 2D trichalcogenides are promising materials for Hydrogen evolution reaction (HER) and Oxygen evolution reaction (OER). Due to their unique properties like layered structure, more active sites, tunable structures, and high abundance on earth, they can be a potential electrocatalyst for water splitting. Hence, the various properties and electrocatalytic activity of these trichalcogenides for HER and OER will be covered in this chapter. Additionally, this chapter emphasizes the various synthesis methods and high catalytic activity observed when transition metals are exfoliated with 2D trichalcogenides. To date, trisulfides and triselenides are mostly used for water splitting from the chalcogen group. The challenges faced and how to overcome those obstacles is explored as well to ensure maximum application in various fields.

Keywords 2D trichalcogenides · Hydrogen evolution reaction · Oxygen evolution reaction · Electrocatalyst

1 Electrocatalytic Activity of Trichalcogenides

Metal trichalcogenides are widely used for splitting water nowadays. The unique features of 2D trichalcogenides, especially having a layer structure with several active sites and low overpotential, permit them to show strong electrocatalytic activity towards water splitting. With a tunable structure, the divacancy configuration offers them high conductivity [1]. A graphical representation of their properties is shown

Md. R. Khan · J. S. Ananna · A. J. S. Ahammad (✉)
Department of Chemistry, Jagannath University, Dhaka 1100, Bangladesh
e-mail: ajsahammad@chem.jnu.ac.bd

P. Jebin
Department of Physics, Jagannath University, Dhaka 1100, Bangladesh

O. P. Pandey et al. (eds.), *2D Materials for Electrochemical Water Splitting*, Engineering Materials, https://doi.org/10.1007/978-981-95-2003-9_6

Fig. 1 Schematic representations of unique properties of 2D trichalcogenides

below in Fig. 1. In the case of water splitting, these electrocatalysts split water in two ways: either through hydrogen evolution or oxygen evaluation reaction.

1.1 Trichalcogenides for Electrocatalytic Hydrogen Evaluation Reaction

By absorbing solar energy, the photocatalyst can splits water. But in electrochemistry, the electrocatalyst offers a more facile and better efficiency for hydrogen evaluation reactions. Noble metal-based electrocatalysts are much more expensive and have some limitations. As a result, researchers have given more attention to develop an efficient chalcogenides-based electrocatalyst that is available in nature and cost-effective. It looks like a semiconductor that is tunable and has appropriate physio-chemical characteristics. Trichalcogenide complexes are also used for the electrocatalytic evolution of hydrogen. Let's consider a general hydrogen evolution reaction mechanism. HER can occur both in acidic and basic media, including adsorption as well as desorption [1].

Adsorption path by Volmer:

$$\text{In acidic medium: } H_3O^+ + e^- + Cat \rightarrow Cat - H + H_2O \tag{1}$$

$$\text{In basic medium: } H_2O + e^- + Cat \rightarrow Cat - H + OH^- \tag{2}$$

Desorption path by Heyrovsky:

$$\text{If the medium is acidic: Cat} - \text{H} + \text{H}_3\text{O}^+ + \text{e}^- \rightarrow \text{H}_2 + \text{H}_2\text{O} + \text{Cat} \quad (3)$$

$$\text{If the medium is basic: Cat} - \text{H} + \text{H}_2\text{O} + \text{e}^- \rightarrow \text{H}_2 + \text{OH}^- + \text{Cat} \quad (4)$$

Desorption path by Tafel:

$$2\text{Cat} - \text{H} \rightarrow \text{H}_2 + 2\text{Cat} \quad (5)$$

Here, the catalyst used is indicated by Cat, and the hydrogen atom adsorbed on the surface of the catalyst is indicated by Cat-H. The desorption step can occur in either the Volmer-Heyrovsky or Volmer-Tafel pathway. So, the HER process is classified into two routes based on desorption mechanisms: either V-H or V-T pathways.

The trichalcogenide with transition metals shows high catalytic activity toward HER. The formation of an efficient electrocatalyst complex between trichalcogenides and transition metals can easily catalysis water to hydrogen by HER process. Recently, Sampath et al. prepared liquid-exfoliated nickel phosphate trisulfide ($NiPS_3$) nanosheets for the HER process in a wide range of pH (1–14) and they used sodium chloride solution (3.5 wt%) for neutral condition. They prepared the crystals of $NiPS_3$ using the chemical vapor transport method. The lateral size of the prepared nanosheets of $NiPS_3$ was 200–400 nm and the thickness of these nanosheets was 0.65–0.7 nm after exfoliation in a liquid phase. Due to exfoliation, the overall electrochemical HER performance was improved than without exfoliated nanosheets under all pH ranges. It was observed that in acidic conditions the overpotential for HER for exfoliated nanosheets of $NiPS_3$ was very much lower than that of basic and neutral media. The values of overpotential in pH 1, 7, 14 were 297 mV, 816 mV and 398 mV at the current density $10\,\text{mA/cm}^2$, respectively. Density functional theory was also used for the further study of HER [2].

In trichalcogenides complex, doping of metals can enhance the HER activity by improving the electrical conductivity of the existing complex. The electrical conductivity of $NiPS_3$ improved drastically when cobalt was doped in the complex due to the changing electronic structure of the nickel phosphate trisulfides. However, cobalt is expensive compared to other commonly used transition metals like iron, nickel, and manganese for the HER process. Iron can also be used for doping because its catalytic property is similar to cobalt and is more abundant in the earth's crust. So, iron can be combined with nickel to enhance the HER activity of the $NiPS_3$-based complexes. Based on this, several series of nanosheets of Ni/Fe phosphate trisulfide were prepared and denoted as $Ni_{1-x}Fe_xPS_3$ where the values of x were 0, 0.05, 0.1, 0.15 and 1. The promising electrocatalytic behavior of these complexes towards the HER process is shown in Fig. 2 [3].

The experimental results showed that the iron-doped nanosheets can perform high electrocatalytic HER activity which is comparable with cobalt doping. A number of

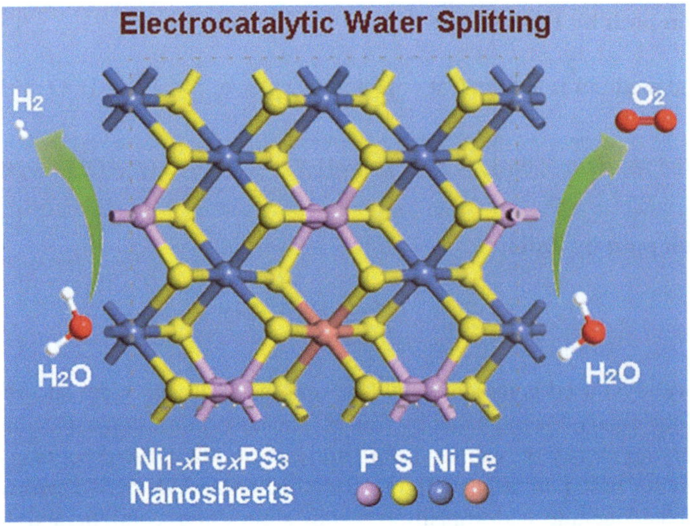

Fig. 2 Mixed nickel iron phosphosulfide nanosheet ($Ni_{0.9}Fe_{0.1}PS_3$) for enhanced HER as well as OER. Adapted with permission from [3] Copyright (2017), ACS

catalysts can be formed from different values of x. Among them, the optimized complex was $Ni_{0.9}Fe_{0.1}PS_3$ with an overpotential only 72 mV at 10 mA/cm^2 current density. The lower Tafel slope value (73 mV/decade) indicates the facile electron transfer.

1.2 Trichalcogenides for Electrocatalytic Oxygen Evaluation Reaction

Trichalcogenides can be used as an efficient electrocatalyst for the OER process. When they form complexes with transition metal they show high catalytic activity towards the OER. In the water-splitting process, HER takes place at the negatively charged electrode, whereas OER takes place at the positively charged electrode. The OER mechanism is completely different from HER since one is the oxidation process and the other is the reduction process. For the OER process, the general mechanism can be written as [1]

$$\text{In basic medium: } OH^- \rightarrow O_2 + 2H_2O + 4e^- \quad (6)$$

$$\text{In acidic medium: } 2H_2O \rightarrow O_2 + 4H^+ + 4e^- \quad (7)$$

It is understandable that the OER process is facile in an alkaline medium while sluggish kinetics are found in the acidic medium. In an alkaline solution, the OER mechanism occurs by the following four steps [1]

$$OH^- + Cat \rightarrow Cat - OH^* + e^- \quad (8)$$

$$OH^- + Cat - OH^* \rightarrow H_2O + Cat - O^* + e^- \quad (9)$$

$$OH^- + Cat - O^* \rightarrow Cat - OOH^* + e^- \quad (10)$$

$$OH^- + Cat - OOH^* \rightarrow O_2 + H_2O + e^- + Cat \quad (11)$$

Here, OH*, O* and OOH* are the free radical species. According to the free energy calculation, the OER reaction need to overcome the 1.23 V versus RHE to produce oxygen.

Recently, R. Dangol et al. prepared liquid exfoliated $NiPS_3$ as well as bulk $NiPS_3$ nanosheets for evaluation of the reaction of oxygen under an alkaline solution [4]. The exfoliated $NiPS_3$ nanosheets show excellent catalytic behavior towards the OER process. Its overpotential value (301 mV) is quite smaller than the bulk $NiPS_3$ (441 mV) at 10 mA/cm^2. This lowering overpotential of the exfoliated $NiPS_3$ from the bulk is due to the increasing effective surface area (ESCA) that is about 3.4 folds larger in exfoliated $NiPS_3$. Schuhmann et al. demonstrated detailed study of the OER process based on exfoliated $NiPS_3$ nanosheets [5].

Similar to the HER, the OER activity is also enhanced by doping some transition elements in the metal trichalcogenide electrocatalyst as a result of increased effective surface area. Song et al. [3] showed that doping the $NiPS_3$ nanosheets with iron metal provides high catalytic activity towards the OER process. Various samples were prepared using the CVT method such as $Ni_{1-x}Fe_xPS_3$ where the values of x were 0, 0.05, 0.1, 0.15 and 1. They showed promising electrocatalytic behavior towards the OER process. The interesting matter is that $FePS_3$ showed almost no catalytic activity but $Ni_{1-x}Fe_xPS_3$ showed higher catalytic activity towards the OER than undoped $NiPS_3$. The doped nanosheets provide an overpotential of about 329 mV at 20 mA cm^{-2}. They also prepared mass production of different metal phosphorus trichalcogenide in a novel way for comparing the nanosheets for OER activity. It was found that the OER performance is better in the case of iron-doped $NiPS_3$ than undoped one. The optimized nanosheets provide a lower overpotential of about 256 mV at 30 mA cm^{-2}. The lower Tafel slope value also indicates facile electron transfer which was about 46 mV/decade and better than that of IrO_2 (56 mV/decade) and RuO_2 (86 mV/decade). Doping of the nanosheets with Fe enhances the ESCA, which is responsible for the high catalytical activity. From the EIS and DFT calculation, it was found that Fe would enhance the conductivity of the electrocatalyst electronically. Also, it would significantly decrease the interaction between the prepared nanosheets and oxygen atom containing intermediates that facile the formation of oxygen molecule by the OER process. The theoretical investigations reveal that when metal is doped in a metal trichalcogenide complex, it lowers the binding energy to achieve intermediate species with improved catalytical performance. The various preparation processes of metal trichalcogenide will be discussed in the following section.

2 Synthesis Strategies

There are various methods to prepare transition metal-based trichalcogenides, that possess good catalytical activity towards the HER as well as the OER. Among various methods, the chemical vapor transport (CVT) method, Liquid Phase Exfoliation (LPE), Mechanical Exfoliation, Electrochemical Exfoliation, and Chemical Vapor Deposition are very common [6].

2.1 Chemical Vapor Transport (CVT)

In a typical CVT growth procedure, the synthesis of $MPCh_3$ in bulk quantity occurs in a double-temperature zone furnace, where stoichiometric amounts of various metals, other nonmetals like phosphorus, and desire chalcogens are vaporized and reacted in a high-temperature zone; then, they are transported into a low-temperature zone for crystallization and deposition in a fused silica ampoule. During this process, two factors, like the heating rate and degree of vacuum, played significant roles in the crystalline quality and defect density of the bulk $MPCh_3$. CVT can achieve the high fidelity and extensive production of $MPCh_3$ materials.

2.2 Liquid Phase Exfoliation

LPE emerges as a notably cost-effective method for producing large-scale ultrathin 2D $MPCh_3$ nanosheets from bulk crystals. In a typical LPE process Route 1 in Fig. 3a–c, the $MPCh_3$ bulk crystals are dispersed into suitable solvents. After undergoing ultrasonication treatment, $MPCh_3$ nanosheets with the desired thickness and size can be collected via centrifugation. The use of ethanol, acetone, and isopropanol as solvents leads to the successful preparation of $CoPS_3$, $NiPS_3$, and $SnPS_3$ nanosheets with ultrathin thickness through the LPE method.

2.3 Mechanical Exfoliation

The methods employed to produce $MPCh_3$ nanoflakes can be categorised in two ways such as micromechanical exfoliation and ball milling techniques. In these methods,

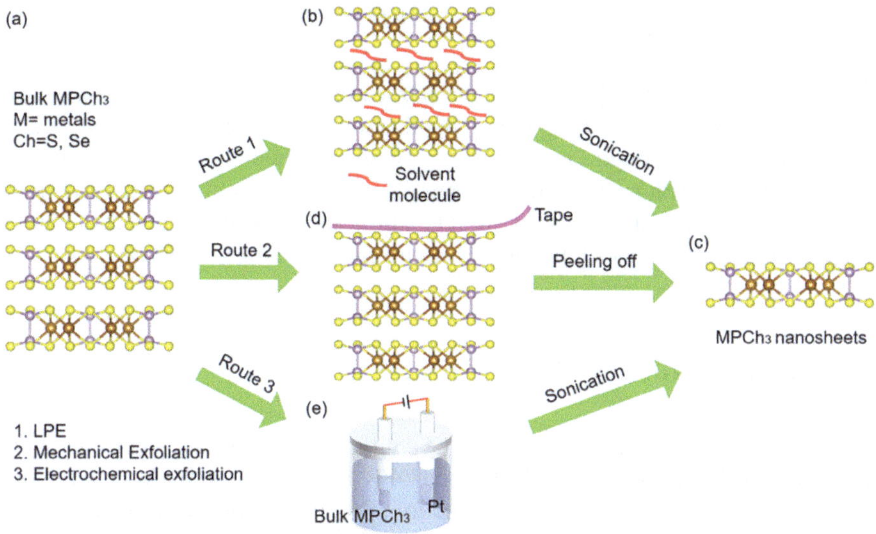

Fig. 3 Graphical representation of the fabricating $MPCh_3$ nanosheets. **a** Bulk $MPCh_3$ crystals. **b** Intercalated bulk $MPCh_3$ crystals. **c** Exfoliated $MPCh_3$ nanosheets. **d** Mechanical exfoliation tape on the bulk $MPCh_3$ crystals. **e** Schematic of the experimental setup for electrochemical intercalation exfoliation. Adapted with permission from [6] Copyright (2024), Wiley–VCH GmbH

between adjacent layers of $MPCh_3$ van der Waals forces are present utilizing both normal and shear forces, respectively. Micromechanical exfoliation involves gradual separation of layered bulk crystals into few-layered or monolayer nanosheets in Route 2, as shown in Fig. 3a, c, d. The dissociation and generation energies associated with lamellar $MPCh_3$ materials are comparatively lower than those of conventional 2D materials, making the exfoliation process thermodynamically favorable. Since only physical alterations occur during micromechanical exfoliation, it can be seen as a non-destructive technique conducive to maintain a pristine surface state.

2.4 Electrochemical Exfoliation

Many existing preparation methods face challenges such as uncontrolled thickness, uneven topography and area distribution, prolonged processing durations, and inadequate quality. In contrast, the electrochemical exfoliation method Route 3, as shown in Fig. 3a, c, e, is typically conducted under mild conditions. This results in large-scale monolayer $MPCh_3$ with superior quality, crystallinity, and yield.

2.5 *Chemical Vapor Deposition (CVD)*

This process is very effective for directly fabricating $MPCh_3$ nanosheets on certain specific substrates for emerging energy applications. In the process of synthesizing $MPCh_3$ nanosheets through vapor deposition, particular amounts of metal precursors, sulfur powder, and phosphorus powder are placed in two ceramic boats; one holds the metal precursors, and the other contain sulfur and phosphorus powders. Out of all the processes, the major three processes are shown below graphically.

3 Trichalcogenide for HER

By providing external energy, water can be split into two parts: H_2 and O_2 gas. The production of H_2 and O_2 are called HER and OER, respectively. It is a key technology for clean and sustainable hydrogen production. Produced hydrogen is a vital energy carrier. It holds great potential for addressing global energy challenges. It can be used as an alternative to conventional fossil fuels. However, the yields of the HER and OER are dependent mainly on the type of electrocatalyst is used. Noble metals like platinum are highly efficient for HER but such metals are costly and scarce. It is essential to explore alternative materials with high efficiency and durability to make HER scalable, cost-effective, and sustainable.

Trichalcogenides with transition metals act as good electrocatalysts. The general formula is MX_3; here, M implies metal and X indicates chalcogen. The d block metals of the periodic table, as well as some post-transition and alkali metals, can be used as an element for preparing electrocatalysts based on trichalcogenide. From the chalcogen atoms, sulfur and selenium are mostly used till now. In a few works the tellurium is used as a promising candidate for splitting water [7].

Trichalcogenides have many advantages for HER due to some unique properties:

(1) It is composed of single or few atomic layers by weak van der Waals forces, exposing abundant active sites for catalytic processes in HER.
(2) Trichalcogenides are a cost-effective and scalable alternative to noble metals.
(3) It is highly stable.
(4) Their tunable electronic properties make them suitable for HER.

2D Trichalcogenides can be composed of single or few atomic layers, as shown in Fig. 4. Monolayer trichalcogenide materials have a large density of exposed active sites and distinct electronic properties [8]. Due to these properties, monolayer trichalcogenide materials are highly efficient for hydrogen evolution reactions (HER). However, they often face a few challenges, like structural instability and limited charge transfer pathways, which makes them less practical for prolonged or large-scale applications. Additionally, multilayer trichalcogenide materials are more cost-effective and scalable. They provide tunable electronic properties through multilayer interactions when interlayer charge-transfer pathways, both basal-plane and edge sites are exposed [9]. It increases catalytic activity. The surface area also increases as a consequence of their porous morphology, boosting their performance in real-world applications. Monolayer trichalcogenide materials are ideal for fundamental studies only due to their high activity. On the other hand, multilayer trichalcogenide materials show great stability, practicality, and balance efficiency, making them more suitable for industrial HER applications [8]. Among various reasons, the main reason for attraction in 2D trichalcogenide is they have some intrinsic properties with unique anisotropic magnetic and electrical properties [10]. Sometimes, MX_3 combines with sulfur or phosphorus materials. The sulfur atoms in these materials provide active sites for the adsorption of hydrogen, a critical step in HER. On the other hand, phosphorus atoms enhance the electronic properties, creating charge polarization that facilitates efficient electron transfer [7]. MPX_3 family has some inherent limitations. Like low intrinsic conductivity and suboptimal electro-activity, etc. To overcome all those problems and improve the electrocatalytic properties of MPX_3 materials, many strategies are being used. These comprise interlayer intercalation [10], integration of conductive materials, by the sulfur, and selenium mutually substituted [11], heteroatom doping [9, 12], forming various metal alloys with a hexagonal shape like lattice [13], and many more.

By tuning their structure, the catalytic activity can be controlled. It enables their applications in various clean energy conversion and storing devices, advanced electronic devices, as well as in sensing systems. The unique combination of catalytic activity, affordability, and versatility highlights 2D trichalcogenides as a potential game-changer in advancing sustainable water-splitting technologies.

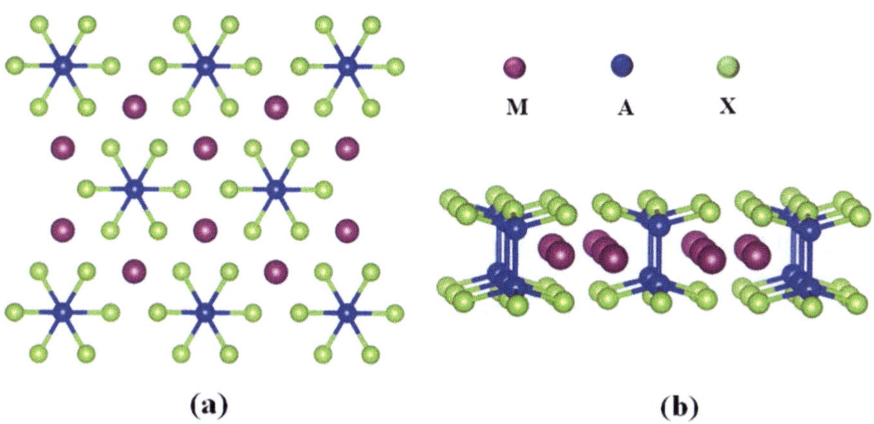

Fig. 4 **a** Top view and **b** side view of monolayer trichalcogenide. Adapted with permission from [8] Copyright (2020), American Chemical Society

3.1 *Sulfur-Based Trichalcogenide for HER*

Sulfur trichalcogenides are one of the most used 2D trichalcogenides as a promising electrocatalyst for catalysis and energy storage. These materials belong to the broader family of transition metal chalcogenides (TMCs). When S (sulfur) atoms are arranged in a layered structure with the transition metals like Ni, Fe, etc., it is known as sulfur trichalcogenide. Molybdenum Trisulfide (MoS_3) and Nickel Phosphorus Trisulfide ($NiPS_3$) are examples of such types. When sulfur is combined with other transition metals, it enhances the performance of HER due to its unique structural and electronic properties. The sulfur atoms in these materials provide more active sites for the adsorption of hydrogen, which is a critical step in HER. On the other hand, phosphorus atoms create charge polarization that facilitates efficient electron transfer [7]. Sulfur atoms are arranged in a layered structure, which facilitates efficient electron transfer by providing a high surface area. Sulfur trichalcogenide is a key player in the development of scalable, eco-friendly hydrogen production technologies for renewable energy systems as:

(1) They are cost-effective and abundant. So, sulfur trichalcogenide is a sustainable alternative to platinum-based catalysts.
(2) These materials are highly stable under acidic and neutral conditions. It is a significant advantage over other non-noble metal catalysts.
(3) A layered structure provides a high surface area and facilitates electron transfer.
(4) The tunable properties make them suitable for HER.
(5) They show a di-vacancy configuration. This lowers the activation energy that is needed for water to split, speeding up the reaction kinetics.

Metal thiophosphates are made of phosphorus (P), sulfur (S), or a mix of P_2S_5 at a high temperature (500 °C) with metal halides. Example: $FePS_3$, $CoPS_3$ and $NiPS_3$ where metal cations (e.g., Fe^{2+}, Co^{2+}, Ni^{2+}) are combined with $P_2S_6{}^{4-}$ anions. These materials exhibit very small band gaps of about only ~1.5 eV, making them potential candidates as catalysts for HER. When tested, these materials show very different results for HER. $CoPS_3$ shows the best catalytic activity and high stability toward HER activity in acidic solutions compared to $FePS_3$, and $NiPS_3$. The reduction process from H^+ to H_2 in $CoPS_3$ occurs at very low applied potentials near 200 mV. $NiPS_3$ shows lower electrocatalytic activity compared to $CoPS_3$. Also, $FePS_3$ shows low HER activity and seems highly resistant to participate in the reaction. When these metals are combined with 2D trichalcogenides, the activity towards HER depending on lower applied potential is Co > Ni > Fe [14]. Moreover, metals like Mn, Fe, Co, and Ni can improve the catalytic properties as monolayer catalyst support.

By the doping process, electrical conductivity dramatically improves. It is well known that the cost of Co is high compared to other common metals like Mn, Fe, and Ni. Fe is earth-abundant and cost-effective than Co. The combination of Fe with nickel can enhance the HER activity of $NiPS_3$-based compounds. Fe-doped $NiPS_3$ can exhibit electrocatalytic HER performance comparable to that of the Co-doped counterpart. For the optimized $Ni_{0.9}Fe_{0.1}PS_3$ compound, the overpotential at

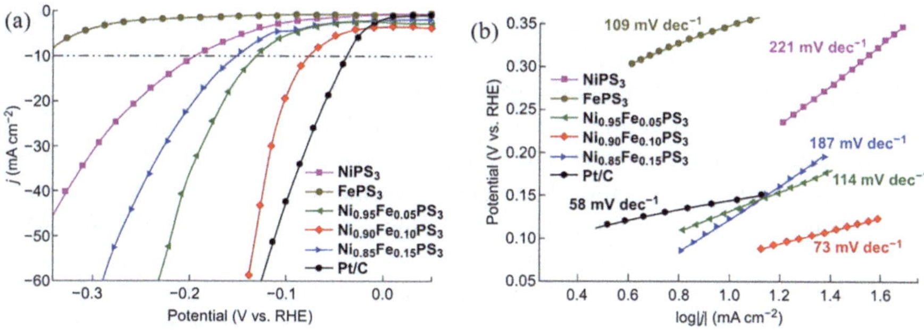

Fig. 5 **a** Current density (j) versus potential (v) curves for $FePS_3$, $NiPS_3$, and several $Ni_{0.9}Fe_{0.1}PS_3$ samples after iR correction in the HER, where 20 wt% Pt/C which is a commercial catalyst used for comparison; **b** Tafel plots for the presenting data in figure **a**. Adapted with permission from [3] Copyright (2017), ACS

10 mA cm^{-2} was 72 mV, and the Tafel slope could reach 73 mV dec^{-1} in Fig. 5a, b. By doping B and C in $NiPS_3$, the behavior of $NiPS_3$ could be altered from a semiconductor to a metal with an enhanced HER performance. If $NiPS_3$ is co-doped with C and N, it displays Pt-like activity in alkaline solution [15].

Two-dimensional (2D) materials like $CuPS_3$ monolayers have poor HER performance due to their high Gibbs free energy ($\Delta G_h = 0.44$ eV). This value is calculated by DFT. But when transition metals strongly bind to the copper (Cu) sites in $CuPS_3$, the binding energy increases. It reduces ΔG_h values and shows great catalytic activity. Anchoring metals like V (Vanadium), Fe (Iron), and Ni (Nickel) significantly reduces ΔG_h values. Especially vanadium, onto $CuPS_3$ shows the best catalytic activity, great stability, and enhanced performance for an affordable HER [16]. Another example of sulfur trichalcogenide is $NiPS_3$. It is antiferromagnetic and shows a di-vacancy configuration. When both nickel and sulfur atoms are missing (NiS di-vacancies), the catalyst works best because these defects make it easier for water to break apart and to make a bond with hydrogen. The NiS di-vacancy configuration lowers the energy that is needed for water to split, speeding up the reaction [17]. Debdyuti et al. have introduced a new promising electrocatalyst, $FePS_3$ for HER. It is a 2D material with thin nanosheet and multiple layers that shows good catalytic behavior in PBS, acidic as well as in basic media. In acidic media, metallic iron is known to be a poor electrocatalyst for HER. Once P and S are combined with Fe, the P and S sites create localized electronic environments within the $[P_2S_6]^{4-}$ unit. During the reaction, it binds with hydrogen or allows hydrogen to detach when required by gaining or losing electrons [18]. This ability makes them highly suitable for catalytic processes like hydrogen evolution reactions. Xinzhe Li et al. have prepared $NiPS_3$ flakes without using ultrasonication and used a very fast cathodic exfoliation method. $NiPS_3$ flakes exhibit a large effective ESCA with poor resistance for reaction. The catalytic active sites of P and S are fully exposed without any contamination of oxygen groups, which delivers promising electrocatalysts for HER and OER performance [19].

The Structure, active sites, and cluster play a vital role in determining the catalytic performance of 2D transition metal trichalchogenides in HER. Additionally, the electrocatalytic activity of metals-based trisulfides can be increased by doping an atom onto the structure. It provides more surface area and more active sites that facilitate the hydrogen evaluation reaction. The N doped $FePS_3$ nanosheet (N-$FePS_3$) exhibits larger values of current densities than $FePS_3$. The introduction of N atoms in $FePS_3$ facilitates the dissociation of the water molecule, which improves the HER performance by providing more catalytically active sites. The EIS plot reveals that they have lower charge-transfer resistance, which indicates faster electron transfer. The EIS data also suggests that the faradaic process is very fast and charge-transfer kinetics is very much facile, indicating better HER performance of prepared electrocatalysts [20].

Sulfur trichalcogenides stand out as versatile, anisotropic materials with unique electrical and magnetic properties driven by their layered structure. Their potential in energy storage makes them a cornerstone for next-generation technologies.

3.2 Selenium-Based Trichalcogenide for HER

Both selenium and sulfur are present in group VIA in the periodic table, which are located in the fourth period and third periods, respectively. Each of them has six valence electrons in the outermost shell, which indicates they have similar oxidation numbers. This electronic configuration gives the idea about the chemical properties of this element in their compound. Like trisulfides, 2D triselenides compounds also show promising behavior towards HER. Due to some unique features of the metal triselenides, they have received a lot of attention for HER. Despite some similarities between sulfur and selenium, they differ in a number of ways which are:

1. Selenium is more metallic than Sulfur.
2. Sulfur has less conductivity than selenium.
3. The ionization energy of selenium is lower than sulfur since it has one more shell.

For these reasons, the triselenides have some unique characteristics when compared with trisulfides toward HER [21]. Also, optimized doping might increase the catalytic performance. It is seen that both P and Se in $FePSe_3$ are significantly catalytically active compared with iron-based trisulfides, phosphides, or other layered structured selenides [15]. HER can also be shown by the $MnPSe_3$, which has lower values of onset potential in both acidic and basic media. It is proved that the onset potential is higher in basic media than in acidic media. Due to the smaller size and thin nanosheets, its performance is enhanced. The major reason is the layered structure with exfoliation $MnPSe_3$ [10].

$MoSe_2$ is the best among the tested metal dichalcogenides ($FeSe_2$, $CoSe_2$, $NiSe_2$) for HER because of its partially filled energy band. This unique electronic structure enhances its ability to interact with hydrogen intermediates during the reaction while improving the catalytic activity. When metal trichalcogenide such as Few-Layer

$FePS_3$ is decorated with thin MoS_2 nanosheets, it shows good catalytic activity toward hydrogen evaluation reaction. The prepared composite MoS_2@$FePS_3$ shows good catalytic activity when compared with $FePS_3$, MoS_2, and commercial Pt/C and all of the cases. The exfoliated $FePS_3$ nanosheets provide both improved electron conductivity and larger effective active sites, which causes faster kinetics than the bulk $FePS_3$ materials. Another reason for this high catalytic activity is that the in-situ growth of MoS_2 on the $FePS_3$ nanosheets causes intimate contact between them. This will facilitate the transfer of electrons as well as protect the nanosheets from agglomeration, providing good catalytic active sites for hydrogen evaluation [9].

4 Trichalcogenides for OER

Group 16 elements of the periodic table are commonly called chalcogens. From them, sulfur and selenium are mostly used as trichalcogenides electrocatalysts for the OER process.

4.1 Sulfur-Based Trichalcogenides for OER

Sulfur is a readily used chalcogen that shows high catalytic activity towards OER when a complex is formed with transition metal or phosphorous. In presence of phosphorous, metal trisulfides can be classified into two types based on their interlayer interactions. One type involves P_2 Pairs and M center in O_h coordination which can be formed into the layered structure by van der Waals interactions (e.g. $FePS_3$, $NiPS_3$). The other types involve P_2 pairs and M center in O_h and trigonal prismatic coordination with non-layered structures (e.g. $SnPS_3$ and $PbPS_3$). The exfoliated and exposed M center in non-layered structures serve as effective reactive sites that enhance the electrocatalytic activity for the OER process. Figure 6 shows a sandwich-like structure for some metal atoms (e.g. $NiPS_3$) while another one shows that the M center is exposed on the surface of nanosheets (e.g. $SnPS_3$). They are highly efficient for energy conversion such as OER [22].

C-F. Du et al. prepared a 2D nanohybrid $Ni_{1-x}Fe_xPS_3$ decorated on the MXene nanosheets surface prepared by in situ solid-state process in low temperature. This optimized nanohybrid shows a lower value of overpotential of 282 mV with a facile electron transfer. The value of the Tafel slope is only 36.5 mV/decade which is lower than that of IrO_2. The iron doping increases the active site which improves electrical conductivity and charge transfer. Here, MXene was used as support and the optimized compounds provided a larger active surface area to enhance catalytic activity. Also, the EIS data shows low charge transfer resistance which increases the overall OER activity [23].

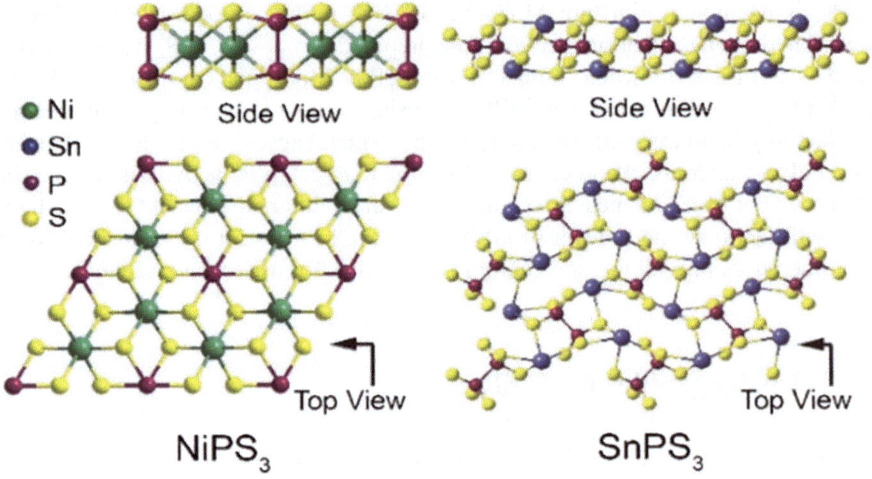

Fig. 6 The upper and side views of the $NiPS_3$, and $SnPS_3$. Adapted with permission from [22] Copyright (2023), Wiley–VCH GmbH

Another promising electrocatalyst for OER uses graphite which is sandwiched with $FeNiPS_3$. It shows a reduced value of overpotential (223 mV) with a fast electron transfer. The value of Tafel slope reaches about 58 mV/decade. The prepared electrode $FeNiPS_3/GL$ shows outstanding performance towards the OER process. Coupling effect, structures like the sandwich and doping with active nickel faster the transfer of electrons as well as inhibited the agglomeration of $FeNiPS_3$. This effect provides a large number of the effective active sites that enhance the activity of the electrocatalyst towards the evolution of oxygen and make this catalyst more efficient [24].

The catalytic performance of the trichalcogenide-based electrocatalyst for the OER process can be enhanced by manipulating the spin exchange interaction through doping and transfer of electrons in selected spin. Between the $FePS_3$ and $Co_xFe_{1-X}PS_3$ crystals, the latter one shows high catalytic activity due to the doping of Co. The $FePS_3$ shows a very high overpotential, which is about 620 mV whereas $Co_xFe_{1-X}PS_3$ crystal shows only 310 mV at 10 mA cm^{-2} current density. The condition can be further improved by using Ni foam since it provides a high surface area with an open porous surface. The lower value of the Tafel slope indicates that the electrocatalyst shows facile electron transfer on Ni foam. $FePS_3$ contains only the Fe^{2+} valency but the Co_x $Fe_{1-X}PS_3$ crystals show a mixed valency of Fe^{2+} and Fe^{3+}. This results from surface defects or creates vacancies during doping, where the doping of cobalt breaks the structural symmetry of $FePS_3$ and enhances its catalytic activity. The mixed valency also improves the electronic conductivity by electronic delocalization. Here, the Co^{2+} are considered a high spin state ($t_{2g}5e_g^2$) when they stabilized in the O_h sites. The EIS data also shows that the cobalt-doped crystals have lower charge transfer resistance than the previous one. The reason for that is the Co^{2+} ($t_{2g}5e_g^2$) cation, which is in a high spin state and contains three unpaired electrons in the d orbital. It may facilitates the bond formation between O H^- and Co atom which advances the OER process [25]. Another promising electrocatalyst is the highly exposed $NiPS_3$/Ni foam which is prepared in situ. It is highly efficient for OER as well as HER. This electrode shows a wide anodic peak due to the conversion of Ni^{2+} to Ni^{3+}, confirming the formation of the oxide or hydroxide of nickel on the surface of electrocatalysts. It generates a large ESCA that enables quicker charge transfer in the OER process. EIS data shows that, the $NiPS_3$/Ni has lower values of R_{ct} than the other composites of nickel which indicates facile charge transfer kinetics. For high catalytic activity towards the OER process, stability of the prepared materials is very important for practical applications. The stability depends on the retention of a nickel oxide or hydroxide layer over the surface of $NiPS_3$/Ni nanosheets. These nanosheets will provide the effective active sites for the oxidation of water in OER process. After several CV analyses, it is confirmed that forming an oxide or a hydroxide layer of nickel facilitates the OER process significantly [26].

The oxygen evolution reaction (OER) with precious metals like Pt has similar electrochemical behavior as 2D materials like TMDs. The world scientific community is always searching for cost-effective materials with high-performance. Martinez et al. [27] prepared some electrocatalysts for Oxygen evolution reaction (OER) at a lower cost than traditional electrocatalysts. This drawback of the traditional electrocatalyst has led the world scientific community to search for alternatives such as metal phosphide-based trichalcogenide. So, the metal trichalcogenides have great importance for OER performance. Among the various sulfur-based trichalcogenides, the $CoPS_3$ are most promising due to a lower value of onset potential 0.84 V at 10 mA/cm^2. However, other prepared materials show poor OER activities. $CoPS_3$ contains cobalt ion which is the main reason for such good catalytic activity. According to the previous work, it is confirmed that when cobalt is stable on the surface of GCE it provides a large surface area and lowers the interaction energy for OER. Li et al. [19] prepared the large-sized and thin-layered flakes of $NiPS_3$ that show high catalytical activity towards OER. The completely exposed multifunctional active sites provide a large ESCA, improved mass-transport as well as electron transport. As a result, the large-sized thin layered $NiPS_3$ serve as a promising bifunctional electrocatalysts for OER [1, 6, 15, 28, 29].

The formation of a strong covalent bonds between metals and metalloids is the primary reason for good catalytic behavior of these metal trichalcogenides towards OER activity. This bond gives them extra stability and the diverse crystal structures. Metal trichalcogenides provide a higher specific active surface area than bulk, which is important for energy conversion devices [30]. They can be synthesized in various desired shapes and sizes such as nanosheets, nanoplates, nanoneedles, and other microstructures as well. Also, they can be a good replacement for precious noble metals which are used as electrocatalysts in the OER process [31].

4.2 Selenium-Based Trichalcogenides for OER

Like sulfur-based trichalcogenides, selenium-based trichalcogenides also show high catalytic activity towards water splitting. This is due to their anisotropic properties that are imposed on by their unique layered structure. They can be used as a host for different intercalation compounds. Because of their magnetic and antiferromagnetic properties, they form a network structure like a honeycomb. Their wide band gap ranges indicate that they can be used in various optoelectronic devices with a large wavelength range [32–35]. Only a few of the metals with various oxidation states, such as M(I/II/III/IV) form an effective composite with chalcogenides for OER activity. The prepared composite shows high catalytic activity towards OER which is shown in Fig. 7 [6].

Owing to the low overpotentials of IrO_2 and RuO_2, they can't be used commercially due to their high cost. It paved the way to prepare other materials for water splitting such as 2D triselenides-based catalysts. Among various metals, $FePSe_3$ shows better OER activity and has a higher onset potential (+1.8 V) than Pt [21].

Another work is done for layered iron selenophosphate ($FePSe_3$) composite with rGO. It shows good catalytic activity towards OER due to a large ESCA and higher effective active sites [36]. There are so many metals in the periodic table but some of them don't form an effective composite with chalcogenides [6].

The exfoliated structures can provide more catalytically active sites than normal nanosheets of $FePSe_3$. The overall electrocatalytic performance is improved as a result of reduced Tafel slope value (58 mV/decade). The optimized exfoliated structure shows the in-plane defects on the nanosheet layers of $FePSe_3$ through X-ray

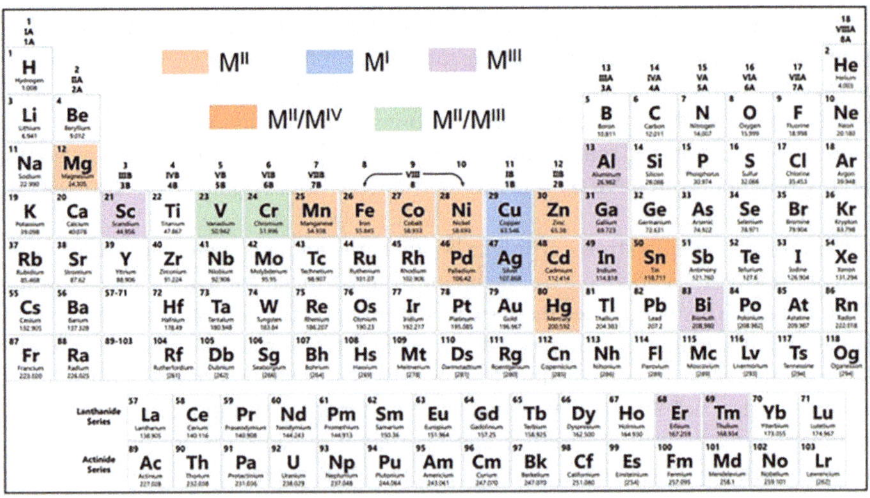

Fig. 7 A schematic diagram of the periodic table in which the colored cell exhibits the elements to prepare various metal-based trichalcogenides materials for OER has been reported. Adapted with permission from [6] Copyright (2022), ACS

spectroscopy and HRTEM. In that case, the catalytic activity is enhanced due to the heterostructure formed between the $FePSe_3$ and FeOOH [37]. Other heterostructures called $NiPS_3$@NiOOH core-shells have been reported as promising electrocatalysts for OER. Because they have very extreme stability for over 160 h and a lower value of onset potential (250 mV). The large number of active metallic edges as well as the introduction of P and S to $NiPS_3$ causes structural disorder and increases defect sites on the structures. This facilitates the transfer of electrons on the NiOOH shell and enhances overall OER activity [5]. When compared with the analogous MPS_3; $MPSe_3$ needed a higher potential for oxidation in most cases indicating its high stability than MPS_3. Due to their promising catalytic activity, they are used in various fields of electrochemistry such as biosensing, and electroactive labels [27]. Now, 2D metal phosphorous trichalcogenides ($MPCh_3$) have gained significant attention in sustainable energy storage and conversion. This is because of their distinct physical and chemical characteristics, such as adjustable energy band gap, significant specific surface area, and abundant active sites [6]. The bandgaps of metal trichalcogenide were also calculated experimentally, which ranges from 1.5 eV to 3.7 eV for $FePSe_3$ and $ZnPS_3$, respectively. $MnPSe_3$ exhibits superior performance with a band gap of 2.9 eV, showing a relation between theoretical and experimental bandgaps [38]. Thus, triselenide can be used as a potential electrocatalyst for OER and will find a broad range of applications in the future.

5 Challenges and Future Perspectives of Trichalcogenide

Trichalcogenides are now used in various fields other than OER or HER. They are very promising electrocatalysts for supercapacitors, metal-ion batteries, electrocatalysis, and Li–S batteries. They are also used in desalination and electrocatalysis. Their high catalytic activity allow them to act as photocatalysts. A schematic diagram for their applications is shown in Fig. 8.

This chapter mainly discusses the 2D trichalcogenide as a promising electrocatalyst for water splitting. Hence, its future perspective discussed in this section dominates. In the past few decades, energy demand has increased and consumption of non-renewable energy sources like fossil fuels has rapidly grown. Therefore, developing new, globally favorable, and high-performing renewable energy storage systems is crucial. From electrolysis, the hydrogen produced from water splitting serves as a sustainable alternative to fossil fuels, aiding in reducing carbon emissions as a clean and renewable energy carrier. It plays a vital role in developing efficient energy storage systems for renewable sources like solar and wind. Advanced OER or HER catalysts help to lower the energy costs and improve the scalability of green hydrogen production. As a cornerstone of hydrogen-based energy systems, HER drives progress in achieving global energy sustainability and decarbonization goals. Trichalcogenides are cost-effective alternatives to precious metal catalysts like platinum. Their earth abundance and chemical stability in harsh acidic and alkaline

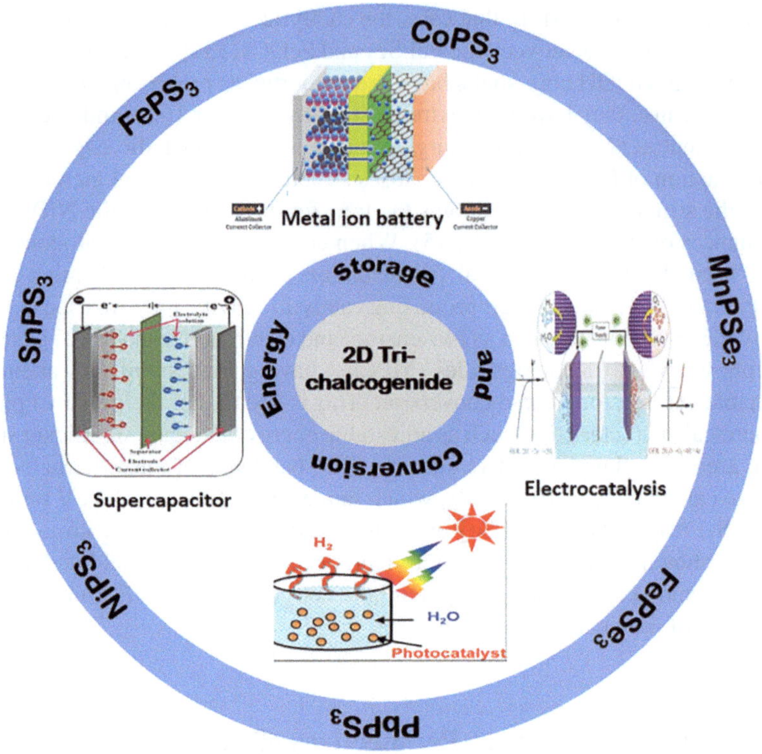

Fig. 8 A schematic diagram of the various applications of metal trichalcogenide for energy conversion and storage

environments make them highly attractive for practical HER applications. Trichalcogenides, a unique family of layered materials, have gained significant attention in the realm of hydrogen evolution reaction (HER) due to their fascinating structural, electronic, and catalytic properties. These materials exhibit excellent conductivity, tunable electronic structures, and abundant active sites, which make them highly efficient catalysts for HER. The tunable bandgap of trichalcogenide allows for enhanced catalytic activity by optimizing hydrogen adsorption energies.

Doping or alloying these materials with other metals or non-metals can further improve their catalytic efficiency and stability. For instance, incorporating transition metals into trichalcogenides creates additional active sites.

Consequently, trichalcogenide compounds are paving the way for efficient, sustainable hydrogen production in green energy technologies by water splitting. Optimizing their catalytic performance through compositional modifications, defect engineering, and heterostructure formation can further improve HER efficiency and scalability. 2D MPX_3 crystals with an ultrathin nature, also contain defect interface as well as an electrochemically active large surface area, tunable spacing between interlayers, enhanced conductivity, and a robust interface. Hence, they are promising alternatives for converting or storage energy [15].

Despite the various applications and advantages of trichalcogenide, there are some challenges as well. Additionally, there are some future perspectives to overcome these challenges. For energy conversion and storage, the major challenges are as follows:

- Poor electrical conductivity of trichalcogenide materials reduces their utilization.
- The low air stability of these materials restricts their uses in energy applications.
- The trichalcogenides are not completely ready yet for practical applications.
- Very few works are done to identify the catalytic active sites and an appropriate catalytic mechanism.
- The accurate synthesis of trichalcogenide nanosheets also faced some difficulties.

Based on these challenges, the future perspective will be:

- The electrical conductivity will be raised by the formation of heterojunction in nanosheets.
- The prepared nanosheets can be modified.
- The practical applicability can be enhanced by analyzing some factors, such as power and loading density.
- Defect engineering enhances the active sites and gives a clear idea about the catalytic mechanism.
- In the future, optimizing synthesis protocols for trichalcogenides nanomaterials will be desirable for practical application.

In summary, these new materials offer a new path for energy storage as well as conversion owing to their unique structure, large active sites, and tunable bandgaps. After covering the major challenges, 2D trichalcogenides will be a prominent material for overall water splitting.

Acknowledgements This work was supported by the Department of Chemistry, Jagannath University, Dhaka-1100, Bangladesh.

References

1. Du, C.-F.: Layered Trichalcogenidophosphate: A New Catalyst Family for Water Splitting. (2018)
2. Jenjeti, R.N., Austeria, M.P., Sampath, S.: Alternate to molybdenum disulfide: a 2D, few-layer transition-metal thiophosphate and its hydrogen evolution reaction activity over a wide pH range. ChemElectroChem **3**, 1392–1399 (2016)
3. Song, B., et al.: Tuning mixed nickel iron phosphosulfide nanosheet electrocatalysts for enhanced hydrogen and oxygen evolution. ACS Catal. **7**, 8549–8557 (2017)
4. Dangol, R., et al.: Few-layer $NiPS_3$ nanosheets as bifunctional materials for Li-ion storage and oxygen evolution reaction. Nanoscale **10**, 4890–4896 (2018)
5. Konkena, B., et al.: Metallic $NiPS_3$ @NiOOH core-shell heterostructures as highly efficient and stable electrocatalyst for the oxygen evolution reaction. ACS Catal. **7**, 229–237 (2017)

6. Wang, H., et al.: 2D metal phosphorous trichalcogenides (MPCh3) for sustainable energy storage and conversion: nanoarchitectonics and advanced applications. Adv. Funct. Mater. **34**, 2407432 (2024)
7. Gusmão, R., Sofer, Z., Pumera, M.: Metal phosphorous trichalcogenides (MPCh3): from synthesis to contemporary energy challenges. Angew. Chem. Int. Ed. **58**, 9326–9337 (2019)
8. Sen, P., Alam, K., Das, T., Banerjee, R., Chakraborty, S.: Combinatorial design and computational screening of two-dimensional transition metal trichalcogenide monolayers: toward efficient catalysts for hydrogen evolution reaction. J. Phys. Chem. Lett. **11**, 3192–3197 (2020)
9. Huang, H., et al.: Few-layer FePS3 decorated with thin MoS2 nanosheets for efficient hydrogen evolution reaction in alkaline and acidic media. Appl. Surf. Sci. **525**, 146623 (2020)
10. Gusmão, R., Sofer, Z., Pumera, M.: Exfoliated layered manganese trichalcogenide phosphite (MnP X3, X = S, Se) as electrocatalytic van der Waals materials for hydrogen evolution. Adv. Funct. Mater. **29**, 1805975 (2019)
11. Alam, K., Das, T., Chakraborty, S., Sen, P.: Finding the catalytically active sites on the layered tri-chalcogenide compounds $CoPS_3$ and $NiPS_3$ for hydrogen evolution reaction. Phys. Chem. Chem. Phys. **23**, 23967–23977 (2021)
12. Guo, Y., et al.: XPS analysis of $Fe_xNi_yPS_3$vdW materials used in the hydrogen evolution processes. ChemPhysChem **25**, e202400039 (2024)
13. Abidi, N., Sahu, A., Raybaud, P., Steinmann, S.N.: Electrochemical potential-dependent stability and activity of MoS_3 during the hydrogen evolution reaction. ACS Catal. **13**, 15290–15300 (2023)
14. Coleman, N., Liyanage, I.A., Lovander, M.D., Leddy, J., Gillan, E.G.: Facile solvent-free synthesis of metal thiophosphates and their examination as hydrogen evolution electrocatalysts. Molecules **27**, 5053 (2022)
15. Wang, F., Sendeku, M.G.: Two-dimensional metal phosphorus trichalcogenide nanostructure for sustainable energy conversion. In: Cheng, Y., Houston, L., Liu, J.L. (eds.) ACS Symposium Series, vol. 1421, pp. 1–25. American Chemical Society, Washington, DC (2022)
16. Sun, Y., Huang, A., Li, Z., Fu, Y.-Q., Wang, Z.: Transition metal atoms anchored on CuPS3 monolayer for enhancing catalytic performance of hydrogen evolution reactions. Electrocatalysis **13**, 494–501 (2022)
17. Han, H.G., Choi, J.W., Son, M., Kim, K.C.: Unlocking power of neighboring vacancies in boosting hydrogen evolution reactions on two-dimensional NiPS3 monolayer. eScience **4**, 100204 (2024)
18. Mukherjee, D., Austeria, P.M., Sampath, S.: Two-dimensional, few-layer phosphochalcogenide, FePS3: a new catalyst for electrochemical hydrogen evolution over wide pH range. ACS Energy Lett. **1**, 367–372 (2016)
19. Li, X., et al.: High-yield electrochemical production of large-sized and thinly layered $NiPS_3$ flakes for overall water splitting. Small **15**, 1902427 (2019)
20. Zhang, H., et al.: Nitrogen-incorporated iron phosphosulfide nanosheets as efficient bifunctional electrocatalysts for energy-saving hydrogen evolution. Ionics **28**, 3927–3934 (2022)
21. Gusmão, R., Sofer, Z., Sedmidubský, D., Huber, Š, Pumera, M.: The role of the metal element in layered metal phosphorus triselenides upon their electrochemical sensing and energy applications. ACS Catal. **7**, 8159–8170 (2017)
22. Wang, H., et al.: Exfoliated 2D layered and nonlayered metal phosphorous trichalcogenides nanosheets as promising electrocatalysts for CO_2 reduction. Angew. Chem. **135**, e202217253 (2023)
23. Du, C., et al.: Self-assemble and in situ formation of $Ni_{1-x}Fe_xPS_3$ Nanomosaic-decorated MXene hybrids for overall water splitting. Adv. Energy Mater. **8**, 1801127 (2018)
24. Wei, L., Huang, S., Zhang, Y., Ye, M., Li, C.C.: Enhancing the coupling effect in a sandwiched $FeNiPS_3$/graphite catalyst derived from graphite intercalation compounds for efficient oxygen evolution reaction. J. Mater. Chem. A **10**, 11793–11802 (2022)
25. Huang, C., et al.: Manipulating spin exchange interactions and spin-selected electron transfers of 2D metal phosphorus trisulfide crystals for efficient oxygen evolution reaction. Adv. Funct. Mater. **33**, 2305792 (2023)

26. Fang, L., et al.: *In situ* formation of highly exposed $NiPS_3$ nanosheets on nickel foam as an efficient 3D electrocatalyst for overall water splitting. Sustain. Energy Fuels **5**, 2537–2544 (2021)
27. Mayorga-Martinez, C.C., et al.: Layered metal thiophosphite materials: magnetic, electrochemical, and electronic properties. ACS Appl. Mater. Interfaces **9**, 12563–12573 (2017)
28. Xie, J., et al.: Zirconium trisulfide ultrathin nanosheets as efficient catalysts for water oxidation in both alkaline and neutral solutions. Inorg. Chem. Front. **1**, 751–756 (2014)
29. Xue, S., Chen, L., Liu, Z., Cheng, H.-M., Ren, W.: $NiPS_3$ nanosheet-graphene composites as highly efficient electrocatalysts for oxygen evolution reaction. ACS Nano **12**, 5297–5305 (2018)
30. Peng, L., Shah, S.S.A., Wei, Z.: Recent developments in metal phosphide and sulfide electrocatalysts for oxygen evolution reaction. Chin. J. Catal. **39**, 1575–1593 (2018)
31. Wu, X., et al.: Morphology-controllable synthesis of Zn–Co-mixed sulfide nanostructures on carbon fiber paper toward efficient rechargeable zinc-air batteries and water electrolysis. ACS Appl. Mater. Interfaces **9**, 12574–12583 (2017)
32. Kuo, C.-T., et al.: Exfoliation and Raman spectroscopic fingerprint of few-layer NiPS3 Van der Waals crystals. Sci. Rep. **6**, 20904 (2016)
33. Zeng, H., Ye, T., Cheng, P., Yao, D., Ding, J.: Raman spectroscopy investigation on the pressure-induced structural and magnetic phase transition in two-dimensional antiferromagnet $FePS_3$. Chin. Phys. B **31**, 056109 (2022)
34. Li, X., Wu, X., Yang, J.: Half-metallicity in $MnPSe_3$ exfoliated nanosheet with carrier doping. J. Am. Chem. Soc. **136**, 11065–11069 (2014)
35. Guo, Y., et al.: Distinctive in-plane cleavage behaviors of two-dimensional layered materials. ACS Nano **10**, 8980–8988 (2016)
36. Mukherjee, D., Muthu Austeria, P., Sampath, S.: Few-layer iron selenophosphate, $FePSe_3$: efficient electrocatalyst toward water splitting and oxygen reduction reactions. ACS Appl. Energy Mater. **1**, 220–231 (2018)
37. Zhu, W., et al.: Exfoliation of ultrathin $FePS_3$ layers as a promising electrocatalyst for the oxygen evolution reaction. Chem. Commun. **54**, 4481–4484 (2018)
38. Oliveira, F.M., Paštika, J., Pires, L.S., Sofer, Z., Gusmão, R.: Photoelectrochemical activity of layered metal phosphorous trichalcogenides for water oxidation. Adv. Mater. Interfaces **8**, 2100294 (2021)

MXenes: A Class of 2D Electrocatalysts for Water Splitting

Mahmoud A. Hefnawy and Shymaa S. Medany

Abstract MXenes, a unique class of two-dimensional in nature (2D) transition metal carbides, carbonitrides, and nitrides, have been identified as promising electrocatalysts for water-splitting purposes. Their exceptional electrical conductivity, high surface area, and tunable surface chemistry make them highly efficient for both the hydrogen (HER) and oxygen (OER) evolution reactions. Various modification strategies, including elemental doping, surface termination engineering, and heterostructure formation, have exhibited considerable improvement in their electrocatalytic activity. MXene-based composites with transition metal oxides, sulfides, and layered double hydroxides have significantly improved catalytic efficiency and stability. However, challenges such as oxidation susceptibility, structural degradation, and limited long-term stability need to be addressed for their practical application. Advances in synthesis techniques, defect engineering, and computational modeling are expected to optimize their performance further. This chapter presents a complete overview of MXene synthesis, properties, and electrocatalytic purposes in water splitting, showing their potential aimed at transforming clean energy technology.

Keywords MXenes · Electrocatalysis · Hydrogen evolution reaction (HER) · Water splitting · Oxygen evolution reaction (OER)

1 Introduction

There are serious worries regarding global energy sustainability due to the increasing world population, the rising demand for energy, and the impact of climate change [1–3]. By employing secure, fossil-free methods to produce vital fuels and chemicals, we can lower CO_2 emissions while still supplying the necessary feedstock for

M. A. Hefnawy (✉) · S. S. Medany
Chemistry Department, Faculty of Science, Cario University, Giza, Egypt
e-mail: maadel@cu.edu.eg; maahefnawy@gmail.com

S. S. Medany
e-mail: shymaasamir80@cu.edu.eg

O. P. Pandey et al. (eds.), *2D Materials for Electrochemical Water Splitting*, Engineering Materials, https://doi.org/10.1007/978-981-95-2003-9_7

the products we use daily [4, 5]. Electrocatalysis is considered a successful strategy to split water into hydrogen and oxygen. The effectiveness of this electrochemical process is influenced by various factors, such as the stability, efficiency, and cost of the electrocatalyst employed [6, 7]. In addition to being stable and inexpensive, an efficient electro-catalyst should also have great charge transferability and many active sites. Practically, the applied voltage for water splitting remains significantly more than the conventional reaction potential. Two primary categories of explanations exist for this high reaction potential: reaction and device limitations. The factors influencing the reaction encompass intricate electron and ion transport mechanisms, the activation energy associated with the electrode material, the entrapment of ions and gas molecules, as well as the dynamics at the electrolyte and electrode interface. The characteristics of a device that splits water encompass wire and the resistance and surface area of the electrode, the amount of solution, the morphology of the electrodes (whether crystalline or amorphous), electrolyte diffusion, among various other factors. All of these influences lead to a rise in reaction voltage, which in turn makes the total response less efficient. Pt is look upon as the most effective electrocatalyst for HER in acidic conditions because it has a low Tafel slope and nil overpotential. Nonetheless, the high cost associated with Pt restricts its industrial application, necessitating the development of a catalyst that contains minimal precious metals while maintaining effective electrocatalytic activity.

Similarly, iridium and ruthenium oxides are considered to be the most efficient electrocatalysts for oxygen evolution reaction (OER), but they are also rather expensive. In the search for alternative catalysts that are more cost-effective, a wide variety of materials have been exploited as electro-catalysts for the process of water splitting. The chemicals that make up these materials are derived from transition metals (Fe, Co, Ni, Mn, and Mo), and they include phosphides, sulfides, hydroxides, oxides, and selenides within their composition. Furthermore, despite the fact that each of these materials has strong electrocatalytic properties, their total efficiency is restricted due to the fact that they have low conductivity and a tiny surface area that is efficient.

Due to the exceptional mechanical, electrical, and physicochemical properties that MXenes possess, they have become increasingly utilized in a wide variety of applications ever since the discovery of $Ti_3C_2T_x$ in 2011. MXenes are an important group of two-dimensional transition metal carbides, nitrides, and/or carbo-nitrides. The typical structure of MXenes is $M_{n+1}X_nT_x$ ($n = 1$–3), where M stand for a transition metal, X stands for carbon or nitrogen, and T_x stands for a surface termination group such as –OH, –O, or –F. MXenes are utilized in a wide variety of applications.

When an A group layer is removed from its parent MAX phase, the result is the formation of MXenes. As a result of the metallic bonding nature of the M and A elements in the Max phase, selective etching may be used to remove the layer A from the phase, which ultimately leads to the production of two-dimensional layered MXene structure. A powerful etchant, such as HF, was used to etch an A component from the MAX phase at the beginning of the design process.

It has been established that MXenes offer a substantial potential for the storage and conversion of energy needs. However, the application of MXenes and related heterostructures in water electrolysis, particularly for hydrogen and oxygen evolution processes, provides prospects for additional research. This is despite the fact that multiple studies have addressed these topics. The enormous surface area of MXenes can be attributed to the presence of layered structures comparable to those seen in MXenes. The transition metals that are a component of their composition are responsible for imparting metallic properties and transitional electronic states. Furthermore, the presence of surface end groups is a factor that adds to the hydrophilic qualities of these substances. Because of all of these characteristics working together, MXenes are extremely effective electrocatalysts for HER, OER, and total water splitting. Alterations to the material's inherent surface characteristics can arise as a result of modifications such as the insertion of functional groups using different etchants. These changes can also influence the work functions of MXenes, which can vary depending on the transition metals that are present in the material. It is possible to vary the spacing between layers by doping with a variety of components, which will result in an increase in the total surface area and the creation of additional catalytic spots. Electrocatalytic performance may be improved by combining a variety of electrocatalysts with MXenes as a conductive substrate. This combination reduces buildup, increases surface area, and produces better results. The presence of functional groups on the surface of MXene enhances the chemical bonding and dissemination capabilities of electrocatalysts. Making use of a variety of electrocatalysts in order to create heterostructures using MXenes has the potential to open up new research pathways and may also significantly improve the electrocatalytic activity of these materials. The progression of hydrogen (H_2) evolution photocatalytically through water dissociation is regarded as an ideal approach for hydrogen production. This process relies on the development of semiconductor photocatalysts. Significant efforts have been made to enhance the efficiency of titanium dioxide (TiO_2) as a photocatalyst. However, TiO_2, which has a band gap of ca. 3.2 eV, falls within the ultraviolet range, limiting its effectiveness. Hence, it is crucial to develop additional photocatalysts with narrower band gaps to facilitate photocatalytic H_2 evolution using visible light. Several challenges hinder the efficacy of photocatalysts. The challenges encompass electron–hole recombination, backward reactions, the limitations of certain photocatalysts in harnessing visible light, and the issue of photo-corrosion. To address these issues, researchers often use precious metals as co-catalysts. However, the scarcity and extraordinary price of noble metals present another challenge. Alternative methods and materials for enhancing photocatalytic activity are being explored. One promising option is the usage of MXenes as co-catalysts, which exhibit several advantageous properties for hydrogen evolution reactions (HER) [8–13].

2 Synthesis of MXenes

MXenes, an innovative category of two-dimensional transition metal nitrides, carbides, and carbonitrides, are synthesized from MAX phases, characterized by the standard expression Mn + 1AXn (where M represents a transition metal, A denotes an A-group element, and X signifies carbon and/or nitrogen) [14–16]. The synthesis of MXenes involves selective engraving of the A element from the MAX phase, typically via strong acids or fluoride-containing solutions. This section details the common synthesis methods, reaction mechanisms, and post-synthesis treatments of MXenes.

2.1 *Selective Etching Methods*

2.1.1 HF Etching Method

The utmost conventional and broadly used means for MXene synthesis involves the use of hydrofluoric acid (HF). In this method, the MAX phase powder is dipped in an aqueous HF solution at controlled temperatures (typically 25–60 °C) for a specific duration (ranging from hours to days). The reaction can be summarized as [17]:

$$M_n + 1AX_n + HF \rightarrow Mn + 1Xn + A_xF_y + H_2 \tag{1}$$

This progression efficiently removes the A element, resulting in formation of exfoliated MXene layers. However, HF-based synthesis poses safety concerns due to the toxicity and corrosiveness of HF. The concentration of HF, reaction temperature, and time suggestively affect the quality and morphology of the obtained MXenes.

Additionally, different MAX phases require specific HF concentrations. For example, Ti_3AlC_2 typically needs a 40–50% HF solution, while Ti_2AlC can be etched at a lower concentration. Control over etching parameters plays a vital role in adapting MXene features for specific purposes.

2.1.2 In-Situ HF Generation (LiF + HCl) Method

To mitigate the hazards associated with concentrated HF, an alternative method involves in-situ generation of hydrogen flouride by reacting lithium fluoride (LiF) with hydrochloric acid (HCl). The general process follows [18, 19]:

$$LiF + HCl \rightarrow HF + LiCl \tag{2}$$

This tactic has been extensively used to manufacture $Ti_3C_2T_x$ MXene, offering better control over etching conditions and reducing HF exposure risks. Moreover, the intercalation of Li^+ ions enhances exfoliation, leading to higher yield and fewer structural defects. The in-situ HF generation method also allows for better dispersion of MXenes in aqueous and organic solvents, enhancing their usability in various applications.

2.1.3 Electrochemical Etching

Electrochemical etching offers a fluoride-free alternative by utilizing anodic oxidation of MAX phases in an acidic electrolyte. In this method, an external voltage is applied, promoting the choosy dissolution of the A component without damaging the underlying MXene layers. This procedure allows exact control over the etching progression and enhances the purity of the resulting MXene [17, 20].

The electrochemical method also provides an environmentally friendly and scalable approach. It can be regulated by fluctuating factors such as applied voltage, current density, and electrolyte composition. A common electrolyte used is sulfuric acid, or a mixture of organic and inorganic acids. This method has gained attention because of its potential for bulky-scale MXene manufacture with reduced environmental impact.

2.1.4 Molten Salt Etching

A recent advancement in MXene synthesis involves using molten salt etching, where the MAX phase is treated by a mixture of halide salts (e.g., $ZnCl_2$, NaCl, KCl) at elevated temperatures (~500–700°C). The molten salts simplify the elimination of the A element while preserving the structural integrity of the MXene. This approach yields MXenes with improved crystallinity and fewer defects compared to traditional etching methods [21–23].

Molten salt etching has advantages such as producing MXenes with higher oxidation resistance and better thermal stability. This creates them appropriate for applications in extreme environments, for example high-temperature catalysis and aerospace applications.

2.2 Post-synthesis Treatment

After etching, MXenes require thorough washing and delamination to obtain high-quality 2D sheets. post-synthesis treatments include [17, 24]:

i. **Washing and Neutralization**: The etched MXene is splashed with deionized water, typically through centrifugation, until the pH reaches neutral (~6–7). This step removes residual acid and reaction byproducts. Repeated washing cycles ensure high-purity MXene sheets are free from contaminants.
ii. **Exfoliation**: Intercalation with organic molecules (e.g., urea, dimethyl sulfoxide) or sonication aids in separating multilayered MXene into single or few-layer sheets. The exfoliation efficiency depends on factors such as the intercalating agent, sonication power, and duration.
iii. **Freeze-Drying**: To prevent restacking of MXene sheets, the material is often freeze-dried, which helps maintain its 2D structure for subsequent applications.

The drying method also influences the MXene's surface chemistry and electrical conductivity.

iv. **Surface Modification**: MXenes tend to oxidize over time, which can degrade their features. Surface functionalization with organic ligands, polymers, or metal ions can enhance stability and tailor possessions for specific uses such as biosensing, energy storage, and catalysis.

2.3 *Effect of Synthesis Parameters on MXene Properties*

The final properties of MXenes depend significantly on the synthesis parameters, including:

i. **Etching Time**: Shorter etching times lead to incomplete A-element removal, while excessive etching can degrade MXene structure.
ii. **Acid Concentration**: High HF concentrations lead to rapid etching but may introduce defects, whereas lower concentrations ensure controlled removal of A-element.
iii. **Temperature**: Elevated temperatures accelerate etching but can also affect the oxidation state and electronic properties of MXenes.
iv. **Post-Synthesis Treatments**: Techniques like ultrasonic exfoliation, intercalation, and drying methods impact MXene stability and functionalization.

Additionally, the selection of etching conditions influences MXene's structural and electronic properties. The presence of residual ions from etching solutions can modify the surface terminations (–OH, –F, –O), which in turn affect electrical conductivity, hydrophilicity, and chemical reactivity. The oxidation resistance of MXenes is extremely hooked on production conditions, where exposure to high-temperature processing or excessive sonication may lead to degradation.

Layer thickness and interlayer spacing also play a critical role in determining the electrochemical performance of MXenes. For applications in energy storage and catalysis, controlling the interlayer spacing through intercalation strategies can enhance ion diffusion and improve performance. Moreover, post-synthesis annealing under a passive or reducing atmosphere can fine-tune the electronic structure of MXenes, optimizing their conductivity and catalytic activity.

Furthermore, defect engineering during synthesis, such as introducing vacancies or dopants, can tailor the microelectronic and electrochemical features of MXenes for specific purposes. By optimizing these synthesis parameters, researchers can design MXenes with targeted functionalities, flagging the approach for their widespread adoption in energy storing, water purification, and biomedical applications.

3 MXene Properties

Materials that are characterized by the general formula $M_xX_əT_o$ (with M as a transition metal, X as carbon or nitrogen, and T as surface terminations like –OH, –O, or –F) demonstrate a distinctive amalgamation of electronic, thermal, mechanical, and chemical features. These properties have led to considerable interest in MXenes for their potential utilization in energy storage, catalysis, electronic devices, and water purification. This document discovers the fundamental properties of MXenes in detail (see Fig. 1).

3.1 Electronic and Conductive Properties

The mostly significant advantage of MXenes is their high electrical conductivity, which is attributed to the existence of transition metals in their structure. Unlike other 2D materials, e.g. graphene oxide, which suffer from low conductivity due to oxygen functional groups, MXenes retain a metallic nature even with surface terminations. The high conductivity of MXenes, reaching values as high as 20,000 S/cm for $Ti_3C_2T_o$, makes them appropriate for solicitations in supercapacitors, batteries, and electromagnetic shielding.

The tunability of MXenes' electronic properties depends on their composition, termination groups, and synthesis methods. The presence of surface terminations can expressively alter the electronic band structure, resulting in either metallic or

Fig. 1 Representing various MXene properties [25]

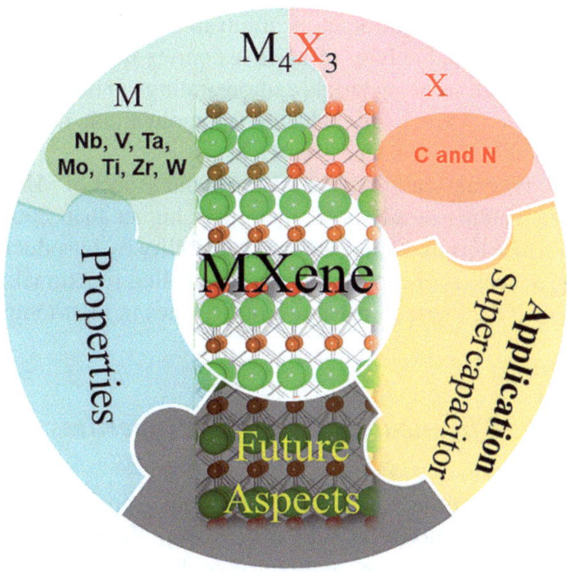

semiconducting behavior. Analytical assessments and empirical investigations have shown that varying the transition metal or modifying surface functional groups can shift the work function and Fermi level, making MXenes highly versatile materials for electronic applications.

3.2 Mechanical Properties

MXenes possess exceptional mechanical properties, such as robustness, versatility, and durability. The mechanical strength is primarily influenced by M-X bond's nature within the layered structure. For instance, Ti_3C_2-based MXenes have been found to possess a Young's modulus of approximately 330 GPa, which is comparable to other strong 2D materials such as graphene and MoS_2.

Despite their high mechanical strength, MXenes maintain a high degree of elasticity, allowing them to be integrated into flexible electronic devices and wearable sensors. The presence of surface terminations can influence the mechanical properties by affecting interlayer interactions. By controlling the interlayer spacing and termination groups, MXenes can be engineered to optimize their mechanical features for definite purposes.

3.3 Thermal Properties

MXenes exhibit remarkable thermal stability and elevated thermal conductivity, which are crucial for applications in transferring heat and thermal management. The thermal conductivity of MXenes varies based on their configuration and construction, with some materials demonstrating conductivities exceeding 1000 W/mK. This feature is principally valuable for applications in thermal interface materials, where efficient heat transfer is crucial.

Thermal stability is another crucial aspect of MXenes, with certain compositions outstanding steady at temperatures higher than 800 °C. However, the presence of surface terminations can impact stability by introducing oxidation pathways. Strategies such as encapsulation and controlled functionalization have been discovered to boost the thermal resistance of MXenes in high-temperature applications.

3.4 Chemical and Surface Properties

Because of the existence of functional groups (–OH, –F, –O), MXenes' rich surface chemistry allows for excellent hydrophilicity and chemical tunability. These properties make MXenes highly dispersible in aqueous solutions, facilitating their use in ink formulations, coatings, and composite materials. Additionally, the surface chemistry of MXenes enables selective adsorption of ions, making them useful for water purification and environmental remediation applications.

The existence of surface terminations also influences the chemical reactivity of MXenes. For example, MXenes can be functionalized through numerous organic and inorganic molecules to adapt their features for definite uses. This functionalization capability has been leveraged in catalysis, where MXenes, owing to their plentiful active sites, function as exceptionally effective electrocatalysts for both hydrogen and oxygen evolution reactions.

3.5 *Optical and Magnetic Properties*

MXenes display exciting optical properties, including tunable captivation in the ultraviolet-visible-near infrared (UV-Vis-NIR) spectrum. This makes them useful for applications in photodetectors, sensors, and optoelectronic devices. The optical features of MXenes can be controlled by altering their configuration, thickness, and surface terminations.

Some MXenes also display unique magnetic properties, principally those incorporating magnetic transition metals such as Cr and Mn. These materials display ferromagnetic or antiferromagnetic performance depending on their atomic configuration and electronic configuration. The tunable magnetism of MXenes has opened new avenues for spintronic applications, where controlling electron spin is essential for developing advanced memory and computational devices [26, 27].

4 HER and OER Mechanisms

4.1 *HER Mechanisms*

At every stage of the water electrolysis process, the cathodic half-reaction, also known as HER, takes place. Furthermore, it encompasses the reduction of protons (H^+) or water molecules (H_2O), which results in the production of hydrogen gas (H_2). This is a significant electrochemical process that serves as a mechanism for the transfer of two electrons and is essential to a wide variety of technologies that convert energy [28]. The routes of this reaction are different depending on the electrolyte's pH and the conditions of the electrode surface. The Volmer reaction is what starts the hydrogen evolution process (HER) in acidic environments. In this reaction, hydronium ions are released at the electrode's surface, leading to the formation of adsorbed hydrogen intermediates (H*). The reaction may be represented as follows: $H_3O^+ + e^- + * \rightarrow H^* + H_2O$. On the other hand, in alkaline conditions, water molecules are the source of protons: $H_2O + e^- + * \rightarrow H^* + OH^-$. The production of molecular hydrogen that follows this is done through two competing methods, which are mostly governed by the surface covering of H*. When the H* coverage is large, the Tafel step becomes the rate-controlling step. This step includes the combining of two neighboring H* atoms to generate H_2, which may be represented by the equation $H^* + H^* \rightarrow H_2$. When it comes to the total response kinetics in these types of situations, this step is frequently seen to be the most important aspect. On the other hand, a low H* coverage is more favorable for the Heyrovský reaction, which involves a single H* interacting with an extra proton and an electron. In acidic settings, this is expressed as $H^* + H_3O^+ + e^- \rightarrow H_2 + H_2O$, whereas in alkaline situations, it is expressed as $H^* + H_2O + e^- \rightarrow H_2 + OH^-$. It is important to recognize the Tafel and Heyrovský mechanisms because the dominance of one of these mechanisms has a major impact on the efficiency and speed of the reaction, which is important for building improved electrocatalysts. In addition, the interaction between these processes and the electrical structure of the electrode's material, surface shape, and catalytic characteristics is essential for maximizing the HER presentation across a diversity of pH levels and operational settings.

4.2 OER Mechanisms

OER, which is the anodic half-reaction in water electrolysis, is more complicated and requires higher energy than HER. The oxidation of water molecules to create O_2 is part of this process. The OER is an significant process in electrochemical water split, but it is also difficult to perform. It is essential for several renewable energy systems. The OER is a slow process that transfers four electrons and includes the creation and transformation of three important intermediates that are deposited on the surface: OOH*, O*, and OH* [29]. This intricacy is a contributing factor to its substantial overpotential need and highlights the need to create effective catalysts in order to overcome these kinetic obstacles. The OER reaction mechanism includes routes that rely on pH. Water molecules are the main reactants in acidic conditions, where they undergo oxidation to form hydrogen ions and oxygen. On the other hand, hydroxyl ions are oxidized in neutral and alkaline environments, producing water and oxygen. Theoretical models suggest that, even though the final reactions are different, the OER goes through comparable intermediary processes in mutually acidic and alkaline settings. It is generally agreed upon that the process that is most frequently recognized is the four consecutive stages of proton-coupled electron transfer (PCET). When it comes to defining the overall kinetics and efficiency of the reaction, the energetics of these intermediate phases, particularly the binding energies of OH*, O*, and OOH* to the surface of the catalyst, are extremely important. In order to lessen the amount of overpotential that is required for each phase, the most effective catalysts should achieve a perfect equilibrium between the binding strengths of these intermediates.

5 Mxenes as Electrocatalyst for HER

MXene-based oxide catalysts have good characteristics and might compete with noble metals. Quantum dots were formed on MXene via Van Der Waals interaction, which decreases catalyst free energy. The large, conductive support with specified surface area hybridizes plentiful Co_3O_4 with mesoporous to create a significant OER active site. Yin et al. thermally treated MXene nanosheets to make amorphous $CoSnO_3$ [30].

The overlap of oxygen's s and p orbitals with d orbital electrons enhances the hydrogen evolution reaction (HER). Nonetheless, oxygen-based MXene catalysts are only effective for hydrogen evolution reactions (HER). Guo et al. enhanced hydrogen evolution reaction (HER) with metallic cobalt [34]. A non-noble metal featuring a dual-active Co-CoO heterojunction MXene was produced. The computational outcomes confirm that HER's active center is Co^0 and OER's is Co^{2+}. The simulation shows that Ti_3C_2 produced the reduced electron density around the Femi level. Ti_3C_2-MXene prevents active Co constituents from aggregating and improves the conductivity and activity. Zhang et al. demonstrate MXene nanosheet-based Ni-NiO with o-vacancy, similar to heterojunction catalysts of Co-CoO/$Ti_3C_2T_x$ [31]. This study presents computational findings and XPS results indicating the disappearance of the Ni^0 peak and the emergence of the Ni^{3+} peak, which signifies the presence of Ni (oxy) hydroxides (NiOOH), identified as the active site for OER. The Ni-NiO heterostructure, along with MXene, O-vacancies, and a novel nanoarray structure exhibiting improved conductivity and aggregation resistance, demonstrated significant interfacial interaction. In-situ development of Ni-NiO with Ovs nanosheets on an open $Ti_3C_2T_x$ substrate inhibits aggregation and promotes mass diffusion and gas escape, boosting catalytic activity.

Wang et al., the author, developed the $Ti_3C_2T_x$-MXenes with $TiOF_2$ nanospheres to enhance hydrogen evolution reactions [32]. $Ti_3C_2T_x$, $TiOF_2$, and $TiOF_2@Ti_3C_2T_x$ composites were all subjected to parallel studies for the purpose of comparison. On the basis of the data shown in Fig. 2a, it can be observed that both $Ti_3C_2T_x$ and $TiOF_2$ demonstrated a low electro-hydrogen evolution activity, with a high onset potential of 718 and 430 mV, individually. To the contrary, the composite material $TiOF_2@Ti_3C_2T_x$ has significantly higher levels of hydrogen evolution activity. In order to get the onset potential down to 103 mV, the two-dimensional substrate that has good stratification and the uniform distribution of the $TiOF_2$ nanospheres are both responsible for this reduction. Additionally, the $Ti_3C_2T_x$, which has a high conductivity, is capable of supplying a large number of electrons for the transfer of interlayer electrons, which is necessary in order to establish the combination of H^+ and O sites [32].

The Tafel slopes of all of the samples are displayed according to the linear sweep voltammetry (LSV) on the Tafel equation in order to conduct a more in-depth analysis of the HER performance of $TiOF_2@Ti_3C_2T_x$. According to the equation $\eta = b \log(j/j_0)$, where η represents the overpotential, b represents the Tafel slope, j represents the current density, and j^0 means the exchange current density. Since the Tafel slope

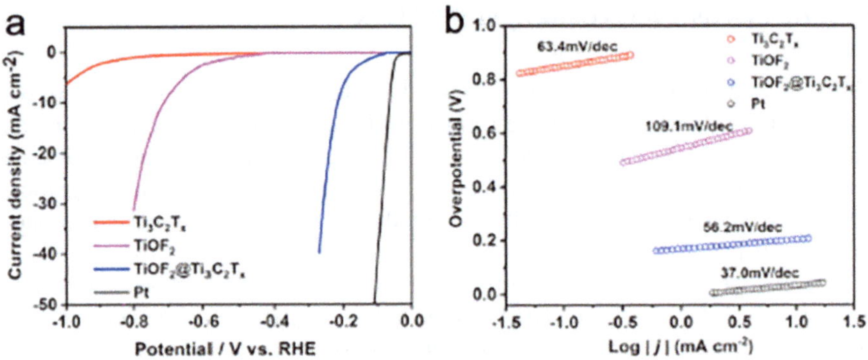

Fig. 2 **a** Shows the HER polarization curves for the four samples. **b** Tafel plots that correspond. Reprinted with permission [32]

of $TiOF_2@Ti_3C_2T_x$ (56.2 mV dec^{-1}) is very near to that of the glassy carbon electrode (37.0 mV dec^{-1}), it can be concluded that $TiOF_2@Ti_3C_2T_x$ exhibits the best electrode catalytic efficiency. On the other hand, the slope values of pure $Ti_3C_2T_x$ and $TiOF_2$ are greater, with values of 63.4 mV dec^{-1} and 109.1 mV dec^{-1}, respectively. This indicates that the composite material enhances the dynamic process of the electrochemical reaction [32].

Not only does the HER activity play a significant part in catalysis, but the stability of the catalyst is also an important factor. The authors carried out a cycle test over an extended period of time to determine the stability of various catalysts, as seen in Fig. 3. Despite the fact that the onset potential of $Ti_3C_2T_x$ did not change considerably after 1000 CV cycles, a considerable drop in current density was detected at the same hydrogen evolution potential. This finding suggests that the catalyst has poor long-term catalytic stability. On the other hand, the composite catalyst did not exhibit a substantial drop-in catalytic activity, which demonstrated that $TiOF_2@Ti_3C_2T_x$ had a high level of electrocatalytic stability. For the composite catalyst, $Ti_3C_2T_x$ and $TiOF_2$, respectively, a long-term constant current test was carried out for a duration of twenty-four hours. The current density was twenty milliamperes per square centimeter. The results of this test are depicted in Fig. 3c. The results of the tests demonstrated that $TiOF_2@Ti_3C_2T_x$ possesses greater long-term catalytic stability than $Ti_3C_2T_x$ and $TiOF_2$. Stability in comparison to $Ti_3C_2T_x$ and $TiOF_2$.

Cui et al. developed Pt immobilized spontaneously on porous MXene/MAX hybrid monolith for hydrogen evolution reaction [33]. Figure 4 displays the Pt immobilization scheme for HER on a porous MXene/MAX hybrid monolith, which was created using a porous Ti_3AlC_2 matrix.

By using the spontaneous redox reaction between $[PtCl_6]_2$ and MXene, we have created a Pt-immobilized partly etched MXene/MAX hybrid monolith as high-performance catalysts for HER. This approach capitalizes on the strong connection between MXene and the MAX phase, high electrical conductivity, and good stability of MAX phases in acidic solutions.

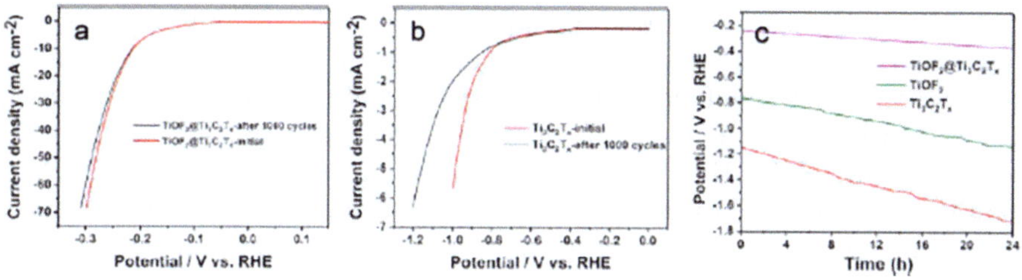

Fig. 3 **a, b** $TiOF_2@Ti_3C_2T_x$ and $Ti_3C_2T_x$ stability test employing linear sweep voltammetry before and after 1000 cycles. **c** E versus t over 24 h at a steady current density of 20 mA cm^{-2}. Reprinted with permission [32]

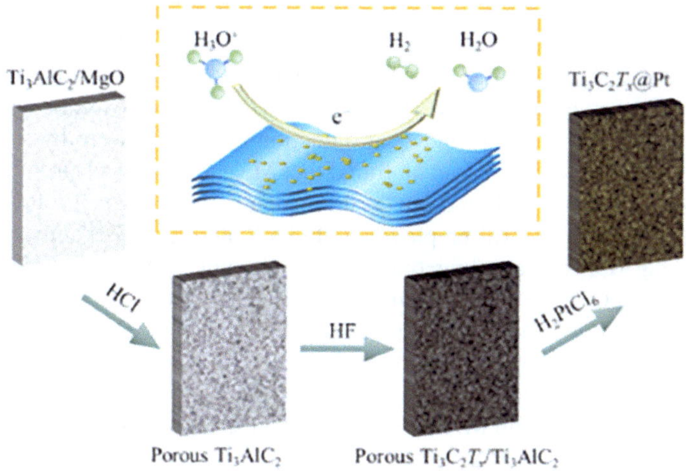

Fig. 4 $Ti_3C_2T_x@Pt$ catalyst preparation schematic. Reprinted with permission [33]

Based on modest Pt loadings (less than 8.9 mg/cm^2), the catalyst exhibits a low overpotential versus RHE (43 mV for about 10 mA/cm^2). In addition to offering a method for creating a Pt-based catalyst for HER, this study opens possibilities for additional research and development of highly effective catalysts based on noble metals and porous MXenes/MAX phase monoliths for other reactions.

Ren et al. found that MXene and MoS_2 nanoflowers had a low overpotential of 176 mV at 10 mA cm^{-2} [34]. The distance between MoS_2 and Mo_2CO_2 heterostructures was determined as 2.77 Å. Compared to MoS_2, $MoS_2@Mo_2CO_2$ has a much lower ΔE_{H*} value. MoS_2 in MXene changes surface adsorption energetics and adds reaction sites. DFT showed that 2H-MoS_2 can modify oxygen-terminated Mo_2CT_x's surface O sites. MXene-derived sulfide catalysts were occasionally synthesized using nickel foam. NF@MXene@Co_3S_4/Ni_3S_2 heterostructure by Li et al. forms a multi-dimensional feather structure [35]. The catalyst performs well in 1.0 M potassium hydroxide with a low overpotential of 186 mV at 10 mA cm^{-2}. The greatest energy input for Co_3S_4's potential is 0.41 V and Ni_3S_2's is 0.37 V in this heterogeneous structure. The density functional active calculation shows that Ni_3S_2's Ni atoms are active sites. In recent times, sulfides doping transition metals that lower kinetic energy improved HER. Liang et al. employed heterostructure engineering to add Co metals to MoS_2 for HER activity, achieving a low overpotential of 112 mV at 10 mA cm^{-2} [36]. Heterostructure catalysts with synergistic effects and enhanced active sites boost electrocatalytic performance. Co-MoS_2 can additionally include MXene to lower the kinetic energy barrier. Electrochemical impedance spectroscopy showed that sulfides catalysts had lower charge transfer resistance. This catalyst also has metallic characteristics and hybridized orbital dispersion in its electronic band.

Many substrates are available in MXene, a vast system. It can make catalysts with other substrates. Zong et al. discovered that the combination of MoS_2 nanoparticles and niobium carbide results in the formation of a buttercup-flower structure following the freeze-drying process. The nanoflowers were consistently formed on the nanosheet surface and had an overpotential of 117 mV at 10 mA cm^{-2} [37]. Because of the comparable lattice structure of Nb_2C and MoS_2, more active spots, Mo edges validated by DFT, and heterojunction constructs were hydrothermally produced. Another work by Xie et al. found that MXene nanosheets in sulfide, cubic phase iron pyrites may give 10 mA cm^{-2} current density at 1.57 V cell voltage [38]. In-situ Raman spectroscopy was employed within a Raman-electrochemistry coupling system during (OER), revealing that a small amount of FeS_2 in FeS_2@MXene transforms to FeOOH. Surface engineering improves nanosheet conductivity and electrical structure [39, 40]. The sulfur species were chemically bonded to MXene terminations. FeS_2 nanoparticles were produced on MXene nanosheets using a careful adsorption-growth approach technique. Surface engineering ensured the close coupling of active FeS_2 and MXene supports and the creation of an interface.

The as-synthesized Ti-MXene nanosheets, featuring oxygen and fluorine functional groups, effectively electrostatically adsorb charged ions, serving as a model for the uniform nucleation of $MoSe_2$. [41]. The $MoSe_2$@Ti-MXene hybrid's electrocatalytic activity for HER in 0.5 M H_2SO_4 was examined. The hybrid has a lower onset potential (61 mV) and overpotential (180 mV at 10 mA cm^{-2}) than $MoSe_2$ (300 mV). Compared to bare $MoSe_2$, the current density rose 6 times. Bare $MoSe_2$ had a 200 mV dec^{-1} Tafel slope, while the hybrid had a 91 mV dec^{-1} slope, indicating fast charge transfer with Ti-MXene. Nano-flowered hybrid with highly conductive MXene has a charge transfer resistance of 16.6 Ω, compared to 242.4 Ω for naked $MoSe_2$.

6 Enhancing the HER Activity of MXenes

Although pristine MXenes have their advantages, they often require further modifications to achieve optimal hydrogen evolution reaction (HER) activity. Scientists have investigated several approaches to improve their catalytic performance.

6.1 Elemental Composition and Doping

The electrocatalytic activity of MXenes can be adapted by varying the transition metal (M) in their structure. Research has demonstrated that Mo-based MXenes, specifically Mo_2CT_x, display loftier performance in HER than Ti-based MXenes, such as $Ti_3C_2T_x$ [42–44]. This loftier performance is owed to their optimal hydrogen binding energy. Furthermore, doping MXenes by non-metallic components such as nitrogen (N), phosphorus (P), or sulfur (S) can enhance their electronic properties and raise the density of active spots.

6.2 Surface Termination Engineering

The presence of functional groups (–O, –OH, –F) on MXene surfaces plays a crucial role in influencing their effectiveness in the hydrogen evolution reaction (HER). Among these, MXenes terminated with oxygen (MXene-O) show the best catalytic activity due to their favorable hydrogen adsorption energy. Researchers have demonstrated that replacing fluorine (–F) terminations with oxygen (–O) can lower the overpotential needed for HER, thereby enhancing efficiency [45–47].

6.3 Defect Engineering

Presenting defects into the MXene lattice can create extra active spots for HER [48, 49]. Point defects, including vacancies in the transition metal or carbon/nitrogen layers, can potentially improve the electrical behavior and catalytic activity of MXenes. For instance, adding molybdenum (Mo) vacancies in $Mo_2TiC_2T_x$ has been demonstrated to improve HER performance by expanding the active spot's number available for hydrogen adsorption.

6.4 MXene-Based Composites

Merging MXenes with additional electrocatalytic materials, such as metal nanoparticles, heterostructures, or single-atom catalysts, has demonstrated an effective strategy for improving HER activity. For example, platinum (Pt) nanoparticles held on $Ti_3C_2T_x$ MXenes improve kinetics for HER with lower Pt loading. Additionally, nickel iron-layered double hydroxides (NiFe-LDH) integrated with MXenes have demonstrated remarkable stability and performance under alkaline conditions for HER [50–53].

7 MXene Electrocatalyst for Oxygen Evolution Reaction

This section summarizes MXene heterostructure alternatives for OER. Kaixin et al. developed FeOOH NSs@$Ti_3C_2T_x$ at room temperature and tested the OER in 1.0 M potassium hydroxide. At 10 mA cm^{-2}, FeOOH NSs@Ti_3C_2 had a 400 mV overpotential and a Tafel slope of 95 mV dec^{-1} [54]. The hierarchical configuration prevented FeOOH aggregation, increasing its surface area. It also created robust interfacial contact, accelerating electrical conduction and increasing electrocatalytic activity [54]. Cobalt oxide with valances like +2 and +3 is a potential OER electrocatalyst due to its tunable morphology and crystallization [55]. Yi Lu et al. created nano-cubic Co_3O_4@Ti-MXene nanosheets using solvothermal methods [55] (see Fig. 5). This circumvented cobalt oxide limitations. The nano-cubic Co_3O_4 synthesized was held down on delaminated Ti-MXene through robust electrostatic interactions and electrical coupling, improving OER performance. The hybrid of Co_3O_4 and Ti-Mxene exhibited 300 mV overpotential at 10 mA cm^{-2} during OER testing in a 1.0 M potassium hydroxide solution. This overpotential was smaller than naked nano-cubic Co_3O_4 and Ti-MXene [55]. Co_3O_4 with conductive Ti-MXene synergistically improved cobalt oxides. This impact increases OER performance, which may improve it.

In a separate piece of research, Jiapeng et al. produced hierarchical cobalt borate@Ti-MXene at room temperature [56]. During the OER tests conducted in an alkaline medium, the hybrid was seen to exhibit an overpotential of 250 mV at a current of 10 mA cm^{-2} [56]. A further point to consider is that the hybrid exhibited a Tafel slope of just 53 mV dec^{-1}. Because of the close contact between Co–B and $Ti_3C_2T_x$, charge transfer was simplified, and the electrostatic attraction of additional anionic intermediates was increased, which resulted in an improvement in OER kinetics.

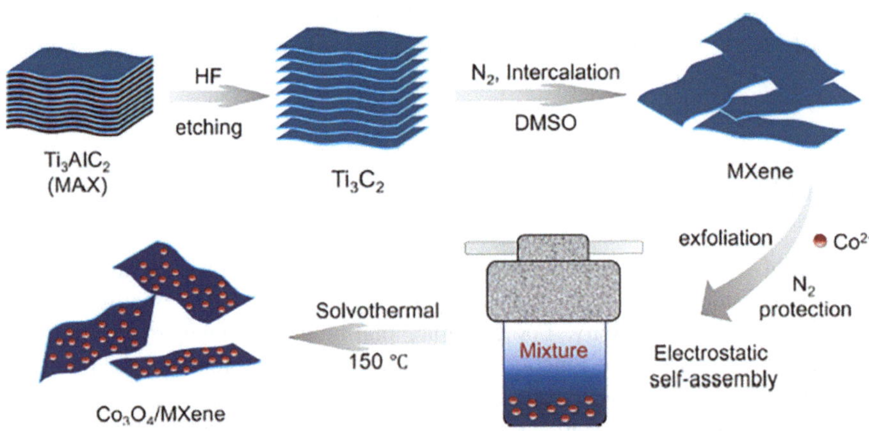

Fig. 5 Diagrammatic representation of the CM hybrids' preparation procedure. Reprinted with permission from [55]

Free-standing electrode films can be synthesized using MXenes and other 2D materials. Tian et al. conducted tests on their hybrid film (TCCN) in an alkaline medium for OER [57]. The hybrid film was made by layering g-C3N4 with Ti MXene. The T-Nx bond phenomena created the porous free-standing film from g-C3N4 and Ti-MXene. Once formed, the structure was water-loving and conducting. The hybrid demonstrated enhanced OER performance, achieving 1.14 V onset potential versus RHE and 165 mV overpotential [57]. This is great proof that g-C3N4 and Ti-MXene worked together. Layer double hydroxides may be OER electrocatalysts, but their humble conductivity restricts their efficacy. Benchakar et al. created an efficient oxygen evolution reaction (OER) electrocatalyst by utilizing an electrochemically active cobalt layer double hydroxide on highly conductive MXene sheets.

At 10 mA/cm^2, the hybrid exhibits a low overpotential of 330 mV. When Co-LDH and MXene worked together synergistically, the hybrid's overall execution performance (OER) improved [58].

The hybrid's OER performance improved when Co-LDH and MXene synergistically collaborated [58]. Similarly, the synthesis of Co-LDH on MXene's surface avoided oxidation and retained conductivity, making it stable over time. They created a CoFe-LDH hybrid on MXene sheets using heating and reflux [59] (see Fig. 6).

A test was conducted in 1.0 M potassium hydroxide to evaluate the OER performance of CoFe-LDH@Ti-MXene. At a current density of 10 mA cm^{-2}, a hybrid with a Tafel slope of 50 mV dec^{-1} exhibited an overpotential of 319 mV. Nevertheless, the overpotential of CoFe-LDH was 352 mV when it was applied at 10 mA cm^{-2}. Due to the conductive tight link between CoFe-LDH and Ti-MXene, electrons were able to flow between the two materials with ease. Similarly, significant electrostatic interactions between CoFe-LDH and Ti-MXene enhanced the stability of the

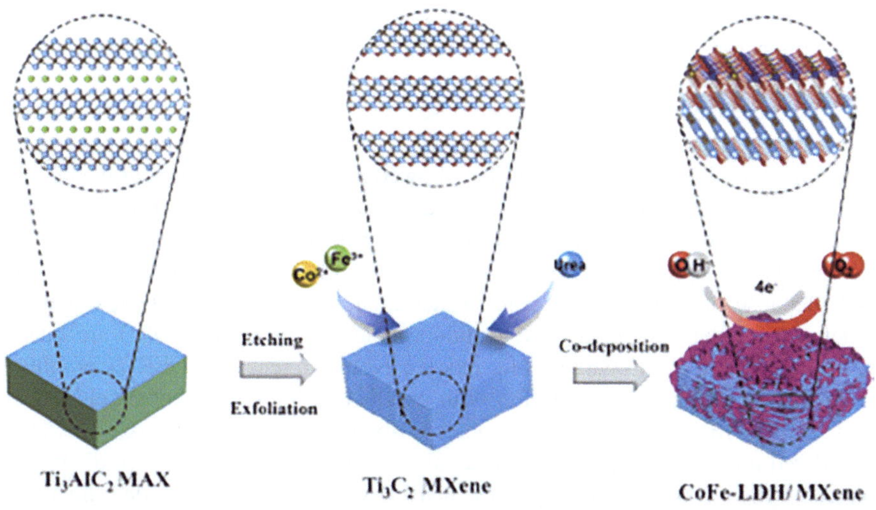

Fig. 6 Diagrammatic representation of the CoFe-LDH/MXene hybrid production process. Reprinted with permission [59]

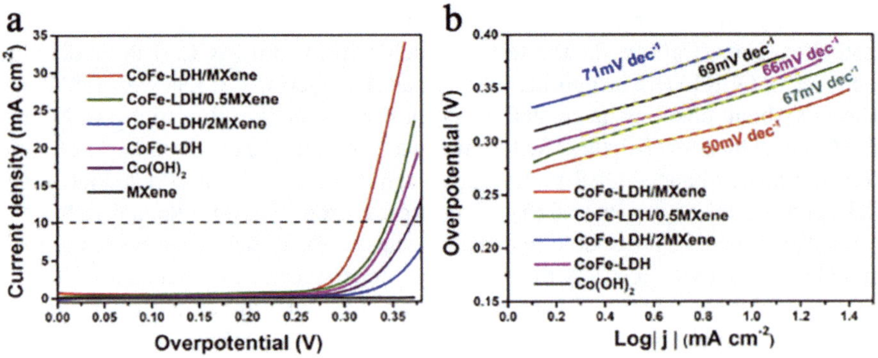

Fig. 7 **a** LSV, **b** Tafel plot for CoFe-LDH/MXene nanohybrid. Reprinted with permission [59]

hetero-structure. The in-situ development of MXene resulted in the creation of three-dimensional architecture. The surface area of the electro-catalyst was expanded because of this architecture, which made it possible for gas to diffuse and be delivered to additional active sites (Fig. 7).

Likewise, the synthesis of Co-LDH on MXene's surface avoided oxidation and retained conductivity, making it stable over time. They created a CoFe-LDH hybrid on MXene sheets using heating and reflux [59]. Chongyan et al. did this. CoFe-LDH@Ti-MXene's OER performance was tested in 1.0 M potassium hydroxide. A hybrid with a 50 mV dec^{-1} Tafel slope showed 319 mV overpotential at 10 mA cm^{-2} current density. However, CoFe-LDH had 325 mV overpotential at 10 mA cm^{-2}. Electrons moved easily between CoFe-LDH and Ti-MXene due to their conductive tight connection. Similarly, substantial electrostatic interactions between CoFe-LDH and Ti-MXene increased hetero-structure stability. MXene's in-situ growth created three-dimensional architecture. This layout increased the electro-catalyst's surface area, allowing gas diffusion and delivery to additional active sites.

In another investigation, Ming et al. described the FeCo-LDH hybrid on a monolayer MXene sheet [60]. The FeCo-LDH @ Ti-MXene electrochemical experiment in 1.0 M potassium hydroxide solution showed a modest overpotential of 268 mV at 10 mA cm^{-2}. The solution's Tafel grade was also 85 mV dec^{-1} [60]. The synergistic result of bimetallic LDH and the quick transfer of electrons caused by the MXene sheet's high conductivity explain the electrocatalyst's low overpotential [60]. Because of their high specific surface area, varied metal centers, functional moiety, and tunable penetrability, metal–organic frameworks (MOFs) are one of the most effective electrocatalysts for oxygen evolution reaction (OER). Combining metal–organic frameworks (MOFs) with conductive substrates boosts electrocatalytic activity [61, 62]. In 1967, 2D cobalt 1,4-benzene dicarboxylate (CoBDC) was hybridized in-situ with $Ti_3C_2T_x$ by Li et al., who then assessed the hybrid's performance for the oxygen evolution process in a 1.0 M potassium hydroxide. The hybrid showed a Tafel gradient of 48.2 mV dec^{-1} and an overpotential of 1.64 V in relation to

RHE [62]. The conductive MXene support further accelerated charge and ion transport transversely the CoBDC and Ti-MXene contact, improving OER performance [62].

Metal sulfides have also shown great promise for OER activity when hybridized with MXenes. Haiyuan et al. synthesized NiCuS on Ti-MXene via MOF [61]. Electrochemical testing for the OER was done in potassium hydroxide (1.0 M). The overpotential of NiCuS@Ti-MXene was 365 mV at 10 mA cm^{-2}, with a Tafel gradient of 58.2 mV dec^{-1} [61]. These data show that the material had good kinetics throughout OER experiments. Strongly bonded NiCuS on Ti-MXene was shown to be accountable for the hybrid's superior OER performance. Continuous macropores formed, allowing electrolyte transport through the open structure [61]. Yangyang et al. [63] fabricated the MXene-supported CoNi-ZIF-67 hybrid via co-precipitation. The hybrid's onset-potential (275 mV) and Tafel grade (65.1 mV dec^{-1}) were negligible [63]. MXene substrate helped create smaller CoNi-ZIF-67 particles and promote Ni/Co oxidation, improving the hybrid's OER [63]. Consequently, the combined effect of the highly electrochemically active CoNi-ZIF-67 and the highly conductive Ti-MXene support enhanced the performance of the oxygen evolution reaction [63].

ABO_3 (rare earth metals and transition metals) perovskite oxides are one of the most effective electrocatalysts for water electrolysis [64–67]. Recent research shows that LSTN, a cation-deficient material, exhibits excellent HER properties [68]. Ramsha et al. proposed connecting the adjustable perovskite structure with greatly conductive and water-loving 2D MXene nanosheets to improve OER kinetics. This would reduce the charge transfer pathway, improving perovskite electrocatalyst performance [68]. Improve charge transfer resistance while retaining electrical conduction was the target. At 1.56 V voltage (vs. RHE), the LSTN@MXene electrode showed a remarkable Tafel slant of 44 mV dec^{-1} for OER and a current density of 10 mA cm^{-2} [68]. The hybrid showed remarkable long-term constancy for 24 h at 100 mA cm^{-2} current density [68]. The hybrid has greater OER activity due to fast ion diffusion, increased exposed area, and strong LSTN-MXene interfacial contact. This contract caused considerable transfers between LSTN and MXene nanosheets, giving the composite structural strength.

For oxygen evolution process (OER), cobalt oxide (Co_3O_4) is a crucial electrocatalyst, however, particle aggregation, volume change, and poor conductivity restrict its efficiency [69]. Anchoring Co_3O_4 on MXene increases conductivity and reduces volume change [69]. Chuang et al. grew Co_3O_4 quantum dots on Ti-MXene by employing a thermal control technique that only required one step. Co_3O_4 forms a strong connection to MXene sheets by the process of van der wall interactions[69]. As spacers, Co_3O_4 quantum dots prevented MXene nanosheets from restacking, preserving active site integrity [69]. The hybrid's electrochemical OER performance was tested using 1.0 M potassium hydroxide. An overpotential of 340 mV at 10 mA cm^{-2} was reported in the hybrid, significantly lower than Co_3O_4 [69]. The hybrid had a Tafel slope of 63.97 mV dec^{-1}, showing a debauched kinetic OER reaction [69].

8 The Difficulties of MXenes in Hybrid Architectures for Water Splitting Applications

While MXene materials demonstrate potential for use in composite applications, they do have several drawbacks that may adversely affect their properties. These limitations primarily stem from structural challenges, compatibility issues with other materials, constraints in processing and manufacturing, and environmental and health concerns associated with MXenes in composite materials.

8.1 Structural Challenges

8.1.1 Layer Stability

Layer stability is one of the main structural challenges MXenes encounter in composite materials. MXenes are known for their multilayered structures, which can easily separate under certain conditions. This instability can settle the truthfulness and performance of the composite material. For instance, research has shown that MXenes are extremely disposed of oxidation, that can lead to the deterioration of their layers. When titanium carbide MXenes are oxidized, titanium dioxide (TiO_2) often forms on their surface, significantly reducing their useful properties [70].

8.1.2 Morphological Control

Regulating their morphology is a significant challenge in synthesizing and producing composites with MXenes. Controlling morphology is vital for adapting the features of MXenes for specific purposes. However, reaching uniform morphology is difficult due to synthesis method variations and environment. For example, the engraving route used to derive MXenes from parent MAX phases often results in flakes that are uneven and non-uniform in size. This lack of uniformity can complicate their integration into a composite matrix. A uniform and well-defined flake morphology is crucial for ensuring efficient load transfer and stress distribution within the composite material [71, 72].

8.1.3 Mechanical Integrity

Another significant issue is maintaining the integrity of MXenes when they are used in composites. While MXenes are known for their inherent mechanical strength, their mechanical properties often degrade when combined with a composite matrix. This degradation typically occurs due to poor adhesion between layers and structural defects. For instance, when MXenes are integrated with polymers or other materials, they tend to restack, which can negatively impact the composite's mechanical integrity. Furthermore, imperfections such as vacancies and edge terminations can create stress concentration points, leading to premature failure when mechanical force is applied [73, 74].

8.2 Challenges of Compatibility with Other Materials

8.2.1 Interfacial Bonding

The major problem in preparing MXene-based composites is that it is difficult to generate strong and long-lasting interfacial bonding between MXenes and other materials. The overall enactment and permanency of the composite are greatly affected by the interfacial bonding. On the other hand, weak bonding, delamination, and decreased mechanical characteristics can result from poor interfacial contacts. For instance, when MXenes are added to polymer matrices, the stress transmission is generally insufficient because of weak van der Waals forces or insufficient chemical reactions at the interface. Interfacial bonding may be improved by using surface modification procedures, such as functionalization with coupling agents or plasma treatments [71, 75].

8.2.2 Chemical Incompatibility

Chemical incompatibility presents a significant challenge when integrating MXenes with certain materials in composite applications. Due to their surface functional groups, MXenes are highly reactive, which can result in undesirable reactions with other components of the composite. For example, the surface hydroxyl, oxygen, and fluorine groups found on MXenes can lead to harmful chemical interactions with metal ions or polymer matrices, negatively impacting the durability and presentation of the composite. Furthermore, the acidic nature of MXenes, a byproduct of their synthesis process, can intensify these chemical incompatibility issues [76, 77].

8.2.3 Dispersion Issues

MXenes must be equally dispersed across a composite matrix in order to produce consistent features and performance. However, MXenes can encounter considerable dispersion difficulties due to their tendency to clump together and restack. The strong van der Waals forces that present between MXene layers are responsible for this behavior. These forces can cause phase separation and unequal distribution in the composite material. If dispersion is poor, it can lead to localized stress concentrations, a decrease in mechanical qualities, and the composite failing before it should. Different methods have been tested in order to improve the dispersion of MXenes. These methods include sonication, mechanical stirring, and the addition of surfactants [78, 79].

8.2.4 Surface Functionalization Challenges

The etching technique used to create MXenes exposes metal atoms, leading to instability and unpredictable surface properties. This presents a challenge when constructing composites with the desired characteristics. A significant obstacle is finding a way to balance surface functional groups with the necessary reactivity [80].

9 Conclusion

In summary, MXenes have developed as talented electrocatalysts for water splitting owing to their extraordinary conductivity, tunable surface chemistry, and structural versatility. Their ability to enhance both hydrogen and oxygen evolution reactions makes them competitive alternatives to traditional precious metal catalysts. Strategies, for instance, elemental doping, surface termination engineering, and composite formation, have significantly improved their catalytic performance. However, challenges remain, including structural instability, oxidation susceptibility, and compatibility with other materials in hybrid architectures. Overcoming these limitations through innovative synthesis and modification techniques will further enhance their potential in energy applications. Upcoming study should emphasis on optimizing MXene-based heterostructures to achieve higher competence and long-term permanency. Additionally, understanding the fundamental reaction mechanisms at the atomic level will be crucial for further advancements. The integration of MXenes into commercial electrolyzers and scalable production methods will be key to their widespread adoption. With continued research efforts, MXenes holds great promise for sustainable hydrogen production and clean energy applications. Moreover, their extraordinary surface area and unique electronic features provide opportunities for multi-functional applications beyond electrocatalysis. Joining MXenes with further 2D materials could yield synergistic effects, further boosting their catalytic activity. Computational modeling and machine learning approaches can speed up the discovery of optimized MXene compositions for specific purposes. Finally, addressing the environmental and economic aspects of MXene synthesis will be critical for their large-scale implementation in energy conversion technologies.

References

1. Adnan, M., Xiao, B., Bibi, S., Xiao, P., Zhao, P., Wang, H., Ali, M.U., An, X.: Known and unknown environmental impacts related to climate changes in Pakistan: an under-recognized risk to local communities. Sustainability **16**, 6108 (2024)
2. Malik, K., Capareda, S.C., Kamboj, B.R., Malik, S., Singh, K., Arya, S., Bishnoi, D.K.: Biofuels production: a review on sustainable alternatives to traditional fuels and energy sources. Fuels **5**, 157–175 (2024)
3. Bilgili, M., Tumse, S., Nar, S.: Comprehensive overview on the present state and evolution of global warming, climate change, greenhouse gasses and renewable energy. Arab. J. Sci. Eng. **49**, 14503–14531 (2024)
4. Vogt, E.T.C., Weckhuysen, B.M.: The refinery of the future. Nature **629**, 295–306 (2024)
5. Janaki, S.T., Madheswaran, D.K., Naresh, G., Praveenkumar, T.: Beyond fossil: the synthetic fuel surge for a green energy resurgence. Clean Energy zkae050 (2024)
6. Munir, A., Nasir, J.A., ul Haq, T., Iqbal, J., Hussain, I., Qurashi, A.: Benchmarking stable electrocatalysts for green hydrogen production: a chemist perspective. Coord. Chem. Rev. **521**, 216112 (2024)
7. Quan, L., Jiang, H., Mei, G., Sun, Y., You, B.: Bifunctional electrocatalysts for overall and hybrid water splitting. Chem. Rev. **124**, 3694–3812 (2024)
8. Tan, P., Liu, Y., Zhu, A., Zeng, W., Cui, H., Pan, J.: Rational design of Z-scheme system based on 3D hierarchical CdS supported 0D Co9S8 nanoparticles for superior photocatalytic H2 generation. ACS Sustain. Chem. Eng. **6**, 10385–10394 (2018)
9. Ding, M., Ao, W., Xu, H., Chen, W., Tao, L., Shen, Z., Liu, H., Lu, C., Xie, Z.: Facile construction of dual heterojunction CoO@ TiO2/MXene hybrid with efficient and stable catalytic activity for phenol degradation with peroxymonosulfate under visible light irradiation. J. Hazard. Mater. **420**, 126686 (2021)
10. Tareen, A.K., Khan, K., Iqbal, M., Zhang, Y., Long, J., Mahmood, A., Mahmood, N., Xie, Z., Li, C., Zhang, H.: Recent advance in two-dimensional MXenes: new horizons in flexible batteries and supercapacitors technologies. Energy Storage Mater. **53**, 783–826 (2022)
11. Gao, G., O'Mullane, A.P., Du, A.: 2D MXenes: a new family of promising catalysts for the hydrogen evolution reaction. ACS Catal. **7**, 494–500 (2017)
12. Rasheed, T.: MXenes as emerging two-dimensional analytical modalities for potential recognition of hazardous environmental contaminants. Mater. Today Chem. **24**, 100859 (2022)
13. Li, X., Zhang, J., Zhang, S., Xu, S., Wu, X., Chang, J., He, Z.: Hexagonal boron nitride composite photocatalysts for hydrogen production. J. Alloys Compd. **864**, 158153 (2021)
14. Gogotsi, Y.: MXenes: From Discovery to Applications of Two-Dimensional Metal Carbides and Nitrides. CRC Press (2023)
15. Deshmukh, K., Muzaffar, A., Kovářík, T., Ahamed, M.B., Pasha, S.K.: Introduction to 2D MXenes: fundamental aspects, MAX phases and MXene derivatives, current challenges, and future prospects. Mxenes Their Compos. 1–47 (2022)
16. Zhang, C., Naguib, M.: Transition Metal Carbides and Nitrides (MXenes) Handbook: Synthesis, Processing, Properties and Applications. Wiley (2024)
17. Kumar, S.: Fluorine-free MXenes: recent advances, synthesis strategies, and mechanisms. Small **20**, 2308225 (2024)
18. Sinha, A., Ma, K., Zhao, H.: 2D Ti3C2Tx flakes prepared by in-situ HF etchant for simultaneous screening of carbamate pesticides. J. Colloid Interface Sci. **590**, 365–374 (2021)
19. Liew, J., Bashir, S., Ramesh, K., Ramesh, S.: A comprehensive characterisation of titanium carbide (Ti3C2) MXene with hydrothermal-assisted in-situ hydrofluoric acid etching. Mater. Sci. Eng. B **314**, 118076 (2025)
20. Siddique, S., Waheed, A., Iftikhar, M., Mehran, M.T., Zarif, M.Z., Arafat, H.A., Hussain, S., Shahzad, F.: Fluorine-free MXenes via molten salt lewis acidic etching: applications, challenges, and future outlook. Prog. Mater. Sci. **139**, 101183 (2023)

21. Defoy, E., Baron, M., Séné, A., Ghoridi, A., Thiaudière, D., Célérier, S., Chartier, P., Brette, F., Mauchamp, V., Portehault, D.: Galvanic replacement and etching of MAX-related phases in molten salts toward MXenes: an in situ study. Chem. Mater. **35**, 8112–8121 (2023)
22. Kruger, D.D., García, H., Primo, A.: Molten salt derived MXenes: synthesis and applications. Adv. Sci. **11**, 2307106 (2024)
23. Mim, M., Habib, K., Farabi, S.N., Ali, S.A., Zaed, M.A., Younas, M., Rahman, S.: MXene: a roadmap to sustainable energy management, synthesis routes, stabilization, and economic assessment. ACS Omega **9**, 32350–32393 (2024)
24. VahidMohammadi, A., Kayali, E., Orangi, J., Beidaghi, M.: Techniques for MXene delamination into single-layer flakes. 2D Met. Carbides Nitrides Struct. Prop. Appl. 177–195 (2019)
25. Hussain, I., Arifeen, W.U., Khan, S.A., Aftab, S., Javed, M.S., Hussain, S., Ahmad, M., Chen, X., Zhao, J., Rosaiah, P., Fawy, K.F., Younis, A., Sahoo, S., Zhang, K.: M4X3 MXenes: application in energy storage devices. Nano-Micro Lett. **16**, 215 (2024). https://doi.org/10.1007/s40820-024-01418-0
26. Xu, H., Ren, A., Wu, J., Wang, Z.: Recent advances in 2D MXenes for photodetection. Adv. Funct. Mater. **30**, 2000907 (2020)
27. Zhang, D.: Optical Properties of MXenes. (2024)
28. Morales-Guio, C.G., Stern, L.-A., Hu, X.: Nanostructured hydrotreating catalysts for electrochemical hydrogen evolution. Chem. Soc. Rev. **43**, 6555–6569 (2014)
29. Suen, N.-T., Hung, S.-F., Quan, Q., Zhang, N., Xu, Y.-J., Chen, H.M.: Electrocatalysis for the oxygen evolution reaction: recent development and future perspectives. Chem. Soc. Rev. **46**, 337–365 (2017)
30. Yin, J., Wei, K., Bai, Y., Liu, Y., Zhang, Q., Wang, J., Qin, Z., Jiao, T.: Integration of amorphous CoSnO3 onto wrinkled MXene nanosheets as efficient electrocatalysts for alkaline hydrogen evolution. Sep. Purif. Technol. **308**, 122947 (2023)
31. Zhang, B., Du, Z., Sun, R., Lai, X., Lan, J., Liu, X., Yan, L.: Tremella-like Ni–NiO with O-vacancy heterostructure nanosheets grown in situ on MXenes for highly efficient hydrogen and oxygen evolution. ACS Appl. Mater. Interfaces **14**, 47529–47541 (2022)
32. Wang, Z., Yu, K., Feng, Y., Qi, R., Ren, J., Zhu, Z.: Stabilizing Ti3C2Tx-MXenes with TiOF2 nanospheres intercalation to improve hydrogen evolution reaction and humidity-sensing performance. Appl. Surf. Sci. **496**, 143729 (2019). https://doi.org/10.1016/j.apsusc.2019.143729
33. Cui, C., Cheng, R., Zhang, C., Wang, X.: Pt immobilized spontaneously on porous MXene/MAX hybrid monolith for hydrogen evolution reaction. Chin. Chem. Lett. **31**, 988–991 (2020)
34. Ren, J., Zong, H., Sun, Y., Gong, S., Feng, Y., Wang, Z., Hu, L., Yu, K., Zhu, Z.: 2D organ-like molybdenum carbide (MXene) coupled with MoS2 nanoflowers enhances the catalytic activity in the hydrogen evolution reaction. CrystEngComm **22**, 1395–1403 (2020)
35. Li, Y., Du, Q.-X., Cui, J., Chen, X., Yang, H.-W., Qian, H.: Identifying the intrinsic active site in bimetallic Co3S4/Ni3S2 feathers on MXene nanosheets as a heterostructure for efficient oxygen evolution reaction. Inorg. Chem. Front. **10**, 184–191 (2023)
36. Liang, J., Ding, C., Liu, J., Chen, T., Peng, W., Li, Y., Zhang, F., Fan, X.: Heterostructure engineering of Co-doped MoS2 coupled with Mo2CTx MXene for enhanced hydrogen evolution in alkaline media. Nanoscale **11**, 10992–11000 (2019). https://doi.org/10.1039/C9NR02085C
37. Zong, H., Hu, L., Gong, S., Yu, K., Zhu, Z.: Flower-petal-like Nb2C MXene combined with MoS2 as bifunctional catalysts towards enhanced lithium-sulfur batteries and hydrogen evolution. Electrochim. Acta **404**, 139781 (2022)
38. Xie, Y., Yu, H., Deng, L., Amin, R.S., Yu, D., Fetohi, A.E., Maximov, M.Y., Li, L., El-Khatib, K.M., Peng, S.: Anchoring stable FeS2 nanoparticles on MXene nanosheets via interface engineering for efficient water splitting. Inorg. Chem. Front. **9**, 662–669 (2022)
39. Su, T., Ma, X., Tong, J., Ji, H., Qin, Z., Wu, Z.: Surface engineering of MXenes for energy and environmental applications. J. Mater. Chem. A **10**, 10265–10296 (2022)
40. Xia, J., Yang, S.-Z., Wang, B., Wu, P., Popovs, I., Li, H., Irle, S., Dai, S., Zhu, H.: Boosting electrosynthesis of ammonia on surface-engineered MXene Ti3C2. Nano Energy **72**, 104681 (2020)

41. Huang, J.-J., Liu, X.-Q., Meng, F.-F., He, L.-Q., Wang, J.-X., Wu, J.-C., Lu, X.-H., Tong, Y.-X., Fang, P.-P.: A facile method to produce MoSe2/MXene hybrid nanoflowers with enhanced electrocatalytic activity for hydrogen evolution. J. Electroanal. Chem. **856**, 113727 (2020)
42. Shuai, T.-Y., Zhan, Q.-N., Xu, H.-M., Zhang, Z.-J., Li, G.-R.: Recent developments of MXene-based catalysts for hydrogen production by water splitting. Green Chem. **25**, 1749–1789 (2023)
43. Hussain, I., Hanan, A., Bibi, F., Kewate, O.J., Javed, M.S., Zhang, K.: Non-Ti (M2X and M3X2) MXenes for energy storage/conversion. Adv. Energy Mater. **14**, 2401650 (2024)
44. Wang, J., Yang, G., Jiao, Y., Yan, H., Fu, H.: Subtle 2D/2D MXene-based heterostructures for high-performance electrocatalytic water splitting. Small Methods 2301602 (2024)
45. Miao, B., Bashir, T., Zhang, H., Ali, T., Raza, S., He, D., Liu, Y., Bai, J.: Impact of various 2D MXene surface terminating groups in energy conversion. Renew. Sustain. Energy Rev. **199**, 114506 (2024)
46. Das, H.T., Balaji, T.E., Dutta, S., Das, N., Maiyalagan, T.: Recent advances in MXene as electrocatalysts for sustainable energy generation: a review on surface engineering and compositing of MXene. Int. J. Energy Res. **46**, 8625–8656 (2022)
47. Kali, S.R., Meghana, V., Madhushree, R., KR, S.D., Pinheiro, D.: Research progress of MXenes and MXenes-based catalysts for photocatalytic water splitting: a systematic review. J. Environ. Chem. Eng. 112748 (2024)
48. Tang, T., Wang, Z., Guan, J.: A review of defect engineering in two-dimensional materials for electrocatalytic hydrogen evolution reaction. Chin. J. Catal. **43**, 636–678 (2022)
49. Wei, Y., Soomro, R.A., Xie, X., Xu, B.: Design of efficient electrocatalysts for hydrogen evolution reaction based on 2D MXenes. J. Energy Chem. **55**, 244–255 (2021)
50. Yan, C., Liu, Y., Zeng, Q., Wang, G., Han, J.: 2D nanomaterial supported single-metal atoms for heterogeneous photo/electrocatalysis. Adv. Funct. Mater. **33**, 2210837 (2023)
51. Kang, Z., Khan, M.A., Gong, Y., Javed, R., Xu, Y., Ye, D., Zhao, H., Zhang, J.: Recent progress of MXenes and MXene-based nanomaterials for the electrocatalytic hydrogen evolution reaction. J. Mater. Chem. A **9**, 6089–6108 (2021)
52. Zubair, M., Hassan, M.M.U., Mehran, M.T., Baig, M.M., Hussain, S., Shahzad, F., Yan, C., Liu, Y., Zeng, Q., Wang, G., Han, J., Kang, Z., Khan, M.A., Gong, Y., Javed, R., Xu, Y., Ye, D., Zhao, H., Zhang, J.: 2D nanomaterial supported single-metal atoms for heterogeneous photo/electrocatalysis. Adv. Funct. Mater. **33**, 2210837 (2023)
53. Zubair, M., Hassan, M.M.U., Mehran, M.T., Baig, M.M., Hussain, S., Shahzad, F.: 2D MXenes and their heterostructures for HER, OER and overall water splitting: a review. Int. J. Hydrogen Energy **47**, 2794–2818 (2022)
54. Zhao, K., Ma, X., Lin, S., Xu, Z., Li, L.: Ambient growth of hierarchical FeOOH/MXene as enhanced electrocatalyst for oxygen evolution reaction. ChemistrySelect **5**, 1890–1895 (2020)
55. Lu, Y., Fan, D., Chen, Z., Xiao, W., Cao, C., Yang, X.: Anchoring Co3O4 nanoparticles on MXene for efficient electrocatalytic oxygen evolution. Sci. Bull. **65**, 460–466 (2020)
56. Liu, J., Chen, T., Juan, P., Peng, W., Li, Y., Zhang, F., Fan, X.: Hierarchical cobalt borate/MXenes hybrid with extraordinary electrocatalytic performance in oxygen evolution reaction. ChemSusChem **11**, 3758–3765 (2018)
57. Ma, T.Y., Cao, J.L., Jaroniec, M., Qiao, S.Z.: Interacting carbon nitride and titanium carbide nanosheets for high-performance oxygen evolution. Angew. Chem. Int. Ed. **55**, 1138–1142 (2016)
58. Benchakar, M., Bilyk, T., Garnero, C., Loupias, L., Morais, C., Pacaud, J., Canaff, C., Chartier, P., Morisset, S., Guignard, N.: MXene supported cobalt layered double hydroxide nanocrystals: facile synthesis route for a synergistic oxygen evolution reaction electrocatalyst. Adv. Mater. Interfaces **6**, 1901328 (2019)
59. Hao, C., Wu, Y., An, Y., Cui, B., Lin, J., Li, X., Wang, D., Jiang, M., Cheng, Z., Hu, S.: Interface-coupling of CoFe-LDH on MXene as high-performance oxygen evolution catalyst. Mater. Today Energy **12**, 453–462 (2019)
60. Tian, M., Jiang, Y., Tong, H., Xu, Y., Xia, L.: MXene-supported FeCo-LDHs as highly efficient catalysts for enhanced electrocatalytic oxygen evolution reaction. ChemNanoMat **6**, 154–159 (2020)

61. Zou, H., He, B., Kuang, P., Yu, J., Fan, K.: Metal–organic framework-derived nickel–cobalt sulfide on ultrathin mxene nanosheets for electrocatalytic oxygen evolution. ACS Appl. Mater. Interfaces **10**, 22311–22319 (2018)
62. Zhao, L., Dong, B., Li, S., Zhou, L., Lai, L., Wang, Z., Zhao, S., Han, M., Gao, K., Lu, M.: Interdiffusion reaction-assisted hybridization of two-dimensional metal–organic frameworks and Ti3C2Tx nanosheets for electrocatalytic oxygen evolution. ACS Nano **11**, 5800–5807 (2017)
63. Wen, Y., Wei, Z., Ma, C., Xing, X., Li, Z., Luo, D.: MXene boosted CoNi-ZIF-67 as highly efficient electrocatalysts for oxygen evolution. Nanomaterials **9**, 775 (2019)
64. Marina, O.A., Canfield, N.L., Stevenson, J.W.: Thermal, electrical, and electrocatalytical properties of lanthanum-doped strontium titanate. Solid State Ion. **149**, 21–28 (2002)
65. Xu, X., Wang, W., Zhou, W., Shao, Z.: Recent advances in novel nanostructuring methods of perovskite electrocatalysts for energy-related applications. Small Methods **2**, 1800071 (2018)
66. Dai, J., Zhu, Y., Zhong, Y., Miao, J., Lin, B., Zhou, W., Shao, Z.: Enabling high and stable electrocatalytic activity of iron-based perovskite oxides for water splitting by combined bulk doping and morphology designing. Adv. Mater. Interfaces **6**, 1801317 (2019)
67. Zhu, Y., Dai, J., Zhou, W., Zhong, Y., Wang, H., Shao, Z.: Synergistically enhanced hydrogen evolution electrocatalysis by in situ exsolution of metallic nanoparticles on perovskites. J. Mater. Chem. A **6**, 13582–13587 (2018)
68. Khan, R., Mehran, M.T., Naqvi, S.R., Khoja, A.H., Baig, M.M., Akram, M.A., Shahzad, F., Hussain, S.: A highly efficient A-site deficient perovskite interlaced within two dimensional MXene nanosheets as an active electrocatalyst for hydrogen production. Int. J. Hydrogen Energy **47**, 37476–37489 (2022)
69. Wang, C., Zhu, X.-D., Mao, Y.-C., Wang, F., Gao, X.-T., Qiu, S.-Y., Le, S.-R., Sun, K.-N.: MXene-supported Co3O4 quantum dots for superior lithium storage and oxygen evolution activities. Chem. Commun. **55**, 1237–1240 (2019)
70. Alam, M.R., Gharami, M., Dev, B., Rahman, M.A., Islam, T.: Next-generation materials for multifunctional applications: design progress and prospects of 2D MXene-enabled hybrid architectures. Adv. Mater. Interfaces 2400681 (2025). https://doi.org/10.1002/admi.202400681
71. AlHadithy, Z.E., AbdulRazak, A.A., Al-Ghaban, A.M.H.A., Alsalhy, Q.F., Meskher, H., Al-Juboori, R.A., Katibi, K.K., Lawal, D.U.: Advancements in water treatment with MXene-enhanced membranes: a review of current progress and future directions. Water Air Soil Pollut. **235**, 809 (2024)
72. Sun, J., Kong, W., Jin, Z., Han, Y., Ma, L., Ding, X., Niu, Y., Xu, Y.: Recent advances of MXene as promising catalysts for electrochemical nitrogen reduction reaction. Chin. Chem. Lett. **31**, 953–960 (2020)
73. Lei, J.-C., Zhang, X., Zhou, Z.: Recent advances in MXene: preparation, properties, and applications. Front. Phys. **10**, 276–286 (2015)
74. Zhang, L., Song, W., Liu, H., Ding, H., Yan, Y., Chen, R.: Influencing factors on synthesis and properties of MXene: a review. Processes **10**, 1744 (2022)
75. Prabhakar Vattikuti, S.V., Shim, J., Rosaiah, P., Mauger, A., Julien, C.M.: Recent advances and strategies in MXene-based electrodes for supercapacitors: applications, challenges and future prospects. Nanomaterials **14**, 62 (2023)
76. Dharmasiri, B., Usman, K.A.S., Qin, S.A., Razal, J.M., Tran, N.T., Coia, P., Harte, T., Henderson, L.C.: Ti3C2Tx MXene coated carbon fibre electrodes for high performance structural supercapacitors. Chem. Eng. J. **476**, 146739 (2023)
77. Qu, D.-Y., Guo, F.-L., Hou, W.-D., Long, J.-F., Li, Y.-Q., Fu, S.-Y.: Cryogenic mechanical properties and liquid oxygen compatibility of MXene/epoxy nanocomposites. Adv. Compos. Hybrid Mater. **7**, 153 (2024)
78. Qi, J., Chen, Z., Jiang, P., Hu, W., Wang, Y., Zhao, Z., Cao, X., Zhang, S., Tao, R., Li, Y.: Recent progress in active mechanical metamaterials and construction principles. Adv. Sci. **9**, 2102662 (2022)

79. Akhter, R., Maktedar, S.S.: MXenes: a comprehensive review of synthesis, properties, and progress in supercapacitor applications. J. Mater. (2023)
80. Im, J.K., Sohn, E.J., Kim, S., Jang, M., Son, A., Zoh, K.-D., Yoon, Y.: Review of MXene-based nanocomposites for photocatalysis. Chemosphere **270**, 129478 (2021)

The Rise of MBenes in Electro-catalysing Water Splitting Reactions

Pratik Shinde, Kassa Belay Ibrahim, Elisa Moretti, and Alberto Vomiero

Abstract Electrocatalytic water splitting has attracted significant attention as a key way to generate clean hydrogen from water without toxic byproducts. The highly efficient electrocatalyst is important to accomplish the energy transfer process during water splitting. To replace currently available precious and noble metal electrocatalysts, nonprecious-metal catalysts have been proposed for large-scale production. Among them, two-dimensional (2D) MBenes class materials attracted curiosity due to the unique characteristics associated with structures. This chapter briefly discusses and presents recent research efforts devoted to exploring 2D MBenes as electrocatalysts for water-splitting reactions. The chapter summarizes different structural aspects of MBenes with their properties, preparation strategies of MBenes, and role of MBenes as an electrocatalyst along with reaction mechanisms. The electrocatalytic properties and performances of MBenes are thoroughly discussed based on theoretical and experimental evidence. A brief outlook on current progress and future challenges in front of 2D MBenes-based electrocatalysts is proposed along with perspectives.

Keywords Water splitting · MBenes · 2D materials · Catalytic activity · Electrocatalyst

P. Shinde · K. B. Ibrahim (✉) · E. Moretti · A. Vomiero
Department of Molecular Sciences and Nanosystems, Ca' Foscari University of Venice, Venezia Mestre, Italy
e-mail: kassabelay.ibrahim@unive.it

A. Vomiero
Division of Materials Science, Department of Engineering Sciences and Mathematics, Luleå University of Technology, Luleå, Sweden

O. P. Pandey et al. (eds.), *2D Materials for Electrochemical Water Splitting*, Engineering Materials, https://doi.org/10.1007/978-981-95-2003-9_8

1 Introduction

The rapidly increasing world's population and its needs demand rapid growth of the industrial sector. This increasing industrialisation and daily use of technologies require energy. Currently, mankind is highly dependent on fossil fuels and coal-based energy resources to mitigate its needs. These resources are available because of thousands of years of environmental processes and are difficult to reproduce as per need. Also, byproducts generated after the use of these fuels are toxic and harmful to the environment. Their use contaminates water resources, pollutes the air and atmosphere, and emits greenhouse gases. This, in turn, results in a rise in global warming and severe changes in climatic conditions. These things are directly impacting the future of living beings and are not good for survival on the planet. To tackle these important challenges, a significant interest has been observed in recent years to find new energy resources and reduce dependency on fossil fuels. In this search, hydrogen enticed attention due to its high gravimetric energy density (120 MJ/kg) and nontoxic behaviour towards nature, which will help us combat environmental pollution and energy shortages for a sustainable future.

Hydrogen is considered a green and clean energy source of energy and can be used as a potential alternative to fossil fuel resources. There is substantial work that has been done and going on in its production, storage, and utilization. At this time, the industrial-scale production of hydrogen is taking place with the use of fossil fuels by methods like steam methane reformation, steam iron process, partial oxidation, and auto-thermal reformation. Though produced hydrogen is a source of clean energy, these methods do not fulfill a vision of zero carbon emission and importantly not reduce reliance on fossil fuels. In alternative to these methods and to meet the goal of clean hydrogen production, the electrocatalytic water splitting method gained attention due to its merits over others. This method helps to produce high-purity and eco-friendly hydrogen without generating any harmful byproducts from abundantly available water resources. It involves two half-reactions after the split of water molecules which are the cathodic hydrogen evolution reaction (HER) and the anionic oxygen evolution reaction (OER). The standard thermodynamic potential of 1.23 V is the minimum required potential for water splitting reaction which corresponds to an energy of 273.1 kJ/mol at 25 °C temperature and one atmospheric pressure. The following reactions are taking place under these conditions:

$$2H^+ + 2e^- \rightarrow H_2 \text{ at } 0 \text{ V versus RHE}$$

and

$$2H_2O \rightarrow O_2 + 4H^+ + 4e^- \text{ at } 1.23 \text{ V versus RHE}$$

The kinetics of these reactions are very sluggish due to the involvement of multi-steps of electron transfer processes and demand additional potential than theoretical values. This limits energy conversion efficiency and subsequently hinders its large-scale production. Therefore, to achieve high performance in water splitting, it is essential to use electrocatalysts with high catalytic activities to overcome these challenges. It will help reduce the energy use required for breaking water molecules and help the process to work in a favourable manner. Thus, the search for electrocatalysts with high current density, low overpotential, and superior longer stability is crucial to developing commercial water-electrolysis systems.

So far, precious metals such as platinum (Pt), ruthenium (Ru), and/or iridium (Ir) based catalysts are recognized as efficient catalysts, due to high reactivity in alkaline and acidic conditions. The HER process shows higher activity with Pt, whereas OER activity becomes better with ruthenium oxide (RuO_2) and/or iridium oxide (IrO_2). However, their low abundance, high costs, and poor durability make these catalysts unsuitable for their large-scale deployment in an economical way. Therefore, it is of great significance to develop low-cost, electroactive, and stable electrocatalysts for water water-splitting method. To achieve this goal, the research community has devoted its efforts to finding and designing electroactive materials.

In the search for strong substitutes and economically affordable catalysts, a wide variety of materials with different dimensions and structures have been used as electrocatalysts for water splitting. Although, zero-dimensional (0D), one-dimensional (1D), and three-dimensional (3D) materials have been widely contributed as electrocatalysts, two-dimensional (2D) materials attracted the greatest attention because of their interesting catalytic properties. The era of these materials started with the exfoliation of monolayer graphene sheets from the graphite structure in 2004. Then, different classes such as transition metal dichalcogenides (TMDs), perovskites, metal monochalcogenides, MXenes, etc., have emerged as potential candidates for the electrolysis of water. These materials exhibit good conductivity, large surface area, tunable band gap, and modifiable structures. Since the discovery of $Ti_3C_2T_x$ MXene in 2011, this class has expanded rapidly in numerous fields due to its interesting physicochemical and optoelectronic properties. The 2D materials, such as transition metal carbides, nitrides, and/or carbo-nitrides, come under the MXene family. These materials are derived from MAX phase material by etching of the A layer due to weaker M-A bonds. Here, M is a transition metal (Ti, Sr, V, Cr, etc.), A is an element from the IIIA and IVA columns (Al, Ga, In, Tl, etc.), and X is carbon or nitrogen. These 2D class materials showed excellent potential in water spitting, yet their performance is unsatisfactory when considering the large-scale purpose. Challenges like large overpotential, substandard, longstanding durability, and difficulty in the formation of controlled structures are still associated with these different families of materials. The advances in MXene structure result in the formation of MBenes, which are mainly due to the smaller molar mass of B as compared to C and N [1–3]. In simple terms, MBenes are boron analogs of MXenes, where X represents boron elements in the MAX phase. The exfoliation of A layer (IIIA and IVA group elements) within the bulk MAB phase results in the formation of the MBenes structure. The bond present between M and A is metallic, while the bond between M and X/B is a mixed covalent/metallic/ionic character [4]. The general chemical formula of MBenes is M_nB_{2n-2} (n = 2–4), where M signifies the transition metal and B is for

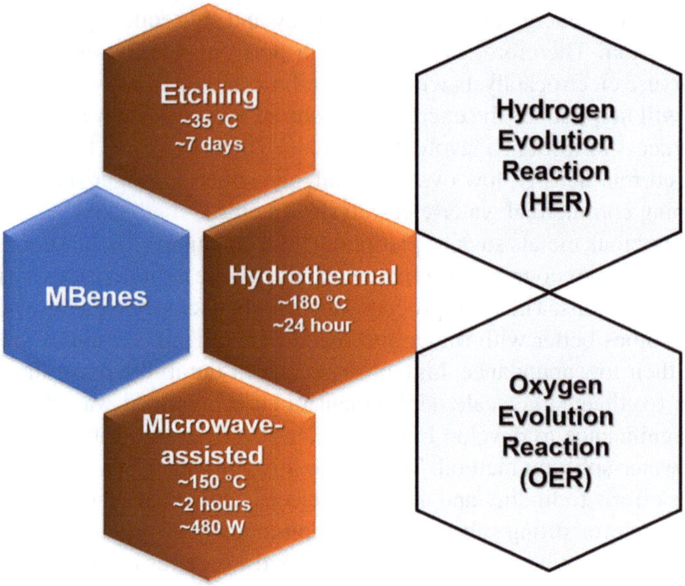

Fig. 1 Schematic representation of the availability of different synthesis methods for MBenes and their applications in water splitting

boron atoms. MBenes exhibit a sandwich kind of layered structure where alternating layers comprising transition metal atoms and boron atoms are observed.

To promote water electrolysis research, many articles are reported focusing on various strategies to utilize 2D material as efficient electrocatalysts. This chapter mainly focuses on current theoretical and practical developments taking place in newly emerged 2D MBenes materials and their use as electrocatalysts in water-splitting reactions (Fig. 1). In the beginning, we introduced different synthesis strategies used to produce MBenes. Then, the mechanism of the water-splitting reaction and the role of MBenes as electrocatalysts a briefly discussed along with their mechanistic insights. In the end, we summarized the current progress, discussed challenges for MBenes used as an electrocatalyst, and pointed out perspectives for their future studies. We believe that this chapter discussion will offer new insights into the development of MBenes-based electrocatalysts for water splitting.

2 Synthesis Strategies for MBenes

There are mainly three different strategies used to prepare MBenes, such as the etching method, hydrothermal synthesis, and microwave-assisted hydrothermal synthesis. This section briefly discusses these methods and their associated parameters.

2.1 Etching

This method is mainly used to produce MBenes structures from MAB structures with the use of solutions of acids or bases. The diluted HCl chosen by researchers to etch out the Al layer from chromium-based MAB phases due to its safe and environmentally friendly behaviour compared to highly acidic etchants [5, 6]. The etching was done at room temperature in a dilute 0.5 mol/L HCl solution for 7 days. In this etching reaction, Cr_2B_2 layers slip relatively 1/2 <a + c> along [101] direction then rebonding takes place (Fig. 2a, b). Then shrining of the structure forms CrB with small lattice constants. The probable chemical reaction happened as follows with the evolution of hydrogen. $Cr_2AlB_2 + 3HCl \rightarrow 2CrB + AlCl_3 + 3/2H_2$ (g). The etching time affects the thickness of the MBene structure during this process. For the preparation of 2D Mo_2B_2 MBene, Jakubczak et al. mixed powder of MoAlB MAB-phase into a solution of 0.9 M HCl and 0.3% H_2O_2 [7]. After 24 h of reaction, the dark grey MAB powder turned to a blue-green colour, which was washed with distilled water until a pH became 6. The etching of Al from the MAB phase results in the formation of an expanded multi-layered structure with open slit-shaped pores (Fig. 2c, d). The partial removal of Al allows for maintaining the structural stability of the MBene because MBenes are unstable and dissolve rapidly. The mixed phase of MoB and Mo_2B in the MBene was obtained after a reaction of MoAlB with 12 M HCl and LiF for 72 h at 35 °C [8]. The deposition of carbon nanotubes (CNT) on MBene changes its multilayer structure to nanosheet structure. The structure of MBene/CNT exhibits a higher surface area of 131.8 m^2/g compared to MBene (41.6 m^2/g) and CNT (94.6 m^2/g). The synthesis of a few layers of MoAlB from MAB material was also achieved with an aqueous hydrofluoric (HF) solution by heating in a water bath at 35 °C for 3 h [9]. Then washed several times with water and added to 20 mL of 20 wt% tetrabutylammonium hydroxide (TBAOH) solution, later centrifuged and shaken for 15 min to get the desired product. Liu et al. prepared multilayered $Mo_{4/3}B_2$ MBenes from $(Mo_{2/3}Y_{1/3})AlB_2$ by LiF and 12 M HCl solution and stirring for 48 h, where etching of Al and Y takes place [10]. The composite of this MBene formed with Fe_3S_4 due to the –OH group available on the MBene structure reacts feasibly with Fe_3S_4 (Fig. 2e, f). The partial etching of Al from MoAlB material was also accomplished at room temperature with NaOH solution [11–13]. The interlayer region can expand when urea is treated on this etched powder. The treatment of undiluted TMAOH, ascorbic acid, and H_2O_2 to etched powder in the autoclave at 140 °C for 24 h, which was further centrifuged, sonicated, and freeze-dried drying produced delaminated $MoAl_{1-x}B$ compound. As a result of this process, the specific surface area of the samples increased, and the surface was enriched with pores, rich basal planes, and molybdenum vacancy defects.

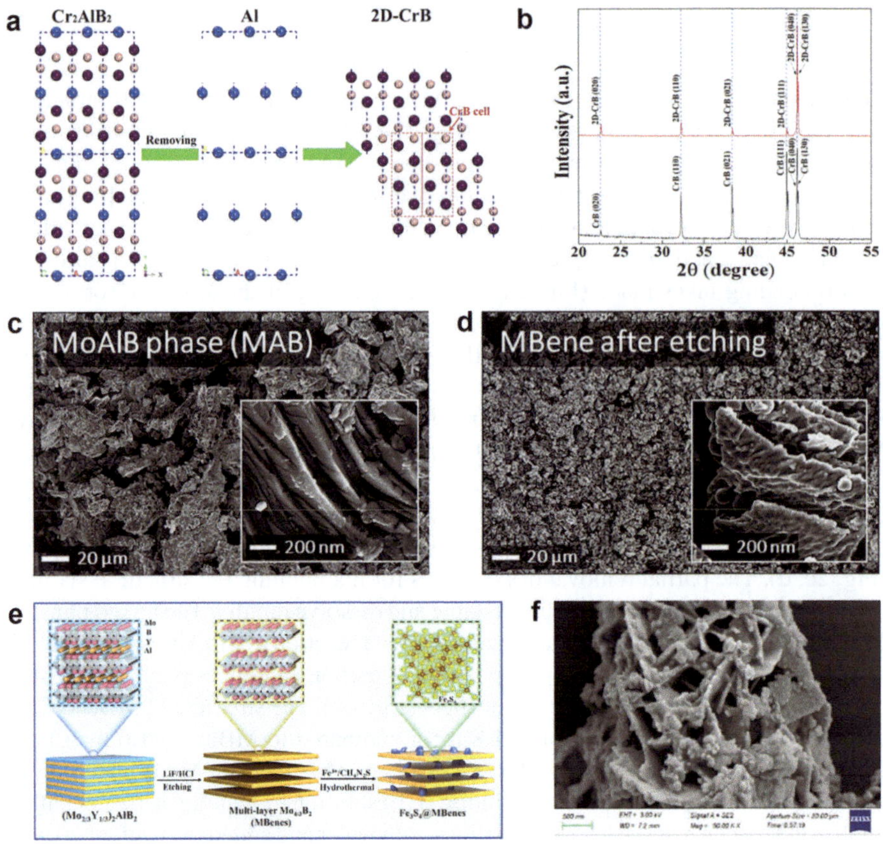

Fig. 2 **a** Schematic represents the transformation from Cr_2AlB_2 structure to 2D-CrB structure. **b** XRD patterns of 2D-CrB nanosheets produced by etching out Al from Cr_2AlB_2 (red) and CrB particles prepared at 1800 °C for 2 h (black). Adapted with permission from Ref. [6] Copyright (2019) Elsevier. **c, d** Scanning electron microscopy (SEM) images of the MoAlB MAB phase and MBene structure after etching. Adapted with permission from Ref. [7] Copyright (2023) Elsevier. **e** Schematic shows preparation of Fe_3S_4@Mbenes composites. **f** SEM image of Fe_3S_4@Mbenes composites. Adapted with permission from Ref. [10] Copyright (2023) Elsevier

2.2 Hydrothermal

The simple and fluorine-free hydrothermal synthesis is also used to synthesize 2D MoB MBene from MAB structure, as shown in Fig. 3a–c [14]. The MoAlB powder was added into 25 wt% NaOH solution and heated at 150 °C for 24 h in an autoclave. Then three- and five-times washed with 1 M NaOH and deionized water respectively during filtration. The MBene powder was collected after drying for 10 h at 80 °C in vacuum conditions. The lamellar-stacking structure of MoAlB changes to an accordion-like layer for MoB after etching treatment. The X-ray diffraction patterns confirm the formation of MoB structure along with residual $MoAl_{1-x}B$ (Fig. 3d).

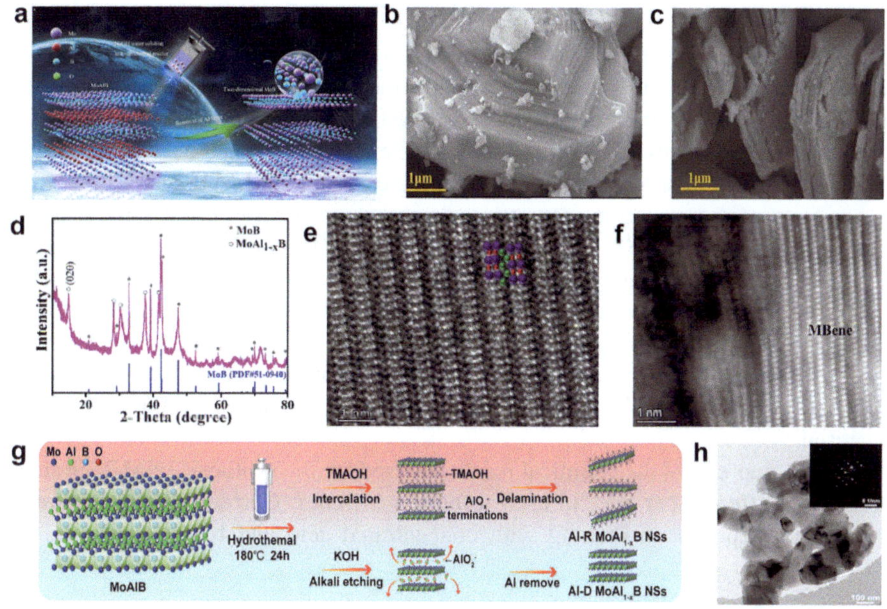

Fig. 3 **a** Synthesis of 2D MoB MBene from MoAlB. **b** SEM image of MoAlB. **c** SEM image of 2D MoB. **d** XRD of 2D MoB Mbene. **e, f** Atomic scale resolution ADF-STEM image of MoAlB and 2D MoB MBene. Adapted with permission from Ref. [14] Copyright (2022) Elsevier. **g** Schematic illustration of the exfoliation method for $MoAl_{1-x}B$ nanosheets in organic and alkali solutions. **h** TEM images of the exfoliated Al-R $MoAl_{1-x}B$ nanosheets. The inset shows SAED patterns. Adapted with permission from Ref. [15] Copyright (2025) American Chemical Society

The percent of MoB Mbene is 73.2% and $MoAl_{1-x}B$ is 26.8%, as per inductively coupled plasma-atomic emission spectrometry. Figure 3e, f shows the annular dark-field scanning transmission electron microscopy (ADF-STEM) images revealing partial etching of Al takes place in 2D MoB. Recently, Chen et al. disintegrated bulk MoAlB powder into 2D MBenes by hydrothermal method with hydrothermal treatment at 180 °C for 24 with tetramethylammonium hydroxide (TMAOH) as an etchant [15]. TMAOH helps to oxidize zero-valence Al atoms into AlO_x^- terminal functional groups and forms Al-R $MoAl_{1-x}B$ nanosheet structure (Fig. 3g, h). Therefore, the hydrothermal approach offers acid-free preparation of MBene structure from MAB phases.

2.3 *Microwave-Assisted Hydrothermal*

Chandel et al. followed a mild microwave-assisted hydrothermal method to prepare a multilayered MoAlB@MBene structure (Fig. 4a–j) [16]. Initially, MoAlB was stirred for 2 h in 0.1 M HCl/NaOH, which was then washed with water until the pH became

6. After that, the residue is transferred to 0.6 M HCl with stirring and sonicated for 2 h. The solution of H_2O_2 (30% (v/v) in H_2O) was added slowly, and the Teflon container was closed. The reaction was performed at 150 °C and 480 W microwave power for 2 h. Then again material was washed with water until pH reached 6. The SEM and TEM images of MoAlB, MAB@MBene, Base-MAB@MBene, and Acid-MAB@MBene are shown in Fig. 4a–h. In the case of acid-MAB@MBene cavities range from nanometer to micrometer scale and show accordion-like structure. The etched regions were ~100 nm wide and parted from each other by <100 nm slabs. The XRD patterns in Fig. 4j confirm the decreased amount of Al atoms in etched products. The bandgaps obtained for MoAlB, MAB@MBene, Acid-MAB@MBene, and Base-MAB@MBene are 3.54, 3.58, 3.65, and 3.88 eV based on the absorption spectra. This method exhibits some advantages like shorter preparation time, less use of harsh chemicals, produced high-quality materials and increased homogeneity of material.

Table 1 represents the chemical resources used for the synthesis of MBene structures and different parameters used to obtain MBenes. Currently, the preparation of MBene depends on etching with concentrated acid and base conditions. Till

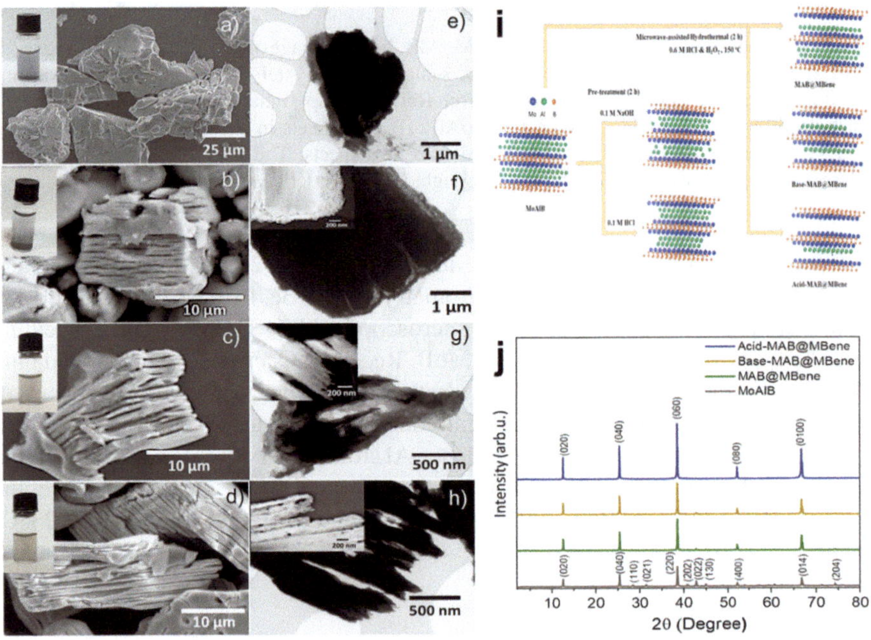

Fig. 4 **a–d** SEM images and **e–h** TEM images of MoAlB, MAB@MBene, Base-MAB@MBene, and Acid-MAB@MBene, respectively. **h** Schematic representation of Al etching from the MoAlB MAB phase for multilayer MoAlB@MBene with pretreatment and microwave-assisted hydrothermal approach. **j** XRD diffraction patterns. Adapted with permission from Ref. [16] Copyright (2025) Royal Society of Chemistry

Table 1 The required chemicals and different synthesis parameters of MBene structures

Method	MBene	MAB	Chemicals	Time	Temp	Refs.
Etching	CrB	Cr_2AlB_2	HCl	8 h	Room temp	[5]
	CrB	Cr_2AlB_2	HCl	7 days	Room temp	[6]
	Mo_2B_2	MoAlB	HCl and H_2O_2	24 h	Cold temp	[7]
	MoB and Mo_2B mixture	MoAlB	HCl and LiF	72 h	35 °C	[8]
	$Mo_{4/3}B_2$	$(Mo_{2/3}Y_{1/3})AlB_2$	HCl and LiF	48 h	Room temp	[10]
	Few layer MoAlB	MoAlB	HF	3 h	35 °C	[17]
	$MoAl_{1-x}B$	MoAlB	NaOH	24 h	Room temp	[11]
Hydrothermal	MoB	MoAlB	NaOH	24 h	150 °C	[14]
	$MoAl_{1-x}B$	MoAlB	TMAOH	24 h	180 °C	[15]

now, experimental evidence is only available for chromium and molybdenum-based MBenes.

3 Water Splitting Reaction

There are two reactions involved in electrochemical water splitting to generate hydrogen and oxygen, which are HER and OER. The selection of catalysts is highly important to get good production. In this section, we have discussed the progress of MBene-based electrocatalysts for both reactions based on experimental and theoretical pieces of evidence.

3.1 Hydrogen Evolution Reaction

This cathodic half-reaction of water splitting involves the transfer of two electrons and on each transmission step catalyst is a key element that reduces the energy barrier. The evolution of hydrogen molecules takes place via Volmer-Heyrovsky or Volmer-Tafel mechanism pathways. In both acidic and basic conditions, hydronium ions and water molecules serve as proton reservoirs, respectively. The HER performance of 2D MoAlB nanosheets prepared using TMAOH (Al-R $MoAl_{1-x}B$) and KOH (Al-D $MoAl_{1-x}B$) is represented in Fig. 5 [15]. The Al-R $MoAl_{1-x}B$ nanosheet catalyst shows an onset potential of 86 mV, an overpotential of 130 mV at 10 mA/cm^2 and a Tafel slope of 58 mV/dec. The pristine MoAlB and Al-D $MoAl_{1-x}B$ nanosheets exhibit a Tafel slope of 128 and 98 mV/dec. The Al-D $MoAl_{1-x}B$ nanosheets show poor linear sweep voltammetry (LSV) performance compared to Al-R $MoAl_{1-x}B$ nanosheets because of the lack of Al sites on the catalyst. The availability of Al^{3+} sites provides a better surface for adsorption, activation, and final deprotonation kinetics (Fig. 5a). The double-layer capacitance (C_{dl}) graphs in Fig. 5d show nearly a 63.5-fold higher value for Al-R $MoAl_{1-x}B$ than for Al-D $MoAl_{1-x}B$ catalyst. This reveals that Al^{3+} atoms with an unsaturated coordination structure act as active sites in the reaction. The Al-R $MoAl_{1-x}B$ catalyst shows stable performance over 70 h at 10 mA/cm^2 current density (Fig. 5f).

The theoretical studies suggest that the bridging oxygen atoms in local –Al–O–Mo– structures possess a strong Brønsted acid nature, which benefits the proton-coupled electron HER process in alkaline conditions. The Pt-$MoAl_{1-x}B$ catalyst requires a lower overpotential in alkaline (32 mV) and acidic (18 mV) conditions to reach 10 mA/cm^2 current than MoAlB and $MoAl_{1-x}B$ catalysts, as shown in Fig. 6 [11]. Its performance in both conditions is also higher as compared to Pt-Mo_2C, Pt-MoS_2, and commercial Pt-C catalysts towards HER. The $MoAl_{1-x}B$ material exhibits strong interactions with single atoms of platinum, which results in enhanced charge density. The Pt single atoms doped into MBene structure by replacing Mo atoms and are favourable for Pt–B or Pt–Mo interactions. The catalyst also exhibits long and stable stability performance in both electrolytes (Fig. 6f). MBene-MoB has better HER catalytic activity than MoAlB and bulk-MoB, which requires a low overpotential of 241.98 mV at 10 mA/cm^2 [18]. Although these catalysts are showing promising results, there are still improvements needed in performance. The performance is still far behind from benchmarked catalyst.

Theoretical studies also predicted that single-atom engineering in MBenes is advantageous to achieve higher HER performance from them [19–21]. The weakening of H–O bonding strength helps to enhance the electrocatalytic activity of MBenes [20]. As shown in Fig. 7a–c, the Gibbs free energy ($|\Delta G_H|$) is closer to zero for S_0 sites than for S_1 and S_2 sites. When a single transition metal atom is embedded in the MBene structure then the H–O bond distance increases, which is useful for H_2 formation. The vanadium atom embedding in $W_2B_2O_2$ decreases its $|\Delta G_H|$ from $|-0.67|$ to 0.013 eV. The transition metals such as Ti, V, Cr, Mn, Fe, Co, Ni and Cu embedded in Mo_2B_2 MBene show improvement in metallic conductivity and enhancement in catalytic performance [21].. The Ni@Mo_2B_2 catalyst shows a lower $|\Delta G_H|$ value of -0.09 eV under 1/4H coverage for the HER. The inclusion of a single transition metal atom in Ti_2B MBene modifies its hydrogen adsorption capacity (Fig. 7d) [19]. The adsorption energy (E_{ads}) values range from -2.85 to -7.05 eV, which are energetically favourable. Also, each single atom-modified Ti_2B system maintains the metallic nature and magnetic states of pure material. The Pd-Ti_2B show ΔG_H of -0.057 eV and exchange current density of 1.25×10^{-4} A cm^{-2}. This ΔG_H value of Pd-Ti_2B is even lower than Pt (111) (-0.089 eV). The monolayer of TiB MBene shows ΔG_H of 0.03 eV, which is nearly close to the ideal 0 eV value,

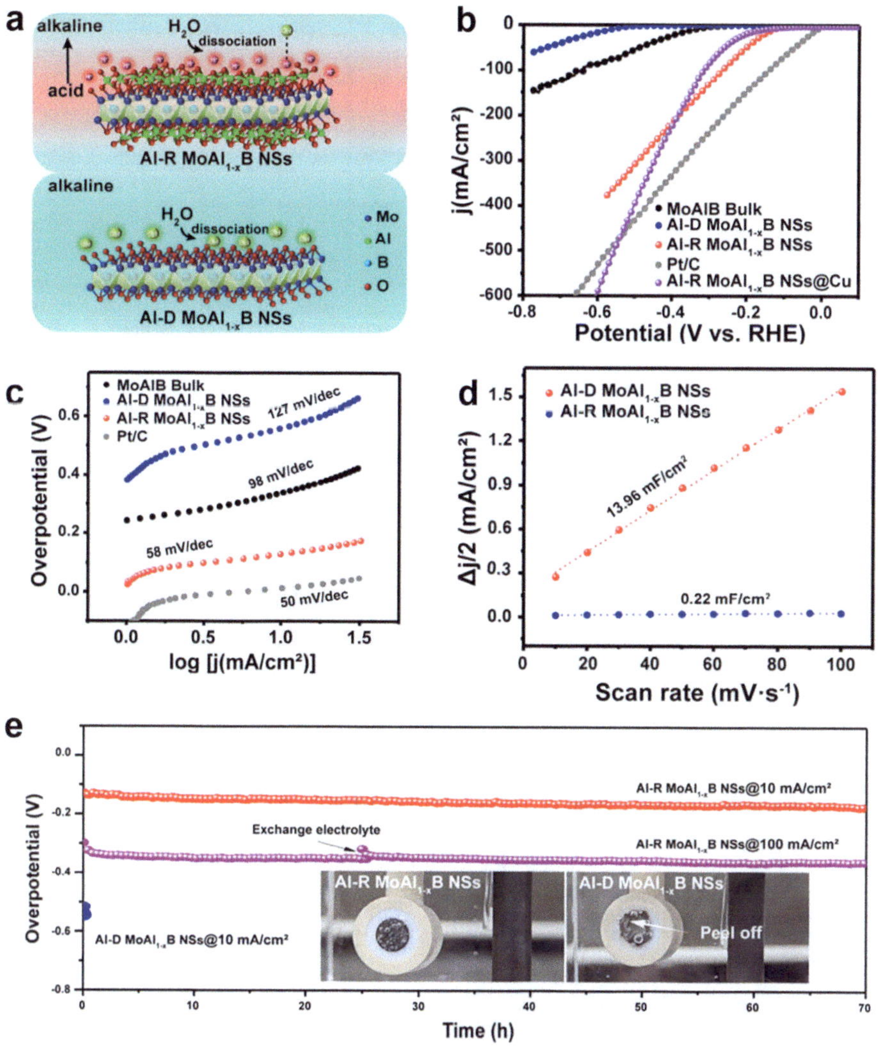

Fig. 5 **a** Schematic illustration of the alkaline HER process. **b** LSV curves. **c** Tafel plots. **d** C_{dl} plots. **e** Chronopotentiometry test at 10 and 100 mA/cm^2. Adapted with permission from Ref. [15] Copyright (2025) American Chemical Society

which signifies its suitability for HER as an efficient electrocatalyst [22]. This thermodynamic activity of TiB material is superior to carbide structures such as MXenes. The hexagonal transition-metal borides like h-ZrBO, h-MoBO, and h-Nb_2BO_2 also show ΔG_H values closer to zero, making them suitable catalysts for HER [23]. The implanting of transition metals such as V, Cr, and Mo in hexagonal Hf_2BO_2 MBene reduces its ΔG_H from 0.939 to −0.04, 0.05 and −0.04 eV, respectively [24]. This happens due to the synergistic effect between them and the activation of the MBene

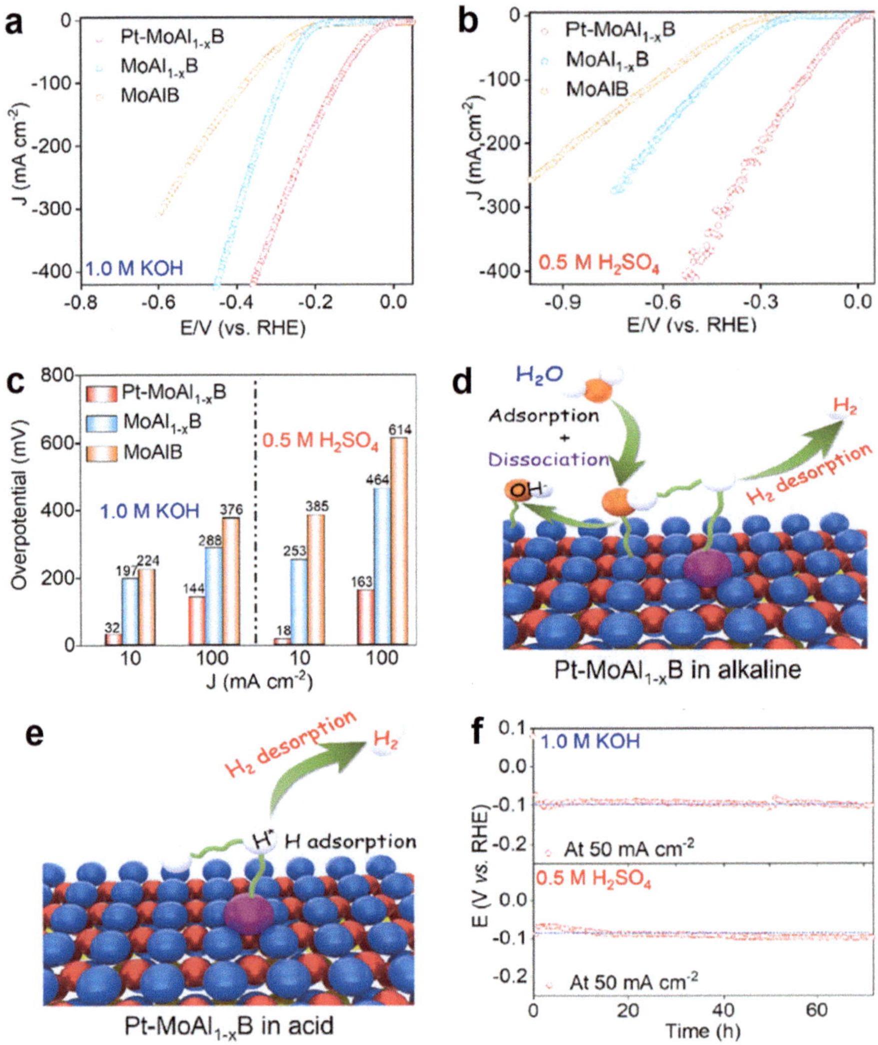

Fig. 6 **a** LSV curves of MoAlB, $MoAl_{1-x}B$, and Pt-$MoAl_{1-x}B$ catalysts in 1.0 M KOH. **b** LSV curves of MoAlB, $MoAl_{1-x}B$, and Pt-$MoAl_{1-x}B$ catalysts in 0.5 M H_2SO_4. **c** Comparison of catalyst performance in alkaline and acid electrolyte. **d** HER mechanism takes place on Pt-$MoAl_{1-x}B$ in an alkaline medium. **e** HER mechanism takes place on Pt-$MoAl_{1-x}B$ in an acidic medium. **f** Stability tests of Pt-$MoAl_{1-x}B$ in different media. Adapted with permission from Ref. [11] Copyright (2023) Royal Society of Chemistry

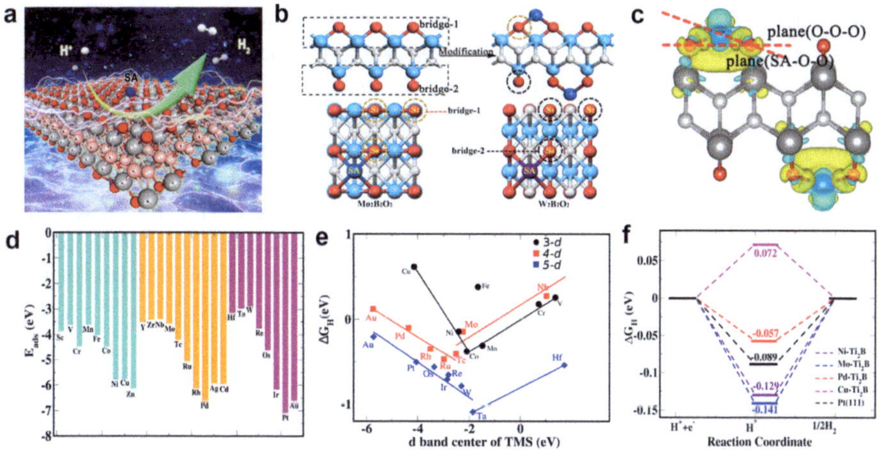

Fig. 7 **a** Schematic representing single-metal atoms supported on MBenes for HER analysis. **b** The O-terminated MBene structure and its surface modification by a single-metal adatom. **c** Three-dimensional configuration of the V-$W_2B_2O_2$. Adapted with permission from Ref. [20] Copyright (2020) American Chemical Society. **d** The calculated adsorption energy of transition metal atoms on Ti_2B. **e** Relation between ΔG_H and the d-band centre value of transition metal on Ti_2B. **f** The ΔG_H of transition metal-Ti_2B. Adapted with permission from Ref. [22] Copyright (2024) Elsevier

surface by broadening O-p orbitals and shifting up the p-band centre. This theoretical and experimental evidence suggests that the embedding of a single atom into the MBenes structure or modification of its surface can produce high-performance HER electrocatalysts.

3.2 *Oxygen Evolution Reaction*

As compared to HER, the anodic OER involves a complicated four-electron multistep transfer process and is more energy-intensive. The electrochemical OER performance of MAB material and a few-layer exfoliated MoAlB (EMAB) in 1 M KOH electrolyte is shown in Fig. 8a–e [9]. The EMAB required an overpotential of 328 mV at 10 mA/cm^2 and a Tafel slope of 70.6 mV. This is better than MAB, which shows an overpotential of 386 mV and a Tafle slope of 85.23 mV/dec. The double-layer capacitance of EMAB (3.61 mF/cm^2) is higher than MAB (2.14 mF/cm^2) material, which suggests EMAB have more active sites. Removing aluminium atoms from MAB material exposed molybdenum atoms, which adsorb the hydroxyl intermediate. The EMAB material revealed better OER performance compared to MAB due to increased hydroxyl adsorption and reduced surface valence of molybdenum. The etching and exfoliation decrease the valence states of molybdenum, which is responsible for better stability performance. The first principle calculations reveal that different MBene-supported single-atom catalyst such as Fe-$Ti_2B_2O_2$, Co-$Ti_2B_2O_2$,

$Fe\text{-}Mo_2B_2O_2$, $Co\text{-}Mo_2B_2O_2$, $Ni\text{-}Mo_2B_2O_2$, $Fe\text{-}W_2B_2O_2$ and $Co\text{-}W_2B_2O_2$ shows an overpotential of 0.51 V, 0.46 V, 0.48 V, 0.41 V, 0.54 V, 0.58 V and 0.30 V, respectively [17]. These values are comparable to precious metal-based catalysts like RuO_2 (0.42 V) and IrO_2 (0.56 V). Another study of dual atom catalysts with MBene such as $Ni\text{-}Mo_2B_2O_2$, $FeCo\text{-}Mo_2B_2O_2$ and $CoNi\text{-}Mo_2B_2O_2$ reveals an overpotential of 0.49, 0.38, and 0.25 V, which is also superior to IrO_2 (110) catalyst [25]. Transition metals (Ni and Cu) with Mo_2B_2 show an overpotential of 0.52 and 0.31 V for OER, where metals act as active sites for reaction. Therefore, these experimental and theoretical discussions show that MBene structures can act as good electrocatalysts for OER reactions.

4 Summary

The discovery of MBenes brought an interesting turn in the search for a suitable electrocatalyst for water-splitting reactions. These materials exhibit superior and tunable electronic properties similar to other 2D materials. The broad range of designing different MBene nanostructures with high surface area and active sites make them efficient electrocatalysts. The research of MBenes is still at the primary stage and several theoretical studies predicted its bright future in catalysis reactions. Most of these predictions are experimentally not proven yet and need more investigations. Very few structures of MBenes are yet reported with experimental evidence as electrocatalysts for water-splitting reactions. Though the performance of studies on MBene catalysts is looking promising, yet not satisfied and not closer to benchmark catalysts like Pt, Ru, and Ir. Moreover, MBene's performance is behind TMDs, oxides, MXenes and other 2D material-based electrocatalyst.

There are still some big obstacles present in utilizing MBenes as electrocatalysts in water splitting. Most of the experimental investigations reported only partial etching of aluminium from MAB structure, instead of its complete etching. The control over the etching process is lacking and also needs strong acidic conditions. Therefore, in the coming time, researchers should keep focused on, control and selective etching of MBene materials from respective MAB phases. There is a need to find out and test other preparation approaches for MBene structure formation. To improve the performance of these materials in the future, most of the focus should be kept on modifying MBene structures with different strategies such as doping, composite formation or expanding interlayer spacing. It will help to make the adsorption and desorption processes of ions easier and improve catalytic activity. Also, need to combine theoretical studies and experimental investigations with each other. This will guide us to understand the mechanism of MBenes as an electrocatalyst and also help with future modifications. The in-situ characterisation techniques will also help understand the mechanisms and role of MBenes in catalysis reactions. We hope that extensive research and indepth analysis of MBene structures will make them suitable electrocatalyst in coming years for water splitting reaction.

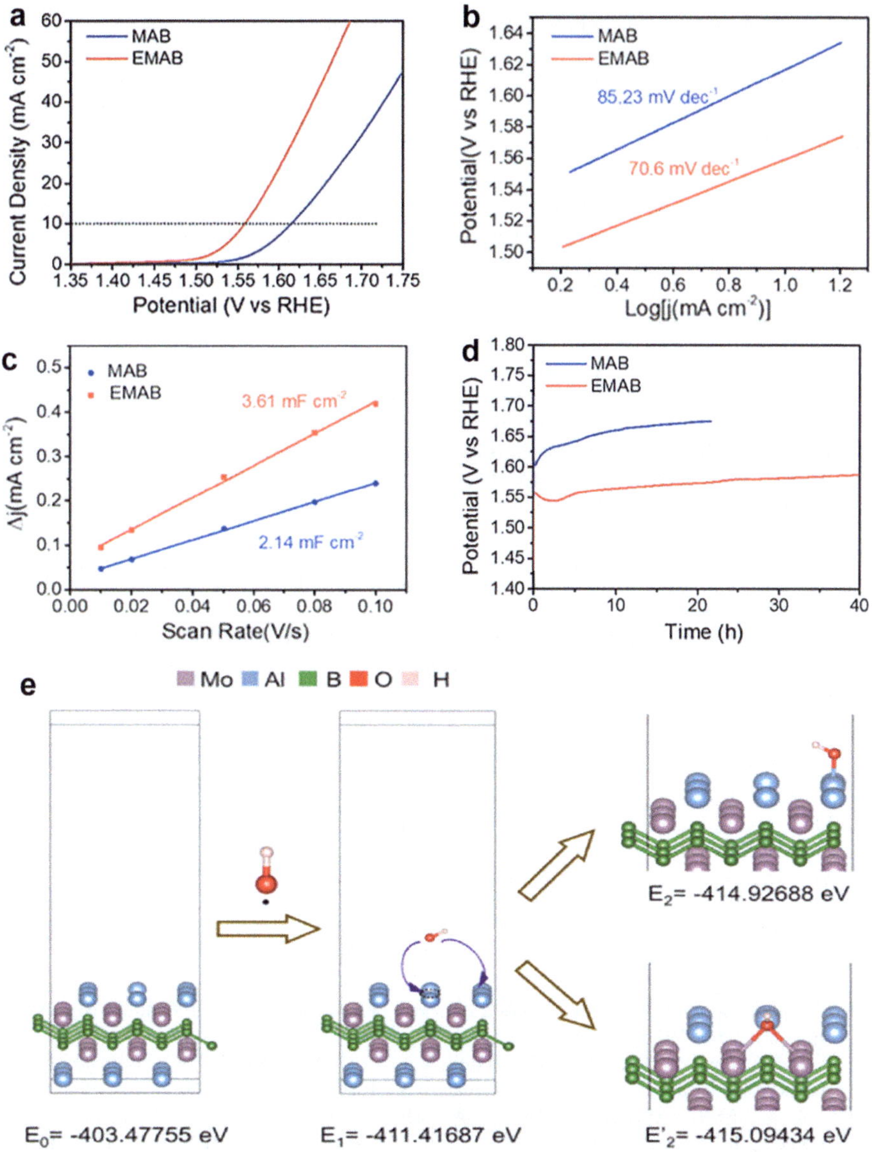

Fig. 8 **a** Polarization curves with scan rate 5 mV/s. **b** Corresponding Tafel plots. **c** Current versus scan rate graph. **d** Chronopotentiometric responses catalyst at 10 mA/cm^2 current density. **e** The side views and total energies of the *OH adsorption surface. Adapted with permission from Ref. [17] Copyright (2024) Elsevier

Conflict of Interest The authors declare no conflict of interest.

References

1. Gao, D., Zhong, J., Li, Y., Zhang, Y., Zhao, S., Guo, Z., Lin, J., Fang, Y., Tang, C.: Efficient polyselenides anchoring and conversion by functionalized Mo_2B_2 MBene for advanced Li-Se battery. J. Alloy. Compd. **15**(932), 167722 (2023)
2. Bo, T., Liu, P.F., Xu, J., Zhang, J., Chen, Y., Eriksson, O., Wang, F., Wang, B.T.: Hexagonal Ti_2B_2 monolayer: a promising anode material offering high rate capability for Li-ion and Na-ion batteries. Phys. Chem. Chem. Phys. **20**(34), 22168–22178 (2018)
3. Ade, M., Hillebrecht, H.: Ternary borides Cr_2AlB_2, Cr_3AlB_4, and Cr_4AlB_6: the first members of the series $(CrB_2)_n$ CrAl with n = 1, 2, 3 and a unifying concept for ternary borides as MAB-phases. Inorg. Chem. **54**(13), 6122–6135 (2015)
4. Rosli, N.F., Nasir, M.Z., Antonatos, N., Sofer, Z., Dash, A., Gonzalez-Julian, J., Fisher, A.C., Webster, R.D., Pumera, M.: MAX and MAB phases: two-dimensional layered carbide and boride nanomaterials for electrochemical applications. ACS Appl. Nano Mater. **2**(9), 6010–6021 (2019)
5. Zhang, H., Xiang, H., Dai, F.Z., Zhang, Z., Zhou, Y.: First demonstration of possible two-dimensional MBene CrB derived from MAB phase Cr_2AlB_2. J. Mater. Sci. Technol. **34**(11), 2022–2026 (2018)
6. Zhang, H., Dai, F.Z., Xiang, H., Wang, X., Zhang, Z., Zhou, Y.: Phase pure and well crystalline Cr_2AlB_2: a key precursor for two-dimensional CrB. J. Mater. Sci. Technol. **35**(8), 1593–1600 (2019)
7. Jakubczak, M., Wojciechowska, A., Zambrano, D.F., Moncada, D., Birowska, M., Moszczyńska, D., Eisawi, K., Naguib, M., Rosenkranz, A., Jastrzębska, A.M.: Synthesis, properties and solid lubrication performance of MoAlB-based MBene. Appl. Mater. Today **1**(35), 101925 (2023)
8. Li, Z., Zeng, Q., Yu, Y., Liu, Y., Chen, A., Guan, J., Wang, H., Liu, W., Liu, X., Liu, X., Zhang, L.: Application of transition metal boride nanosheet as sulfur host in high loading Li-S batteries. Chem. Eng. J. **15**(452), 139366 (2023)
9. Chen, H., Wang, Z., He, H., Chen, J., Yin, H., Yu, D., Liang, J., Qin, L., Huang, Y., Chen, D.: Few-layer MoAlB nanosheets with Al vacancies enhanced hydroxyl adsorption for improved water oxidation kinetics. Renew. Energy **1**(225), 120268 (2024)
10. Liu, F., Zhang, Y., Wang, S., Lu, Y., Hu, B.: Elimination mechanisms of U (VI) and Se (IV) by MBenes supported Fe_3S_4 micro-crystal: synergy of adsorption, reduction and transformation. Appl. Surf. Sci. **1**(636), 157813 (2023)
11. Park, S.J., Nguyen, T.H., Tran, D.T., Dinh, V.A., Lee, J.H., Kim, N.H.: Delaminated MBene sheets beyond usual 2D transition metal materials for securing Pt single atoms to boost hydrogen evolution. Energy Environ. Sci. **16**(9), 4093–4104 (2023)
12. Alameda, L.T., Lord, R.W., Barr, J.A., Moradifar, P., Metzger, Z.P., Steimle, B.C., Holder, C.F., Alem, N., Sinnott, S.B., Schaak, R.E.: Multi-step topochemical pathway to metastable Mo_2AlB_2 and related two-dimensional nanosheet heterostructures. J. Am. Chem. Soc. **141**(27), 10852–10861 (2019)
13. Alameda, L.T., Holder, C.F., Fenton, J.L., Schaak, R.E.: Partial etching of Al from MoAlB single crystals to expose catalytically active basal planes for the hydrogen evolution reaction. Chem. Mater. **29**(21), 8953–8957 (2017)
14. Xiong, W., Feng, X., Xiao, Y., Huang, T., Li, X., Huang, Z., Ye, S., Li, Y., Ren, X., Wang, X., Ouyang, X.: Fluorine-free prepared two-dimensional molybdenum boride (MBene) as a promising anode for lithium-ion batteries with superior electrochemical performance. Chem. Eng. J. **15**(446), 137466 (2022)

15. Chen, Z., Wang, H., Zhang, C., Gou, Y., Gong, Z., Jiang, Y., Zeng, H., Wang, J., Meng, F., Cui, Y.: MBene Brønsted acid catalyst for hydrogen evolution reaction in alkaline electrolyte. ACS Catal. **3**(15), 2885–2895 (2025)
16. Chandel, M., Purbayanto, M.A., Kowal, D., Moszczynska, D., Wójcik, A., Birowosuto, M.D., Naguib, M., Jastrzębska, A.: Multilayered MoAlB@MBene using mild microwave-assisted etching and their optical properties. Mater. Horiz. (2025)
17. Wang, E., Zhang, B., Zhou, J., Sun, Z.: High catalytic activity of MBenes-supported single atom catalysts for oxygen reduction and oxygen evolution reaction. Appl. Surf. Sci. **1**(604), 154522 (2022)
18. Wei, S., Kale, G., Lai, X.: Unlocking enhanced electrochemical performance of MBene-MoB through controlled aluminum dissipation from MoAlB. Small **20**(38), 2401573 (2024)
19. Shukla, A.: Tailoring hydrogen evolution reaction with single-atom catalysts on Ti_2B MBenes: insights from computational screening. Int. J. Hydrogen Energy **5**(77), 479–485 (2024)
20. Li, B., Wu, Y., Li, N., Chen, X., Zeng, X., Arramel, Zhao, X., Jiang, J.: Single-metal atoms supported on MBenes for robust electrochemical hydrogen evolution. ACS Appl. Mater. Interfaces **12**(8), 9261–9267 (2020)
21. Zhang, T., Zhang, B., Peng, Q., Zhou, J., Sun, Z.: Mo_2B_2 MBene-supported single-atom catalysts as bifunctional HER/OER and OER/ORR electrocatalysts. J. Mater. Chem. A **9**(1), 433–441 (2021)
22. Li, F., Tang, Q.: First-principles calculations of TiB MBene monolayers for hydrogen evolution. ACS Appl. Nano Mater. **2**(11), 7220–7229 (2019)
23. Miao, N., Duan, Z., Wang, S., Cui, Y., Feng, S., Wang, J.: H-MBenes: promising two-dimensional material family for room-temperature antiferromagnetic and hydrogen evolution reaction applications. ACS Appl. Mater. Interfaces **16**(5), 5792–5802 (2024)
24. Feng, S., Miao, N., Wang, J.: Hexagonal MBene (Hf_2BO_2): a promising platform for the electrocatalysis of hydrogen evolution reaction. ACS Appl. Mater. Interfaces **13**(47), 56131–56139 (2021)
25. Mou, Y., Wang, Y., Wan, J., Yao, G., Feng, C., Zhang, H., Wang, Y.: Rational design of 2D MBene-based bifunctional OER/ORR dual-metal atom catalysts: a DFT study. Phys. Chem. Chem. Phys. **25**(42), 29135–29142 (2023)

Graphdiyne-Supported Catalysts for Enhanced HER and OER Performance

Prakash Chandra

Abstract Graphdiyne (GDY) has gained recognition as an effective base for electrocatalytic systems owing to its distinctive sp–sp^2 carbon structure, expansive surface, and superior electrical conductivity. This review examines how GDY-supported catalysts boost the efficiency of the hydrogen evolution reaction (HER) and oxygen evolution reaction (OER). Combining GDY with transition metal oxides (e.g., $NiFe_2O_4$, Co_3O_4, MnO_2), single-atom catalysts (e.g., Ru, Ir, Fe), and bifunctional systems (e.g., CoFe-GDY, Ni-GDY) enhances stability and charge movement, elevating catalytic performance. Additionally, hybrid materials integrating GDY with MXenes, metal–organic frameworks (MOFs), and perovskites show improved electrocatalytic effectiveness, durability, and potential for large-scale use. The discussion covers recent synthesis advancements, obstacles in mass production, and future opportunities for refining GDY-based catalysts in green energy technologies.

Keywords GDY · Electrocatalysis · HER · OER · Transition metal oxides · Single-atom catalysts · Hybrid materials · Water splitting

1 Introduction

The Industrial Revolution marked a surge in fossil fuel use, fueling economic progress but also accelerating their depletion due to escalating global energy needs. The burning of these fuels has worsened environmental problems, such as climate change and pollution, with atmospheric CO_2 levels now nearly twice their pre-industrial levels. This has spurred the search for affordable, eco-friendly energy solutions. Electrocatalysis stands out as a viable option, though developing high-performance catalysts remains a hurdle. Traditional carbon supports like graphene and carbon nanotubes have been employed to improve catalyst stability and activity, tackling issues such as low conductivity and insufficient active sites.

P. Chandra (✉)
Department of Chemistry, School of Energy Technology, Pandit Deendayal Energy University, Gandhinagar, Gujarat, India
e-mail: prakash.chandra@sot.pdpu.ac.in

O. P. Pandey et al. (eds.), *2D Materials for Electrochemical Water Splitting*, Engineering Materials, https://doi.org/10.1007/978-981-95-2003-9_9

Unlike typical carbon materials reliant on sp^2 hybridization, GDY combines sp- and sp^2-hybridized carbon, offering unique electronic, mechanical, and optical traits. Theorized by Baughman et al. in the late 1980s as part of the graphyne family, GDY features a highly conjugated π-system, a structured porous design, and adjustable electronic properties. Its first synthesis, achieved by Li et al. in 2010 via an in-situ cross-coupling technique, opened doors to applications in batteries, catalysis, solar energy, biomedicine, gas separation, and water purification. GDY shares Dirac cone characteristics with graphene and exhibits a tunable bandgap with charge mobility rivaling graphene and carbon nanotubes. Its mechanical properties can be modified by varying acetylenic linkages and stacking patterns, providing a high surface area and nanoporous structure that supports efficient charge transport [1–5]. GDY demonstrates a high specific surface area and unique sp- and sp^2-hybridized carbon atom networks, capable of naturally adjustable bandgap, uniformly nanoporous structure, rapid carrier mobility activated by inhomogeneous π-orbitals, and accessible on/off ratios [6, 7]. Unlike traditional carbonaceous supports, GDY's diacetylenic linkages (–C≡C–) enhance its ability to conduct electricity efficiently and transfer of electrons. These advantages, coupled with the direct binding ability of metal atoms resulting from *sp*-hybridization, enable GDY to be applied across various research fields, especially as a substrate to support and further improve the performance of metals and metal compounds. One of the most compelling advantages of GDY is its high surface area and porous architecture, which provide abundant anchoring sites for metal and non-metal catalysts.

Lately, well-regulated synthesis, structural alteration, and real-world applications of GDY with dissimilar morphologies and 3D assemblies have been effectively produced; the functionalization approaches for GDY are also significantly augmented, and electrocatalysts based on GDY have also been advanced robustly (see Fig. 1) [8–19].

GDY's benefits arise from its large surface area, hybrid carbon framework, and adaptable bandgap, which enhance charge transfer and catalytic efficiency. The diacetylenic (–C≡C–) bonds improve electrical conductivity and electron flow in electrochemical processes, distinguishing it from conventional carbon supports. Its sp-hybridized sites allow direct metal atom attachment, making it an ideal platform for boosting metal and metal compound performance. GDY also shows robust chemical and electrochemical stability in harsh conditions, resisting oxidation and degradation better than traditional carbon materials due to its strong carbon–carbon bonds and π-conjugated network, ensuring long-term catalytic reliability [16]. Additionally, vacancy engineering and structural modifications can tailor GDY's interaction with metal catalysts, improving their adsorption energy for key reaction intermediates. These tunable properties position GDY as a highly versatile platform for developing advanced electrocatalysts. More recently, the application of GDY as an electrocatalyst support, significantly enhanced energy conversion efficiency in processes like the HER, OER, oxygen reduction reaction (ORR), and overall water splitting (OWS).

GDY's extensive surface area and nanoporous 2D structure prevent catalyst clumping and ensure even distribution of active species, maximizing site exposure for high reaction rates. Its porous design also aids mass transport of reactants and

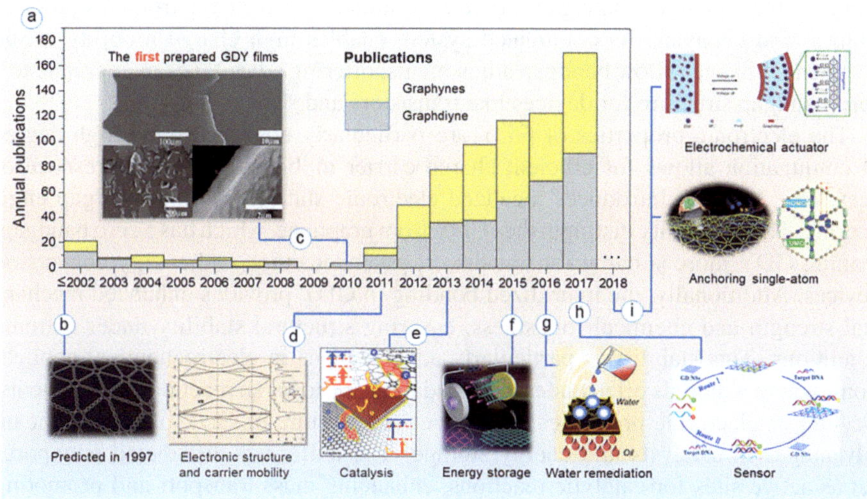

Fig. 1 An overview of graphyne and graphdiyne research trends from 1987 to 2017, derived from Web of Science data. Reprinted with authorization [20]. Copyright 2018 The Royal Society of Chemistry

products, reducing diffusion barriers in electrocatalysis. The π-conjugated system strengthens interactions with metal or non-metal catalysts, stabilizing single-atom sites and optimizing their electronic environment for intermediate adsorption. This synergy enhances charge transfer, prevents metal leaching, and boosts durability, positioning GDY-supported catalysts as promising options for energy conversion processes like water splitting [21–24].

In this book chapter, a basic understanding of GDY along with a comprehensive overview of the synthesis and applications of GDY-supported metals and metal compounds is presented. Moreover, the applications of GDY-supported electrocatalysts in energy conversion are described. Furthermore, based on current understanding, the key challenges and prospects of GDY and GDY-based materials are outlined to facilitate their continued development. The potential of GDY-supported materials is also discussed, offering diverse perspectives and insights for future research directions [25].

2 Structure of GDY

GDY, a 2D carbon allotrope, differs from graphene with its mix of sp^2- and sp-hybridized bonds, featuring benzene rings linked by diacetylenic units (–C≡C–C≡C–). This creates a honeycomb-like porous network with a high surface area, supporting efficient charge movement. The sp-hybridization imparts semiconducting behavior with a tunable bandgap (~0.46 eV), suitable for nanoelectronics, optoelectronics, and catalysis. Its conjugated system enables high charge mobility, while acetylenic linkages allow bandgap adjustments, offering advantages over graphene's zero-bandgap structure for devices like transistors and sensors.

The electronic properties of GDY are particularly noteworthy. Its high degree of conjugation allows for efficient charge carrier mobility, while the presence of acetylenic linkages introduces localized electronic states, enabling bandgap engineering. This tunability distinguishes GDY from graphene, which has a zero bandgap, making GDY more suitable for transistors, photodetectors, and energy conversion devices. Additionally, the hybridized bonding in GDY provides enhanced mechanical strength and chemical robustness, ensuring structural stability under extreme conditions. This stability is particularly advantageous in electrochemical applications, where materials often undergo degradation due to harsh reaction environments. Beyond its electronic properties, the porous architecture of GDY offers significant advantages in catalysis and energy storage. The uniformly distributed nanopores act as active sites for catalytic reactions, enhancing mass transport and promoting efficient adsorption and diffusion of reactants. This makes GDY an excellent candidate for electrocatalysis, particularly in the HER, OER, and CO_2 reduction reaction (CO_2RR). Additionally, the presence of triple bonds in the GDY framework provides anchoring sites for metal atoms, dopants, and functional groups, allowing for precise material modifications that can optimize catalytic activity and selectivity. These functionalization capabilities make GDY highly versatile for applications in fuel cells, metal-air batteries, and supercapacitors. Overall, GDY, unlike traditional catalysts such as platinum (Pt) for HER and iridium oxide (IrO_2) or ruthenium oxide (RuO_2) for OER, GDY offers a cost-effective, stable, and tunable alternative with high catalytic activity when properly engineered.

3 Synthesis Techniques of Graphdiyne (GDY)

The synthesis of GDY is challenging due to the complexity of its sp–sp^2 hybridized structure, but several successful approaches have been developed to fabricate high-quality GDY with well-defined morphologies. The most widely used techniques include bottom-up chemical synthesis, surface-mediated polymerization, and chemical vapor deposition (CVD).

3.1 Bottom-Up Chemical Synthesis

The bottom-up chemical synthesis of graphdiyne (GDY) is a well-established approach that involves the controlled polymerization of small molecular precursors through cross-coupling reactions. This method enables the precise design and assembly of GDY at the molecular level, ensuring structural uniformity and controlled functionalization. Among the various coupling strategies, Glaser-Hay coupling and Eglinton coupling are the most commonly employed reactions to form the characteristic diacetylenic (–C≡C–C≡C–) linkages in GDY. These reactions facilitate the formation of extended conjugated frameworks, allowing GDY to exhibit its unique electronic, mechanical, and chemical properties.

The polymerization of graphdiyne (GDY) typically starts with 1,3,5-triethynylbenzene (TEB), a highly conjugated molecular precursor containing three terminal ethynyl (–C≡C–H) groups. These ethynyl groups act as reactive sites for oxidative coupling, enabling the formation of an extended GDY network. In the presence of an oxidative coupling catalyst, such as Cu(I) or Cu(II) salts, the terminal alkynes undergo polymerization through the formation of diacetylene (–C≡C–C≡C–) linkages, which are essential for constructing the highly conjugated sp–sp^2 hybridized carbon structure of GDY [26–29]. Thus far, the synthesis of GDY has predominantly relied on four copper-mediated coupling strategies: Glaser, Glaser-Hay, Eglinton, and alkynylsilane couplings, as illustrated in Fig. 2A. Gaining insight into the fundamental catalytic systems and reaction mechanisms is essential for advancing more efficient coupling strategies for GDY.

3.2 Glaser-Hay Coupling for GDY Synthesis

The Glaser-Hay coupling reaction relies on Cu(I) salts (e.g., CuCl or CuBr) as catalysts, in the presence of molecular oxygen (O_2) or air to regenerate the active Cu(I) species [30–33]. The reaction proceeds through the formation of Cu(I)-acetylide intermediates, which subsequently undergo oxidative dimerization, resulting in the formation of diacetylenic linkages. One critical aspect of this reaction is the requirement of a stabilizing ligand, such as N,N,N′,N′-tetramethyl ethylenediamine (TMEDA), which enhances the solubility and stability of the Cu(I) intermediates. TMEDA prevents undesired side reactions and promotes selective coupling, leading to the efficient growth of GDY layers. Despite its effectiveness, the Glaser-Hay coupling method has some limitations in GDY synthesis. One major issue is the competition between polymerization and oxidative degradation, as Cu(I) catalysts can promote unwanted oxidation of the ethynyl groups under excessive oxygen exposure. Additionally, the reaction often yields small, fragmented GDY sheets due to the difficulty in achieving long-range order and uniform coupling. To overcome these challenges, researchers have explored alternative oxidative coupling methods, such as Eglinton coupling, which employs Cu(II) salts (e.g., $Cu(OAc)_2$ or $CuCl_2$) instead of Cu(I) and eliminates the need for an external ligand. This modification improves reaction selectivity, minimizes side reactions, and facilitates the formation of more structurally uniform GDY networks.

In conclusion, Glaser-Hay coupling remains a widely used approach for GDY synthesis due to its ability to efficiently construct diacetylene-linked carbon networks. However, its reliance on Cu(I) catalysts and ligands introduces challenges related to reaction control, polymerization efficiency, and structural integrity. To address these

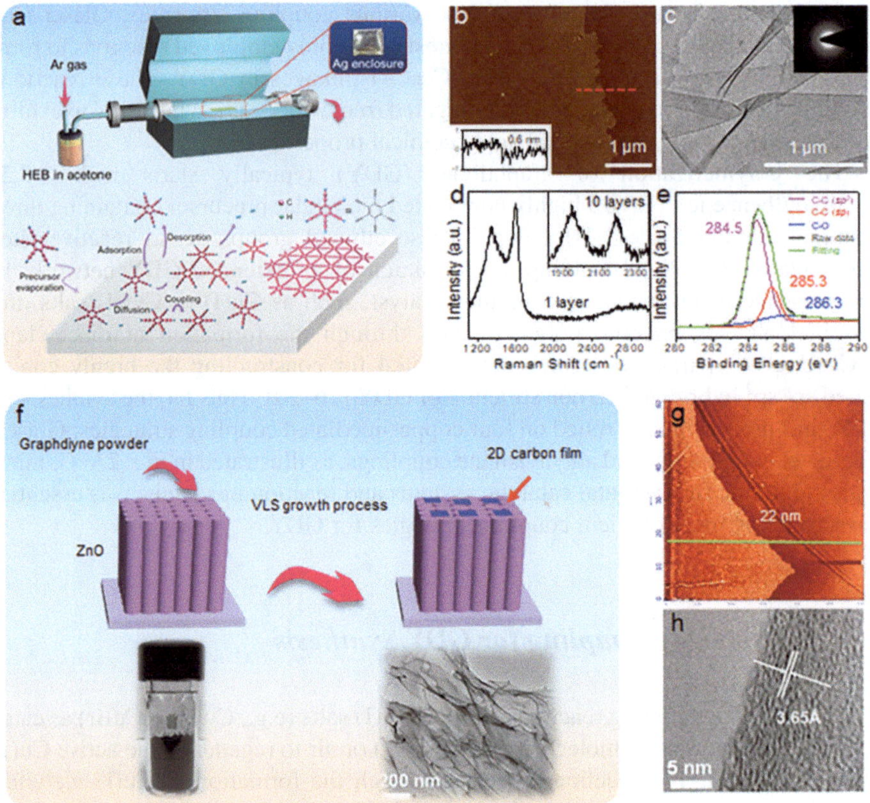

Fig. 2 Synthesis and characterization of GDY by CVD technique [1]

issues, researchers continue to refine the synthesis process by optimizing reaction conditions, exploring alternative catalysts, and developing new polymerization strategies that enhance the yield and quality of GDY for its diverse applications in energy storage, catalysis, and electronics.

3.3 The Eglinton Coupling for GDY Synthesis

The Eglinton coupling reaction is a copper(II)-mediated oxidative coupling method commonly used in the synthesis of GDY, a 2D allotrope of carbon possessing unique sp–sp^2 hybridized carbon structures. Unlike the Glaser-Hay coupling, which uses copper(I) catalysts and requires external ligands, Eglinton coupling makes use of Cu(II) salts like $Cu(OAc)_2$ or $CuCl_2$ to act as oxidants and catalysts simultaneously., providing a more efficient and selective pathway for polymerization. This method helps form the characteristic diacetylenic linkages (–C≡C–C≡C–) that define GDY's structure, leading to the development of extended 2D networks of conjugated carbon atoms. The absence of external ligands eliminates side reactions, thus enhancing the overall selectivity and improving the quality of the resulting GDY material. In the Eglinton coupling reaction, 1,3,5-triethynylbenzene (TEB), a commonly used precursor, undergoes oxidative polymerization in the presence of Cu(II) salts. The reaction starts with the activation of the terminal alkyne (–C≡C–H) by Cu(II), forming copper-acetylide intermediates. These intermediates then undergo oxidative coupling, resulting in the formation of diacetylenic linkages between the molecules, which ultimately leads to the polymerization of TEB into the desired GDY framework. The process efficiently forms conjugated carbon networks, ensuring a high degree of order in the final product. Moreover, the reaction regenerates Cu(II) throughout the process, maintaining the catalytic cycle and enhancing overall efficiency.

One of the main advantages of the Eglinton coupling is its high reaction selectivity. By using Cu(II) salts, the reaction avoids the unwanted side reactions commonly associated with copper(I)-mediated processes, such as overoxidation or side-product formation. This makes the reaction particularly useful for the controlled synthesis of GDY with fewer structural defects. Furthermore, the elimination of external ligands simplifies the reaction setup and reduces the steric hindrance that might otherwise affect the coupling efficiency. Eglinton coupling facilitates the controlled synthesis of 2D GDY structures, essential for use in electronics, energy devices, and catalysis. Despite its advantages, the Eglinton coupling method is not without challenges. Limited solubility of copper(II) acetylide intermediates in common solvents can hinder the polymerization process, reducing reaction efficiency and limiting the growth of the GDY network. To address this, researchers have optimized solvent systems, such as pyridine, DMF, and toluene, which enhance solubility and facilitate better reaction kinetics. Additionally, there is a risk of structural defects and incomplete coupling, which may arise from steric hindrance or uneven reactivity between the alkyne groups. However, by fine-tuning the reaction conditions (e.g., temperature, concentration, and time), these issues can be minimized. Despite these challenges, Eglinton coupling remains a promising and versatile approach for producing high-quality GDY, especially as future developments aim to enhance its scalability and applicability for large-scale synthesis. Both coupling reactions enable the formation of sp–sp^2 hybridized carbon networks, but Glaser-Hay coupling is more commonly used due to its relatively mild reaction conditions and better compatibility with large-scale synthesis. However, controlling the reaction parameters such as solvent composition, catalyst concentration, and reaction time-is critical to optimizing polymerization efficiency and achieving high-quality GDY films.

3.4 Structural and Morphological Challenges

Although bottom-up chemical synthesis offers precise control over the molecular structure of graphdiyne (GDY), several challenges hinder its large-scale production and structural quality. One major limitation is the low yield and restricted lateral dimensions of the synthesized GDY sheets. The formation of long-range ordered networks is constrained by kinetic and thermodynamic factors, often leading to incomplete polymerization and fragmented nanosheets. Steric hindrance between reactive sites further limits the expansion of the GDY framework, making it difficult to achieve large, defect-free structures suitable for practical applications.

Another significant challenge is the presence of structural defects and irregular growth patterns in GDY synthesis. Side reactions, including cyclization and oxidation of terminal alkynes, can introduce unwanted imperfections, negatively impacting the electronic and mechanical properties of the material. Achieving a uniform and defect-free network requires precise reaction control and careful tuning of catalyst-substrate interactions to minimize undesired transformations. Researchers are actively refining synthesis conditions to enhance the structural integrity of GDY, ensuring improved performance in advanced applications. Additionally, poor solubility and limited processability pose obstacles in the practical utilization of GDY. The material's inherent insolubility in common organic solvents makes it difficult to process, transfer, or integrate into various devices. To address this, researchers are exploring functionalization strategies such as heteroatom doping and substrate-assisted growth, which improve both solubility and structural stability. Novel synthesis approaches, including templated polymerization, solvothermal methods, and catalyst engineering, are being developed to enhance the crystallinity and scalability of GDY. Moreover, defect engineering and heteroatom doping during polymerization can further optimize its catalytic and electronic properties, broadening its applicability in energy conversion, electronics, and optoelectronic technologies.

3.5 Surface-Mediated Polymerization

On-surface polymerization is a highly controlled synthesis approach that enables the fabrication of ultrathin, well-ordered GDY films by using a solid substrate such as copper (Cu), gold (Au), or highly oriented pyrolytic graphite (HOPG). The substrate acts as a template, guiding the polymerization of ethynyl-containing monomers like 1,3,5-triethynylbenzene (TEB). Through strong substrate-monomer interactions, the monomers align in a structured manner, leading to the formation of a continuous and uniform GDY layer. This approach significantly improves the long-range order and crystallinity of GDY compared to traditional solution-phase polymerization, which often leads to fragmented nanosheets with irregular structures. One of the primary benefits of this technique is the ability to engineer high-quality GDY monolayers with minimal defects. By precisely controlling the reaction parameters—such as temperature, precursor concentration, and surface energy of the substrate—researchers can tailor the thickness, morphology, and conductivity of GDY films. This is particularly important for applications in nanoelectronics, catalysis, and optoelectronic devices, where the presence of defects can significantly impact performance. Furthermore, the strong interaction between the GDY layer and the substrate promotes efficient charge transfer, which is crucial for electrocatalytic and sensing applications.

The polymerization process generally follows a stepwise reaction mechanism. First, the ethynyl monomers are adsorbed onto the substrate surface in an ordered fashion due to π–π stacking interactions and metal coordination effects. Once properly aligned, oxidative coupling occurs through Glaser-Hay or Eglinton coupling, facilitating the formation of diacetylene (–C≡C–C≡C–) linkages. This leads to the growth of an extended sp–sp^2 hybridized carbon network with a high degree of conjugation, which is essential for efficient electron transport and catalytic activity. The result is a highly crystalline, substrate-supported GDY thin film with enhanced mechanical stability and electronic properties.

3.6 *Chemical Vapor Deposition (CVD)*

Chemical Vapor Deposition (CVD) is an advanced technique for synthesizing high-quality GDY by decomposing gaseous precursors on a heated substrate, followed by a controlled polymerization reaction. Unlike traditional solution-based synthesis, CVD enables the direct growth of continuous GDY films on substrates such as copper, gold, or silicon, offering superior control over thickness, crystallinity, and morphology. This method utilizes gaseous carbon sources, often hydrocarbon precursors like ethynyl-containing compounds, which react under high-temperature and low-pressure conditions to form extended sp–sp^2 hybridized GDY networks.

One of the key advantages of CVD-based GDY synthesis is its potential for large-scale production while maintaining high purity and structural integrity. This technique enables precise control over reaction parameters, including temperature, precursor concentration, and gas flow rate, which are critical for achieving uniform GDY layers with minimal defects. Unlike conventional bottom-up polymerization methods, which often result in low yield and small lateral dimensions, CVD offers a more scalable approach for fabricating continuous GDY sheets with superior electronic, optical, and catalytic properties.

Recent advancements in Plasma-Enhanced CVD (PECVD) have further enhanced GDY synthesis by introducing high-energy plasma to activate gaseous precursors at lower temperatures. This not only accelerates polymerization but also reduces the formation of structural defects, leading to highly uniform and well-ordered GDY films. PECVD allows for tunable film thickness, better adhesion to substrates, and improved electronic performance, making it an attractive method for applications in nanoelectronics, electrocatalysis, and flexible optoelectronics. Additionally, PECVD enables the direct functionalization of GDY by incorporating heteroatoms such as nitrogen, sulfur, or boron, further enhancing its catalytic and semiconducting properties.

With its ability to produce high-quality, large-area GDY films, CVD-based synthesis—especially PECVD-is emerging as a promising technique for industrial applications. The integration of GDY into electronic devices, energy storage systems, and catalysis platforms requires precisely engineered thin films, which CVD can provide with superior reproducibility. Future research efforts will focus on optimizing precursor design, improving substrate compatibility, and scaling up production to enable the widespread adoption of GDY-based technologies. By leveraging advanced deposition techniques and in-situ functionalization strategies, CVD and PECVD can drive next-generation applications in hydrogen production, energy conversion, and high-performance electronic devices (see Fig. 2).

3.7 Synthesis Approaches Confined to Interfacial Regions

The formation of 2D-polymeric materials at miscellaneous crossing points, including solid–liquid, liquid–liquid, and gas–liquid, is an extensively accepted method, predominantly for making surface covalent organic frameworks (COFs) and metal–organic frameworks (MOFs). Applying interfacial zones for making of graphdiyne (GDY) offers three key advantages: (1) it provides a wide, flat, and consistent surface for growth; (2) the underlying liquid layer acts as a storage medium for both monomers and catalysts; and (3) it enables the straightforward and non-destructive transfer of the resulting films to various substrates. To overcome the drawbacks associated with solution-based methods, an interface-constrained technique is suggested to fabricate few-layer GDY films by limiting the reaction to a specific interfacial area [34–40].

Nishihara and colleagues reported liquid–liquid and gas–liquid interfacial synthesis of GDY. They used a nearly immiscible system of dichloromethane containing HEB monomers and an aqueous phase with Cu(II) acetate and pyridine as catalysts. GDY films formed at the liquid–liquid interface, exhibiting sheet morphology (25–100 μm lateral size, 24 nm thickness). In the gas–liquid method, a dilute monomer solution was added dropwise onto an aqueous catalyst phase under argon protection, leading to GDY nanosheet formation at the interface. TEM confirmed hexagonal sheets (~1–2 μm). Hu and co-workers later modified this method in an air atmosphere, enabling the synthesis of large-area, freestanding GDY ultrathin films, which can be easily transferred onto solid substrates for device applications [41].

4 GDY for HER: Enhancing Proton Adsorption and Hydrogen Evolution

Porous and high-surface-area structure in GDY improve the exposure of active sites and facilitate efficient mass transport of reactants and intermediates. GDY demonstrates excellent corrosion resistance in both acidic and alkaline electrolytes, outperforming many metal-based catalysts. The hybridized sp/sp^2 bonding allows for precise bandgap engineering, optimizing charge transfer for both HER and OER. The triple bonds in GDY enable strong metal–carbon interactions, enhancing the stability of metal-functionalized catalysts.

HER involves the reduction of protons (**H^+**) into molecular hydrogen (**H_2**) at the cathode through a two-step reaction in acidic media:

1. **Volmer step**: $H^+ + e^- \rightarrow H_{ads}$ (proton adsorption)
2. **Tafel or Heyrovsky step**: $2H_{ads} \rightarrow H_2$ or $H_{ads} + H^+ + e^- \rightarrow H_2$

GDY's sp-hybridized carbon sites provide high-density active centers that enhance proton adsorption and facilitate electron transfer. Additionally, the delocalized π-electron network in GDY improves charge transport, reducing reaction over potential. However, pristine GDY has moderate HER activity due to its relatively low intrinsic catalytic activity. To overcome this, researchers have employed various modification strategies like heteroatom doping via incorporating heteroatoms such as N, B, P, or S introduces electron density variations, optimizing hydrogen adsorption energy and boosting HER kinetics. For example, N-doped GDY enhances the binding affinity of hydrogen intermediates, improving catalytic efficiency. Anchoring transition metals (e.g., Mo, Co, Fe, Ni, or Pt) onto GDY creates highly active HER sites by synergistically modifying the electronic structure [42–47]. Mo-doped GDY exhibits high HER activity due to Mo's ability to reduce the activation energy for hydrogen evolution. Introducing vacancies or structural defects in GDY provides additional reactive sites, enhancing proton adsorption and catalytic efficiency.

GDY has a π-conjugated structure that ensures excellent electron transport. This facilitates rapid charge transfer between the catalyst and the electrolyte, accelerating HER. The high surface area of GDY allows for uniform dispersion of MoP, CoP, and MoS_2, preventing catalyst aggregation and ensuring maximum active site exposure. A major challenge in electrocatalysis is the structural degradation of catalysts during long-term operation. GDY helps stabilize MoP, CoP, and MoS_2 by preventing their dissolution, oxidation, and aggregation, leading to enhanced durability and reusability. This ensures that the composite catalyst maintains high HER activity over prolonged cycles. GDY in corroboration with MoP, CoP, and MoS_2 is combined with these catalysts, the resulting hybrid material exhibits faster HER kinetics and energy-efficient hydrogen production because it provides more active sites, leading to increased hydrogen evolution rates. Moreover, GFY provides better long-term stability to catalyst degradation in acidic or alkaline media. GDY-based transition metal phosphide and sulfide catalysts offer an exciting alternative to noble metals like Pt for HER. The strong interactions between GDY and MoP, CoP, and MoS_2 lead to enhanced charge transfer, better hydrogen adsorption properties, and improved long-term stability, making these materials highly promising for sustainable hydrogen production.

Graphdiyne (GDY)-based hybrid catalysts have emerged as promising materials for HER applications due to their unique sp–sp^2 hybridized carbon structure, excellent conductivity, and tunable electronic properties. However, pristine GDY often faces challenges in catalytic efficiency, which has led researchers to explore its integration with other advanced materials such as MXenes, metal–organic frameworks (MOFs), and perovskites. These hybrid catalysts capitalize on the strengths of each component, creating synergistic effects that enhance HER performance. By combining GDY with these materials, researchers aim to improve charge transfer kinetics, increase active site availability, and enhance overall catalytic stability.

MXenes, a class of two-dimensional transition metal carbides and nitrides, are particularly attractive for GDY-based hybrid catalysts due to their high electrical conductivity, hydrophilicity, and excellent chemical stability. When incorporated with GDY, MXenes can facilitate faster charge transport, reducing overpotential and improving catalytic activity for HER. Similarly, MOFs—highly porous materials with tunable metal centers and organic linkers—offer a high density of active sites that can interact effectively with GDY. This combination provides a larger electrochemically active surface area and better control over catalytic processes, leading to enhanced hydrogen evolution efficiency.

Xue's team first used NGDY to stabilize MoS_2 nanosheets, significantly improving charge transfer and exposure of active sites, leading to a remarkable reduction in overpotential to 186 mV for a current density of 10 mA cm^{-2}. To further enhance performance, they developed an eGDY/MoS_2 heterojunction, which required an even lower overpotential of 128 mV under acidic conditions. Additionally, He and colleagues synthesized GDY-coated MoS_2 nanosheets on a carbon fiber network (GDY-MoS_2 NS/CF) using a hydrothermal approach. This structure significantly reduced the free energy barrier (ΔGH) and bandgap, improving conductivity. As a result, it demonstrated ultra-low overpotentials of 90 mV and excellent stability across thousands of cycles, outperforming conventional Pt-based catalysts [47, 48].

Wang's group introduced a 3D Cu@GDY core–shell heterojunction with a self-catalyzed growth process, achieving overpotentials of 79 and 162 mV for 10 and 100 mA cm^{-2} current densities, respectively. The GDY shell provided stability by preventing material degradation, allowing it to maintain performance even after 20 h of continuous operation [8]. Similarly, Yang's team developed CoNC/GDY, which exhibited high HER activity across different pH environments, outperforming 10% Pt/C catalysts in stability tests. Meanwhile, Liu and colleagues synthesized a 2D/2D WS_2/GDY heterojunction via hydrothermal methods. Sulfur vacancies introduced defects in WS_2, acting as active sites, and a built-in electric field facilitated charge transfer. This structure resulted in a low onset potential of 140 mV and a Tafel slope of 54 mV dec^{-1}, showcasing exceptional catalytic efficiency. Li's team investigated GDY-supported zero-valence transition-metal atoms (Ni^0 and Fe^0), achieving superior HER performance. Fe^0/GDY demonstrated the lowest onset potential (9 mV) and high TOF values, surpassing other transition-metal-based catalysts, including noble-metal ones. Additionally, Fe^0/GDY and Ni^0/GDY retained stability after 5000 cycles. These advancements highlight the potential of GDY-based electrocatalysts for sustainable hydrogen production, suggesting further innovations in electrochemical energy applications.

Recently, SACs are particularly beneficial for the HER, as they can significantly lower the reaction energy barrier while ensuring high selectivity and stability. Graphdiyne (GDY) serves as an excellent platform for anchoring SACs because

GDY consists of benzene rings connected by acetylenic (–C≡C–) linkages, creating an extended π-conjugated system that enhances electronic conductivity and charge mobility. Unlike graphene, GDY contains both sp^2 (graphene-like) and sp (linear carbon chains) hybridized carbon atoms. The presence of sp-hybridized carbon atoms provides high-density binding sites for metal atoms, making it an ideal host for SACs. The electronic structure of GDY can be easily modified by doping or functionalization, allowing for further enhancement of catalytic activity when combined with SACs. One of the major challenges in SACs is preventing metal atoms from aggregating into clusters, which would reduce catalytic efficiency. GDY provides a strong metal–carbon interaction, preventing metal atom migration and aggregation due to the metal atoms form coordination bonds with the sp/sp^2 carbon sites, ensuring stable anchoring. The electron density from GDY can interact with the d-orbitals of metal atoms, stabilizing them against movement. The porous structure of GDY helps trap metal atoms, further preventing aggregation. This strong interaction ensures that SACs on GDY remain highly active and durable, even under harsh electrochemical conditions.

An ultrathin GDY-supported Pd electrocatalyst was used for HER reaction. Theoretical and experimental analyses confirmed the successful and uniform distribution of Pd atoms on GDY without aggregation. Remarkably, with just 0.2% Pd loading, the catalyst achieved an overpotential of only 55 mV at 10 mA cm^{-2} surpassing commercial Pt/C catalysts in both efficiency and stability. The superior performance stems from strong p–d coupling between Pd and GDY, enabling effective charge transfer and enhanced hydrogen adsorption. Electrochemical tests, such as Tafel analysis, mass activity, turnover frequency (TOF), and long-term durability, further demonstrated the catalyst's outstanding HER performance. Pd^0/GDY exhibited a TOF of 16.7 s^{-1} and excellent stability over 72 h, with minimal performance degradation. These results highlight the potential of GDY as a robust and conductive platform for stabilizing single-atom catalysts and emphasize the promise of atomic Pd^0 for scalable, cost-effective hydrogen production technologies [49].

Transition-metal single atoms anchored on graphdiyne (TM@GDY) as electrocatalysts for key energy conversion reactions such as the HER, OER, and ORR. Using density functional theory (DFT) calculations, the study investigates 11 different transition metals (from Sc to Zn and Pt) supported on a GDY monolayer. Furthermore, single atoms exhibit strong binding to specific sites on GDY, particularly the corners of acetylenic rings, which provide structural stability and prevent aggregation into metal clusters. Among the studied configurations, Ti@GDY and V@GDY showed outstanding HER activity with both metal and carbon atoms acting as active sites. Similarly, Ni@GDY and Pt@GDY demonstrated impressive bifunctional capabilities, performing efficiently in both OER and ORR. Furthermore, the study revealed that the electronic structure and adsorption energies of the metal atoms can be tuned by selecting different transition metals, enabling optimization for specific reactions. The stability of the atoms on GDY and the low overpotentials achieved make these catalysts highly attractive for practical applications. Importantly, the volcano plots based on Gibbs free energy descriptors confirmed that Ni@GDY and Pt@GDY are among the most promising electrocatalysts, combining high activity with low cost and excellent stability [50].

Perovskites, known for their unique electronic structures and defect-tolerant nature, further expand the capabilities of GDY-based catalysts. Their ability to modulate charge transfer and stabilize active sites makes them excellent candidates for hybrid systems. The integration of perovskites with GDY enhances catalytic performance by improving electron mobility and facilitating synergistic interactions between different components. Ultimately, GDY-based hybrid catalysts incorporating MXenes, MOFs, or perovskites show superior stability, efficiency, and scalability for HER applications, making them strong contenders for next-generation sustainable hydrogen production technologies.

5 GDY for OER: Improving Oxygen Evolution Kinetics

OER is the **anodic reaction** in water splitting, where water molecules are oxidized to generate oxygen (**O_2**). The reaction follows a four-electron transfer process:

1. $H_2O \rightarrow OH^- + H^+ + e^-$ (Water dissociation)
2. $OH^- \rightarrow OOH^- + e^-$ (Formation of oxyhydroxide intermediate)
3. $OOH^- \rightarrow O_2 + H^+ + e^-$ (Oxygen formation and release)

OER is significantly more challenging than HER due to its higher kinetic barriers and sluggish reaction rates. Pristine GDY alone has limited OER activity, but its electronic structure can be tuned to enhance its performance through incorporating Ni, Fe, Co, Mn, or Ru into GDY creates highly active OER centers. To enhance OER performance, reducing the overpotential, accelerating reaction kinetics, and improving stability are essential. In 2018, Yu and colleagues synthesized a GDY/NiFe-LDH catalyst through an in situ growth method, demonstrating improved water oxidation performance. Density functional theory (DFT) calculations indicated a stronger chemical interaction and higher electron transfer efficiency between GDY and NiFe-LDH compared to RGO/NiFe-LDH. As a result, GDY/NiFe-LDH exhibited outstanding OER performance, requiring only 260 mV to achieve 10 mA cm^{-2}, outperforming other catalysts, including RuO_2. It also displayed the smallest Tafel slope and the highest turnover frequency (TOF), along with excellent durability under long-term testing. Further advancements in GDY-based catalysts include the development of a metalloporphyrin-based GDY analogue (Co-PDY) by Zhang's group. This nanomaterial, synthesized via Glaser-Hay coupling, featured a broadened pore structure (2.34 nm) that facilitated rapid mass transport, improving catalytic efficiency. Additionally, the Co atoms provided numerous active sites, while butadiyne linkages enabled efficient electron transfer. Consequently, Co-PDY demonstrated superior OER performance with a low overpotential of 270 mV at 10 mA cm^{-2}, a small Tafel slope of 99 mV dec^{-1}, and remarkable stability.

Besides these examples, several other GDY-based materials have been developed, consistently exhibiting exceptional catalytic performance for OER. Their unique structural properties, high stability, and cost-effectiveness highlight the significant potential of GDY-based electrocatalysts in energy conversion technologies, paving the way for further innovations in sustainable energy applications. GDY-based electrocatalysts have shown significant promise for oxygen evolution reaction (OER) enhancement, particularly when combined with transition metal oxides such as $NiFe_2O_4$, Co_3O_4, and MnO_2. These oxides serve as active sites for OER by stabilizing reaction intermediates and facilitating charge transfer processes. Their strong interaction with GDY improves electron mobility and structural stability, reducing overpotential and increasing overall catalytic efficiency. By anchoring these oxides onto GDY, researchers achieve a highly conductive and durable hybrid system suitable for sustainable water oxidation applications.

In addition to transition metal oxides, single-atom catalysts (SACs) based on Ru, Ir, and Fe supported on GDY have emerged as highly efficient OER electrocatalysts. These SACs optimize the adsorption energy of oxygen intermediates, thereby enhancing intrinsic catalytic activity. GDY's rich sp–sp^2 hybridized carbon network provides excellent electronic conductivity and strong metal coordination sites, ensuring uniform dispersion of single atoms and minimizing aggregation. This atomic-scale engineering of active centers significantly improves catalytic selectivity and reaction kinetics, making GDY-SAC hybrids competitive for high-performance electrolysis systems.

γ-GDY possessing sp and sp^2 hybridized carbon atoms with unique structural, optical, and electronic properties. Its high electron/hole mobility, tunable band gap, and uneven electronic structure make it a promising photoelectrocatalyst. Among miscellaneous synthesis techniques discussed in previous sections, the liquid–solid method is highlighted as the most effective due to its ability to yield high-crystallinity GDY films with moderate output on diverse substrates. However, the solid–gas approaches may support large-scale production, but they typically result in GDY with poor crystallinity.

GDY enhances light absorption and charge separation, especially when combined with semiconductors like TiO_2 and ZnO, significantly improving degradation rates of organic pollutants and efficiency in photoelectrochemical (PEC) water splitting. In electrocatalysis, its porous structure and electronic properties facilitate the anchoring of single-atom catalysts, particularly for ORR and HER/OER. Doping and metal loading on GDY further elevate its performance, in some cases rivaling or surpassing commercial Pt-based catalysts. The review concludes by emphasizing the need for further research in scalable synthesis, structural optimization, and mechanistic understanding to fully exploit GDY's catalytic potential in sustainable energy technologies [51].

Nitrogen-doped triphenylene-graphdiyne (N@TP-GDY) has been investigated as a promising metal-free, bifunctional catalyst for both electrochemical (EC) and photoelectrochemical (PEC) overall water splitting. Using density functional theory (DFT) and nonadiabatic molecular dynamics (NAMD), the authors show that doping nitrogen into TP-GDY dramatically enhances its catalytic activity. The doped structure exhibits low overpotentials for the HER and OER as low as 0.06 V and 0.49 V, respectively. This enhanced performance arises from the altered electronic structure, increased charge redistribution, and more favorable adsorption energies of reaction intermediates. Importantly, N@TP-GDY also demonstrates high thermal and chemical stability, making it a suitable candidate for sustainable hydrogen production.

Beyond EC applications, N@TP-GDY also shows great potential in PEC water splitting. Several doped structures meet the energetic and optical requirements to act as effective photoelectrodes, with improved carrier lifetimes (up to 729 ps) and enhanced light absorption in the visible spectrum. Notably, N_2@TP-GDY and N_4@TP-GDY stand out for their superior PEC performance under both acidic and alkaline conditions, driven entirely by solar energy. These monolayers also maintain their efficiency when stacked into multilayers or when partially oxidized, confirming their robustness and practical applicability. The work offers a strategic pathway to design next-generation, metal-free, multifunctional catalysts for clean energy applications (see Fig. 3) [52].

The computational study of the catalytic properties of transition metal (TM) clusters comprising 1–4 atoms (TM_{1-4}) anchored on graphdiyne (GDY). Computational investigations reveal that the performance of these TM clusters—specifically those of V, Cr, Mn, Fe, Co, Ni, Cu, Ru, Rh, Pd, Ir, and Pt—for key electrochemical reactions: OER, ORR, and HER. The findings indicate that certain configurations, such as Pd_2, Pd_4, and Pt_1 anchored on GDY, exhibit trifunctional catalytic activity with low overpotentials, making them promising candidates for applications in energy conversion and storage devices.

Moreover, the study highlights that precise control over the number of atoms in TM clusters significantly influences their catalytic efficiency. For instance, Pd_1 and Pt_2 clusters on GDY demonstrate excellent bifunctional performance for water

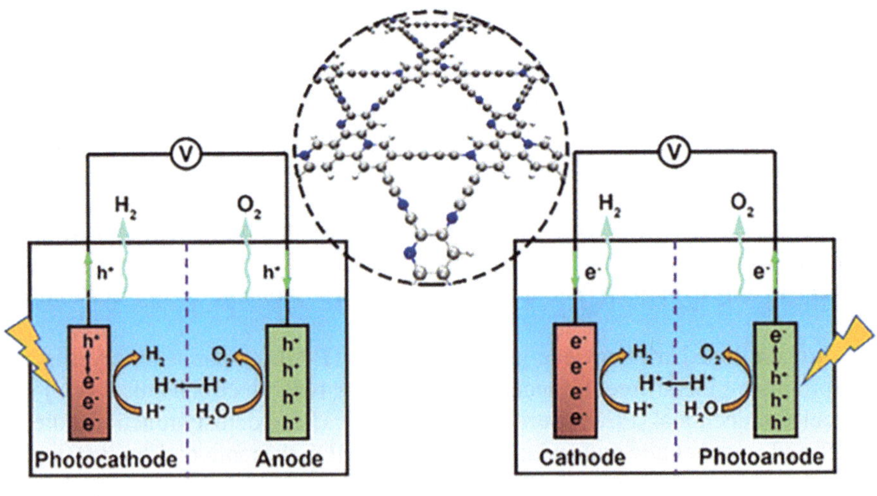

Fig. 3 Application of N@TP-GDY for PEC water splitting

splitting, while Ni_1 and Pd_3 clusters are effective for metal–air and fuel cells. This research underscores the potential of atomically precise TM clusters on GDY as versatile and efficient electrocatalysts, offering a strategic approach to designing advanced materials for sustainable energy technologies [53].

Graphdiyne (GDY)-supported nickel-iron layered double hydroxide (NiFe LDH) nanosheets for the OER in alkaline media. GDY functions as a conductive and porous support to improve the conductivity and active site exposure of the NiFe LDH catalyst. The composite was fabricated by first synthesizing a GDY film on copper foil, followed by electrodeposition of NiFe LDH. The hierarchical structure formed allowed uniform deposition and intimate contact between GDY and NiFe LDH, leading to enhanced charge transport and catalytic performance. Electrochemical analysis showed that the GDY@NiFe composite exhibited a low overpotential of 260 mV to achieve a current density of 10 mA cm^{-2} and superior long-term durability, outperforming both pristine GDY and NiFe LDH alone. The enhanced performance was attributed to the increased electrochemical surface area, efficient charge transfer, and improved mass transport facilitated by the hierarchical porous structure. Additional tests, including electrochemical impedance spectroscopy (EIS), X-ray photoelectron spectroscopy (XPS), and transmission electron microscopy (TEM), confirmed the synergistic interaction between GDY and NiFe LDH (see Fig. 4) [54].

Cobalt nanoparticles wrapped in nitrogen-doped carbon (CoNC) and supported on graphdiyne (GD) synthesized using scalable and cost-effective method and demonstrated remarkable HER catalytic activity and durability across a wide pH range (acidic, neutral, and alkaline conditions). Compared to commercial Pt/C catalysts, CoNC/GD not only offered competitive performance with lower onset potentials and overpotentials but also exhibited significantly greater stability-retaining activity after 36,000 cycles in basic, 38,000 in acidic, and 9,000 in neutral media. The unique structure of graphdiyne, with its sp and sp^2 carbon framework and high conductivity, facilitated efficient electron transfer, while the nitrogen-doped carbon layer

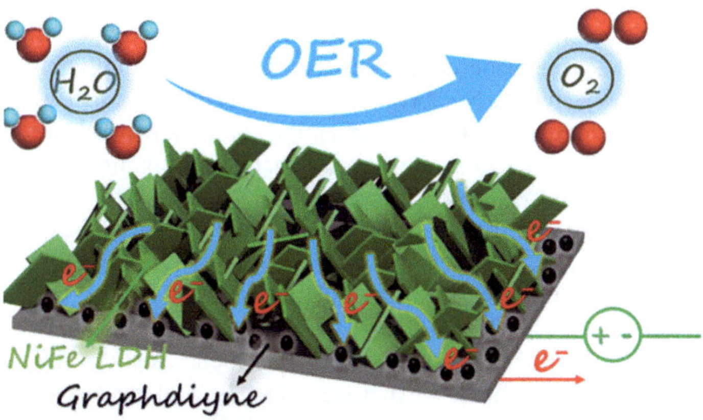

Fig. 4 Application of GDY@NiFe composite for OER

protected the cobalt nanoparticles from corrosion and aggregation. Extensive characterization confirmed the successful formation and high porosity of the CoNC/GD material, which contributed to its high surface area and abundance of catalytically active sites. Electrochemical tests showed superior HER performance due to the synergistic effects between the GD support, cobalt nanoparticles, and nitrogen-doped carbon shell. Importantly, the material maintained its structural integrity and catalytic activity over prolonged operation, which is rarely achieved with non-precious metal catalysts [55].

Nickel–iron layered double hydroxides (NiFe-LDH) supported on a graphdiyne/graphene (GDY/G) heterostructure were used as electrocatalyst for OER reaction. This hybrid material leverages the high conductivity of graphene and the unique electronic structure of graphdiyne to significantly enhance the performance of NiFe-LDH, a known OER catalyst. The in-situ hydrothermal growth of NiFe-LDH on the GDY/G substrate results in a well-anchored, uniformly distributed nanostructure with a high electrochemical surface area and low charge-transfer resistance. The composite exhibits an impressively low overpotential of 215 mV at 10 mA cm^{-2} and a small Tafel slope of 22 mV dec^{-1}, indicating fast OER kinetics and efficient electron transport. Detailed physicochemical characterizations confirm the robust interaction between the GDY/G support and the NiFe-LDH catalyst, facilitating rapid charge mobility and improved wettability. Electrochemical tests demonstrate that this hierarchical structure not only enhances intrinsic catalytic activity but also provides remarkable stability over extended cycling and continuous operation. The study concludes that the few-layered GDY/G heterojunction effectively regulates the electronic environment of active Ni/Fe sites and serves as a superior conductive platform, making the NiFe-LDH/GDY/G composite a promising candidate for large-scale water splitting and other energy-related applications [56].

Cobalt- and nitrogen-doped graphdiyne (Co–N-GDY) electrocatalyst demonstrated high bifunctional activity for both the ORR and HER. Co–N-GDY was prepared through a simple two-step high-temperature calcination process, using a small amount of cobalt precursor and melamine as the nitrogen source. The optimized composition, with 1% cobalt content, demonstrated outstanding catalytic activity due to the synergistic effects between Co, N, and the highly conductive graphdiyne (GDY) network. Notably, this material exhibited a half-wave potential for ORR that was comparable to commercial Pt/C and a low onset overpotential of 97 mV for HER, with excellent durability across acidic and alkaline media. Structural analysis confirmed that Co and N were successfully doped into the GDY framework, enhancing the number of active sites and the overall electronic conductivity. The Co–N-GDY displayed a high electrochemically active surface area (ECSA), low charge transfer resistance, and remarkable methanol tolerance and cycling stability. In both 1 M KOH and 0.5 M H_2SO_4, 1% Co–N-GDY retained its activity after extended operation, outperforming many other non-precious metal catalysts [16].

A combined computational and experimental analysis of various nitrogen-doped graphdiyne (NGDY)-supported transition metals (TMs) revealed a catalytic activity trend of Pd@NGDY > Ni@NGDY > Co@NGDY > Fe@NGDY. A detailed structure–property correlation study further identified key descriptors that effectively capture intermediate binding strength and ORR performance across these TM single-atom catalysts (TM-SACs). These descriptors include the free energy of *OH adsorption* (ΔGOH), the d-orbital center position ($\varepsilon_d/\varepsilon_{d-a}$), the extent of charge transfer before and after reactant adsorption (ΔQTM/ΔQTM-a), and the TM–O bond length (LTM–O). Notably, an inverse relationship was observed between catalytic activity and the binding strength of reaction intermediates, indicating that excessively strong adsorption of *OOH, *O, and *OH species impairs ORR efficiency [57].

6 Conclusions and Future Prospects

To conclude, GDY-supported catalysts have demonstrated exceptional potential in advancing HER and OER technologies, offering a pathway toward efficient and sustainable water-splitting systems. The integration of GDY with various catalytic materials has led to enhanced charge transfer, increased active site availability, and improved stability. GDY's intrinsic properties, including its tunable electronic structure and high surface area, make it a versatile platform for developing next-generation electrocatalysts. Despite these advantages, challenges remain in achieving large-scale synthesis, precise structural control, and long-term electrochemical stability. By addressing key challenges in scalability, active site optimization, and stability, GDY could become a cost-effective and durable solution for water electrolysis and fuel cell applications. Ongoing research in synthetic techniques, computational modeling, and real-world electrolysis trials will be crucial in transforming GDY from a laboratory innovation to an industrial-scale catalyst, accelerating the transition toward a hydrogen-based economy.

Despite its advantages, scalability remains a major challenge in practical GDY applications. Existing fabrication methods, such as on-surface polymerization, chemical vapor deposition (CVD), and solvothermal synthesis, often result in low yields, non-uniform films, or high production costs. Ensuring high crystallinity, uniform doping, and large-area deposition is difficult due to the complex reaction conditions required for GDY formation. Researchers are exploring scalable techniques such as electrochemical polymerization and plasma-enhanced deposition, which offer improved control over structure and cost-effectiveness for industrial applications. In its pristine form, GDY exhibits moderate catalytic activity. To enhance HER and OER efficiency, researchers employ heteroatom doping (e.g., N, B, P, S) and metal functionalization (e.g., Ni, Co, Fe, Mo, Ru, Ir) to introduce active sites that improve charge transfer and reaction kinetics. Theoretical modeling and Density Functional Theory (DFT) calculations help predict optimal dopants and defect engineering strategies for tuning GDY's electronic structure. Additionally, defect engineering-creating vacancies and edge terminations-can increase the density of active sites, enhancing overall electrocatalytic efficiency. While GDY demonstrates high chemical stability, long-term electrolysis performance under harsh conditions remains a concern. Prolonged exposure to extreme pH environments (acidic or alkaline electrolytes) can cause catalyst degradation, structural deterioration, or metal leaching in hybrid catalysts.

Strategies such as surface passivation (thin oxide coatings), strong metal anchoring, and defect stabilization are being explored to enhance GDY's durability. Real-world testing in industrial electrolysis setups is essential to evaluate its long-term catalytic performance and commercial viability.

To overcome these challenges, future research should focus on scalable production methods, advanced doping techniques, and long-term stability testing. AI-driven machine learning models can accelerate material discovery by predicting the most efficient doping strategies and structural modifications. Additionally, hybrid catalyst systems—combining GDY with metal oxides, hydroxides, or sulfides—can create synergistic effects to further enhance HER and OER efficiency. These advancements will pave the way for GDY-based catalysts to serve as viable alternatives to precious metal catalysts like platinum (Pt) and iridium (Ir).

References

1. Qian, X., Zheng, Y., Chen, S., Xu, J.: Atomic-level functionalized graphdiyne for electrocatalysis applications. Catalysts **10**(8), 929 (2020)
2. Li, J., Gao, X., Zhu, L., Ghazzal, M.N., Zhang, J., Tung, C.-H., Wu, L.-Z.: Graphdiyne for crucial gas involved catalytic reactions in energy conversion applications. Energy Environ. Sci. **13**(5), 1326–1346 (2020)
3. Chen, X., Jiang, X., Yang, N.: Graphdiyne electrochemistry: progress and perspectives. Small **18**(24), 2201135 (2022). https://doi.org/10.1002/smll.202201135
4. Zheng, Z., Xue, Y., Li, Y.: A new carbon allotrope: graphdiyne. Trends Chem. **4**(8), 754–768 (2022). https://doi.org/10.1016/j.trechm.2022.05.006
5. Hayat, A., Sohail, M., Moussa, S.B., Al-Muhanna, M.K., Iqbal, W., Ajmal, Z., Raza, S., Al-Hadeethi, Y., Orooji, Y.: State, synthesis, perspective applications, and challenges of Graphdiyne and its analogues: a review of recent research. Adv. Coll. Interface Sci. **319**, 102969 (2023). https://doi.org/10.1016/j.cis.2023.102969
6. Li, X., Li, X., Sun, Q., He, J., Yang, Z., Xiao, J., Huang, C.: Synthesis and applications of graphdiyne derivatives. Acta Phys. Chim. Sin. **39**(1), 2206029 (2023)
7. Zhao, N., Li, J., Chang, X., Zheng, W., Zhang, J., Liu, X.: Synthesis and application of graphdiyne-based materials for advanced chemical sensors. Coord. Chem. Rev. **521**, 216171 (2024)
8. Xue, Y., Guo, Y., Yi, Y., Li, Y., Liu, H., Li, D., Yang, W., Li, Y.: Self-catalyzed growth of Cu@graphdiyne core–shell nanowires array for high efficient hydrogen evolution cathode. Nano Energy **30**, 858–866 (2016). https://doi.org/10.1016/j.nanoen.2016.09.005
9. Yu, H., Hui, L., Xue, Y., Liu, Y., Fang, Y., Xing, C., Zhang, C., Zhang, D., Chen, X., Du, Y.: 2D graphdiyne loading ruthenium atoms for high efficiency water splitting. Nano Energy **72**, 104667 (2020)
10. Qi, L., Chen, Z., Luan, X., Zheng, Z., Xue, Y., Li, Y.: Atomically dispersed Mn enhanced catalytic performance for overall water splitting on graphdiyne-coated copper hydroxide nanowire. Chin. J. Struct. Chem. **43**(1), 100197 (2024). https://doi.org/10.1016/j.cjsc.2023.100197
11. Zheng, Z., Qi, L., Gao, Y., Luan, X., Xue, Y., He, F., Li, Y.: Ir_0/graphdiyne atomic interface for selective epoxidation. Natl. Sci. Rev. **10**(8) (2023). https://doi.org/10.1093/nsr/nwad156. Accessed 3 Feb 2025
12. Qi, L., Zheng, Z., Xing, C., Wang, Z., Luan, X., Xue, Y., He, F., Li, Y.: 1D nanowire heterojunction electrocatalysts of $MnCo_2O_4$/GDY for efficient overall water splitting. Adv. Func. Mater. **32**(11), 2107179 (2022). https://doi.org/10.1002/adfm.202107179

13. Fu, X., Zhao, X., Lu, T.-B., Yuan, M., Wang, M.: Graphdiyne-based single-atom catalysts with different coordination environments. Angew. Chem. Int. Ed. **62**(16), e202219242 (2023). https://doi.org/10.1002/anie.202219242
14. Jin, Z., Li, T., Zhang, L., Wang, X., Wang, G., Hao, X.: Construction of a tandem S-scheme GDY/CuI/CdS-R heterostructure based on morphology-regulated graphdiyne (g-C_nH_{2n-2}) for enhanced photocatalytic hydrogen evolution. J. Mater. Chem. A **10**(4), 1976–1991 (2022). https://doi.org/10.1039/D1TA09347A
15. Lu, T., Wang, H.: Graphdiyne-supported metal electrocatalysts: from nanoparticles and cluster to single atoms. Nano Res. **15**(11), 9764–9778 (2022). https://doi.org/10.1007/s12274-022-4157-1
16. Wang, X., Yang, Z., Si, W., Shen, X., Li, X., Li, R., Lv, Q., Wang, N., Huang, C.: Cobalt-nitrogen-doped graphdiyne as an efficient bifunctional catalyst for oxygen reduction and hydrogen evolution reactions. Carbon **147**, 9–18 (2019). https://doi.org/10.1016/j.carbon.2019.02.033
17. Chen, X., Wang, M., Yuan, M.: Recent advances of graphdiyne: synthesis, functionalization, and electrocatalytic applications. Mater. Chem. Front. **5**(22), 7964–7981 (2021). https://doi.org/10.1039/D1QM01079D
18. Gao, X., Li, J., Du, R., Zhou, J., Huang, M.-Y., Liu, R., Li, J., Xie, Z., Wu, L.-Z., Liu, Z., et al.: Direct synthesis of graphdiyne nanowalls on arbitrary substrates and its application for photoelectrochemical water splitting cell. Adv. Mater. **29**(9), 1605308 (2017). https://doi.org/10.1002/adma.201605308
19. Li, X., Cui, X., Zhang, L., Du, J.: Review of graphdiyne-based nanostructures and their applications. ACS Appl. Nano Mater. **6**(22), 20493–20522 (2023). https://doi.org/10.1021/acsanm.3c03038
20. Gao, X., Liu, H., Wang, D., Zhang, J.: Graphdiyne: synthesis, properties, and applications. Chem. Soc. Rev. **48**(3), 908–936 (2019). https://doi.org/10.1039/C8CS00773J
21. He, T., Matta, S.K., Will, G., Du, A.: Transition-metal single atoms anchored on graphdiyne as high-efficiency electrocatalysts for water splitting and oxygen reduction. Small Methods **3**(9), 1800419 (2019)
22. Yu, H., Jin, X., Li, Y., Zhang, L., Yang, M., Li, J.: Principal strategies for designing graphdiyne-based catalyst toward green hydrogen production from water electrolysis. J. Phys.: Energy **5**(2), 021001 (2023)
23. Sun, C., Zhang, S., Wang, P., Wei, M., Wang, S., Shi, X.-R.: High-throughput screening of efficient graphdiyne supported transition metal single atom toward water electrolysis and oxygen reduction. J. Catal. **439**, 115773 (2024). https://doi.org/10.1016/j.jcat.2024.115773
24. Li, B., Lai, C., Zhang, M., Zeng, G., Liu, S., Huang, D., Qin, L., Liu, X., Yi, H., Xu, F., et al.: Graphdiyne: a rising star of electrocatalyst support for energy conversion. Adv. Energy Mater. **10**(16), 2000177 (2020). https://doi.org/10.1002/aenm.202000177
25. Xu, S., Kang, B., Li, H., Yoo, P.J., Lee, J.Y.: Graphdiyne-supported metals and metal compounds: synthesis, applications, and future prospects. Adv. Func. Mater. **34**(39), 2403095 (2024). https://doi.org/10.1002/adfm.202403095
26. Haley, M.M., Bell, M.L., English, J.J., Johnson, C.A., Weakley, T.J.: Versatile synthetic route to and DSC analysis of dehydrobenzoannulenes: crystal structure of a heretofore inaccessible [20] annulene derivative. J. Am. Chem. Soc. **119**(12), 2956–2957 (1997)
27. Marsden, J.A., Haley, M.M.: Carbon networks based on dehydrobenzoannulenes. 5. Extension of two-dimensional conjugation in graphdiyne nanoarchitectures. J. Org. Chem. **70**(25), 10213–10226 (2005)
28. Sonogashira, K., Tohda, Y., Hagihara, N.: A convenient synthesis of acetylenes: catalytic substitutions of acetylenic hydrogen with bromoalkenes, iodoarenes and bromopyridines. Tetrahedron Lett. **16**(50), 4467–4470 (1975)
29. Diederich, F., Rubin, Y.: Synthetic approaches toward molecular and polymeric carbon allotropes. Angew. Chem., Int. Ed. Engl. **31**(9), 1101–1123 (1992)
30. Kong, Y., Li, J., Zeng, S., Yin, C., Tong, L., Zhang, J.: Bridging the gap between reality and ideality of graphdiyne: the advances of synthetic methodology. Chem **6**(8), 1933–1951 (2020). https://doi.org/10.1016/j.chempr.2020.06.011

31. Li, J., Xie, Z., Xiong, Y., Li, Z., Huang, Q., Zhang, S., Zhou, J., Liu, R., Gao, X., Chen, C., et al.: Architecture of β-graphdiyne-containing thin film using modified Glaser-Hay coupling reaction for enhanced photocatalytic property of TiO_2. Adv. Mater. **29**(19), 1700421 (2017). https://doi.org/10.1002/adma.201700421
32. Hou, J., Sun, L., Pan, Q., Zhao, Y.: Recent synthetic methods towards graphdiyne. ChemNanoMat **9**(9), e202300273 (2023). https://doi.org/10.1002/cnma.202300273
33. Zhou, J., Li, J., Liu, Z., Zhang, J.: Exploring approaches for the synthesis of few-layered graphdiyne. Adv. Mater. **31**(42), 1803758 (2019). https://doi.org/10.1002/adma.201803758
34. Zwaneveld, N.A., Pawlak, R., Abel, M., Catalin, D., Gigmes, D., Bertin, D., Porte, L.: Organized formation of 2D extended covalent organic frameworks at surfaces. J. Am. Chem. Soc. **130**(21), 6678–6679 (2008)
35. Dmitriev, A., Spillmann, H., Lin, N., Barth, J.V., Kern, K.: Modular assembly of two-dimensional metal-organic coordination networks at a metal surface. Angew. Chem. Int. Ed. **42**(23), 2670–2673 (2003)
36. Takada, K., Sakamoto, R., Yi, S.T., Katagiri, S., Kambe, T., Nishihara, H.: Electrochromic bis(terpyridine)metal complex nanosheets. J. Am. Chem. Soc. **137**(14), 4681–4689 (2015). https://doi.org/10.1021/ja510788b
37. Sakamoto, R., Hoshiko, K., Liu, Q., Yagi, T., Nagayama, T., Kusaka, S., Tsuchiya, M., Kitagawa, Y., Wong, W.Y., Nishihara, H.: A photofunctional bottom-up bis(dipyrrinato)zinc(II) complex nanosheet. Nat. Commun. **6** (2015). https://doi.org/10.1038/ncomms7713
38. Makiura, R., Motoyama, S., Umemura, Y., Yamanaka, H., Sakata, O., Kitagawa, H.: Surface nano-architecture of a metal–organic framework. Nat. Mater. **9**(7), 565–571 (2010)
39. Dai, W., Shao, F., Szczerbiński, J., McCaffrey, R., Zenobi, R., Jin, Y., Schlüter, A.D., Zhang, W.: Synthesis of a two-dimensional covalent organic monolayer through dynamic imine chemistry at the air/water Interface. Angew. Chem. **128**(1), 221–225 (2016)
40. Bauer, T., Zheng, Z., Renn, A., Enning, R., Stemmer, A., Sakamoto, J., Schlüter, A.D.: Synthesis of free-standing, monolayered organometallic sheets at the air/water interface. Angew. Chem. Int. Ed. **50**(34), 7879–7884 (2011)
41. Li, W., Liu, J., Yu, Y., Feng, G., Song, Y., Liang, Q., Liu, L., Lei, S., Hu, W.: Synthesis of large-area ultrathin graphdiyne films at an air–water interface and their application in memristors. Mater. Chem. Front. **4**(4), 1268–1273 (2020)
42. Tang, T., Jiang, W.J., Niu, S., Liu, N., Luo, H., Zhang, Q., Wen, W., Chen, Y.Y., Huang, L.B., Gao, F.: Kinetically controlled coprecipitation for general fast synthesis of sandwiched metal hydroxide nanosheets/graphene composites toward efficient water splitting. Adv. Func. Mater. **28**(3), 1704594 (2018)
43. Gao, Z.W., Liu, J.Y., Chen, X.M., Zheng, X.L., Mao, J., Liu, H., Ma, T., Li, L., Wang, W.C., Du, X.W.: Engineering NiO/NiFe LDH intersection to bypass scaling relationship for oxygen evolution reaction via dynamic tridimensional adsorption of intermediates. Adv. Mater. **31**(11), 1804769 (2019)
44. Zhang, R., Shao, M., Xu, S., Ning, F., Zhou, L., Wei, M.: Photo-assisted synthesis of zinc-iron layered double hydroxides/TiO_2 nanoarrays toward highly-efficient photoelectrochemical water splitting. Nano Energy **33**, 21–28 (2017)
45. Wang, Y., Zhang, Y., Liu, Z., Xie, C., Feng, S., Liu, D., Shao, M., Wang, S.: Layered double hydroxide nanosheets with multiple vacancies obtained by dry exfoliation as highly efficient oxygen evolution electrocatalysts. Angew. Chem. Int. Ed. **56**(21), 5867–5871 (2017)
46. Lv, L., Yang, Z., Chen, K., Wang, C., Xiong, Y.: 2D layered double hydroxides for oxygen evolution reaction: from fundamental design to application. Adv. Energy Mater. **9**(17), 1803358 (2019)
47. Hui, L., Xue, Y., He, F., Jia, D., Li, Y.: Efficient hydrogen generation on graphdiyne-based heterostructure. Nano Energy **55**, 135–142 (2019)
48. Yu, H., Xue, Y., Hui, L., Zhang, C., Zhao, Y., Li, Z., Li, Y.: Controlled growth of MoS_2 nanosheets on 2D N-doped graphdiyne nanolayers for highly associated effects on water reduction. Adv. Func. Mater. **28**(19), 1707564 (2018). https://doi.org/10.1002/adfm.201707564

49. Yu, H., Xue, Y., Huang, B., Hui, L., Zhang, C., Fang, Y., Liu, Y., Zhao, Y., Li, Y., Liu, H. et al.: Ultrathin nanosheet of graphdiyne-supported palladium atom catalyst for efficient hydrogen production. iScience **11**, 31–41 (2019). https://doi.org/10.1016/j.isci.2018.12.006
50. He, T., Matta, S.K., Will, G., Du, A.: Transition-metal single atoms anchored on graphdiyne as high-efficiency electrocatalysts for water splitting and oxygen reduction. Small Methods **3**(9), 1800419 (2019). https://doi.org/10.1002/smtd.201800419
51. Sun, T., Gao, F.-Y., Tang, X.-L., Yi, H.-H., Yu, Q.-J., Zhao, S.-Z., Xie, X.-Z.: The preparation and use of γ-graphdiyne, a superb new photoelectrocatalyst. New Carbon Mater. **36**(2), 304–321 (2021). https://doi.org/10.1016/S1872-5805(21)60021-5
52. Fan, Y., Wang, J., Zhang, Z., Ma, X., Zhao, M.: Nitrogen-doped triphenylene-graphdiyne as metal-free multifunctional (photo)electrocatalysts for overall water splitting. ACS Appl. Nano Mater. **6**(14), 13037–13047 (2023). https://doi.org/10.1021/acsanm.3c01701
53. Liu, X.-Y., Liu, J.-W., Li, G., Zhao, J.-X.: Transition metal clusters with precise numbers of atoms anchored on graphdiyne as multifunctional electrocatalysts for OER/ORR/HER: a computational study. Rare Met. **43**(7), 3107–3117 (2024). https://doi.org/10.1007/s12598-023-02611-7
54. Shi, G., Yu, C., Fan, Z., Li, J., Yuan, M.: Graphdiyne-supported NiFe layered double hydroxide nanosheets as functional electrocatalysts for oxygen evolution. ACS Appl. Mater. Interfaces **11**(3), 2662–2669 (2019). https://doi.org/10.1021/acsami.8b03345
55. Xue, Y., Li, J., Xue, Z., Li, Y., Liu, H., Li, D., Yang, W., Li, Y.: Extraordinarily durable graphdiyne-supported electrocatalyst with high activity for hydrogen production at all values of pH. ACS Appl. Mater. Interfaces **8**(45), 31083–31091 (2016). https://doi.org/10.1021/acsami.6b12655
56. Shi, M., Sun, X., Bai, Q., Zhang, Y., Yu, S., Liu, M., Wang, L., Yu, W.W., Sui, N.: Graphdiyne/graphene heterostructure supported NiFe layered double hydroxides for oxygen evolution reaction. Colloids Surf., A **637**, 128217 (2022). https://doi.org/10.1016/j.colsurfa.2021.128217
57. Zou, M., Yang, J., Yue, X., Yuan, Y., Che, Z., Li, M., Li, B., Cui, J., Hu, W., Wang, S., et al.: Design of efficient oxygen reduction reaction catalysts with single transition metal atom on N-doped graphdiyne. J. Phys. Chem. Lett. **14**(43), 9624–9632 (2023). https://doi.org/10.1021/acs.jpclett.3c02649

Engineering of MOF-Based Heterostructures for Enhanced HER and OER Performance

Abdul Majid, Amina Shehbaz, Iram Shahzadi, and Bisma Wasim

Abstract The production of hydrogen through the water splitting process is a practical way to address issues with energy sources and environment. Using two-dimensional materials is the most efficient way to address existing environmental problems because of their outstanding qualities. Therefore, it is necessary to search for a photocatalyst that is efficient, economical and has effective electrical and optical features in addition to a suitable band gap and band off sets. Though, the progress of a cost-effective photocatalyst through exceptional catalytic activity and prolonged stability to perform the hydrogen evolution reaction (HER) and oxygen evolution reaction (OER). In the available spectrum of multifunctional materials, multiporous metal–organic frameworks (MOFs) are capable of offering tunable morphology, electronic structural, transport, and electrochemical characteristics. The MOF based heterostructures comprising of components with different morphology, structures, and electronic properties are potential candidates for enhanced catalytic performance. The efforts on engineering of MOF-Based Heterostructures for Enhanced HER and OER Performance are thoroughly described in this chapter.

Keywords MOF · Heterostructure · Catalysis · HER · OER

1 Introduction

The world's energy landscape is rapidly changing due to rising fossil fuel consumption, industrialization, and environmental contamination, necessitating the adoption of a more efficient, clean, and sustainable energy storage system [1, 2]. Fossil fuel consumption severely negatively affects our planet's ecosystem, temperature, and overall health, as demonstrated by its devastating impacts. The global population is projected to grow by 26% to 9.7 billion by 2050, while primary energy consumption is predicted to rise by nearly 50%. The current rate of global warming is expected to enhance worldwide temperature by 1.5 °C between 2030 and 2052, leading to

A. Majid (✉) · A. Shehbaz · I. Shahzadi · B. Wasim
Department of Physics, University of Gujrat, Gujrat, Pakistan
e-mail: abdulmajid40@uog.edu.pk

O. P. Pandey et al. (eds.), *2D Materials for Electrochemical Water Splitting*, Engineering Materials, https://doi.org/10.1007/978-981-95-2003-9_10

a sea level rise of nearly 3.8 m [3, 4]. The promotion of clean and renewable energy sources is vital for an ecological forthcoming and addressing environmental concerns. Clean energy solutions aim to improve efficiency, resource utilization, cost-effectiveness, and environmental sustainability, as renewable energy growth is crucial for meeting growing needs without compromising future decades [5, 6]. Hydrogen is a highly competitive and sustainable green energy solution, with extensive research focusing on its use as an environmentally friendly energy carrier in post-fossil fuel era. Hydrogen is a promising alternative energy source, with a vast potential that helps to address exhaustion, pollution, and climate change's effects by providing a sustainable and eco-friendly alternative [7].

Hydrogen energy provides a carbon-free, sustainable, efficient and eco-friendly energy origin that uses various raw materials to produce clean fuel for practical energy applications. Hydrogen has high-energy-density fuel (142 MJ/kg) with high combustion efficiency, non-toxicity, clean air pollutants, storage capacity, and renewable nature [8, 9]. The electrochemical hydrogen evolution process (HER) is an efficient and eco-friendly method for producing hydrogen by eliminating the need for hydrocarbons as the primary source [10]. Electrochemical water-splitting is a procedure which utilizes electrical energy for decomposing water into H_2 and O_2 using an aqueous electrolyte, cathode, and anode. The applied potential difference among electrodes drives the generation of oxygen and hydrogen gas at anode and cathode [11, 12]. The water electrolysis process exhibits a kinetically lethargic oxygen evolution reaction (OER) due to energy accumulation in each step, with a significant over-potential in contrast to the HER. Additionally, OER's behavior in basic and acidic media significantly depends on pH of the electrolytic solution. Hydrogen and oxygen are produced in acidic and neutral conditions when two H_2O molecules are oxidized at 0.404 V of potential, and the hydroxyl group is converted into oxygen and water at 1.230 V (vs RHE) in an alkaline media [13, 14]. The process function performs more smoothly in an alkaline environment than in an acidic one, as illustrated in equations [2, 15].

In acidic solution:

$$2H_2O \rightarrow O_2 + 4H^+ + 4e^- \quad E^o = -1.23\ \text{V} \tag{1}$$

$$4H^+ + 4e^- \rightarrow 2H_2 \quad E^o = 0.00\ \text{V} \tag{2}$$

$$2H_2O \rightarrow 2O_2 + 2H_2 \quad E^o = -1.23\ \text{V} \tag{3}$$

In alkaline solution:

$$4OH^- \rightarrow O_2 + 2H_2O + 4e^- \quad E^o = -0.40\ \text{V} \tag{4}$$

$$4H_2O + 4e^- \rightarrow 2H_2 + 4OH^- \quad E^o = -0.83\ \text{V} \tag{5}$$

$$2H_2O \rightarrow 2H_2 + 2O_2 \quad E^o = -1.23\ V \tag{6}$$

The main factors that control electrocatalyst activity are active sites and intrinsic catalytic activity. The availability of catalytic active sites was enhanced through various strategies including morphological modifications, nanostructure fabrication, and large surface areas [16]. To enhance HER and OER activity, the most effective ways include altering bulk and slab characteristics by selecting appropriate selection of anion and cation, optimizing morphology using various synthesis techniques, and improving charge transfer through surface electronic structure modification [17, 18]. To achieve effective electrocatalytic activity, catalysts must be capable of transfer charge, which may be achieved by coupling functional materials like MOF-based electrocatalysts with reduced graphene oxide (RGO) [19, 20]. The HER process involves the contest amid chemical adsorption and desorption of hydrogen atoms. An effective HER catalyst and adsorbed H must have a strong enough bond to facilitate proton-electron transfer, weak bond breaking, and the release of H_2 gas [16].

Recently, metal–organic frameworks (MOF), unique category of porous materials have drawn significant interest for their potential in water splitting, electrochemical energy conversion, and storage field [21]. During recent decade, researchers have been exploring potential of MOF as a catalyst owing to its exceptional properties including high crystallinity, porosity and pore size [22]. The designable frameworks of these materials are created through a modular self-assembly technique due to their broad range of inorganic and organic elements [23, 24]. MOFs are homogeneous catalysts, generally transition metal complexes, that can be considered as a single isolated unit within the unlimited MOF group [25]. The periodic structure of these materials ensures consistent active site distribution, while their porosity permits easy transportation of substrates and products for catalytic purposes within the MOF structure [26]. Thus, MOFs exhibit homogeneous catalytic characteristics and function similarly to metal complex catalysts. Heterogeneous catalysts offer the unique feature of long-term recycling of MOFs post-catalytic operation due to their porosity nature [27]. MOFs and MOF-based compounds offer numerous advantages for use as catalysts, such as excellent electrochemical characteristics, dynamic structure, electrical conductivity, and durable stability [28]. Recently, MOF has been widely explored as an efficient catalyst for an electrochemical process for the generation of H_2, with potential implications in numerous domains like gas sensing [29], biomedicine, adsorption [30], and CO_2 capture [31].

MOFs are promising water-splitting catalysts owing to their ability to retain structural properties and functionality even after post-synthetic alteration procedures, and their exceptional electrical, chemical, and physical properties [7, 32]. Furthermore, various techniques for fabricating pore structures, such as pore space partition, are crucial for the catalytic potentials of MOFs. They exhibit enhanced host–guest interactions due to their reduced pores during pore partitioning, potentially enhancing their catalytic efficiency [33]. The pore-partition strategy can introduce unique heterometallic elements into MOFs, that improves the catalytic efficiency. As a result, MOFs are unique solid catalysts and hold great promise in catalysis applications [34, 35]. However, MOFs exhibit weak metal utilization and ion diffusion efficiency, causing significant interfacial resistance in electrocatalysis due to their unfavorable properties [36]. The discovery of novel 2D MOFs offers a promising class of porous material with exceptional physicochemical and topological characteristics and has sparked research into functional non-noble metal-based catalysts. The invention and research on MOF composite materials have led to evolution of heterogeneous structures regarding to MOFs [37, 38]. Currently, the integration of multiple distinct MOF structures has led to the formation of various multi-component heterostructures. MOFs exhibit adaptability by integrating with various multifunctional nanomaterials like metallic substances, carbon-based compounds, and organic polymers [39]. The combination of these advancements provides numerous design methods for creating intricate and unique heterostructures. The study reveals the complex interactions among MOFs and various materials, laying the foundations for establishing new functional materials having precise characteristics [40].

This chapter explores the fascinating frontier of designing advanced heterostructures for enhanced electrochemical water splitting using MOFs. Water splitting offers hygienic and renewable hydrogen fuel source, but requires the development of reliable electrocatalysts owing to lethargic kinetics of the HER and OER. This study will delve into challenges of effective water splitting, including significant overpotentials required, and the crucial role of electrocatalysts in overcoming these obstacles, focusing on thermodynamics and kinetics. Subsequently, we will explore the MOF's intrinsic properties and potential as electrocatalyst building blocks, including synthetic approaches and post-synthetic modification methods to enhance their catalytic efficiency. The core of this chapter will focus on designing MOF's heterostructure with tailored properties by highlighting the role of interfacial engineering and enhancing synergetic interaction. The study will thoroughly examine the influence of various heterostructure engineering techniques on catalytic efficiency of MOF-based materials for both HER and OER. The key performance indicators depend on overpotential, exchange current density, long-term stability, and Tafel slope. The mechanistic explanations of MOF-based heterostructures for enhanced catalytic performance, such as active site exposure, charge transfer kinetics, and more efficient reaction pathways, will be thoroughly discussed. The study explores the potential applications of advanced MOF-based heterostructures in practical water-splitting devices like fuel cells and electrolyzers, evaluating current obstacles and constraints.

1.1 MOF Structure

MOFs are crystalline porous materials based on inorganic and organic materials, also known as porous coordination polymers (PCPs) [41]. They comprise a network of organic molecules linked to metal ions in one, two, or three dimensions. Metal ions serve as nodes in these structures, while organic ligands act as linkers, exhibiting exceptional crystallinity, versatility, and long-order range [42]. MOFs are synthesized and developed using reticular chemistry principles, which aim to fabricate complex structures from different molecular building elements. Porous MOFs have open pores that allow organic molecules to regularly enter and exit the channel without affecting the framework's structural integrity, allowing the incorporation of gases like carbon dioxide and hydrogen [43–45]. Research on metal–organic materials has led to advancements in several scientific domains including nonlinear optics, gas separation and storage, and catalysis. The initial research focused on phosphates and aluminosilicates but later found that developed MOFs exhibit excellent thermal stability and may replace mesoporous silica in regard to surface area and guest molecule loading capacity [46]. Porous materials like MOFs can be prepared for catalytic reactions by inserting specific-sized and shaped molecules into their pores, particularly aluminosilicate zeolites, as widely published. Zeolite structure hinders the utilization of channels by molecules due to their stiff internal pore surface [47, 48]. In pristine MOFs, catalytic sites are typically located in Lewis's acid core, active groups on organic linkers, and co-ordinately unstable metal locations. However, MOFs are restricted to a specific range of catalytic reactions due to this situation, but two strategies can effectively address this limitation. (a) Functionalized modification, a multivariate method known as post-synthetic adaptation, is used to extract specific active locations from metallic ions or organic linkers [49, 50]. (b) In pore confinement/encapsulation, MOFs can host various active substances in pore space, such as organic materials, inorganic nanoparticles, and enzymes, operating as nanoreactors for catalytic activities [51].

MOF-on-MOF is a novel heterostructure that has garnered significant attention in this field owing to its intricate morphology and unique topologies [52]. Typically, lattice mismatching is used to arrange multiple MOF materials to form MOF-on-MOF structures [53]. Secondary MOFs successfully nucleate and develop on surface of original MOF substrates in the case of strong lattice functionality [54]. On the other hand, higher Lattice mismatch in MOF-on-MOF heterostructure hinders its formation by increasing the energy barrier at interfaces and severely affecting epitaxial formation [55]. Furthermore, nucleation can be influenced by several approaches including thermodynamics and kinetics management, while lattice mismatch barriers across MOFs can be removed using intermolecular forces and related methods. This method offers precise assembly of MOF structures with diverse lattice properties, expanding design options and enabling intricate multi-component assemblies [56, 57]. Chemical reactions can be categorized into homogeneous and heterogeneous forms, with heterogeneous reactions occurring at interfaces affected by surface characteristics and interim interactions [58]. Hydrothermal synthesis is widely used for simulating reactions between solid and liquid phases, forming MOF-on-MOF heterostructures. Heterogeneous reactions allow an additional organized and precise pathway for developing secondary MOF, while homogeneous may result in disorganized MOF development [59, 60].

This study delves into the complex aspects of MOF-on-MOF interface development, examining all intrinsic and extrinsic elements [61, 62]. Firstly, intrinsic factors are significantly influenced by role of metal ions and organic ligands. For instance, Tsung and his team utilized a solid–liquid hydrothermal synthesis technique to combine ZIF-67 particles with a mixture of 2-methylimidazole and zinc metal [63, 64]. Since, ZIF-67 and ZIF-8 have distinct metallic ions but identical ligands, resulting in comparable interaction at the interface when metal ions attempt to bind locations simultaneously. The method renders ZIF-67's intrinsic instability, making it sensitive to etching, to create multi-shell ZIF-67@ZIF-8 materials [65]. However, the degree of lattice matching across MOF structures is a vital factor in determining the outcomes of MOF-on-MOF development [66, 67]. The addition of structure-directing factors can significantly alter interfacial development and interfacial energy. For Example, Yu and his team used a lattice-matching technique to combine two MOF structures with distinct metals and ligands. The large lattice matching between the (020) and (110) facets of ZIF-8 and c-MIL-125 led to ZIF-8 nucleating and developing selectively at the four corner sites. On the other hand, ZIF-8 developed primarily along the b-MIL-125 border due to a tight lattice match and extrinsic factors impacting MOF-on-MOF development [68–70]. Figure 1 illustrates that MOFs can be improved by altering their internal structures, incorporating nanoparticle catalysts, and adding elements such as metal sulfides or carbon-based compounds. These versatile materials have great potential for applications such as gas storage, separation processes, catalysis, and energy storage due to their unique structure and large surface area.

1.2 MOF for Electrocatalysis

The use of MOF-on-MOF heterostructure in electrocatalysts requires identification of catalytically dynamic regions and processes. The MOF porous structure provides frameworks for introducing more active elements with adjustable size, shape, portion, and site [72]. The catalytic efficiency of these catalysts can be enhanced by fostering synergistic interactions between the functional units. These structures possess unique characteristics beyond their active sites, which may contribute to their exceptional electrocatalytic activity. The excellent permeability of MOF channels facilitates efficient mass movement in electrochemical processes [73]. MOFs are highly effective electrocatalysts due to their substrate specificity, which enhances the electrochemical response and sensitivity, making them a promising candidate for various applications. Three primary methods have been suggested to increase the electrocatalytic properties of MOFs: creating metallic or electrochemically active MOFs, supporting MOFs on conductive substrates, and combining MOFs via active substances [74, 75]. The electrocatalytic properties of MOF can be increased by adding redox or active positions of the catalyst. The electrochemically reactive MOFs can be created using organic ligands having π–π^* bonding or elongation of connected systems, like thiophene, pyridine, and benzene derivatives [76]. All of these organic ligands can be combined to redox metal ions to develop electrocatalytically effective MOFs. Wang et al. created an OER electrocatalyst using a cobalt-based zeolitic imidazolate structure (Co-ZIF-9) by combining redox cobalt cores with benzimidazole linkers. The

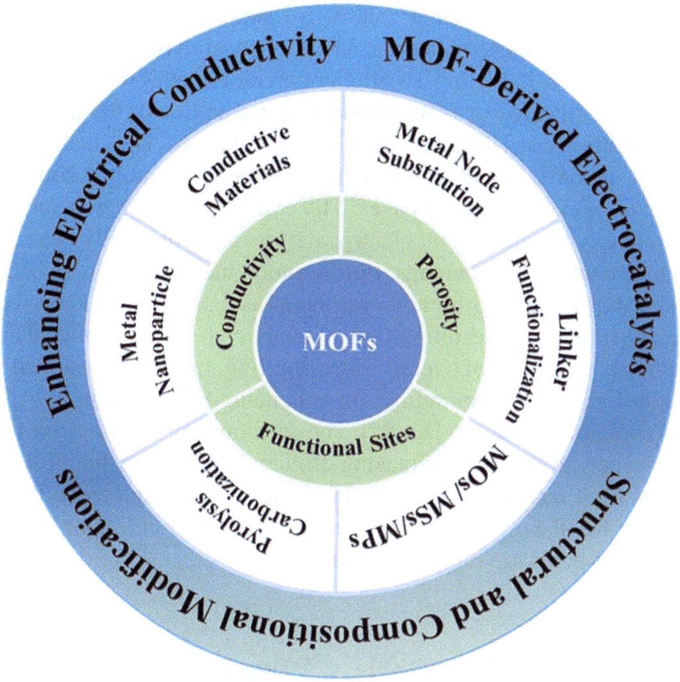

Fig. 1 The diagram illustrates recent strategies to enhance the electroactivity of MOF and MOF-based materials for electrochemical water splitting [71]. Reprinted with permission from Open Access license

synthesized Co-ZIF-9 demonstrated significant OER catalytic activity across a broad pH range [77]. In another research, Gong and his colleagues fabricated two innovative Co(II) coordination polymers using 1,4-bis (3-pyridylaminomethyl) benzene, resembling the significantly efficient OER catalysts based on ruthenium [78]. These two structures demonstrated strong OER electrocatalytic capabilities due to reduced overpotential and increased densities of current, but their structures could collapse due to electrochemically oxidation and reduction reactions.

MOFs are often deemed ineffective as electrocatalysts for OER and HER, the primary mechanisms for electrochemical water splitting [79, 80]. The most advanced MOFs, like OER, perform at higher energy costs than thermodynamic specifications due to larger overpotential and lower turnover frequency throughout OER [81]. The effectiveness of an electrocatalyst is influenced by various factors such as electrode shape, electrical conductivity, and the availability of active centers. It's crucial to optimize all aspects, but it can be challenging to achieve one without compromising others. Therefore, despite their inherent metallic locations, MOFs are not commonly utilized for electrocatalysis due to their low electrical conductivity and tiny pores [82, 83]. The recently reported techniques like elevated temperature calcination may compromise the metallic locations of molecules in MOFs [84]. The hybridization method, which includes additional conductive supports, can potentially block the inherent small pores in these materials. The electrocatalytic water-splitting process faces several obstacles, including existing nonprecious catalyst synthesis techniques often result in complicated or mismatched integration due to the differences in the production conditions of HER and OER electrocatalysts. The interaction between HER and OER in conventional water electrolysis can potentially result in hazardous mixtures of hydrogen and oxygen gases. The combination of gases and the production of reactive oxygen species by electrocatalysts can potentially cause membrane failure and premature device damage. Furthermore, the production of OER is a sluggish reaction, that requires a higher overpotential than HER to keep the identical current density, making the resulting O_2 less beneficial. To overcome all these obstacles, and increase efficiency of overall water-splitting, rationally designed innovative heterostructures can play a vibrant part in enhancing binding affinities of hydrogen and oxygen-containing products [11, 85]. The development of innovative heterostructure presents a new avenue for increasing efficiency of electrochemical water splitting. The structural factors and various key reactions that influence the interfacial development of MOF-on-MOF heterostructures and their emerging applications in different practical fields are represented in Fig. 2.

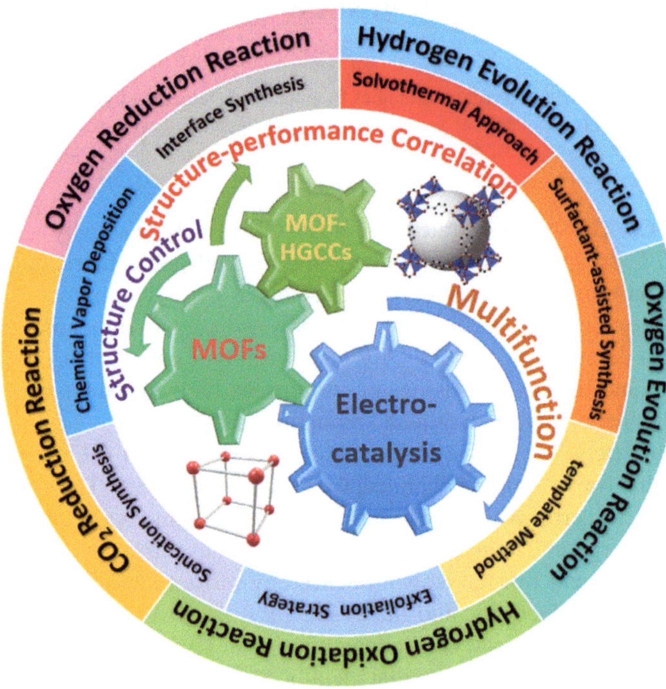

Fig. 2 Overview of various approaches that influence the interfacial development of MOF-on-MOF heterostructures and emerging applications of MOF-on-MOF heterostructures for numerous practical applications [86]. Reprinted with permission from Open Access license

1.3 Synthesis Methods for MOF-Based Electrocatalytic Water Splitting

The demand of MOF-on-MOF heterostructures significantly increased owing to their increased intrinsic charge transport and conductance, making the synthesis of new MOFs advantageous for electrochemical characteristics. The electrocatalyst's morphology significantly influences its performance, as it increases surface adsorption capacity and facilitates the existence of an active region. The process of preparing MOF-based electrocatalysts involves modifying their morphology. For instance, Yang et al. utilized a straightforward electrochemical synthesis method with zinc nitrate and terephthalic acid to create an innovative MOF-5 with a distinctive flower-shaped structure [87]. The extensive structure of MOF-5 is attributed to the binding effect of Zn^{2+} ions and terephthalic acid. Qin et al. developed a wide structure of 3-dimensional MOF-based porous polyoxometalate (POM), joining POM fragments as nodes via organic ligands [76]. These catalysts can be developed by utilizing functional groups with organic ligands like carbon, nitrogen, boron, sulfur, and oxygen. Additionally, the post-synthesis approach allows MOFs to incorporate guest species like metal centers, nanoparticles [88], and POMS [89] by utilizing their inherent properties. The MOFs can be utilized as templates for pyrolysis of coordinate complexes in inert environments like air, ammonia, and N_2 at elevated temperatures [90, 91]. The ideal framework of MOFs is effected by the complex functioning of selected MOF precursors and their synthesis. These strategies are compatible with each other. These specially designed MOF precursors are efficiently converted into comparable inorganic metal complexes via hydrothermal or pyrolysis processes. On the other hand, modifying the reactants and reaction conditions can lead to the development of an extensive range of nanostructures with unique topologies. Typically, MOFs do not have sufficient electrical conductance to be effective. Metal oxides are often combined with conductive substances like graphene, carbon nanotubes, and conduction polymers, and occasionally pyrolyzed in an inert atmosphere to develop the carbon networks of decorated metal oxide [92]. MOF-composited substances exhibit superior electrocatalysis and good electrical conductivity compared to their pristine counterparts. Duan et al. used a single-stage chemical deposition method involving an organic ligand and aqueous metal salt to generate an extremely thin nanosheet of 2D-MOF by using an in-situ growth strategy [93].

The framework of crystalline MOF was formed from inorganic metal–oxygen and organic hydrocarbons, with metal ions arranged octahedrally. In order to regulate the width and structure of electrode on the microscopic level, Cao et al. developed an independent electrode based on extremely thin nanosheets of Ni-rich Ni(Fe)-MOF employing Ni–Fe composite as the metallic sources and substrates utilizing a semi-sacrificial pattern technique [94]. Moreover, earlier research has classified fabrication techniques into three distinct types: (a) In-situ nitrogen doping is a method that utilizes MOFs including nitrogen, like zeolitic imidazolate frameworks (ZIFs); (b) ex-situ nitrogen doping is a technique utilized to treat gaseous host molecules like ammonia; (c) The third process involves using a solid substance in combination with

MOFs as a nitrogen source. Additionally, MOF can serve as a direct precursor for creating porous carbon electrocatalysts doped with heteroatoms. The use of templates in the structural precursor transformation for MOFs containing nitrogen can result in the production of nitrogen-doped carbon [95]. MOFs are often used as sacrificial precursors for pyrolysis, producing carbon-based compounds like amorphous carbon or carbon nanotubes, or metal-based nanoparticles from metal nodes. The produced carbon substrate can prevent the accumulation of more nanoparticles for metal-based. The pyrolysis environments and modifying MOFs can also enhance the electrocatalytic efficiency of pyrolysis products [96]. The bimetallic MOF substances were successfully utilized to control atom-type restriction of an MOF precursor and develop robust catalysts through carbonization. These Bimetallic electrocatalysts often exhibit superior effectiveness compared to individual atom MOF due to the synergistic effect of bimetal doping [97].

2 Mechanisms Enhancing HER Performance in MOFs Heterostructures

The primary determinants of electrocatalyst activity are often intrinsic catalytic activity and accessible active sites. Creation of active sites, surface modifications including heterostructure formation, surface decoration etc., fabrication of nanostructures and synergistic effects can all amplify the amount of catalytic sites and thus, catalytic activity [16]. Optimizing morphology through the adaption of unconventional synthesis procedures, improving charge transfer process through the purposeful alteration of electronic structure like composite synthesis or employing hybrid catalysts are the most productive procedures to maximize rates of HER and OER. Moreover, this can also be done by the modification of bulk and surface characteristics by utilizing an appropriate combination choice of cations and anions [17, 18]. The efficiency of electrocatalytic activity relies on the capacity of the catalyst to transfer charge and employing combination of functional materials fulfills this requirement. For instance, the charge transfer ability and hence, the conductivity of MOF based electrocatalysts can be amplified by utilizing reduced graphene oxide (RGO) [19, 20]. The processes of chemical adsorption and desorption of hydrogen atoms take place simultaneously and competitively in case of HER. A sufficiently strong bond should exist between an effective HER catalyst and adsorbed hydrogen to facilitate the phenomenon of electron-proton transfer. Meanwhile, it should be weak sufficient to ensure easy bond breakage and discharge of H_2 gas.

Analysis of stability and activity of the catalyst is crucial for assessing its feasibility for water splitting and its application-specific efficiency. The benchmark catalysts are often compounds based on noble metals. Because of the low Tafel slope and nearly zero overpotential, Platinum (Pt) is chosen as a standard since it is thought to be the highly efficient catalyst. The best catalyst in terms of activity and instability for OER is thought to be IrO_2. Despite having greater catalytic activity than IrO_2,

RuO_2 has a limited range of application due to its severe corrosion susceptibility [98–100]. A variety of evaluation criterion can be used to quantify the water splitting capability of a catalyst and Tafel slope is a crucial metric in this regard. The current generated in an electrochemical cell is plotted graphically against electrode potential of a specific metal. The Tafel plot, in short, shows how well an electrode can generate current when exposed to an applied potential shift. The reaction rate is linked to the overpotential through Tafel slope. A smaller overpotential is desirable for attaining significant amount of current if the slope value is small because a lower Tafel slope implies the electronic transport to be faster [101]. RuO_2 is said to produce a Tafel slope of 89 mV/dec [102] while electrode of Pt/C is said to have a Tafel slope of 30.0 mV/dec [103].

Overpotential is also an indispensable metric to estimate the potential of a catalyst for water splitting. Overpotential is termed as the potential above equilibrium potential in order to start the reaction and discharge the entity at electrodes. Although water must be separated in to its components by an applied voltage of at least 1.23 V, a substantial overpotential is typically required because of complex reaction kinetics [104]. A catalyst with a small Tafel slope and great current density is appropriate for water splitting [105]. The Tafel slope and overpotential for water splitting are influenced by the electron transfer, which is further affected by the crystallinity of the catalyst, its size, dimensionality and band gap. Thus, these evaluation criteria are immensely improved by the existence of MOF catalysts. The inclusion of metal centers in MOFs improves their charge transfer characteristics. Moreover, the charge carriers are easily transported due to their small size, tunable band gap, high surface area and multi-dimensionality.

Since the chemical environment provided by the electrolyte influences the characteristics of electron affinity of cations, the performance of electrode material is also impacted by it. In contrast to acidic or neutral medium, catalysts portray optimal performance in alkaline medium because of its high oxidation potential. Hence, the current issue is to develop a stable and effective material over the entire pH range [106]. The chemical instability of an MOF in an electrolyte may cause the failure of framework, dissolution of MOF, formation of MOF derived metal oxides during or prior to the electrochemical reaction [107].

The catalyst activity can be significantly improved by a procedure recognized as synergistic interaction between several material kinds. It is characterized by the sequential or supplemental manifestation of nonlinear collective effects of the various features of two or more active substances. In the context of manufacture of water splitting catalyst, a synergistic effect is characterized through the combination of catalyst material along with an effective assisting substance which results in formation of hybrids or composites having improved active sites and boosted activity. MOF based catalysts are more likely to have synergistic effects which improve catalytic activity. MOF based composite catalysts are thought to be successful due in large part to the synergistic effects found among MOF and conductive substrates. Synergistic impacts can be engineered to act at multiple levels, like the coexistence of synergistic effects of two well-blended functional MOFs or MOF/substrate hybrids. On the other hand, distinct metal clusters within a multi-metallic MOF exhibit intramolecular synergistic effects [73].

2.1 Role of Active Sites

Significant efforts have been made to transform MOFs into inorganic functional materials in a controlled manner in addition to attempting to modify MOF materials for use in energy conversion systems [24]. The derivatives of MOFs exhibit considerable potential for use in electrocatalysis, thanks to the rich compositions, high specific area and porous nature, all innate from MOFs. There are two ways to categorize the several heterostructure derivatives that have been fabricated so far, i.e., by composition and by structural geometry. In particular, the final electrocatalytic performance is significantly influenced by the compositions and structural geometry. Classification by composition involves metal hydroxide, metal oxide, metal sulphide, metal phosphide, metal selenide, metal nitride heterostructures and carbon containing heterostructures. On the other hand, categories on the basis of structural geometry include geometric morphology, shell architecture and heterointerface.

Shalom et al. conducted a thorough experimental and theoretical examination of active sites participating from phosphides and sulphides for the better comprehension of the involved phenomena [108]. Nickel phosphide (Ni_5P_4), sulphide (Ni_3S_2) and nitride of nickel (Ni_3N) were synthesized, and the (0001) hexagonal basal plane was taken into consideration as the primary factor influencing the kinetics of HER. Additionally, dominantly electrolyte-oriented Ni_5P_4 $(10\bar{1}0)$ prism planes were taken into account. While P-termination was more suitable for HER conditions in case of Ni_5P_4 and S-termination for Ni_3S_2, they found that the (0001) surface with Ni termination was more favourable for Ni_3N.

Ni_3S_2 (0001) is found at the top of the volcano plot in Fig. 3a according to DFT calculations, suggesting that it is more active than nitride and phosphide. However, Fig. 3b shows that its activity has close resemblance with that of the pristine Ni. The group noted that the S termination lies near the Ni-terminated surface's boundary and they are more likely to shift into one another. Ni termination appears when the redox reaction causes the uppermost S atoms to be removed. As a result, the structure becomes unstable and lacks any ΔGH^* sites around 0. The active sites on each of the four surface terminations of Ni_5P_4, however, were nearly zero. In both Ni-rich and P-rich terminations, nearly thermally neutral H adsorption makes Ni_5P_4 highly stable. Based on the aforementioned observations, it is concluded that development of more stable and active electrocatalyst having greater number of sites along with having nearly zero hydrogen adsorption energy can be realized in case of interfaces containing phosphides.

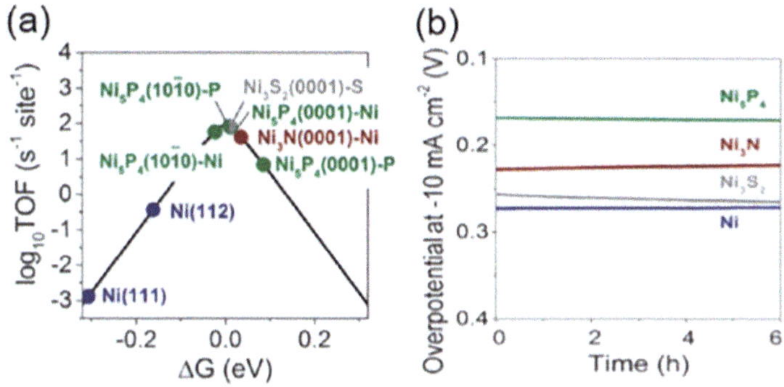

Fig. 3 **a** Volcano plot showing HER activity. **b** Prolonged stability for up to 6 h [109]

2.2 Surface Engineering for Enhanced Catalytic Activity

The electrical structure can be tuned through several surface engineering techniques including heterostructure fabrication, doping, alloying, defect engineering etc. to further boost the electrocatalytic activity. The electrical conductivity and water adsorption capacity of catalysts, mostly dependent on their electronic configurations, have a significant impact on their HER and OER activity.

A partially controlled phosphorization technique was used by Qiling Li et al. to fabricate a 2D porous CoP/Co-MOF nanosheet array [110]. Building heterostructures in 2D nanosheets has been shown to be a successful method of increasing HER activity throughout a broad pH range. The current density of -10 mA cm^{-2} in 0.5 M H_2SO_4, 1.0 M PBS, and 1.0 M KOH medium may be driven by the 2D CoP/Co-MOF nanosheet array with overpotentials of 52, 106, and 26 mV, respectively. Additionally, CoP/Co-MOF exhibits favourable endurance, particularly in neutral media and generally in acidic as well as basic electrolytes. The swift charge transfer rate and many accessible active sites owing to altered electronic structure along with improved specific surface area, are responsible for remarkable HER activity. This work offers a practical method for maximizing HER performance of MOF-based electrocatalysts.

In order to boost slow kinetics of HER and OER in basic media, it is essential to perform in situ topological modificationss of electrocatalysts. This approach results in heterostructures of mixed-metal oxides or hydroxides, which are extremely active for OER, combined with metals that have minor barrier for hydrogen adsorption, thereby promoting efficient HER [111]. It is established that the in situ oxidation of a micropillar-like heterostructured Fe–Co–O/Co@NC/NF precatalyst results in a freestanding, defect-rich, and in situ oxidized Fe–Co–O/Co metal@N-doped carbon (Co@NC) mesoporous nanosheet (mNS) heterostructure on Ni foam (Fe–Co–O/Co@ NC-mNS/NF). The presence of oxygen vacancies, advanced particular surface area, enlarged electrochemical active surface area, and in situ created new metal (oxyhydr) oxide phases are responsible for the long-term durability of the in situ oxidized Fe–Co–O/Co@NC-mNS/NF, which demonstrates exceptional bifunctional characteristics through requiring only respective low overpotentials of 257 and 112 mV for OER and HER with current density of 10 mA cm^{-2}. Furthermore, a current density of 10 mA cm^{-2} may be acquired using the Fe–Co–O/Co@NC-mNS/NF (+/−) electrolyzer with just a small cell potential of 1.58 V. Thus, a new avenue for increasing the overall alkaline water splitting is made possible by this work.

2.3 Synergistic Effects in MOF Heterostructures

Significantly increased electron conductivity can be achieved by pairing active ingredients with conductive surfaces including carbon, nickel foam and MXene [112–116]. Electron transfer in the heterostructures is assisted by interfacial bonding present in them. In addition to conductivity, other adjustable parameters involve mass transfer efficiency, mechanical property, active site density, hydrophilicity or aerophobicity and electrochemical stability. A type of 2-dimensional (2D) CoP nanosheets/1D carbon nanotubes heterostructure (CoP NSs/CNTs), for instance, was described by Zhang et al. [117] and shown noticeably enhanced HER activity. An HER current density of 10 mA cm^{-2}, a small Tafel slope of 57 mV dec^{-1}, and an excellent stability of 24 h were all attained by the CoP NSs/CNTs composite at a low overpotential of 68 mV. The synergistic effect of oxygen functional groups on CNTs is responsible for the improved wettability of CoP NSs/CNTs as compared to CoP NSs, as demonstrated by the examination of water contact angle. In addition to facilitating water adsorption, high wettability also makes water breakdown in HER easier. Similarly, by adding aerophobic WO_3, Peng et al. [118] recently created a composite catalyst, Pt_2W/WO_3/rGO (reduced graphene oxide) possessing high-performance. The Pt_3W/WO_3/rGO with customized aerophobicity is significantly more active than the 20% Pt/C available commercially. According to DFT calculations, each constituent of the heterostructure cooperating synergetically may attain excellent catalytic activity, resulting in rapid and continuous desorption of H_2. To bring the Gibbs free energy for hydrogen adsorption closer to the optimal value. i.e., zero, the synergistic effect in Pt_2W/WO_3 can significantly reduce hydrogen adsorption energy on the surface of the catalyst. Furthermore, the synergetic effect can also improve endurance, which is one of the most important aspects of electrocatalytic reactions. An intricate internal structure and densely interfaced NiFe LDH@NiCoP/NF electrodes were developed by Zhang et al. [119]. Over a prolonged analysis of more than 100 h, the 3D hierarchical heterostructure NiFe LDH@NiCoP/NF maintains the crystalline phase and morphological structure of nanowire due to the strong interfacial interaction and syenrgic effects within the heterostructure. Different components generally have distinct activities for several basic electrochemical reaction steps and ultimate kinetics of the reaction are enhanced under the driving force of synergetic effects. Guan et al.

suggested using FeOOH as a co-catalyst to improve Ni_3N's electrochemical performance [120]. With a current density of 10 mA cm^{-2}, the composite portrayed excellent activity at overpotential values of 244 mV and 67 mV for OER and HER, respectively. Throughout the reaction, the negatively charged Ni_3N preferably adsorbed the H atom, whereas the positively charged FeOOH had great tendency to make bond with O atom. Hence, the stimulation of water molecule and expedition of H-OH bond cleavage is the outcome of the synergetic effects existing between FeOOH and Ni_3N. Although FeOOH can act as a co-catalyst to assist the water molecule in adsorption or dissociation, but its excessive amounts can cause the downfall of electrochemical activity.

3 Mechanism Enhancing OER Performance in MOFs Heterostructures

The oxygen evolution reaction (OER) is the electrochemical reaction for the production of oxygen gas, proton and electrons which requires an effective catalyst for lowering the overpotential in water splitting. MOFs being porous in structure offer several benefits for the OER including high surface area, tunability such as metal' s redox activity and versatility for having multiple catalytic sites or coordinating easily with other materials [121]. As the heterostructure consisting of Co-based MOFs and NiO have improved performance because of transition metal ions work synergistically to optimize the OER. To understand the enhanced OER performance different factors such as design principles, characterization, catalytic activity and mechanism is important to consider which are further discussed.

The first key factor leading to enhance the OER performance is to shift synthesizing bulk powder which have poor electronic conductivity and electrochemical response leading to blockage of active sites by the presence of organic ligands. The bulk MOFs should be converted into ultrathin layer in nanometer thickness for overcoming decline in OER activity [122]. The ultrathin film supports the rapid mass transport, charge transfer rate during catalysis and interaction of unsaturated metal centers with respective reaction substrate is easy. For understanding let's consider and NiCo ultrathin bimetal-organic framework which was synthesized by Tang and his collaborators. For visualizing the OER performance in alkaline solution a rotating ring-disk electrode was placed with copper foam. The excellent overpotential of about 180 mV at 10 mA cm^{-2} current density was reported because of unsaturated Ni and Co sites acting as dominating center for OER compared to bulk NiCo-MOFs [123]. Further Wang et al. developed NiFe-UMNs [124] by ultrasonic oscillation reaching the overpotential of 260 mV reaching current density of 10 mA cm^{-2} because of the synergistic effect between Ni and Fe which was proved by DFT calculation. The energy barrier for the OH^* adsorption on Ni sites can be tuned by the synergy of Fe and Ni with the help of DFT calculations [125]. This effect may be described as the partial electron transfer from the Ni to Fe atoms. Rather than ultrathin films the heterostructure modulate the properties to benefit the OER.

3.1 Synergistic Effect Benefiting the OER

The existence of heterointerface boost the electron conductivity with the help of conductive substrate like carbon, nickel foam and MXene. In heterostructure the interfacial bonding act as a channel for electron transfer. Peng and his coworkers recently reported the composite $Pt_2W/WO_3/rGO$ [126] which shows 20% higher activity than the commercial Pt/C by the fast and continuous H_2 desorption and the synergistic cooperation in lowering the adsorption energy over the surface of catalyst result in maintaining the adsorption Gibbs free energy approaches to zero value which is the ideal case. The overall reaction kinetics is effected by multiple components which possess certain activity for various fundamental steps in an electrochemical reaction which would be influenced by the synergistic effect. For example, the heterostructure FeOOH and Ni_3N exhibits excellent performance with an overpotential of 244 mV for OER and 67 mV for HER at 10 mA cm^{-2}. The reaction carried out in such a way that the FeOOH behaves as positively charged reactant forming the bond with O atom and Ni_3N being negative have preference to adsorb H atom. In this way the synergistic effect between these two materials accelerate the H-OH band while FeOOH promoting the adsorption/dissociation of water molecule. The thermodynamic overpotential is non-zero because of multiple reaction intermediates during OER this may be negotiate by heterostructure electrocatalysts and their underlying mechanism.

3.2 Controllable Morphology and Hierarchical Assembly

The mechanism can be enhanced by controlling the morphology of MOF-based heterostructures via hierarchical assembly. The hierarchical assembly of MOFs maximizes functionality by engineering the material with the desired specific shape, size and internal structure. It basically refers to the process that construct the multi-scale structures from molecular to larger macroscopic structures. The individual MOF units are arranged at different length scale with controlled integration to form complex structures [127]. The multiple length scale includes Nano, Micro and Macroscopic scales and different assembly methods to attain it can be Self-assembly resulting because of the favorable metal center or organic ligands interactions, Template-assisted assembly which use polymers or other frameworks to guide the formation and the Bottom-up or top-down strategies. The former is the widely used strategy as either starting with small units and assembling into the larger systems, or breaking down larger ones into smaller MOF units and then according to the application reorganizing into hierarchical assembly.

The regulation of morphology is carried out by the following main classifications:

i. By functionalizing the organic ligands i.e. varying the approach of growth which results in the change of overall heterostructure.
ii. The number of layers in the heterostructure can be controlled by the number of assembly times for regulating the morphology.
iii. Another important factor is the order of assembly that leads to change in heterostructure morphology.
iv. Growth time for the hierarchical assembly can be changed to control morphology for certain application.
v. The insertion of ion such as Fe^{3+} as indicator can also control the morphology of the heterostructure.

The heterostructure for the advancement in the OER activity can be constructed by different strategies having benefits over the other. The liquid phase epitaxy (LPE), layered spray LPE, internal extended growth and anti-epitaxial growth are some of them that advantages of regulating the construction of MOFs based heterostructure for electrochemical reaction. The porous nature of MOFs makes it more effective to be an electrocatalyst and the porosity can also be tailored. The pore size and the pore structure of MOF based heterostructure may be controlled which help in determining the performance. The atomic or molecular units of the MOFs based heterostructures help in defining pore structure and size which directly affect successful synthetization of larger pore size MOF behaving as core. For example, there was a fruitful synthetization of hybrid membranes of core/shell structured pitches from two individual different MOFs in which larger pore sized MOF (UiO-66-NH_2) worked as core for providing the transport pathway [128]. The smaller and larger overlap in disproportionality at the interface to create the smaller pores that behave as molecular filters enhancing molecular performance. Such as recently its studied that the zeolite pores can by effectively reduced by the microporous coating on the outer surface of zeolite by the deposition. Similar techniques have been implied for desired pore size and for improving selectivity.

3.3 Interfacial Charge Transfer

The formation of heterostructure significantly improves the performance of OER as electron transfer, bond structure and density of states (DOS) observed to change at the interface. The interfacial charge transfers on the synthesized heterostructure FeNi-MOF/caffeic acid (CA) $NiMoO_4$ is directed by the electron research. There is two electron transfer regions i.e. electron accumulation and electron depletion at the interface which is the boundary joining both the surfaces. The electron transfer can be further explained by the charge density difference (CDD). There was seen that the bridging agent CA act as a source for the transfer of electrons from FeNi-MOF to $NiMoO_4$ [128]. The catalytic activity is increased because of the electron interactions with reactants and intermediates. The active sites in the oxidation process are

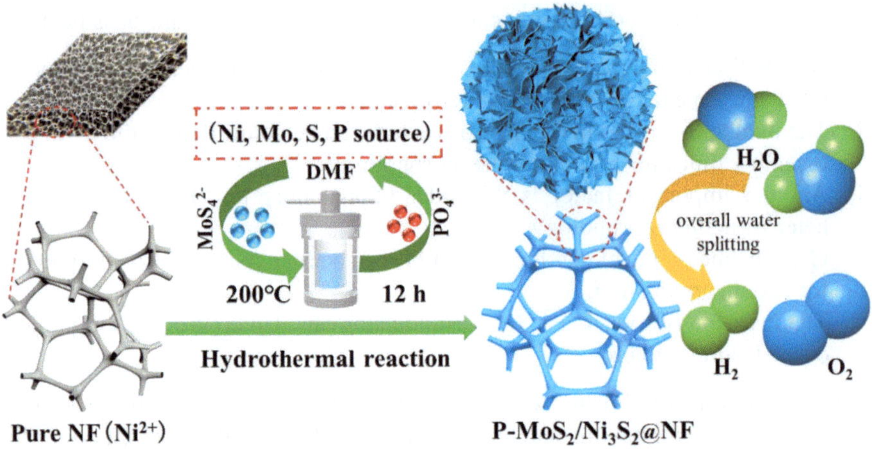

Fig. 4 The diagram representing the process of how the electrocatalyst $P\text{-}MoS_2/Ni_3S_2@NF$ has been prepared using the strategy to promote the synergistic effect in heterostructure with abundant active sites [129]. Reprinted with permission from Open Access license

explained by charge distribution on the respective internal surfaces of heterostructure with the accumulation of electrons. The active sites at 0.81 internal surface of Mxene's where the charge transfer process majorly occurs. Studying the electronic states at the interface shows that near the Fermi level more active states are present leading to the enhancement of adsorption capacity. It was seen that there is hole transfer mechanism from valance band of bismuth vanadate to the adjacent interfacial sites of the MOF. There will be prominent enhancement of e^-–h^+ separation inside the heterostructure, the transfer of electrons inside the respective layers produces the built-in electric field resulting in the charge accumulation on MOF. Hence the interface serves as a redistribution of charges and enables built in electric field by enrichment of positive charges and promoting electron transfer and eventually increasing the performance. Figure 4 illustrates the hydrothermal synthesis and application of $P\text{-}MoS_2/Ni_3S_2@NF$, a hybrid catalyst for water splitting. Pure nickel foam treated with dimethylformamide (DMF), PO_4^{3-}, and other sources for 12 h at 200 °C, yielding a blue colored heterostructure. The resulting $P\text{-}MoS_2/Ni_3S_2@NF$ is capable of electrolyzing water, to produce hydrogen and oxygen gases.

3.4 *Tafel Slope and Free Energy*

The Tafel slope basically quantifies the rate of overpotential with the current density which is important parameter in context of electrochemical kinetics. The Tafel equation is used to draw the slop and it's the connection between overpotential η and current density j as:

$$\eta = b \cdot \log\left(\frac{j}{j_o}\right) \quad (7)$$

whereas b is the Tafel slope and it's unit is mV dec^{-1}, j and j_o are the current density and exchange current density (intrinsic activity of electrode) respectively [130]. Let's consider a novel OER electroatalyst which is decorated by IrNi nanoalloys over nanosheets forming the heterostructure IrNi@NiFe-MOFs. The active species of heterostructure which formed by the complete transformation into IrNi@NiFeOOH participate for the highly OER catalytic activity. The electrocatalytic mechanism for enhanced performance estimated by measuring the electrochemical parameters in 1 M KOH using a three electrode system for all individual materials IrO_2, IrNi heterostructure NiFe-MOFs and the decorated heterostructure IrNi@NiFe-MOFs. At the scan rate of 5 mVs^{-1}, the decorated heterostructure shows higher performance having only low overpotential of 228 mV superior to 255 mV, 287 mV and 343 mV for IrNi, IrO_2 and NiFe-MOFs respectively to attain a current density of 10 mA cm^{-2} [131]. Hence the decoration and the heterostructure aligned to notably increase the catalytic performance. For the further demonstration the Tafel slope which achieved from respective linear sweep voltametry as shown in Fig. 5a can be used to understand the reaction kinetics. The slope represented by the purple dotted line has the less value symbolizing the applicability of the modifying techniques. Typically, the smaller Tafel slope for OER range in 40–120 mV dec^{-1} points towards faster reaction and effective electron transfer indicating favorable reaction mechanism [132]. This also lead to the lower internal charge transfer resistance of about only 7.6 Ω hence at electrode the fast charge transfers kinetics.

For the more insight into the enhanced OER activity the exploration of electronic properties via spin-polarized DFT calculations should carried out. The behavior near the fermi level give the information about bonding and anti-bonding states which points towards the hybridization and the covalent bonding which is more effective

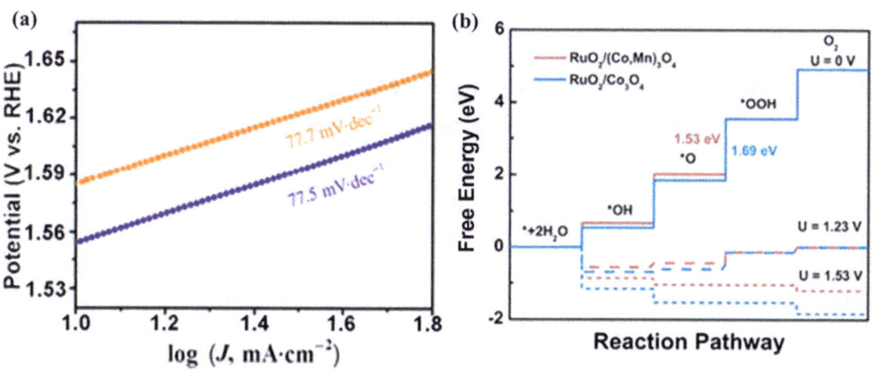

Fig. 5 **a** The Tafel plots for Co-MOF-NK and Co-MOF-K electrocatalyst representing by purple and orange lines respectively in 1.0 M KOH [134]. **b** Gibbs free energy diagram for RuO_2/Co_3O_4 heterostructure and $RuO_2/(Mn,Co)_3O_4$ [135]. Reprinted with permission from Open Access license

for the catalyst [133]. There is also a strong declaration by Hammer and Norskov regarding the reaction between catalyst and the intermediates. They demonstrated that the interaction is directly correlated with the position of d-band center of catalyst i.e. ε_d being closer to fermi level the adsorption reaction will be strong on the active metal sites if fewer occupied antibonding orbitals (d-σ) exists [130]. Hence this position of d-band center can be effectively constructed with the help of heterostructures, which is the crucial condition for adsorption and desorption of oxygen comprising intermediates in OER process.

The Sabatier's principle state the for avoiding of over and under adsorption it's necessary to achieve the optimal adsorption value of intermediates such as OH^*, O^* and OOH^* [124]. The general analysis depicts that the adsorption energies for the heterostructure and the modified heterostructures such as doped, decorated or defect created have low values for all the oxygen-based intermediates in comparison with the pure materials. Figure 5 explains the detailed comparison of the electrocatalytic performance of two ruthenium-based catalysts, RuO_2/[Co(Mn)$_3$]O_4 and RuO_2/Co_3O_4, for OER during water splitting. The Tafel plot (a) illustrates the relationship between the overpotential (η) versus logarithmic current density (log J), revealing the catalytic efficiency of the materials. The plot shows that RuO_2/[Co(Mn)$_3$]O_4 (orange line) has a slightly lower Tafel slope of 77.7 mV dec^{-1} compared to RuO_2/Co_3O_4 (purple line) at 77.5 mV dec^{-1}, indicating that both catalysts exhibit similar kinetics for OER, with the former being marginally more efficient in reducing the energy barrier for the reaction. Figure 5b represents a free energy diagram outlining the reaction pathway for OER on these catalysts. It highlights key intermediates ($^*O$, $^*OH$, and $^*OOH$) and their respective energy levels. For RuO_2/[Co(Mn)$_3$]O_4 (red line), the overpotential required to drive the reaction is 1.53 eV, while for RuO_2/Co_3O_4 (blue line), it is slightly higher at 1.69 eV. The diagram also indicates the potential at which oxygen (O_2) is evolved (U = 0 V) and the overpotentials needed to overcome the energy barriers (U = 1.233 V and U = 1.53 V), demonstrating that RuO_2/[Co(Mn)$_3$]O_4 is more effective for OER due to its lower energy requirement.

4 Key Strategies for Improving MOF Based Electrochemical Water Splitting

According to theoretical perspective, the energies related to the thermodynamic adsorption or desorption of reaction intermediates and kinetic reaction barriers impede rate of electrocatalytic reaction. It has been observed that the electrocatalytic reactions can be enhanced considerably in terms of their thermodynamics and kinetics by the molecular scale alteration of the electrocatalysts [12, 136]. One of the most practical and most efficient ways of manipulating the electronic structure and enhance the electrocatalytic performance of heterostructure materials is through chemical doping. Three categories of chemical doping can be distinguished on the basis of doped elements: noble metals, early or non-3d transition metals and non-metallic elements.

A phase transition from cubic to orthorhombic phases in $CoSe_2$ was observed by Zheng et al. as a result of phosphorus doping [137]. The modified orthorhombic $CoSe_2$ having 8 wt% doped phosphorus was found to be the most optimal sample as it offered 10 mA cm^{-2} of current density with a minimum value of 104 mV overpotential in 1 M KOH. The onset potential for HER was found out to be a minimum of 31 mV. Following 20 h of catalysis, the P-doped $CoSe_2$ showed minimum degradation. Phosphorus dopant caused the optimization of the electronic structure along with activation of local reaction which further led to the provision of remarkable combination for reaction intermediate, and ultimately a significant energy change was observed, as demonstrated by computed and experimental outcomes. Moreover, a straight forward and adaptable cation exchange method for fabrication of Fe doped $Ni(OH)_2$ and $Co(OH)_2$ nanostructures with greater number of active sites for improved OER was described by Zhou et al. with regard to doping of metallic elements [138]. Doping Ni with Fe can increase conductivity of $Ni(OH)_2$ and speed up charge transfer towards Ni active center from the Fe site, causing the ultimate enhancement of catalytic activity. Furthermore, charge redistribution and a synergistic impact between various components can be accomplished through chemical doping to improve the charge transfer, which in turn raises the heterostructure's stability and catalytic performance. For instance, Jin et al. described bottom-up approach to creating a CoO/CoP heterostructure doped with Indium (In) [139].

Additionally, due to their ability to maximize the Gibbs free energy of creation of electrocatalytic intermediate, non-3d metals with high valency like W, Mo etc. were considered excellent dopants [140]. A molybdenum (Mo)-doped $NiCo_2O_4/Co_{5.47}N$ heterostructure catalyst that provides outstanding electrocatalytic activity in both HER and OER was synthesized using the method we described in earlier study. Using pyrolysis, a Mo-doped porous $NiCo_2O_4/Co_{5.47}N$ heterostructure after preparing a Mo, Ni and Co trimetallic metal–organic framework (TMOF). By lowering the chemisorption free energy of H_2 and O_2 intermediates, fruitful doping of Mo gives Mo-$NiCo_2O_4/Co_{5.47}N$ an optimal electronic structure. Additionally, a plenty of active sites and excellent conductivity is delivered by distinct structure of Mo-$NiCo_2O_4/Co_{5.47}N$. Consequently, at the current density of 50 mA cm^{-2}, the Mo-$NiCo_2O_4/Co_{5.47}N$ shows better electrocatalytic efficacy with lower overpotentials of 310 mV and 170 mV for OER and HER, respectively.

N [141–143] and F [144, 145], two highly electronegative non-metallic elements of main group, could be employed as dopants in addition to metallic ones. Hao et al. used a hydrothermal technique to introduce NH_4F, highlighting an F-doped $Ni(OH)_2/Ni_3S_2$ heterostructure [145]. The addition of NH_4F produced a $Ni(OH)_2/Ni_3S_2$ heterostructure rather than a single Ni_3S_2 constituent under similar parameters. The development of $Ni(OH)_2/Ni_3S_2$ heterostructure is thus facilitated by the doping of F, in addition to improving electroconductivity and charge transport efficiency. With a low overpotential of 360 mV at 100 mA cm^{-2}, the F-doped $Ni(OH)_2/Ni_3S_2$ heterostructure demonstrated sufficient electrocatalytic activity.

5 Challenges and Future Perspective

The heterostructures based on MOFs are multicomponent systems that combine the properties of each system and hence produce unique novel properties consisting of different functional materials. These heterostructures combining multiple adsorbent or catalytic competencies have potential for the use in different applications such as storage, catalysis, sensing or hybrid functions that's why have range of application in the electrochemical field. Even though for constructing the MOF-based heterostructure a lot of advancement have been made but still there are many challenges to overcome listed below:

i. As recently the assembly method is mostly used for constructing the MOFs based structures and the criteria of low lattice mismatch cause the limitation for the random epitaxial growth which is attained by the method of adjusting the solution concentrations. Thus for the fine structure the controlled space is restricted and development mechanism is not established so far resulting in the limited application in different fields [128].
ii. There are promising methods to synthesize these MOFs based heterostructure and they have been developed but still there is reliable and universal method for the construction of heterostructure on a larger scale.
iii. Also there is a great challenge to precisely control the interaction between the interface and the synergy between the MOFs and the other doped or defect functional material.
iv. For chemical doping the selected atom plays a part in optimizing Gibbs free energy of the intermediates and wrong choice and unmonitored addition of the dopant such as >10% leads to the destroying of material structure [146]. This will lead to the new crystal phase, blocking the active sites which results into the decrease of catalytic performance. Similarly, if one is interested for the multielement doping the difficulty in finding the real active sites because of the uncertainty in the crystal structure makes indistinct understanding. That's why it is difficult to continue the process without troubling the original structure of MOF.
v. With the help of nanostructure engineering, favorable structures for ion diffusion and adsorption/desorption of gases can possibly constructed. There is limited cycle reversibility caused by the inefficient mass diffusion and electron transfer because of the structural anisotropy.
vi. Theoretically the electron transfers visualization near the interface observed to have a change that cause the misunderstanding, hence there is a need to contract gaps between theoretical model and practical one. The interfacial charge transfer study for the electrocatalysis application is still remain a challenge for its real mechanism cycle.

There are some predictable solutions that can be implemented to address the problems. For instance, one should first carefully understand the thermodynamic process along with assembly dynamics under the action of complex condition. It's possible to have selective modification over the surface by the use of certain direction agent for example PVP and the mismatched components synthesized by the layered assembly [147]. Before work, explore the relationship between the individual structures for deciding the application of the constructed heterostructure. The crisis of thermal stability can be minimized by using an electric circuit type packaging method, such as the electrolytic process in the presence of harsh electrolyte to improve the stability and sustainability. Furthermore, the more in depth understanding can be possible if more advanced characterization tools, in-situ such as TEM and ex-situ techniques should be applied beside the computational simulations [148]. This will help to identify the active site and growth of unstable interfaces could be attained which proves valuable for explaining longevity of electrocatalyst.

The stability challenge may be controlled by synthesis of advanced catalytic materials using the MOFs as precursors and it's proving beneficial. The MOFs is transformed by the thermal decomposition which lead to new field of the heterogeneous catalysis. The MOFs as precursors have good tunable porosity, controlled particle size, dispersion and good metal loading and also the final structure can be modified by the simple methods such as by external templates and carbonization atmosphere etc [149]. The revolution of MOFMSs which have the incomparable features made them center of attention for the multiple catalytic reactions [150]. This will have a commercial stance if the base metals and low cost ligands are utilized with no solvent being used for the MOF activation. However, certain type of transformation also observed during the pyrolysis which form the derivatives of MOFs. All these problems will be discussed in near future by the catalysis community as MOFMSs will continue to receive the attention.

Summary

Some of the most intriguing approaches to utilize MOFs for energy storage in electrochemical devices have been highlighted in this chapter. The porosity, stability and synthetic tenability of MOFs offer a wealth of design criteria to address certain obstacles in the fabrication and development of electrodes and electrolyte. MOFs would be differentiated from porous carbons, polymers and inorganic materials by future identification, consumption and development of approaches to assist charge storage and its effective transmission.

Slurries of MOF composites, which may have few charge conduction channels, are utilized in several devices. This can be addressed by forming homogeneous MOF films by deposition techniques, thereby enhancing the interfacial contact. Homogeneity and thorough control over the thickness of the layer are guaranteed via direct deposition onto a substrate. Moreover, emergent flexible electrodes will be made possible by the utilization of soft materials such as polymers as substrates. MOF-based devices are anticipated to benefit from design strategies for devices and processes employing polymers, porous carbons and metal oxides as direct deposition techniques have been investigated through other categories of materials. Attempts to evolve frameworks having organized structures at meso and macroscale lengths have been prompted by the enormous potential for additional capabilities in MOFs. The porosity and accessibility of metal sites can be improved by crystallographic control of MOF constituents, which will eventually improve device performance. Hence, MOF materials might open up a new market for electronic applications that prioritize rapid charge mobility, increased flexibility and transparency.

The stability of the interested reactions which take place in the high acidic or alkaline conditions is still challenge and it's significant. There is stability of MOF-based heterostructure which consist on the first row of transition metals being coordinated with the hydroxide species. The careful choice of the linker, dopant atom and metal is necessary for the explicit activity. There are two recent trends observed in this field i.e. firstly, a lot of efforts have been made to control the derivation process such as catalyst with specific morphology and active sites via the application of heating. Secondly, 2D MOF-based catalysts are in greater use owing to their enlarged surface area and anisotropic structure. There will be a massive diversity of innovative MOF-based heterostructure catalyst or MOF-assisted photocatalytic systems having much enhanced results that are already supported. Therefore, the objective is not to increase quantity of instances but rather to use MOF-centric chemistry to analyze concepts and create novel, appealing catalysts or light absorbers that could be hard or impractical to make or evaluate using conventional methods with enhanced mechanism for the catalysis application.

References

1. Li, X., et al.: Water splitting: from electrode to green energy system. Nano-Micro Lett. **12**, 1–29 (2020)
2. You, B., Sun, Y.: Innovative strategies for electrocatalytic water splitting. Acc. Chem. Res. **51**(7), 1571–1580 (2018)
3. Atalan, R.A.D.Y.A.: The comparison of energy resources in the context of global energy supply security. In: Experimental and Numerical Advances in Science, Engineering and Technology, p. 3. (2021)
4. Gu, D., Andreev, K., Dupre, M.E.: Major trends in population growth around the world. China CDC Weekly **3**(28), 604 (2021)
5. Zhai, P., et al.: Global Warming of 1.5 C. An IPCC Special Report on the impacts of global warming of 1.5 C above pre-industrial levels and related global greenhouse gas emission pathways, in the context of strengthening the global response to the threat of climate change. In: Sustainable Development, and Efforts to Eradicate Poverty, p. 32 (2018)
6. Shi, Y., Zhang, B.: Recent advances in transition metal phosphide nanomaterials: synthesis and applications in hydrogen evolution reaction. Chem. Soc. Rev. **45**(6), 1529–1541 (2016)
7. Wang, W., et al.: Recent progress in metal-organic frameworks for applications in electrocatalytic and photocatalytic water splitting. Adv. Sci. **4**(4), 1600371 (2017)
8. Roy, S., et al.: Electrocatalytic hydrogen evolution from a cobaloxime-based metal–organic framework thin film. J. Am. Chem. Soc. **141**(40), 15942–15950 (2019)
9. Huang, D.-D., et al.: In situ synthesis of a Fe 3 S 4/MIL-53 (Fe) hybrid catalyst for an efficient electrocatalytic hydrogen evolution reaction. Chem. Commun. **55**(31), 4570–4573 (2019)
10. Bakuru, V.R., DMello, M.E., Kalidindi, S.B.: Metal-organic frameworks for hydrogen energy applications: advances and challenges. ChemPhysChem **20**(10), 1177–1215 (2019)
11. Yan, Y., et al.: A review on noble-metal-free bifunctional heterogeneous catalysts for overall electrochemical water splitting. J. Mater. Chem. A **4**(45), 17587–17603 (2016)
12. Wang, J., et al.: Recent progress in cobalt-based heterogeneous catalysts for electrochemical water splitting. Adv. Mater. **28**(2), 215–230 (2016)

13. Lan, Z.-A., Zhang, G., Wang, X.: A facile synthesis of Br-modified g-C3N4 semiconductors for photoredox water splitting. Appl. Catal. B **192**, 116–125 (2016)
14. Busch, M., et al.: Beyond the top of the volcano?—A unified approach to electrocatalytic oxygen reduction and oxygen evolution. Nano Energy **29**, 126–135 (2016)
15. Gong, M., Dai, H.: A mini review of NiFe-based materials as highly active oxygen evolution reaction electrocatalysts. Nano Res. **8**, 23–39 (2015)
16. Zheng, Y., et al.: Advancing the electrochemistry of the hydrogen-evolution reaction through combining experiment and theory. Angew. Chem. Int. Ed. **54**(1), 52–65 (2015)
17. Huang, Z.-F., et al.: Hollow cobalt-based bimetallic sulfide polyhedra for efficient all-pH-value electrochemical and photocatalytic hydrogen evolution. J. Am. Chem. Soc. **138**(4), 1359–1365 (2016)
18. You, B., et al.: Bimetal–organic framework self-adjusted synthesis of support-free nonprecious electrocatalysts for efficient oxygen reduction. ACS Catal. **5**(12), 7068–7076 (2015)
19. Hou, Y., et al.: Metal−organic framework-derived nitrogen-doped core-shell-structured porous Fe/Fe3C@ C nanoboxes supported on graphene sheets for efficient oxygen reduction reactions. Adv. Energy Mater. **4**(11), 1400337 (2014)
20. Lin, H.-W., et al.: Bi-metallic MOFs possessing hierarchical synergistic effects as high performance electrocatalysts for overall water splitting at high current densities. Appl. Catal. B **258**, 118023 (2019)
21. Antony, R.P., et al.: MOF derived nonstoichiometric NixCo3−xO4−y nanocage for superior electrocatalytic oxygen evolution. Adv. Mater. Interfaces **3**(20), 1600632 (2016)
22. Sun, Y., et al.: Modulating electronic structure of metal-organic frameworks by introducing atomically dispersed Ru for efficient hydrogen evolution. Nat. Commun. **12**(1), 1369 (2021)
23. Lu, Y., et al.: Construction of complex Co3O4@ Co3V2O8 hollow structures from metal–organic frameworks with enhanced lithium storage properties. Adv. Mater. **30**(1), 1702875 (2018)
24. Lu, X.F., et al.: Metal–organic frameworks based electrocatalysts for the oxygen reduction reaction. Angew. Chem. **132**(12), 4662–4678 (2020)
25. Lee, J., et al.: Metal–organic framework materials as catalysts. Chem. Soc. Rev. **38**(5), 1450–1459 (2009)
26. Corma, A., Garcia, H., Llabrés i Xamena, F.: Engineering metal organic frameworks for heterogeneous catalysis. Chem. Rev. **110**(8), 4606–4655 (2010)
27. Liu, J., et al.: Applications of metal–organic frameworks in heterogeneous supramolecular catalysis. Chem. Soc. Rev. **43**(16), 6011–6061 (2014)
28. Yu, L., Yang, J.F., Lou, X.W.: Formation of CoS2 nanobubble hollow prisms for highly reversible lithium storage. Angew. Chem. **128**(43), 13620–13624 (2016)
29. Li, H., et al.: Recent advances in gas storage and separation using metal–organic frameworks. Mater. Today **21**(2), 108–121 (2018)
30. Zhang, Y., et al.: Luminescent sensors based on metal-organic frameworks. Coord. Chem. Rev. **354**, 28–45 (2018)
31. Jiao, L., et al.: Metal–organic frameworks as platforms for catalytic applications. Adv. Mater. **30**(37), 1703663 (2018)
32. Meyer, K., Ranocchiari, M., van Bokhoven, J.A.: Metal organic frameworks for photocatalytic water splitting. Energy Environ. Sci. **8**(7), 1923–1937 (2015)
33. Zhai, Q.-G., et al.: Pore space partition in metal–organic frameworks. Acc. Chem. Res. **50**(2), 407–417 (2017)
34. Zhao, X., et al.: Pore space partition by symmetry-matching regulated ligand insertion and dramatic tuning on carbon dioxide uptake. J. Am. Chem. Soc. **137**(4), 1396–1399 (2015)
35. Zhao, X., et al.: Multivariable modular design of pore space partition. J. Am. Chem. Soc. **138**(46), 15102–15105 (2016)
36. Hu, A., et al.: Epitaxial growth and integration of insulating metal–organic frameworks in electrochemistry. J. Am. Chem. Soc. **141**(28), 11322–11327 (2019)
37. Chai, L., et al.: Rational design and growth of MOF-on-MOF heterostructures. Small **17**(36), 2100607 (2021)

38. Zhang, Y., et al.: MOF-on-MOF nanoarchitectures for selectively functionalized nitrogen-doped carbon-graphitic carbon/carbon nanotubes heterostructure with high capacitive deionization performance. Nano Energy **97**, 107146 (2022)
39. Chai, L., et al.: Fe7C3 nanoparticles with in situ grown CNT on nitrogen doped hollow carbon cube with greatly enhanced conductivity and ORR performance for alkaline fuel cell. Carbon **174**, 531–539 (2021)
40. Wang, X., et al.: MOF derived N-doped carbon coated CoP particle/carbon nanotube composite for efficient oxygen evolution reaction. Carbon **141**, 643–651 (2019)
41. Zheng, F., et al.: Immobilizing Pd nanoclusters into electronically conductive metal-organic frameworks as bi-functional electrocatalysts for hydrogen evolution and oxygen reduction reactions. Electrochim. Acta **306**, 627–634 (2019)
42. Zheng, F., et al.: Advanced electrocatalysts based on metal–organic frameworks. ACS Omega **5**(6), 2495–2502 (2019)
43. Hayati, P., et al.: Survey of temperature, reaction time and ultrasound irradiation power on sonochemical synthesis of two new nanostructured lead (II) coordination supramolecule compounds. Ultrason. Sonochem. **35**, 81–91 (2017)
44. Souri, B., et al.: The effects of altering reaction conditions in synthesis of a novel lead (II) 2D coordination polymer as well as in achieving the same morphologies of lead (II) oxide micro-nano structures via solid-state process. Polyhedron **154**, 443–456 (2018)
45. Fard, M.J.S., et al.: Synthesis and characterization of a new nano lead (II) 0-D coordination supramolecular compound: a precursor to produce pure phase nano-sized lead (II) oxide. Ultrason. Sonochem. **39**, 129–136 (2017)
46. Davis, M.E.: Ordered porous materials for emerging applications. Nature **417**(6891), 813–821 (2002)
47. Mohammadi, M.K., et al.: Sonication-assisted synthesis of a new rod-like metal-organic coordination polymer compound; novel precursor to produce pure phase nano-sized lead (II) oxide. J. Mol. Struct. **1176**, 434–446 (2019)
48. Souri, B., et al.: The effects of modifying reaction conditions in green sonochemical synthesis of a copper (II) coordination polymer as well as in achieving to different morphologies of copper (II) oxide micro crystals via solid-state process. Inorg. Chim. Acta **483**, 516–526 (2018)
49. Deng, H., et al.: Multiple functional groups of varying ratios in metal-organic frameworks. Science **327**(5967), 846–850 (2010)
50. Cohen, S.M.: The postsynthetic renaissance in porous solids. J. Am. Chem. Soc. **139**(8), 2855–2863 (2017)
51. Dhakshinamoorthy, A., Asiri, A.M., Garcia, H.: Metal–organic framework (MOF) compounds: photocatalysts for redox reactions and solar fuel production. Angew. Chem. Int. Ed. **55**(18), 5414–5445 (2016)
52. Shekhah, O., et al.: MOF-on-MOF heteroepitaxy: perfectly oriented [Zn 2 (ndc) 2 (dabco)] n grown on [Cu 2 (ndc) 2 (dabco)] n thin films. Dalton Trans. **40**(18), 4954–4958 (2011)
53. Wu, S., et al.: Shape control of core–shell MOF@ MOF and derived MOF nanocages via ion modulation in a one-pot strategy. J. Mater. Chem. A **6**(37), 18234–18241 (2018)
54. Wang, F., Wang, H., Li, T.: Seaming the interfaces between topologically distinct metal–organic frameworks using random copolymer glues. Nanoscale **11**(5), 2121–2125 (2019)
55. Chai, L., et al.: Stringing bimetallic metal–organic framework-derived cobalt phosphide composite for high-efficiency overall water splitting. Adv. Sci. **7**(5), 1903195 (2020)
56. Wang, F., et al.: Uncovering two kinetic factors in the controlled growth of topologically distinct core–shell metal–organic frameworks. Chem. Sci. **10**(33), 7755–7761 (2019)
57. Yang, X., et al.: One-step synthesis of hybrid core-shell metal–organic frameworks. Angew. Chem. **130**(15), 3991–3996 (2018)
58. Niu, H., et al.: Heterogeneous reactions between toluene and NO2 on mineral particles under simulated atmospheric conditions. Environ. Sci. Technol. **51**(17), 9596–9604 (2017)
59. Roginsky, S., Rosenkewitsch, L.: The quantum theory of the kinetics of homogeneous and heterogeneous reactions. Nature **125**(3149), 347–348 (1930)

60. Wang, Y., et al.: Electrocatalytic oxygen reduction to hydrogen peroxide: from homogeneous to heterogeneous electrocatalysis. Adv. Energy Mater. **11**(15), 2003323 (2021)
61. Hong, D.H., et al.: MOF-on-MOF architectures: applications in separation, catalysis, and sensing. Bull. Korean Chem. Soc. **42**(7), 956–969 (2021)
62. Wu, M.-X., et al.: Core–shell MOFs@ MOFs: diverse designability and enhanced selectivity. ACS Appl. Mater. Interfaces **12**(49), 54285–54305 (2020)
63. Liu, X.Y., et al.: Using a multi-shelled hollow metal–organic framework as a host to switch the guest-to-host and guest-to-guest interactions. Angew. Chem. Int. Ed. **57**(8), 2110–2114 (2018)
64. Wu, F., et al.: Fabrication of binary MOF-derived hybrid nanoflowers via selective assembly and their microwave absorbing properties. Carbon **182**, 484–496 (2021)
65. Ikigaki, K., et al.: MOF-on-MOF: oriented growth of multiple layered thin films of metal-organic frameworks. Angew. Chem. **131**(21), 6960–6964 (2019)
66. Kwon, O., et al.: Computer-aided discovery of connected metal-organic frameworks. Nat. Commun. **10**(1), 3620 (2019)
67. Liu, C., et al.: Site-specific growth of MOF-on-MOF heterostructures with controllable nano-architectures: beyond the combination of MOF analogues. Chem. Sci. **11**(14), 3680–3686 (2020)
68. Zhang, M., et al.: 3D-agaric like core-shell architecture UiO-66-NH2@ ZIF-8 with robust stability for highly efficient REEs recovery. Chem. Eng. J. **386**, 124023 (2020)
69. Chen, D., et al.: MOF-on-MOF-derived hollow FeNi3/N-doped carbon nanorods for efficient oxygen evolution. Chem. Eng. J. **470**, 144418 (2023)
70. Huang, H.B., et al.: Engineering hierarchical architecture of metal-organic frameworks for highly efficient overall CO2 photoreduction. Small **18**(16), 2200407 (2022)
71. Wei, R., et al.: Recent strategies to improve the electroactivity of metal–organic frameworks for advanced electrocatalysis. Inf. Funct. Mater. **1**(2), 181–206 (2024)
72. Wang, H., et al.: Metal-organic frameworks for energy applications. Chem **2**(1), 52–80 (2017)
73. Liu, W., Yin, X.-B.: Metal–organic frameworks for electrochemical applications. TrAC, Trends Anal. Chem. **75**, 86–96 (2016)
74. Cheng, M., et al.: Metal-organic frameworks for highly efficient heterogeneous Fenton-like catalysis. Coord. Chem. Rev. **368**, 80–92 (2018)
75. Wu, H.B., Lou, X.W.: Metal-organic frameworks and their derived materials for electrochemical energy storage and conversion: promises and challenges. Sci. Adv. **3**(12), eaap9252 (2017)
76. Qin, J.-S., et al.: Ultrastable polymolybdate-based metal–organic frameworks as highly active electrocatalysts for hydrogen generation from water. J. Am. Chem. Soc. **137**(22), 7169–7177 (2015)
77. Wang, S., et al.: Water oxidation electrocatalysis by a zeolitic imidazolate framework. Nanoscale **6**(17), 9930–9934 (2014)
78. Gong, Y., et al.: Two novel Co (ii) coordination polymers based on 1, 4-bis (3-pyridylaminomethyl) benzene as electrocatalysts for oxygen evolution from water. Dalton Trans. **42**(34), 12252–12259 (2013)
79. Flügel, E.A., et al.: Homonuclear mixed-valent cobalt imidazolate framework for oxygen-evolution electrocatalysis. Chem. A Eur. J. **22**(11), 3676–3680 (2016)
80. Gong, Y., et al.: Metal (II)-induced coordination polymer based on 4-(5-(Pyridin-4-yl)-4H-1, 2, 4-triazol-3-yl) benzoate as an electrocatalyst for water splitting. Cryst. Growth Des. **14**(2), 649–657 (2014)
81. Miner, E.M., et al.: Electrochemical oxygen reduction catalysed by Ni3 (hexaiminotriphenylene) 2. Nat. Commun. **7**(1), 10942 (2016)
82. Sun, L., Campbell, M.G., Dincă, M.: Electrically conductive porous metal–organic frameworks. Angew. Chem. Int. Ed. **55**(11), 3566–3579 (2016)
83. Sheberla, D., et al.: Conductive MOF electrodes for stable supercapacitors with high areal capacitance. Nat. Mater. **16**(2), 220–224 (2017)

84. Zhong, H.X., et al.: ZIF-8 derived graphene-based nitrogen-doped porous carbon sheets as highly efficient and durable oxygen reduction electrocatalysts. Angew. Chem. Int. Ed. **53**(51), 14235–14239 (2014)
85. Zhang, J., et al.: Interface engineering of MoS2/Ni3S2 heterostructures for highly enhanced electrochemical overall-water-splitting activity. Angew. Chem. **128**(23), 6814–6819 (2016)
86. Yang, Y., et al.: Metal–organic frameworks for electrocatalysis: beyond their derivatives. Small Sci. **1**(12), 2100015 (2021)
87. Yang, H., et al.: Electrochemical synthesis of flower shaped morphology MOFs in an ionic liquid system and their electrocatalytic application to the hydrogen evolution reaction. RSC Adv. **4**(30), 15720–15726 (2014)
88. Yu, J., et al.: Nanoparticle/MOF composites: preparations and applications. Mater. Horiz. **4**(4), 557–569 (2017)
89. Freire, C., et al.: POM & MOF-based electrocatalysts for energy-related reactions. ChemCatChem **10**(8), 1703–1730 (2018)
90. Hu, M.-L., Masoomi, M.Y., Morsali, A.: Template strategies with MOFs. Coord. Chem. Rev. **387**, 415–435 (2019)
91. Liu, J., et al.: Design strategies toward advanced MOF-derived electrocatalysts for energy-conversion reactions. Adv. Energy Mater. **7**(23), 1700518 (2017)
92. Gopi, S., Kathiresan, M., Yun, K.: Metal-organic and porous organic framework in electrocatalytic water splitting. J. Ind. Eng. Chem. **126**, 127–136 (2023)
93. Duan, J., Chen, S., Zhao, C.: Ultrathin metal-organic framework array for efficient electrocatalytic water splitting. Nat. Commun. **8**(1), 15341 (2017)
94. Cao, C., et al.: Semisacrificial template growth of self-supporting MOF nanocomposite electrode for efficient electrocatalytic water oxidation. Adv. Funct. Mater. **29**(6), 1807418 (2019)
95. Zhou, W., et al.: N-doped carbon-wrapped cobalt nanoparticles on N-doped graphene nanosheets for high-efficiency hydrogen production. Chem. Mater. **27**(6), 2026–2032 (2015)
96. Huang, X., et al.: Highly dispersed Pt clusters encapsulated in MIL-125-NH2 via in situ auto-reduction method for photocatalytic H2 production under visible light. Nano Res. **14**, 4250–4257 (2021)
97. Stelmachowski, P., et al.: Carbon-based composites as electrocatalysts for oxygen evolution reaction in alkaline media. Materials **14**, 4984 (2021). 2021, s Note: MDPI stays neu-tral with regard to jurisdictional claims in ….
98. Li, X., et al.: Nanostructured catalysts for electrochemical water splitting: current state and prospects. J. Mater. Chem. A **4**(31), 11973–12000 (2016)
99. Lee, Y., et al.: Synthesis and activities of rutile IrO2 and RuO2 nanoparticles for oxygen evolution in acid and alkaline solutions. J. Phys. Chem. Lett. **3**(3), 399–404 (2012)
100. Cherevko, S., et al.: Oxygen and hydrogen evolution reactions on Ru, RuO2, Ir, and IrO2 thin film electrodes in acidic and alkaline electrolytes: a comparative study on activity and stability. Catal. Today **262**, 170–180 (2016)
101. Kear, G., Walsh, F.C.: The characteristics of a true Tafel slope. Corros. Mater. **30**(6), 51–55 (2005)
102. Gao, M., et al.: Facile electrochemical preparation of self-supported porous Ni–Mo alloy microsphere films as efficient bifunctional electrocatalysts for water splitting. J. Mater. Chem. A **5**(12), 5797–5805 (2017)
103. Chen, X., et al.: Polyvinyl alcohol protected Mo2C/Mo2N multicomponent electrocatalysts with controlled morphology for hydrogen evolution reaction in acid and alkaline medium. Electrochim. Acta **273**, 239–247 (2018)
104. Bard, A.J., Faulkner, L.R.: Electrochemical methods: fundamentals and applications. Surf. Technol. **20**(1), 91–92 (1983)
105. Guo, Y., et al.: Nanoarchitectonics for transition-metal-sulfide-based electrocatalysts for water splitting. Adv. Mater. **31**(17), 1807134 (2019)
106. Tahir, M., et al.: Electrocatalytic oxygen evolution reaction for energy conversion and storage: a comprehensive review. Nano Energy **37**, 136–157 (2017)

107. Li, J.-H., et al.: Metal–organic frameworks toward electrocatalytic applications. Appl. Sci. **9**(12), 2427 (2019)
108. Ledendecker, M., et al.: Experimental and theoretical assessment of Ni-based binary compounds for the hydrogen evolution reaction. Adv. Energy Mater. **7**(5), 1601735 (2017)
109. Lalwani, S., et al.: Advances in interfacial engineering and their role in heterostructure formation for HER applications in wider pH. ACS Appl. Energy Mater. **5**(12), 14571–14592 (2022)
110. Li, Q., et al.: Building CoP/Co-MOF heterostructure in 2D nanosheets for improving electrocatalytic hydrogen evolution over a wide pH range. J. Electroanal. Chem. **895**, 115514 (2021)
111. Singh, T.I., et al.: Alkaline water splitting enhancement by MOF-derived Fe–Co–oxide/Co@NC-mNS heterostructure: boosting OER and HER through defect engineering and in situ oxidation. Small **17**(29), 2101312 (2021)
112. Han, L., et al.: Stable and efficient single-atom Zn catalyst for CO_2 reduction to CH_4. (2020)
113. Liu, W., et al.: Bioinspired interfacial engineering of a CoSe2 decorated carbon framework cathode towards temperature-tolerant and flexible Zn–air batteries. Nanoscale **13**(5), 3019–3026 (2021)
114. Sha, L., et al.: A heterogeneous interface on NiS@ Ni3S2/NiMoO4 heterostructures for efficient urea electrolysis. J. Mater. Chem. A **8**(35), 18055–18063 (2020)
115. Jiang, L., et al.: Iron-cluster-directed synthesis of 2D/2D Fe–N–C/MXene superlattice-like heterostructure with enhanced oxygen reduction electrocatalysis. ACS Nano **14**(2), 2436–2444 (2020)
116. Liu, J., et al.: 2D MXene-based materials for electrocatalysis. Trans. Tianjin Univ. **26**, 149–171 (2020)
117. Zhang, Y., et al.: Heterostructure of 2D CoP nanosheets/1D carbon nanotubes to significantly boost the alkaline hydrogen evolution. Adv. Mater. Interfaces **7**(2), 1901302 (2020)
118. Peng, Y.W., et al.: Polyoxometalate-derived ultrasmall Pt2W/WO3 heterostructure outperforms platinum for large-current-density H2 evolution. Adv. Energy Mater. **9**(26), 1900597 (2019)
119. Zhang, H., et al.: Bifunctional heterostructure assembly of NiFe LDH nanosheets on NiCoP nanowires for highly efficient and stable overall water splitting. Adv. Funct. Mater. **28**(14), 1706847 (2018)
120. Guan, J., et al.: FeOOH-enhanced bifunctionality in Ni3N nanotube arrays for water splitting. Appl. Catal. B **269**, 118600 (2020)
121. Yaqoob, L., et al.: Electrochemical synergies of Fe–Ni bimetallic MOF CNTs catalyst for OER in water splitting. J. Alloy. Compd. **850**, 156583 (2021)
122. Ma, Y.: Highly enhanced oer performance by er-doped fe-mof nanoarray at large current densities. Nanomaterials **11**(7), 1–11 (2021)
123. Li, Z., et al.: MOF-derived hollow heterostructures for advanced electrocatalysis. Coord. Chem. Rev. **439**, 213946 (2021)
124. Li, F., et al.: Enhanced OER performance of composite Co–Fe-based MOF catalysts via a one-pot ultrasonic-assisted synthetic approach. Sustain. Energy Fuels **5**(4), 1095–1102 (2021)
125. Gaur, A., et al.: Intense nano-interfacial interactivity stimulates the OER in a MOF-derived superhydrophilic CuO–NiO heterostructure. Sustain. Energy Fuels **5**(21), 5505–5512 (2021)
126. Zheng, D., et al.: Structural advantages and enhancement strategies of heterostructure water-splitting electrocatalysts. Cell Rep. Phys. Sci. **2**(6) (2021)
127. Zha, Q., et al.: Cobalt-based MOF-on-MOF two-dimensional heterojunction nanostructures for enhanced oxygen evolution reaction electrocatalytic activity. Inorg. Chem. **59**(2), 1295–1305 (2020)
128. He, Q., et al.: Emerging pristine MOF-based heterostructured nanoarchitectures: advances in structure evolution, controlled synthesis, and future perspectives. Small **19**(50), 2303884 (2023)
129. Jia, F., et al.: Synergistic effect of P doping and Mo-Ni-based heterostructure electrocatalyst for overall water splitting. Materials **16**(9), 3411 (2023)

130. Kang, T., Kim, J.: Optimal cobalt-based catalyst containing high-ratio of oxygen vacancy synthesized from metal-organic-framework (MOF) for oxygen evolution reaction (OER) enhancement. Appl. Surf. Sci. **560**, 150035 (2021)
131. Jia, J., et al.: Boosting OER performance of NiFe-MOFs via heterostructure engineering: promoted phase transformation and self-optimized dynamic interface electron structure. Adv. Funct. Mater. 2500568 (2025)
132. Du, J., Li, F., Sun, L.: Metal–organic frameworks and their derivatives as electrocatalysts for the oxygen evolution reaction. Chem. Soc. Rev. **50**(4), 2663–2695 (2021)
133. Zhou, J., et al.: Discovery of quantitative electronic structure-OER activity relationship in metal-organic framework electrocatalysts using an integrated theoretical-experimental approach. Adv. Funct. Mater. **31**(33), 2102066 (2021)
134. Li, C., et al.: Recent progress in metal-organic frameworks (MOFs) for electrocatalysis. Ind. Chem. Mater. **1** (2023)
135. Xu, X., et al.: Modulating metal–organic frameworks for catalyzing acidic oxygen evolution for proton exchange membrane water electrolysis. SusMat **1**, 460–481 (2021)
136. Zhao, G., et al.: Heterostructures for electrochemical hydrogen evolution reaction: a review. Adv. Funct. Mater. **28**(43), 1803291 (2018)
137. Zheng, Y.-R., et al.: Doping-induced structural phase transition in cobalt diselenide enables enhanced hydrogen evolution catalysis. Nat. Commun. **9**(1), 2533 (2018)
138. Zhou, Q., et al.: Active-site-enriched iron-doped nickel/cobalt hydroxide nanosheets for enhanced oxygen evolution reaction. ACS Catal. **8**(6), 5382–5390 (2018)
139. Jin, W., et al.: Oxygen vacancy–rich in-doped CoO/CoP heterostructure as an effective air cathode for rechargeable Zn–air batteries. Small **15**(46), 1904210 (2019)
140. Liu, W., et al.: Non-3d metal modulation of a 2D Ni–Co heterostructure array as multi-functional electrocatalyst for portable overall water splitting. Small **16**(10), 1906775 (2020)
141. Zhou, S., et al.: Heterostructures of MXenes and N-doped graphene as highly active bifunctional electrocatalysts. Nanoscale **10**(23), 10876–10883 (2018)
142. Saha, S., et al.: A review on the heterostructure nanomaterials for supercapacitor application. J. Energy Storage **17**, 181–202 (2018)
143. Zhang, B., et al.: Recent progress on carbonaceous material engineering for electrochemical hydrogen peroxide generation. Trans. Tianjin Univ. **26**, 188–196 (2020)
144. Fu, L., et al.: A fluorine-modulated bulk-phase heterojunction and tolerance factor for enhanced performance and structure stability of cesium lead halide perovskite solar cells. J. Mater. Chem. A **6**(27), 13263–13270 (2018)
145. Hao, P., et al.: Morphology and electronic structure modulation induced by fluorine doping in nickel-based heterostructures for robust bifunctional electrocatalysis. Nanoscale **10**(43), 20384–20392 (2018)
146. Zhang, T.S., et al.: Toward rational design of ordered heterostructures for energy and environmental sustainability: a review. Adv. Energy Sustain. Res. **4**(6), 2200204 (2023)
147. Bavykina, A., et al.: Metal–organic frameworks in heterogeneous catalysis: recent progress, new trends, and future perspectives. Chem. Rev. **120**(16), 8468–8535 (2020)
148. Radwan, A., et al.: Design engineering, synthesis protocols, and energy applications of MOF-derived electrocatalysts. Nano-Micro Lett. **13**, 1–32 (2021)
149. Yang, L., et al.: Recent progress in MOF-derived, heteroatom-doped porous carbons as highly efficient electrocatalysts for oxygen reduction reaction in fuel cells. Adv. Funct. Mater. **28**(7), 1704537 (2018)
150. Guo, Y., et al.: CoMo carbide/nitride from bimetallic MOF precursors for enhanced OER performance. Int. J. Hydrogen Energy **46**(43), 22268–22276 (2021)

High Entropy 2D Materials for Electrochemical Water Splitting

Pratiksha Singh, Kuldeep Kumar Kushwaha, Pranay Ranjan, Devendra Singh Negi, and Prashant Kumar

Abstract High-entropy two-dimensional materials (HE2DMs) have emerged as a transformative class of electrocatalysts for efficient and sustainable electrochemical water splitting. By incorporating five or more principal elements in near-equimolar ratios within atomically thin 2D frameworks, HE2DMs uniquely combine the advantages of configurational entropy stabilization with the high surface area and tunable electronic properties of 2D materials. These materials demonstrate enhanced catalytic activity and stability in both hydrogen evolution reaction (HER) and oxygen evolution reaction (OER), outperforming many conventional and noble-metal-based catalysts. The high entropy effect introduces lattice distortions and abundant active sites, while the synergistic interactions among multicomponent elements optimize the adsorption energies of reaction intermediates. This chapter comprehensively explores the synthesis strategies, structural and electronic design principles, and mechanistic insights governing HE2DM performance in water splitting. We also highlight recent experimental and theoretical advancements, including density functional theory (DFT) and machine learning-guided material discovery. Despite current challenges related to scalability, phase control, and long-term durability, HE2DMs hold significant promise for large-scale, cost-effective green hydrogen production. This work provides a foundational understanding to guide future research in the

Pratiksha Singh and Kuldeep Kumar Kushwaha have contributed equally to this work.

P. Singh · K. K. Kushwaha · P. Ranjan (✉) · D. S. Negi (✉)
Department of Metallurgical and Materials Engineering, Indian Institute of Technology Jodhpur, Karwar, India
e-mail: pranay.ranjan@iitj.ac.in

D. S. Negi
e-mail: devendra@iitj.ac.in

P. Kumar (✉)
Global Innovative Center for Advanced Nanomaterials, College of Engineering, Science and Environment (CESE), The University of Newcastle, Callaghan, NSW, Australia
e-mail: Prashant.Kumar@newcastle.edu.au

Department of Applied Sciences, School of Advanced Engineering, University of Petroleum and Energy Studies, Energy Acres, Bidholi Campus, Dehradun 248007, Uttarakhand, India

O. P. Pandey et al. (eds.), *2D Materials for Electrochemical Water Splitting*, Engineering Materials, https://doi.org/10.1007/978-981-95-2003-9_11

rational design and development of HE2DMs as next-generation electrocatalysts for clean energy technologies.

1 Introduction

High entropy two-dimensional materials (HE2DMs) are two-dimensional (2D) materials that incorporate multiple elements in a near equimolar ratio and have enhanced structural stability, unique electronic and mechanical properties, and thermal performance. HE2DMs leverage configurational entropy, enhanced by incorporating multiple elements to stabilize disordered atomic arrangements, enabling robust catalytic activity for hydrogen evolution (HER) and oxygen evolution (OER) reactions. Their 2D geometry further amplifies surface reactivity and mass transport, addressing key challenges in energy-intensive water electrolysis [1]. Hydrogen (H_2) production as a clean energy carrier produces no emissions during operation. The dominant methods of H_2 production through steam methane reforming (SMR) and coal gasification continue to be unsustainable because they depend on fossil fuels and generate significant CO_2 emissions. Water electrolysis is a sustainable choice for H_2 and O_2 separation through renewable electricity sources [2]. The clean process of water electrolysis encounters obstacles that affect its economic feasibility. Water electrolysis demands high energy by operating a hydrogen evolution reaction (HER) at the cathode and an oxygen evolution reaction (OER) at the anode. The reactions require energy loss, inducing overpotentials that limit attainable reaction rates. Water electrolysis needs efficient electrocatalysts for economic feasibility because these electrocatalysts lower the energy barriers throughout HER and OER. Different effective catalysts exist for the two reactions based on recent scientific advancements. The HER has benefited from electrocatalysts that include platinum on carbon (Pt/C), palladium on carbon (Pd/C), molybdenum disulfide (MoS_2), and graphitic carbon nitride (g-C_3N_4) [3]. The OER benefits from catalytic activity provision by iridium dioxide (IrO_2), layered double hydroxides (LDHs), and nickel sulfide (Ni_3S_2). Using platinum-iridium or ruthenium noble metals, researchers have achieved the best results in catalysis [3]. The high price and scarcity of these rare materials limit their adoption across the market. Cost-effective advanced catalyst development remains vital since it will reduce the manufacturing expenses of green hydrogen while making it match traditional fossil fuel-based production methods. Alternative sustainable hydrogen production methods require intense ongoing research on solutions that match this specific challenge. Researchers work to lower production costs through the development of alternative catalyst materials without or reducing noble metal content, which includes metals and alloys along with intermetallic compounds, hydroxides, oxides, halides, carbides, nitrides, sulfides, phosphides, and metal–organic frameworks (MOFs) [4]. The development of nanotechnology generated substantial scrutiny of high-entropy nanostructures because they demonstrate superior catalytic operations for electrochemical and thermochemical procedures. Nanomaterials demonstrate excellent activity capabilities for generating hydrogen

(HER), sustaining oxygen-related operations (OER/ORR) etc. The development of two-dimensional (2D) architectural systems now lead to bulk materials in performance because surface reactivity and adaptable physical properties outshine their capabilities. Since the two-dimensional sheets exist at an atomic scale, engineers can fuse these materials with various substances to build advanced functional composite systems. Scientists today extend research by introducing stable ultrathin 2D metal frameworks with five or more metallic elements to link high-entropy alloy principles with nanotechnology. Research scientists use natural crystalline layering to develop these high-entropy 2D constructs through exfoliation processes that separate atomic layers. This method allows researchers to design catalysts with exact electronic configurations and surface properties, as shown in Fig. 1.

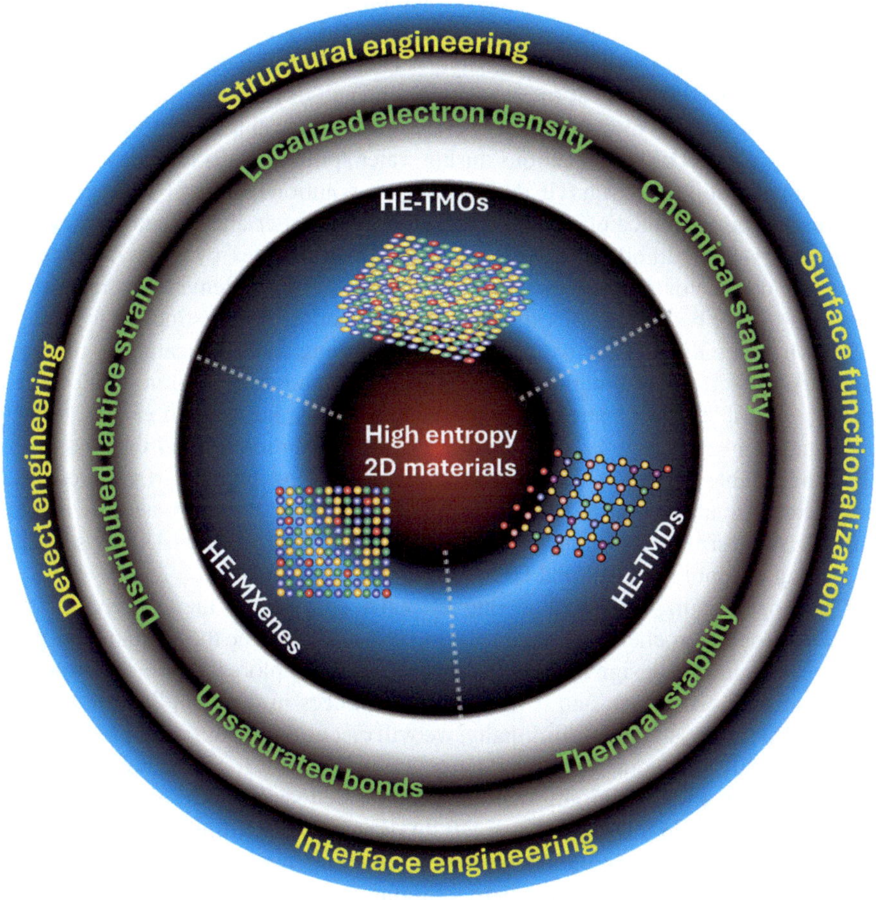

Fig. 1 The schematic illustrations showcase high-entropy two-dimensional materials' unique properties and material manipulation strategies

The development of high-entropy materials has evolved from bulk alloys (2004) and ceramics (2015) to 2D systems (2020–2021), where their combination of high surface area and entropy-driven stabilization offers transformative potential for electrochemical water splitting [5], as shown in Fig. 2a. HE2DMs are groundbreaking material that connects concepts between high entropy alloys and 2D nanomaterials. This novel design integrates high entropy material entropy stabilization methods with atomically thin 2D layer properties. A single-phase 2D structure of HE2DMs maintains five or more primary elements in ratios that are mostly equal. The coexistence of multiple structural elements produces three distinct effects that lead to dramatic changes in material properties [6]. The configurationally high entropy strengthens the material stability, which enables 2D solid solution formation between incompatible elements that were previously immiscible. HE2DMs provide extensive possibilities for adjusting electronic, catalytic, and physical properties that exceed traditional 2D materials [7]. Electrochemical water-splitting is a promising method for clean hydrogen fuel production. The hydroelectric cells function as water separators utilizing electrical power to create hydrogen and oxygen gas, which becomes a green fuel production method when operated from renewable energy resources [8]. The reaction speed of half-reactions during water splitting leads to slow kinetics for hydrogen evolution reaction (HER) and oxygen evolution reaction (OER) processes that restrict performance efficiencies, which require better electrocatalyst development [9]. At the cathode, the HER produces hydrogen gas by reducing protons. Water oxidation for oxygen evolution occurs as the most demanding reaction at the anode while both processes simultaneously function. The distinctive features of HE2DMs make them attractive candidates for serving as electrocatalysts that split water. An effective combination of lattice distortion and defects, along with diverse elements, enables the creation of multiple active sites due to high entropy effects and enables binding energy optimization [9]. The two-dimensional morphological structure enables effective charge transport by offering broad surface areas that boost catalytic efficiency. Modern research demonstrates that purposefully created HE2DMs perform electrocatalysis better than traditional noble metal catalysts while delivering exceptional OER and HER results [10].

This chapter aims to provide a comprehensive overview of the emerging field of HE2DMs for electrochemical water splitting. We will examine the fundamental principles governing the design and synthesis of HE2DMs, explore their unique structural and electronic properties, and review recent advances in their application as HER and OER electrocatalysts. Additionally, we will discuss characterization techniques, mechanistic insights, and theoretical modeling approaches for these complex materials. Finally, we will highlight the remaining challenges and future research directions to fully realize the potential of HE2DMs in water splitting and beyond. This work seeks to guide the rational development of next-generation HE2DM electrocatalysts for clean energy applications by bridging the gap between materials science and catalysis.

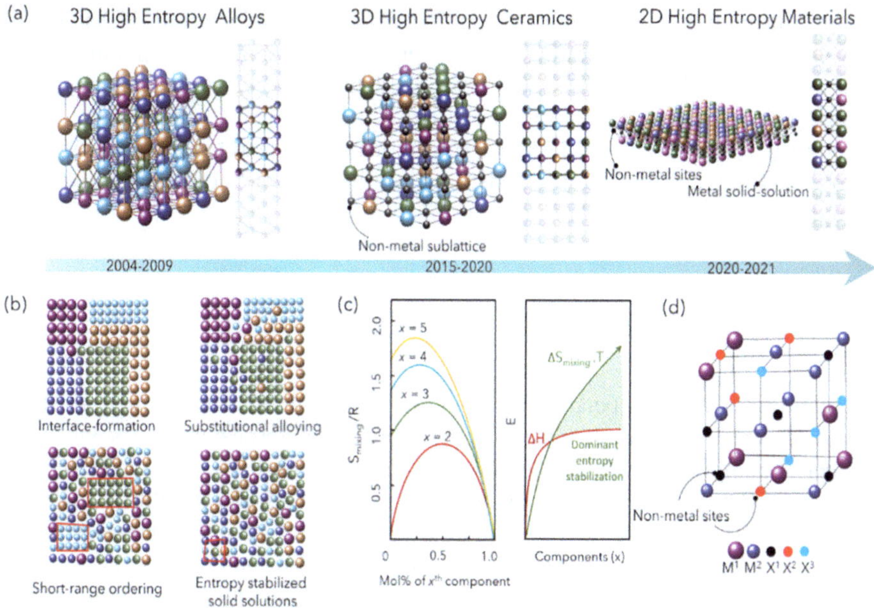

Fig. 2 **a** The timeline tracing advancement in high entropy systems for decades and, now shift toward HE2DMs. **b** Illustrates varying order and disorder between phase-segregated structures and ideal entropy-stabilized solid solutions. **c** The configurational entropy changes as a function of species number (N) and mole fraction, ranging from N = 2 to N = 5, with maximum entropy at equimolar compositions. Based on S/R, materials are categorized as low, medium, or high entropy. The right panel depicts the energy (E) range where entropy effects dominate (shaded area), with the red line representing the enthalpic contribution (ΔH) and the green line indicating the entropic contribution (ΔS). **d** An example of entropy stabilization involves a multi-principal element system containing two metallic and three non-metallic species within the crystal lattice. The entropic contribution arises from both metal and non-metal sublattices and is expressed as: $\Delta S_{mix} = -R\left[\left(\sum_{i=1}^{N} x_i \ln x_i\right)_{cation-site} + \left(\sum_{i=1}^{N} x_i \ln x_i\right)_{anion-site}\right]$. Adapted with permission from Ref. [7].

2 Fundamentals of High Entropy 2D Materials

HEMs contain multiple principal elements of almost equal proportion. HEMs comprising at least five main elements generate distinctive properties beyond traditional materials that contain only one or two dominant elements [11]. The HEM concept emerged from high entropy alloys (HEAs) in the 2000s, and ceramic materials, including oxides, carbides, nitrides, and two-dimensional materials, have been added recently in years [12]. The main attribute of HEMs is their characteristic high configurational entropy, which delivers both material stability and exceptional properties unavailable in traditional materials [13]. The scientific idea of high entropy emerged through investigations of alloys. Yeh et al. showed the world HEAs through their 2004 research by integrating five or more metallic elements with almost identical compositions [14]. These systems exhibited distinctive characteristics with high strength and resistance to corrosion and thermal effects because their configuration entropy stayed high. The success of HEAs motivated researchers to develop high-entropy ceramics and other compounds, which collectively formed the broad category of High Entropy Materials (HEMs).

2.1 Entropy Stabilization

The stability of HEM depends mainly on configurational entropy, which measures how disordered systems become. The stability of materials improves when the number of components within materials made from multiple elements increases due to rising configurational entropy [15] as shown in Fig. 2b–d. The stability of materials depends on the Gibbs free energy equation where both enthalpy (ΔH) and entropy (ΔS) terms are included together. HEMs can sustain their material stability because the high configurational entropy term (ΔS) dominates over enthalpic conditions that would otherwise be unfavorable [16]. The configurational entropy (ΔS) calculation requires the application of this formula [17]:

$$\Delta S = -R\sum_{i=1}^{N} x_i \cdot \ln x_i$$

The HEMs show maximum stability when the mole fractions become equal because of mathematical relationships between the gas constant R and component number N and the ith mole fraction x_i. The system with maximal configurational entropy occurs when moles are distributed equally, thus creating the most stable HEMs [18].

HEMs exist in various types, which are differentiated by their structure and composition. A group of metallic compositions called high entropy alloys (HEAs) results from five or more principal elements when each element exhibits near-equimolar concentrations. HEAs distinguish themselves through exceptional mechanical qualities, the ability to withstand wear and corrosion, high resistance, and excellent flexibility [14]. High Entropy Ceramics (HECs) incorporate oxides, carbides, nitrides, and borides as base elements. High entropy ceramics show application potential in thermal barrier coatings because they combine specialized electrical, thermal, and mechanical characteristics suitable for energy storage devices and catalyst systems [19]. High Entropy Oxides (HEOs) are ceramic materials that incorporate various oxygen anions together with multiple metal cations. Scientists study HEOs because their high thermal stability alongside adjustable electronic features make them suitable for energy storage applications and serve as catalysts and thermal barrier materials [13]. Scientific research about High Entropy 2D Materials now incorporates 2D materials, including MXenes, transition metal dichalcogenides (TMDs), and hydroxides. Combining HEMs and 2D materials enhances specific surface area and unusual electronic properties, thus achieving better functionality for storage systems and catalytic and electronic applications [5].

Multiple elements in HEMs interact to generate unique properties that appear as lattice distortion and slow diffusion mechanisms alongside the "cocktail effect" because the combined properties surpass the total properties of individual elements [15]. The strong stability and adjustable electronic properties in HEMs provide a suitable foundation for electrode applications in batteries and supercapacitors [20]. Multiple active sites arising from HEM elements convert the materials into strong catalysts, which drive hydrogen evolution reactions, oxygen evolution reactions, and carbon dioxide reduction reactions [21, 22]. Research is going on about thermal barrier coatings utilizing high entropy ceramics with oxides and carbides because these materials show excellent resistance to thermal deterioration and chemical reactions [23]. The mechanical sector adopts HEAs to manufacture components that require exceptional strength and wear and corrosion resistance for applications in the aerospace and automotive industries [24].

2.2 Unique Properties of 2D Materials

2D materials exhibit a maximum surface-to-volume proportion, allowing scientists to establish optimized applications for sensing, catalysis, and energy storage systems. The adjustable electronic features of 2D materials become possible by controlling the layers and implementing doping strategies and external stimuli such as strain and electric field. Device developers in electronics and optoelectronics fields should focus on working with adjustable features for their applications. 2D materials preserve their superb mechanical properties at their thin dimensions. Graphene exhibits the greatest tensile strength ever observed in all known materials because its measure surpasses 100 GPa. Graphene and MXenes showcase exceptional conductivity and superior thermal properties, making them suitable candidates for electronic and thermal management applications. Transparent 2D materials possess adjustable absorption spectra that power photovoltaic devices and generate light emissions [25].

2.3 Advantages of 2D Materials in High Entropy Systems

The catalyst functionality of 2D materials is enhanced because they have significant contact areas and adjustable electronic properties. High entropy stabilization strategies with these materials generate advanced catalytic behavior by activating multiple catalytic site locations [16]. 2D materials have extensive applications in developing batteries and supercapacitors that store energy. HE2DMs demonstrate improved energy density and enhanced cycling stability because of their high entropy stabilization methods. High entropy stabilization as a mechanism improves the mechanical properties of 2D materials, making them useful for flexible electronics and structural applications [7]. The thermal stability of HE2DMs is enhanced when stabilized through high entropy methods, which makes them applicable to thermal barrier coating applications and high-power electronics [5, 26].

2.4 Synthesis Strategies for High Entropy 2D Materials (HE2DMs)

The increase in entropy within 2D materials can be achieved through doping and alloying strategies to enhance elemental diversity. During synthesis process, high temperature processing promotes uniform mixing while kinetic control of nucleation and growth to prevent phase separation. Several synthesis techniques have been developed to introduce and stabilize multiple elements in 2D materials as shown in Fig. 3.

Wet-chemical synthesis involves dissolving metal salts in a solvent, followed by precipitation, reduction, or crystallization to form 2D MHEMs. The reaction conditions (temperature, pH, time, solvent) are crucial in determining material quality. Co-precipitation includes mixing multiple metal precursors in a solution to form uniform nanoparticles, often used for high-entropy oxides. The solvothermal/hydrothermal synthesis route uses high-pressure, high-temperature conditions to improve crystallinity and phase stability. Molten salt synthesis provides a medium for metal ion exchange and better control over nanostructure formation as shown in Fig. 3a, b. Wet-chemical synthesis has advantages, such as a simple, cost-effective, versatile route for different compositions and can produce layered structures. With all the mentioned advantages, there are some limitations, such as difficulty in achieving atomic-level mixing and phase separation risk in slow nucleation.

The ***chemical vapor deposition*** (CVD) method involves the thermal decomposition of precursors in a carrier gas at high temperatures (~600–1200 °C). This method is suitable for producing single-layer and few-layered high-entropy 2D materials with excellent crystallinity. ***Calcination doping*** includes Metal precursors that are mixed and annealed at high temperatures to form a solid solution. It synthesizes metal sulphides, oxides, and carbides with entropy-driven stability. CVD produces highly crystalline HE2DMs and allows precise control over phase purity, but it requires specialized high-temperature equipment.

Electrodeposition doping includes the metal ions that are reduced and deposited onto an electrode in a controlled manner. Multi-metallic films with high entropy can be synthesized by adjusting voltage, current, and electrolyte composition as shown in Fig. 3e. It is a scalable, cost-effective, low-temperature synthesis and high compositional control. However, in this method, obtaining single-phase materials is very difficult and limited to conductive substrates only.

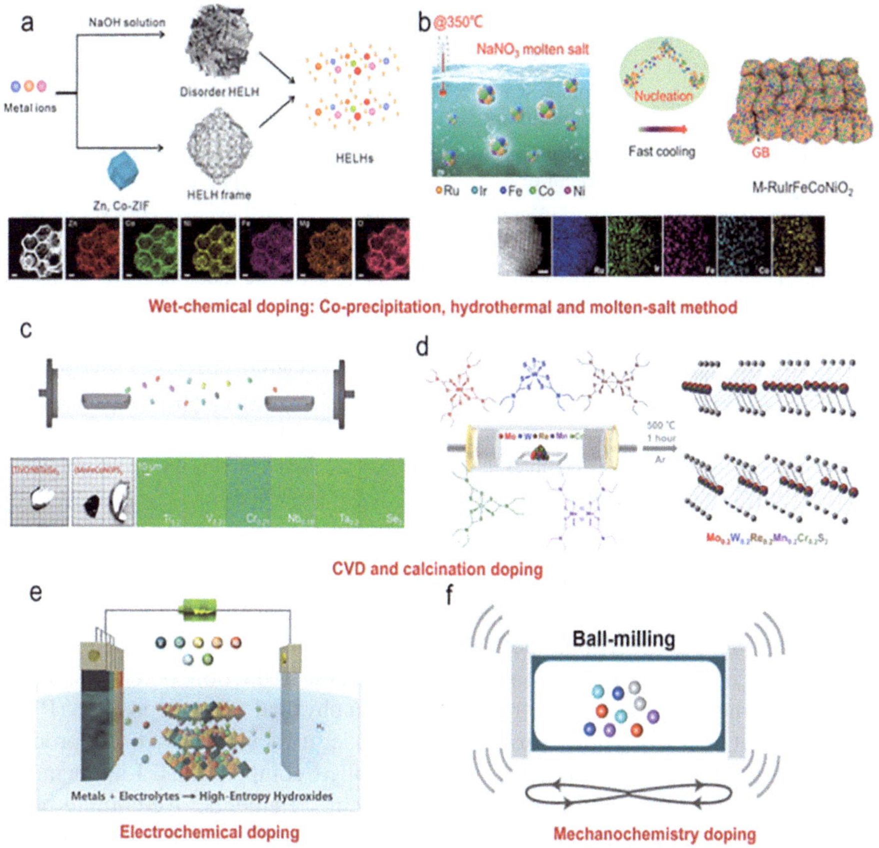

Fig. 3 Doping synthesis to increase the entropy of 2D materials, including wet-chemical doping, chemical vapor deposition doping, electrochemical doping, and mechanochemistry doping. **a** Wet-chemical etching of Zn/Co-MOF to prepare high-entropy hydrotalcite. **b** Molten-salt synthesis of 2D high-entropy $RuIrFeCoNiO_2$. **c** The CVD synthesis of HEMs monocrystals includes Metal dichalcogenides, halides, and phosphorus trisulfides. **d** Calcination synthesis of 2D high-entropy $Mo_{0.2}W_{0.2}Re_{0.2}Mn_{0.2}Cr_{0.2}S_2$. **e** Direct electrodeposition synthesis of high-entropy hydrotalcite. Adapted with permission from Ref. [40]. Copyright (2024) Copyright The Authors, some rights reserved; exclusive licensee Wiley–VCH GmbH. Distributed under a Creative Commons Attribution License 4.0 (CC BY) https://creativecommons.org/licenses/by/4.0/

Mechanochemistry doping exploits mechanical energy (ball milling, high-pressure grinding) to mix elements at the atomic level. It is often used for solid-state reactions to incorporate multiple components into a single-phase 2D material. It has advantages like eco-friendly (solvent-free process), is scalable, cost-effective, and enables high-entropy mixing. However, additional heating/annealing is required for crystallization, and the risk of amorphous phase formation is always present.

Atomic exchange alloying includes a liquid-phase ion exchange process where metal cations replace existing atoms in a 2D material. It is based on the reduction of

Table 1 Comparison based on advantages and limitations of synthesis methods

Method	Advantages	Limitations
Wet-chemical synthesis	Simple, scalable, and tunable compositions	Phase separation, low atomic control
CVD and calcination	High purity and controlled structure	Requires high temperatures, costly equipment
Electrodeposition	Cost-effective and scalable	Limited to conductive substrates, phase control
Mechanochemistry	Eco-friendly and good atomic mixing	Requires additional processing
Atomic exchange alloying	Maintains 2D structure while effective at nano-scale	Requires precise ion exchange control
Magnetron sputtering	Thin films with precise control	Expensive equipment, limited scalability
Liquid metal-assisted	Overcomes immiscibility and has phase control	Requires specific processing conditions

potential differences, allowing controlled element substitution. This technique has advantages such as efficient nano-scale doping and maintaining the 2D structure during alloying. However, this method requires precise control over ion exchange kinetics and is limited to elements with compatible reduction potentials. Other synthetic methods include magnetron sputtering, a physical vapor deposition (PVD) technique that deposits atoms onto a substrate under a vacuum, which can produce ultrathin high-entropy 2D materials with controlled compositions. Liquid metal-assisted synthesis uses a liquid metal medium to mix elements, overcoming phase immiscibility issues that are beneficial for high-entropy oxides and intermetallic.

In conclusion, the synthesis of 2D MHEMs requires a balance of thermodynamics and kinetics to achieve high entropy while maintaining the 2D structure. Each synthesis method offers distinct advantages depending on the material type, application, and scalability. CVD and electrodeposition are promising for thin films. Wet-chemical and mechanochemistry provide scalable synthesis options as shown in Table 1. Atomic exchange alloying is emerging as a method for nano-engineered MHEMs. Selecting the appropriate synthesis method is crucial for tailoring the electronic, catalytic, and structural properties of 2D MHEMs for various advanced applications.

3 Electrochemical Water Splitting: A Brief Overview

The electrical splitting of water shows itself as an effective method to produce hydrogen (H_2) and oxygen (O_2) from water (H_2O). Sustainable splitting represents a fundamental technology for manufacturing clean energy that merges hydrogen storage with electricity applications [9]. The water-splitting process executes two parallel reactions at the cathode using hydrogen evolution reaction (HER), while oxygen evolution reaction (OER) occurs at the anode. Fast OER and HER reactions with minimal overpotential requirements can be reached using effective electrocatalysts [3]. The process of water electrolysis decomposes water into hydrogen and oxygen gases through the application of electricity [3]. The overall reaction is:

$$2H_2O_{(l)} \rightarrow 2H_{2(g)} + O_{2(g)}$$

Under standard pressure at 25 °C, the endothermic reaction requires at least 1.23 V thermodynamic voltage for operation. The operational voltage requirements for this reaction increase because of the kinetic barriers alongside overpotential effects caused by OER and HER processes. Water electrolyzers feature an anode and a cathode as their three principal parts. The electrolyte substance facilitates ionic transfer between electrodes while the device operates in one of three chemical states: acidic, alkaline, or neutral. Electrocatalysts act as materials that lower the activation energy of OER and HER reactions to improve catalytic speed and operational efficiency [27]. Half-reactions in different electrolyte mediums are as follows.

In Acidic Electrolyte:

$$\text{Cathode (HER)} : 2H^+ + 2e^- \rightarrow H_2$$

$$\text{Anode (OER)} : \ 2H_2O \rightarrow O_2 + 4H^+ + 4e^-$$

In Alkaline Electrolyte:

$$\text{Cathode (HER)} : 2H_2O + 2e^- \rightarrow H^2 + 2OH^-$$

$$\text{Anode (OER)} : 4OH^- \rightarrow O_2 + 2H_2O + 4e^-$$

3.1 Catalysts in HER and OER

The function of electrocatalysts in water splitting depends on their ability to reduce activation energy barriers for HER and OER, which produces faster and more efficient reactions [2]. The process of HER requires both proton (H^+) and water molecule reductions to generate hydrogen gas. The ideal HER catalyst needs to exhibit high electrically conductive properties. The catalyst requires excellent performance in hydrogen intermediate adsorption–desorption functions while maintaining stability during reducing conditions [2]. Acidic HER applications use Platinum (Pt) as the catalyst because of its best hydrogen binding capability, yet Nickel (Ni) and cobalt (Co) function as alkaline HER catalysts along with their alloyed materials (e.g., NiMo, CoP) [3]. OER requires water or hydroxide ions to oxidize, producing oxygen gas. OER catalysts must support multiple proton-electron transfer steps and possess strong oxygen intermediate binding characteristics and oxidation resistance. The acidic OER catalysts comprise Iridium oxide (IrO_2) and ruthenium oxide (RuO_2). In contrast, transition metal oxides (e.g., $NiFeO_x$, CoO_x) and perovskites are the prevalent catalysts for alkaline media due to their high activity and stability characteristics [2, 27–29].

3.2 Efficiency Metrics

The performance of electrocatalysts is evaluated using several key metrics such as overpotential (η), Tafel slope, stability, durability, faradaic efficiency, and turnover frequency (TOF). ***Overpotential*** is the additional voltage required to drive a reaction at a specific current density. Lower overpotentials indicate higher catalytic efficiency. For example, a suitable HER catalyst should achieve a current density of 10 mA/cm^2 at an overpotential close to 0 V [2]. ***The Tafel slope*** measures the catalyst's kinetic performance and is derived from the Tafel equation: $\eta = a + b.\log(j)$, where b is the Tafel slope. A lower Tafel slope indicates faster reaction kinetics. For HER, a Tafel slope of ~30 mV/decade is ideal, while for OER, slopes of ~40–60 mV/decade are typical for efficient catalysts [30]. ***Stability*** refers to the ability of the catalyst to maintain its performance over time under operating conditions. Stability is often tested by measuring the overpotential or current density change over extended periods (e.g., 24–100 h) [22]. ***Durability*** assesses the catalyst's resistance to degradation under harsh electrochemical conditions. Key factors include resistance to corrosion, dissolution, and structural changes. ***Faradaic efficiency*** measures the electrical current fraction contributing to the desired reaction (HER or OER). Ideally, the Faradaic efficiency should be close to 100%, indicating minimal side reactions [31]. ***TOF*** quantifies the number of reactions catalyzed per active site per unit of time. High TOF values indicate highly active catalysts [30, 32, 33].

3.3 Challenges in Traditional Electrocatalysts

Overpotential is the extra voltage required beyond the thermodynamic minimum (~1.23 V) to drive the reactions. High overpotentials lead to energy losses and reduced efficiency [2]. OER is particularly challenging due to its complex four-electron transfer process, requiring higher overpotentials than HER. Many catalysts degrade under harsh electrochemical conditions, especially at high potentials and in acidic or alkaline environments. Corrosion, dissolution, and surface oxidation are common issues. Precious metals like platinum (Pt) for HER and iridium (Ir) or ruthenium (Ru) oxides for OER are highly effective but expensive and scarce [22]. Developing low-cost, earth-abundant alternatives is crucial for large-scale applications. The kinetics of HER and OER are inherently slow, requiring efficient catalysts to accelerate the reactions. Catalyst performance varies significantly with pH, limiting their versatility across electrolytes [22]. Electrochemical water splitting is vital for sustainable hydrogen production, but its efficiency and scalability depend on developing advanced electrocatalysts. Understanding the principles of HER and OER, addressing the challenges of traditional catalysts, and optimizing efficiency metrics are critical for advancing this field. Ongoing research focuses on designing cost-effective, stable, and highly active catalysts to enable large-scale implementation of water electrolysis [3, 34].

4 Mechanistic Insights into HE2DMs for Water Splitting

The emerging high-entropy two-dimensional materials (HE2DMs) attract extensive interest in electrocatalysis because of their distinctive structural and electronic features. High configurational entropy and exceptional functional properties emerge from two-dimensional materials with equal proportions of multiple principal elements. The adjustable electronic framework and surface composition of HE2DMs, combined with their ability to engineer defects, mainly benefit Water Splitting by optimizing both hydrogen evolution reaction (HER) and oxygen evolution reaction (OER) [32].

4.1 Electronic Structure of Materials

Electronic structure is the key determining factor for what happens during catalysis. The electronic structure of HE2DMs becomes unique because multiple elements interact in their lattice structure to enable tailored electrochemical reaction characteristics. Changes in the band structure and DOS of HE2DMs occur because different elements produce orbital hybridization within the material structure [35, 36]. A catalytically beneficial electronic structure requires that the Fermi level positions optimally for reaction intermediates like H* in HER or O* in OER. A high density of states near the Fermi level facilitates electron transfer. Transition metals inside HE2DMs have their d-band centers serving as a critical parameter that determines the adsorption energies of reaction intermediates based on the d-band Center Theory [37]. An appropriate placement of the d-band center lies between its upper and lower limits, ensuring balanced interaction between intermediate molecules and the catalyst surface. Multiple constituent elements in HE2DMs function together to improve charge transfer capabilities. The electrocatalytic process requires high electrical conductivity because it reduces the losses caused by electrical resistance. HE2DMs enable tuning of their electronic structure when using different ratios between component elements. The ability to adjust properties through this method facilitates optimizing activities required to perform HER and OER [38, 39].

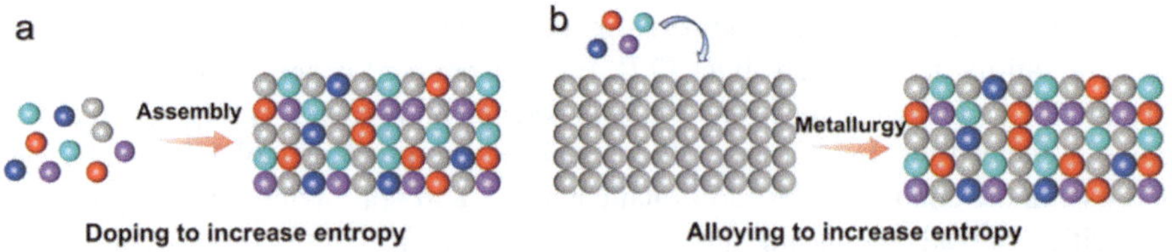

Fig. 4 **a** Doping strategy to increase the entropy of 2D materials. **b** Alloying reaction for achieving medium/high entropy. Adapted with permission from Ref. [40]. Copyright (2024) Copyright The Authors, some rights reserved; exclusive licensee Wiley–VCH GmbH. Distributed under a Creative Commons Attribution License 4.0 (CC BY) https://creativecommons.org/licenses/by/4.0/

4.2 High-Entropy Stabilization

It is the main characteristic of HE2DMs, which emerges when multiple elements mix uniformly at near-equal ratios. Multiple effects occur on electrocatalytic performance due to this stabilizing mechanism. The crystal structure of HE2DMs receives stabilizing effects from high configurational entropy, which protects them from electrochemical degradations, thus enabling long-term water-splitting applications. Different elements randomly distributed through the lattice with doping and alloying to increase the entropy of a materials, cause local structural deformations and strain as shown in Fig. 4. The structural distortions create locations with unique electronic properties, boosting catalytic properties. When multiple elements coexist in the system, they create interactive effects that enhance reactor performance. Among the elements in the catalyst structure, some aid intermediate adsorption, and others aid desorption. The high-entropy stabilization procedure enables scientists to generate exceptional entropy-based phases that existing materials cannot achieve. The phases demonstrate outstanding catalytic functions because their unique electronic structure amalgamated with atomic arrangement works together [10, 12, 18, 19, 40, 41].

4.3 Modeling and Computational Methods

These are essential mechanisms to explain how HE2DMs split water. Such approaches help scientists understand both electronic structure and catalytic activity and well' the reaction pathways of HE2DMs. The electronic structure and reaction pathways, along with the adsorption energies of HE2DMs, are calculated using DFT as one of the main computational tools. Through DFT calculations, researchers can determine chemical activity levels between composition variations and choose promising design options for experimental laboratory verification. By learning their properties from available information, researchers can find new HE2DMs through ML approaches. High-performance catalysts become attainable through ML models for composition-structure–activity relationship discovery. Researchers observe real-time HE2DM behavior dynamics while they operate under electrochemical conditions. The computer simulations display information about HE2DM stability while showing the transformation of their surface and the movement of their structural defects. The reaction kinetics for both HER and OER occur on HE2DMs through simulation based on microkinetic models. These algorithms assist researchers in determining controlling reaction steps and choosing optimal operational parameters. Computational analysis reveals the complete process of HER and OER through the identification of essential molecular intermediates as well as transient states. The obtained knowledge enables the development of catalysts that show decreased activation energies. DFT calculations determine the adsorption bond strength of reaction intermediates (H*, O*, OH*) when placed on various active sites. The equilibrium between intermediate adsorption and desorption requires specific adsorption energy levels. The electronic structure of HE2DMs undergoes modification through computational studies of composition, surface functionalization, and defects. The obtained analytical findings support the development of electronic property tuning methods to boost catalytic performance. Computational methods serve as a tool to develop new HE2DMs with properties optimized for applications. The prediction capability of computational studies speeds up the process of finding high-performance catalysts by analyzing hypothetical material properties [34, 42].

HE2DMs demonstrate great potential for water splitting because they contain distinctive electronic structures, high-entropy stabilization features, and adjustable surface chemistry properties. Researchers must establish electronic structure relationships with catalytic activities, entropy stabilization methods, and surface defect optimization for HE2DM optimization. Theoretical models, together with computational insights, function as guidelines for the development of new HE2DMs with superior catalytic functions. The advancement of HE2DM research will significantly improve sustainable electrochemical water-splitting technology for hydrogen production.

5 High Entropy 2D Materials for Hydrogen Evolution Reaction (HER)

HE2DMs have recently become promising electrochemical water-splitting materials, especially activating the hydrogen evolution reaction (HER). These materials integrate high-entropy alloy properties with 2D structure benefits to enhance electrocatalytic activity and stability [21]. Researchers currently focus on high-entropy transition metal dichalcogenides (HE-TMDs) because these materials provide significant advancement through veterinary high-entropy metal disulfides and high-entropy metal phosphorus tri-chalcogenide systems [43]. Studies of HE2DM performance for HER analyze their behavior through four main criteria, combining overpotential measurements with Tafel slope analysis and exchanging current density characterization and stability tests. Catalytic activity measurements show that $(MoWReMnCr)S_2$ behaves as a superior catalyst than MoS_2 because it operates at 229 mV overpotential at 10 mA cm^{-2} current density in 0.5 M H_2SO_4 solution [2]. Multiple transition metals incorporated in the 2D structure create synergistic effects that boost the catalytic activity levels.

HE2DMs demonstrate improved HER performance through doping and compositional tuning techniques. Doping methods produce modifications in electronic structures and create more active sites while lowering hydrogen adsorption-free energy levels [7, 40]. The optimal hydrogen evolution activity occurred when using a precise Co-molar ratio, which verifies that precise material composition matters significantly. Research indicates that HE2DMs achieve competitive or superior performance results compared to Pt and MoS_2 catalysts regarding overpotential and Tafel slope and stability factors. HE2DMs demonstrate overpotential levels near those of platinum despite its benchmark position. HE2DMs demonstrate more advantageous Tafel slopes than their binary equivalents because certain HE-TMDs doped structures can reach below 50 mV dec^{-1} while approaching the optimal level of Pt (-30 mV dec^{-1}) [20]. Various advantages contributing to HE2DMs superior performance include double active sites from multicomponent incorporation and optimum electronic arrangements for lower hydrogen adsorption energies and increased conductivity in some instances. HE2DMs possess attractive characteristics for large-scale HER applications because of their cost-effectiveness, abundance over platinum, and favorable properties. Continued advancement of research in this area should lead to improvements that will optimize the structure and composition of HE-2D catalysts to boost their efficiency in HER processes. The development of these materials marks a substantial advance in sustainable clean hydrogen production through water splitting because they enable progress toward better energy technologies [21, 44, 45].

6 High Entropy 2D Materials for Oxygen Evolution Reaction (OER)

High entropy 2D materials have led to an advanced electrocatalytic method for oxygen evolution reaction (OER) within electrochemical water splitting systems for hydrogen production. High-entropy 2D materials benefit from two-dimensional structural elements like expanded surface area ratios, numerous accessible active sites, and electronic property flexibility, and they harness special features of entropy-enabled systems, which generate structural distortions along with configuration-based stability and compound synergy effects. The merger of these materials produces outstanding catalytic performance, extending to long-term operation and functional adaptability across different electrochemical domains [21, 44].

The development of HE2DMs marks a revolutionary step in electrocatalysis because they integrate the compositional complexity of multi-metal systems with

2D architectural advantages. These materials reach outstanding oxygen evolution reaction (OER) performance through four main mechanisms. The optimized electronic structures decrease their overpotential response. Spinels made of FeCoNiCrMn show a remarkable OER overpotential of 288 mV when operating at 10 mA cm^{-2} in alkaline solutions while surmounting the overpotential of binary MoS_2 at 362 mV. This confinement structure boosts material electrical conductivity and active site numbers. The electrons within quantum-confined HE2DMs undergo faster transfer, resulting in minimal Tafel slopes down to 34.7 mV dec^{-1}. The perovskite composition $LaCr_{0.2}Mn_{0.2}Fe_{0.2}Co_{0.2}Ni_{0.2}O_3$-δ displays stronger OER currents than isolated-metal materials because of its heterogeneous surface sites with inherent high catalytic potential. Adapter devices made from multi-metal structures achieve improved performance because they modulate reaction intermediate adsorption energies (ΔG_OOH and ΔG_OH) toward the theoretical peak position through effects of d-band center modifications caused by lattice strain [46]. The composition of multiple metals leads to three beneficial effects that maximize the kinetics of OER reactions. The d-band centers in $La_{1-x}Sr_x(CrMnFeCoNi)_{0.2}O_3$ rise by 0.35 eV through electronic modulation via Sr doping, leading to better *OH adsorption and lower overpotential values by 330 mV [37]. Strained equatorial metal-O bonds whose length decreased by 0.08 Å function as the basis for low Tafel slopes under 35 mV dec^{-1} since atomic radius differences (4–15% mismatch) cause compressive strain. The redox pair Fe^{3+}/Fe^{4+} and Co^{3+}/Co^{4+} present in $FeCo(PO_4)_4$@rGO composites accelerates electron transfer, resulting in 0.54 s^{-1} turnover frequency while maintaining an overpotential of 237 mV. Performance benchmarks demonstrate that HE2DMs operate equally well compared to precious metal benchmarks. Testing of FeCoNiCrMn high-entropy oxide shows 307 mV overpotential at 10 mA cm^{-2} with a Tafel slope of 34.7 mV dec^{-1}, thus matching IrO_2 performance in a superior manner due to extended operating time of 95 h compared to IrO_2's less than 20 h runtime [47]. Earth-abundant metals in the catalyst compose a cost-effective system with 5–10 times more efficient operation than Ir/Ru-based technologies. Energy conversion technologies of the future can benefit from HE2DMs, which unite excellent activity with industrial-level stability attributes and scalability aspects [2, 21, 48].

The research investigation presented in Fig. 5 deepens the knowledge about how two-dimensional (2D) materials enhance oxygen evolution reaction (OER) electrocatalysis. The text explains OER's fundamental operational principles while showing recent achievements in constructing catalysts with two-dimensional structures. The new strategies described in this work focus on three points for improving catalyst behavior: (a). Structural Development, (b). Electronic property optimization, and (c). deliberate defect implementation. The research explores existing difficulties in the field while recommending innovative methods for improving material performance potential. An analysis of these elements forms the foundation to provide essential information about building advanced electrocatalysts for efficient OER systems. Figure 6 summarizes contemporary achievements in high-entropy materials development for water electrolysis by analyzing their capacity to activate both hydrogen evolution reaction (HER) and oxygen evolution reaction (OER). The investigation

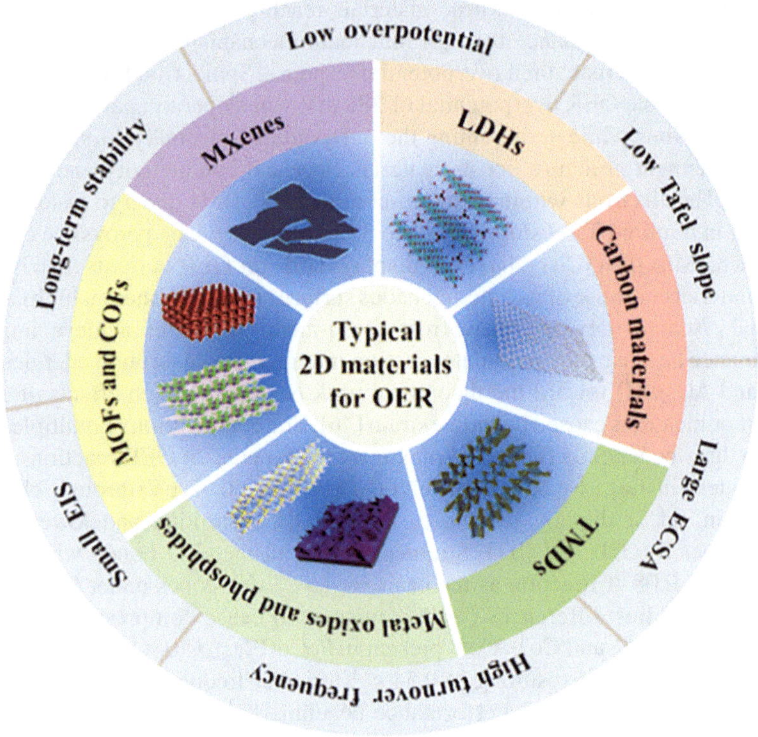

Fig. 5 A schematic overview of typical 2D materials and critical factors in developing efficient electrocatalysts for OER. Adapted with permission from Ref. [48]. Copyright (2022) Copyright The Authors, some rights reserved; exclusive licensee Wiley–VCH GmbH. Distributed under a Creative Commons Attribution License 4.0 (CC BY) https://creativecommons.org/licenses/by-nc/4.0/

includes a detailed analysis of high-entropy catalysts, including high-entropy alloys (HEAs), (oxy)hydroxides, oxides, and other upcoming materials.

7 Challenges and Strategies for Performance Optimization

The electrochemical water splitting applications use novel high-entropy materials (HEMs) because these materials offer distinct features to boost catalytic performance. Multiple important issues must be resolved before HEMs can achieve their full practical potential. HEMs present major stability obstacles because of their complex combination of multiple elements that aid catalytic function yet reduce operational stability. The complicated HEM structure requires advanced methods to improve its operational stability and performance characteristics.

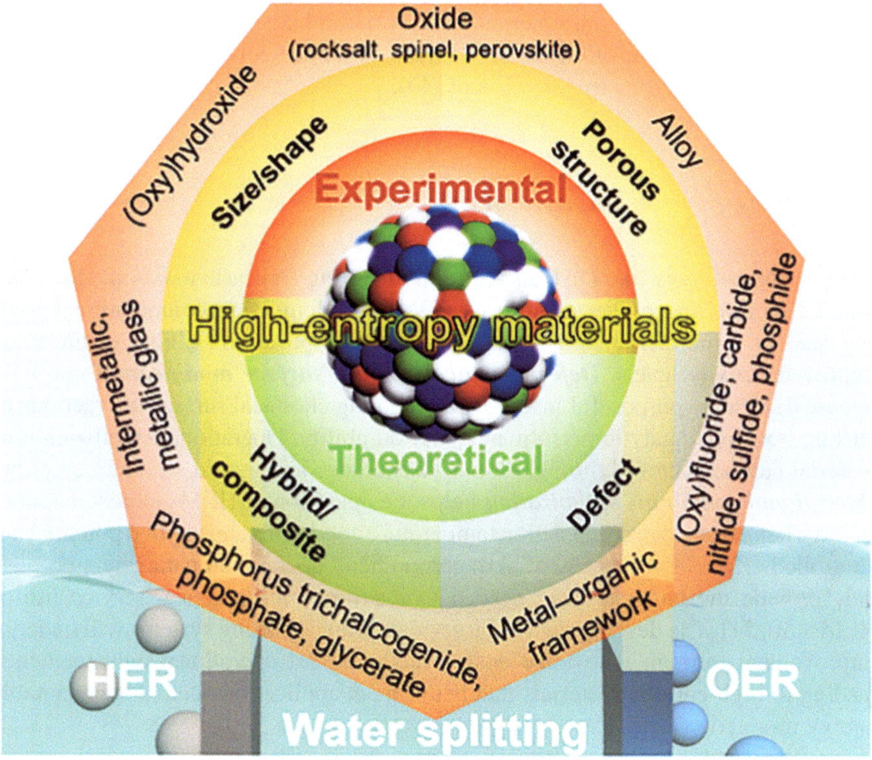

Fig. 6 The schematic provides a detailed representation of high-entropy materials and their key design strategies for water electrolysis applications, including hydrogen evolution reaction (HER) and oxygen evolution reaction (OER). Adapted with permission from Ref. [48]. Copyright (2022) Copyright The Authors, some rights reserved; exclusive licensee Wiley–VCH GmbH. Distributed under a Creative Commons Attribution License 4.0 (CC BY) https://creativecommons.org/licenses/by-nc/4.0/

7.1 Challenges

Complex composition and stability as extended-term operations of HEMs become complicated due to their five-element composition with near-equiatomic concentrations. This composition necessitates complex maintenance of structural and chemical stability. The multiple elements in HEM materials generate operational challenges that cause element segregation or phase separation, thus reducing their catalytic performance and service life [12]. ***The scalability and cost*** of the production of HEMs prove affordable compared to noble metal catalysts; however, the main obstacle lies in their large-scale manufacturing without sacrificing performance consistency. Manufacturing operations requiring extensive batch processing require consistent quality standards for successful industrial implementation. ***Interfacial dynamics*** of the active surface of HEMs will change when exposed to electrolyte solutions because these interactions cause both functional and structural deterioration of their electrochemical features. Controlling these interface dynamics between molecules proves necessary for performance enhancement [49].

7.2 Strategies for Improvement

Computational tools like DFT and machine learning forecasts maximize the electronic structure and phase stability of HEMs. Implementing computational analysis leads to a more efficient materials design with optimized stability features and improved activity levels. ***Defect engineering and surface modification*** can help release defects by purposeful insertion or changing chemical surface interactions of HEMs, boosting catalytic performance and durability. Operational stabilization of material can be achieved through doping techniques and hybrid structure creation. ***Operational conditions optimization*** helps the operation of HEMs to become more stable when electrolyte composition temperature conditions and applied potential are controlled effectively. It is essential to comprehend operational parameters because they provide the understanding needed to maximize the practical service life of HEMs [50, 51]. The development of large-scale water-splitting systems will undergo improvement through innovative design methods and operational enhancements, leading to high-entropy materials that better meet application needs for sustainable energy production [15].

8 Future Perspectives and Emerging Trends

Researchers have discovered that high-entropy materials (HEMs) show excellent electrochemical water-splitting characteristics and continue to present promising opportunities across numerous scientific fields. Research progress leads to developing multiple promising trends and future perspectives [44]. Fuel cells will benefit from HEMs because they function as highly efficient catalysts for two key reactions: oxygen reduction reaction (ORR) and hydrogen oxidation reaction (HOR). High-entropy alloy nanoparticles consisting of PtFeCoNiCu show exceptional Oxygen Reduction catalytic activity, which surpasses the performance of commercial Pt/C catalysts [15]. Multilayered high-entropy two-dimensional materials possess structures that prove difficult for material designers but offer new design prospects. Research has been rising toward the combined utilization of artificial intelligence (AI) and machine learning (ML) systems to speed up HE2DM discovery and optimization processes for multiple applications. DFT calculations and high-throughput screening methods enable researchers to predict stable compositions and HE2DM properties. This advanced technological system enables quick evaluation of large numbers of potential compound iterations, which decreases lab testing periods and practical requirements. Machine learning algorithms currently identify the relationship patterns between HE2DM composition and structure and their catalytic properties. The designed materials become optimal for electrochemical reactions through computational tools, enabling the identification of specific properties [9, 52, 53]. The continued research into HEMs has led scientists to create advanced synthesis methods that handle scalability alongside cost-effectiveness issues. Large-scale production of high-entropy alloys can be achieved through the promising CCC method, which combines casting with comminution. Atom-up synthetic techniques receive increasing recognition because they create perfect homogeneity through element blending at the atomic scale. High-entropy materials and other nanoparticles can be synthesized with versatility through rapid thermal treatment methods involving metal salt solutions on carbon supports. Industries strive to achieve industrial viability by shifting their attention to earth-abundant elements that reduce costs while preserving performance metrics. Scientists are developing high-energy metals that perfectly fulfill the integration needs of current electrolyzer systems to overcome process integration obstacles.

8.1 Environmental Impact and Sustainability Consideration

The commercial deployment of HEMs demands a thorough evaluation of their sustainability practices because they are approaching industrial use. The multiple ingredients used in HEMs present environmental concerns regarding raw material sustainability and proper disposal options in the life cycle of catalyst development. Researchers must study the entire lifespan of HEMs, starting from their extraction of raw materials through disposal or potential recycling options. Earth-abundant materials used in HEM construction help lower production expenses and match sustainability objectives by promoting resource conservation. HEMs promise to boost the operational efficiency of electrochemical processes, enabling industries to decrease their energy costs and environmental footprints. More efficient water-splitting catalysts developed from HEMs will speed up the establishment of sustainable green hydrogen economies worldwide while reducing greenhouse gas emissions. High entropy materials for electrochemical applications show quick progress through recent developments in material fabrications, synthesis techniques, and potential utilization areas. To achieve their complete industrial potential, research must address scalability problems while ensuring sustainability and cost-effectiveness. AI-driven design methods combined with modern characterization methods are anticipated to enhance the development of clean energy breakthroughs by delivering transformative technological achievements [20].

9 Conclusion

The investigation inside this chapter discusses how high entropy two-dimensional materials (HE2DMs) will be used for electrochemical water splitting, which is a necessary step toward sustainable hydrogen generation. This section starts by explaining the composition of HE2DMs through mixed elements in equal proportions, generating greater durability and stability while improving catalytic activity. An introduction describes HE2DM principles, chemical vapor deposition (CVD) synthesis approaches, and solution-based procedures that enable electrocatalytic property modification. The text reviews electrochemical water splitting reactions, starting with an analysis of hydrogen evolution reaction (HER) and oxygen evolution reaction (OER) while exploring overpotential and Tafel slope and stability metrics. The article explores HE2DMs through structural mechanics, showing that their electronic design and entropy balance with defect construction produces optimal catalytic capability. This chapter explains how high entropy transition metal dichalcogenides (HE-TMDs) achieve better catalytic efficiency through their multiple metal components while replacing platinum-based catalysts. HE2DMs function as promising large-scale hydrogen production candidates during OER applications because they provide enhanced oxidation resistance and stability properties. The exceptional characteristics of HE2DMs still come with obstacles regarding scalability, high expense, and limited operational endurance and next-generation energy storage technology while evaluating their industrial impact and environmental benefits. The chapter validates HE2DMs as a promising electrocatalyst category with extensive potential for sustainable energy technologies and green hydrogen production that guides future research development in this field.

References

1. Cavin, J., et al.: 2D high-entropy transition metal dichalcogenides for carbon dioxide electrocatalysis. Adv. Mater. **33**, 2100347 (2021)
2. Wang, S., Lu, A., Zhong, C.-J.: Hydrogen production from water electrolysis: role of catalysts. Nano Converg **8**, 4 (2021)
3. Kazemi, A., Manteghi, F., Tehrani, Z.: Metal electrocatalysts for hydrogen production in water splitting. ACS Omega (2024). https://doi.org/10.1021/acsomega.3c07911
4. Zhai, W., et al.: Recent progress on the long-term stability of hydrogen evolution reaction electrocatalysts. InfoMat **4**, 12357 (2022)
5. Cui, Y., et al.: A perspective on high-entropy two-dimensional materials. SusMat **2**, 65–75 (2022)
6. Han, L., et al.: Multifunctional high-entropy materials. Nat. Rev. Mater. **9**, 846–865 (2024)
7. Nemani, S.K., Torkamanzadeh, M., Wyatt, B.C., Presser, V., Anasori, B.: Functional two-dimensional high-entropy materials. Commun. Mater. **4**, 16 (2023)
8. Ursua, A., Gandia, L.M., Sanchis, P.: Hydrogen production from water electrolysis: current status and future trends. Proc. IEEE **100**, 410–426 (2012)
9. Zhang, H.-M., et al.: Recent advances of high-entropy electrocatalysts for water electrolysis by electrodeposition technology: a short review. Rare Met. **43**, 2371–2390 (2024)

10. Löffler, T., Ludwig, A., Rossmeisl, J., Schuhmann, W.: What makes high-entropy alloys exceptional electrocatalysts? Angew. Chem. Int. Ed. **60**, 26894–26903 (2021)
11. Wang, Q., Velasco, L., Breitung, B., Presser, V.: High-entropy energy materials in the age of big data: a critical guide to next-generation synthesis and applications. Adv. Energy Mater. **11**, 2102355 (2021)
12. Fu, M., Ma, X., Zhao, K., Li, X., Su, D.: High-entropy materials for energy-related applications. iScience **24**, 102177 (2021)
13. Sen, S., Palabathuni, M., Ryan, K.M., Singh, S.: High entropy oxides: mapping the landscape from fundamentals to future vistas. ACS Energy Lett. **9**, 3694–3718 (2024)
14. Liu, H., et al.: High-entropy alloys and compounds for electrocatalytic energy conversion applications. SusMat **1**, 482–505 (2021)
15. Bolar, S., Ito, Y., Fujita, T.: Future prospects of high-entropy alloys as next-generation industrial electrode materials. Chem. Sci. **15**, 8664–8722 (2024)
16. Sun, Y., Dai, S.: High-entropy materials for catalysis: a new frontier. Sci. Adv. **7**, 1600 (2021)
17. Nataraj, C.M., van de Walle, A., Samanta, A.: Temperature-dependent configurational entropy calculations for refractory high-entropy alloys. J. Phase Equilibria Diffus. **42**, 571–577 (2021)
18. Hanke, T., Upterworth, A.L., Sebastiani, D.: Explicit configurational entropy of mixing in molecular dynamics simulations. J. Phys. Chem. Lett. **15**, 11320–11327 (2024)
19. Li, L., et al.: High entropy ceramics for electromagnetic functional materials. Adv. Funct. Mater. (2024). https://doi.org/10.1002/adfm.202416673
20. Han, L., Mu, W., Wei, S., Liaw, P.K., Raabe, D.: Sustainable high-entropy materials? Sci. Adv. **10**, 3926 (2024)
21. Yang, X., et al.: Engineering high-entropy materials for electrocatalytic water splitting. Int. J. Hydrogen Energy **47**, 13561–13578 (2022)
22. Raveendran, A., Chandran, M., Dhanusuraman, R.: A comprehensive review on the electrochemical parameters and recent material development of electrochemical water splitting electrocatalysts. RSC Adv. **13**, 3843–3876 (2023)
23. Wang, Y., Zhang, J., Wu, T., Huang, G.: Full-scale insight into high-entropy ceramics from basic concepts, synthesis technologies, structural characteristics, and properties to application prospects. J. Mater. Res. Technol. **33**, 398–430 (2024)
24. Lee, J., Seo, J.H., Gao, B., Jang, H.W.: Transition metal-based high-entropy materials for catalysis. MetalMat (2025). https://doi.org/10.1002/metm.31
25. Shanmugam, V., et al.: A review of the synthesis, properties, and applications of 2D materials. Part. Part. Syst. Charact. **39**, 2200031 (2022)
26. Lu, D., et al.: Challenges and opportunities in 2D high-entropy alloy electrocatalysts for sustainable energy conversion. SusMat **3**, 730–748 (2023)
27. Yu, M., Budiyanto, E., Tüysüz, H.: Principles of water electrolysis and recent progress in cobalt-, nickel-, and iron-based oxides for the oxygen evolution reaction. Angew. Chem. Int. Ed. **61**, e202103824 (2022)
28. Li, W., et al.: Unlocking efficiency: minimizing energy loss in electrocatalysts for water splitting. Adv. Mater. (2024). https://doi.org/10.1002/adma.202404658
29. Zhu, Y., et al.: Transition metal single-atom catalysts for water splitting: unravelling coordination strategies and catalytic mechanisms for sustainable hydrogen generation. Next Mater. **6**, 100491 (2025)
30. van der Heijden, O., Park, S., Vos, R.E., Eggebeen, J.J.J., Koper, M.T.M.: Tafel slope plot as a tool to analyze electrocatalytic reactions. ACS Energy Lett. **9**, 1871–1879 (2024)
31. Slobodkin, I., Davydova, E., Sananis, M., Breytus, A., Rothschild, A.: Electrochemical and chemical cycle for high-efficiency decoupled water splitting in a near-neutral electrolyte. Nat. Mater. **23**, 398–405 (2024)
32. Binod Raj, K.C., Kumar, D., Bastakoti, B.P.: Enhancing electrocatalytic performance of RuO2-based catalysts: mechanistic insights, strategic approaches, and recent advances. J. Phys. Energy **7**, 022001 (2025)
33. Costentin, C., Passard, G., Savéant, J.-M.: Benchmarking of homogeneous electrocatalysts: overpotential, turnover frequency, limiting turnover number. J. Am. Chem. Soc. **137**, 5461–5467 (2015)

34. Cho, S.C., et al.: Expanding the frontiers of electrocatalysis: advanced theoretical methods for water splitting. Nano Converg. **12**, 4 (2025)
35. Rose, M.J.: Semiconductor band structure, symmetry, and molecular interface hybridization for the chemist. J. Am. Chem. Soc. **146**, 5735–5748 (2024)
36. Dallaev, R., Spusta, T., Allaham, M.M., Spotz, Z., Sobola, D.: Synthesis and band gap characterization of high-entropy ceramic powders. Crystals (Basel) **14**, 295 (2024)
37. Bhattacharjee, S., Waghmare, U.V., Lee, S.-C.: An improved d-band model of the catalytic activity of magnetic transition metal surfaces. Sci. Rep. **6**, 35916 (2016)
38. Chen, M.: Strong metal-support interaction of Pt-based electrocatalysts with transition metal oxides/nitrides/carbides for oxygen reduction reaction. Microstructures (2023). https://doi.org/10.20517/microstructures.2023.12
39. Sher Shah, Md.S.A., Jang, G.Y., Zhang, K., Park, J.H.: Transition metal carbide-based nanostructures for electrochemical hydrogen and oxygen evolution reactions. EcoEnergy **1**, 344–374 (2023)
40. Mei, H., et al.: Entropy engineering of 2D materials. Adv. Sci. **11**, 2409404 (2024)
41. Xiao, W., et al.: Synthesis of high entropy and entropy-stabilized metal sulfides and their evaluation as hydrogen evolution electrocatalysts. Chem. Mater. **35**, 7904–7914 (2023)
42. Chen, C., Zheng, Y., Qiao, S.-Z.: Understanding the local environment in electrocatalysis. Natl. Sci. Rev. **11**, nwae250 (2024)
43. Wu, L., Hofmann, J.P.: High-entropy transition metal chalcogenides as electrocatalysts for renewable energy conversion. Curr. Opin. Electrochem. **34**, 101010 (2022)
44. Ruiz Esquius, J., Liu, L.: High entropy materials as emerging electrocatalysts for hydrogen production through low-temperature water electrolysis. Mater. Futur. **2**, 022102 (2023)
45. Wang, Q., et al.: Recent progress in high-entropy alloy electrocatalysts for hydrogen evolution reaction. Adv. Mater. Interfaces **11**, 2301020 (2024)
46. Liu, D., et al.: Development of perovskite oxide-based electrocatalysts for oxygen evolution reaction. Small **17**, 2101605 (2021)
47. Duan, C., et al.: Nanosized high entropy spinel oxide (FeCoNiCrMn)3O4 as a highly active and ultra-stable electrocatalyst for the oxygen evolution reaction. Sustain. Energy Fuels **6**, 1479–1488 (2022)
48. Xu, X., Shao, Z., Jiang, S.P.: High-entropy materials for water electrolysis. Energy Technol. **10**, 2200573 (2022)
49. Calero, C., Faraudo, J.: The interaction between electrolyte and surfaces decorated with charged groups: a molecular dynamics simulation study. J. Chem. Phys. **132**(2), 024704 (2010)
50. Lopata, J.S. et al.: An experimentally validated three-dimensional computational fluid dynamics model for polymer electrolyte membrane water electrolyzers. In: ECS Meeting Abstracts MA2021-02, pp. 1227–1227 (2021)
51. Kim, Y., Kim, Y., Lee, D., Kim, J.: Computational modeling of anion exchange membrane water electrolysis. In: ECS Meeting Abstracts MA2024-02, pp. 5097–5097 (2024)
52. Li, H., Ling, L., Li, S., Gao, F., Lu, Q.: High entropy materials—emerging nanomaterials for electrocatalysis. Energy Adv. **2**, 1800–1817 (2023)
53. He, C., et al.: Fresh perspectives and insights into the challenges and opportunities in the emerging high-entropy electrocatalysts. Coord. Chem. Rev. **531**, 216496 (2025)

Challenges in Scaling Up 2D Material-Based Electrocatalysts

K. C. Seetha Lakshmi and Balaraman Vedhanarayanan

Abstract Two-dimensional (2D) material-based electrocatalysts, including graphene, transition metal dichalcogenides (TMDs), and MXenes, have demonstrated exceptional catalytic performance for energy conversion and storage applications. However, their large-scale implementation faces significant challenges related to synthesis, structural engineering, device integration, and long-term stability. This chapter provides a comprehensive analysis of the major obstacles in scaling up 2D material-based electrocatalysts for industrial applications. Key synthesis challenges include batch-to-batch variability, purity control, and the scalability of techniques such as chemical vapor deposition (CVD) and liquid-phase exfoliation. Structural limitations, such as aggregation, surface engineering inconsistencies, and doping challenges, hinder their catalytic efficiency at scale. Additionally, the integration of 2D materials into electrochemical devices is complicated by electrode fabrication issues, poor adhesion, and binder-induced performance degradation. Long-term stability concerns, including corrosion, electrochemical cycling degradation, and environmental effects, further restrict their commercialization. Despite these challenges, emerging solutions such as, incorporation of hybrid nanostructures, roll-to-roll manufacturing, and sustainable synthesis approaches offer promising pathways for large-scale production. The chapter concludes by highlighting the importance of standardization efforts to facilitate the transition of 2D material-based electrocatalysts from laboratory research to real-world applications.

K. C. Seetha Lakshmi · B. Vedhanarayanan (✉)
Department of Applied Chemistry and Biotechnology, Graduate School of Engineering, Chiba University, 1-33 Yayoi-cho, Inage-ku, Chiba 263-8522, Japan
e-mail: vedhanab@srmist.edu.in

B. Vedhanarayanan
Department of Chemistry, Faculty of Engineering and Technology, SRM Institute of Science and Technology, Chengalpattu 603203, Tamil Nadu, India

O. P. Pandey et al. (eds.), *2D Materials for Electrochemical Water Splitting*, Engineering Materials, https://doi.org/10.1007/978-981-95-2003-9_12

1 Introduction

1.1 Overview of 2D Material-Based Electrocatalysts

Two-dimensional (2D) materials have emerged as highly promising candidates in electrocatalysis, especially for energy conversion reactions like the hydrogen evolution reaction (HER) and oxygen evolution reaction (OER; Fig. 1) [1–3]. Their structural uniqueness—featuring atomic-scale thickness, a large surface area, and abundant accessible active sites—makes them ideal platforms for facilitating electron transfer and catalytic reactions. In particular, 2D carbon-based materials have shown exceptional performance in OER processes by minimizing metal usage while enhancing catalytic efficiency through nanoconfinement effects [4, 5]. The confined microenvironments in these materials shorten the diffusion paths of intermediates, reduce the limiting potential for water oxidation, and expose a greater number of active sites, contributing to overall improved performance [6, 7]. Nevertheless, the practical deployment of these materials still demands the development of standardized evaluation protocols and stability metrics for long-term operation [7].

Emerging classes of 2D materials—such as graphene derivatives and transition metal dichalcogenides (TMDs)—offer intrinsic electronic properties and high atomic utilization, making them suitable for advanced electrocatalysis [8]. These materials ensure that a significant fraction of catalytic sites are accessible during electrocatalysis, thus enhancing reaction kinetics and electron transport [9, 10]. Continued investigation into their electronic structures has yielded deeper insights into electrocatalytic mechanisms, aiding the rational design of high-efficiency catalysts [11]. Traditional synthetic routes for these materials often rely on toxic organic solvents

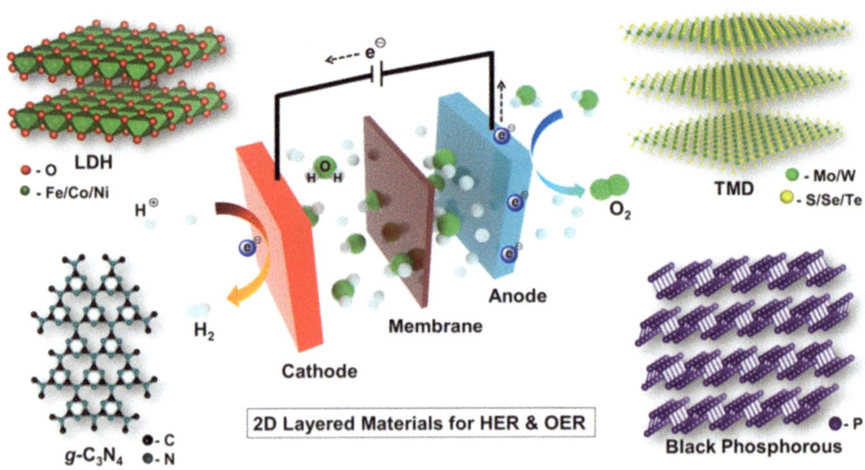

Fig. 1 Pictorial representation of various 2D-layered materials utilized for electrocatalytic water splitting (HER and OER). Adapted with permission from [3]. Copyright (2023) Elsevier

and tend to localize active sites at material edges. An innovative approach introduced a green and self-assembling method for preparing oxygen evolution catalysts under alkaline conditions, yielding materials with uniformly distributed active sites on a stable support [12]. Further, 2D metal–organic frameworks (MOFs) and their derivatives have emerged as a new category of electrocatalysts with tunable structures [13]. By precisely engineering their coordination environments, it is possible to tailor their electronic and catalytic properties to optimize their performance in HER and OER [14]. Incorporation of transition metals and the construction of heterostructures within these MOFs further enhance catalytic outcomes [15].

Hybridization effects in these systems also yield synergistic benefits, improving catalytic activity and stability, as demonstrated in recent studies on 2D composites [16]. Furthermore, on the synthetic side, controlled fabrication techniques that manipulate layer thickness, chemical composition, and defect structures have emerged as key tools for tuning electrocatalytic performance [17]. In particular, defect engineering—through doping with transition metals—has been shown to greatly enhance catalytic efficiency and opens the door to scaling up for real-world applications [18]. Collectively, 2D material-based electrocatalysts present an exciting frontier in clean energy technologies. Their remarkable properties, combined with continued progress in synthetic strategies and device integration, can pave the way for widespread industrial adoption.

1.2 Significance of Scaling Up

The transition of 2D material-based electrocatalysts from laboratory-scale synthesis to industrial-scale implementation is essential for their commercial viability (https://doi.org/10.1002/aenm.202303451) [19]. Large-scale deployment in renewable energy systems such as water electrolyzers demands materials that are not only efficient but also economically viable and environmentally benign [20]. The importance of scaling up lies in the unique physical and chemical properties of 2D materials, such as high surface area, tunable electronic structures, and low-cost compositional flexibility. Materials like graphene, TMDs, and MOFs hold immense promise for large-scale applications due to these attributes. For example, 2D carbon-based materials are notable for their excellent conductivity and low production costs, which are advantageous for industrial water splitting technologies [7, 21].

The move toward industrial-scale synthesis also requires scalable and cost-effective production methods. Recent developments in fabricating layered double hydroxides (LDHs) and transition metal composites at scale, while preserving their electrocatalytic characteristics, mark important milestones [17, 22]. In practice, however, materials optimized under controlled laboratory conditions often experience performance degradation in real-world operational environments. Therefore, robust evaluation metrics and reliability frameworks are needed to assess activity and stability under industrial conditions [23]. Strategies such as substrate hybridization and structural reinforcement have been shown to mitigate degradation, enhance resistance to oxidation, and prolong operational lifespans [24]. Ambient-condition synthesis of hierarchically porous 2D structures is another promising strategy, capable of lowering energy inputs and eliminating the need for toxic solvents, thus making the process more sustainable. For example, exfoliation of MoS_2 layers is performed with the help of NiFe nanosheets in aqueous solution to form NiFe/MoS_2 hybrid in an environmental friendly synthetic process (Fig. 2) [12, 25].

Moreover, the modular design of 2D MOFs allows for the integration of multiple functionalities, including enhanced electrical conductivity, structural flexibility, and

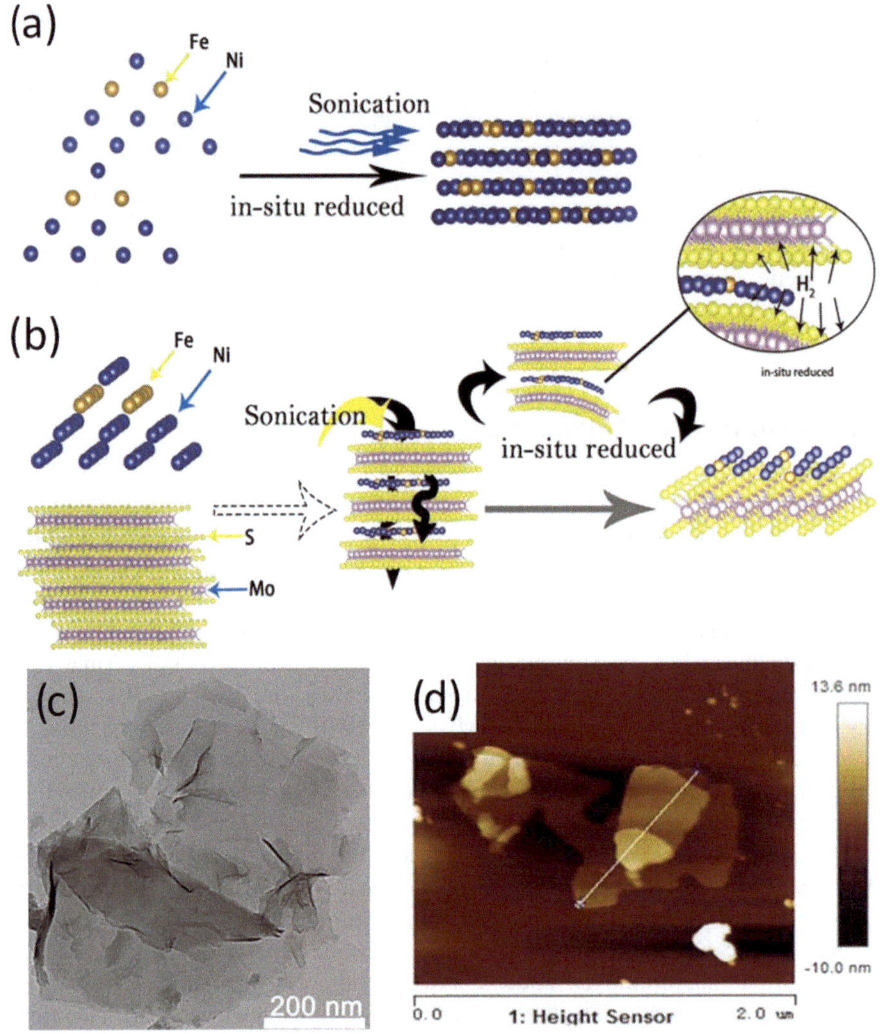

Fig. 2 **a**, **b** Synthesis, **c** TEM and **d** AFM images of NiFe/MoS_2-based 2D heterostructures. Adapted with permission from [12]. Copyright (2018) Wiley

catalytic selectivity [26, 27]. Their superior electronic and catalytic properties are critical for a variety of large-scale electrocatalytic systems [28]. The successful scale-up of 2D material-based electrocatalysts is vital for advancing sustainable energy technologies, ensuring that lab-based innovations translate effectively into practical, industry-ready systems.

1.3 Key Challenges in Scaling Up

Despite their promising potential, scaling up 2D electrocatalysts introduces several scientific and engineering challenges that must be overcome to realize their full impact on commercial energy systems [29]. A major hurdle lies in the controlled synthesis of high-quality 2D materials with reproducible features such as surface defects, edge configurations, and elemental composition. Achieving atomic-level precision across large-scale production batches remains technically demanding, as even slight inconsistencies can drastically affect catalytic performance [30].

Durability is another pressing issue. Carbon-based 2D materials are especially susceptible to structural degradation, aggregation, or chemical instability when subjected to prolonged electrochemical cycling or harsh conditions in real applications [7]. Preventing restacking and ensuring long-term operational stability remain central goals. Moreover, many high-performance synthesis methods—such as CVD [31] and solvothermal techniques—are energy-intensive or involve hazardous reagents, making them unsuitable for industrial-scale implementation. Greener, milder methods that preserve material performance while reducing costs and ecological impact are urgently needed [32]. The integration of 2D electrocatalysts into real-world energy devices introduces another layer of complexity. These materials must operate compatibly with electrolytes, separators, and other device components under fluctuating thermal and electrochemical conditions [21]. Efforts to design hybrid catalysts—such as bimetallic 2D MOFs—hold great promise due to synergistic interactions between multiple metallic elements. However, such systems are often plagued by synthetic challenges related to interfacial energy barriers and nucleation control [33]. Addressing these challenges through innovative materials design, environmentally friendly synthesis routes, and strategic integration frameworks is key to enabling the industrial adoption of 2D electrocatalysts in the energy sector.

2 Synthesis Challenges

2.1 Scalability of Synthesis Methods

The scalability of synthesis methods for two-dimensional (2D) material-based electrocatalysts remains one of the most significant barriers to their large-scale deployment in real-world electrochemical applications [34]. Despite their immense potential in catalyzing vital reactions such as hydrogen evolution, and oxygen evolution, the reliable and cost-effective production of these materials at industrial scale continues to be a major hurdle. The challenge arises from the delicate interplay between structural integrity, material consistency, and desired functional properties, all of which are often compromised during scale-up processes. Widely employed synthesis techniques such as CVD [31], liquid-phase exfoliation [35], and wet chemical synthesis [36], although effective at laboratory scale, face severe constraints when translated to large-scale manufacturing. For instance, CVD is renowned for producing high-quality 2D films with excellent crystallinity and uniformity; however, it frequently suffers from complications like wrinkle formation due to thermal expansion mismatch between the substrate and the deposited film. Additionally, the transfer process associated with CVD-grown films often introduces contamination, delamination, or structural defects, compromising the material's electrocatalytic performance.

Liquid-phase exfoliation, on the other hand, offers a simpler and more scalable route but is limited by its inability to control the lateral size, thickness, and homogeneity of the resulting nanosheets [37]. This leads to batch-to-batch variability and poor production yields. As a result, the resultant material often exhibits inconsistent catalytic activity, limiting its practical deployment. Wet chemical synthesis, while versatile and conducive to functionalization, also faces drawbacks such as the introduction of surface inhomogeneity and impurities during the synthesis process [36]. These defects may degrade the catalytic activity or reduce the chemical and structural stability of the material in harsh electrochemical environments. The need to balance material quality with scalable synthesis has led researchers to explore a broader spectrum of synthetic routes. Across various studies, two-dimensional carbon-based materials have attracted significant attention due to their inherent advantages—high surface area, tunable electronic properties, and abundant active edge sites [38]. These properties make them particularly attractive for catalytic applications like water splitting and oxygen evolution reactions. However, ensuring that these features are retained in mass-produced materials requires precise control over synthesis parameters, including temperature, pressure, chemical precursors, and reaction time.

The push toward non-precious metal-based electrocatalysts, especially those incorporating carbon matrices, has introduced an additional layer of complexity [39]. Performance evaluation standards, including benchmarks for activity, durability, and long-term operation, remain poorly defined, particularly under realistic conditions such as fluctuating temperature and electrolyte concentration [40]. This ambiguity underscores the importance of not just scalable synthesis but also the development of rigorous protocols to ensure reproducibility and performance validation across different labs and production batches. Moreover, the development of two-dimensional transition metal-based materials, such as MOFs and LDHs, poses specific synthesis-related challenges [41]. These materials are structurally complex, often requiring multi-step procedures and precise stoichiometric control. During scale-up, the risk of forming undesired phases or incomplete reactions increases significantly. This may lead to inhomogeneous compositions and deviations in crystal morphology, both of which can adversely affect electrocatalytic activity and selectivity. Therefore, ensuring reproducibility and maintaining thin, defect-free layers over large areas remains an unresolved issue for many of these materials.

Emerging synthetic techniques are showing promise in mitigating these challenges. One notable example is the molten salt synthesis route, which provides a relatively straightforward, scalable, and thermodynamically favorable environment for the formation of layered transition metal nitrides (TMNs) [42]. The molten salt medium facilitated the controlled growth of single-crystal layered materials under mild conditions (Fig. 3). This method not only preserved the intrinsic properties of the 2D TMNs but also enhanced their electrocatalytic activity due to improved crystallinity and minimized structural defects. The use of molten salts also contributed to a unique microstructural arrangement that promoted high conductivity and surface reactivity, thereby making these materials well-suited for HER application.

Such advancements emphasize the role of innovative processing environments in overcoming synthesis-related limitations. The above-mentioned 'molten salt synthesis' is particularly notable for allowing uniform nucleation and growth while avoiding aggregation—a common problem in colloidal or aqueous-based methods [43]. The scalability of this approach, coupled with its relatively low cost and energy requirements, positions it as a strong candidate for future industrial adoption. Alongside this, solution-processed methods for the synthesis of elemental 2D materials have been explored for specific applications like HER and OER [44]. These techniques offer distinct advantages such as low-temperature operation, minimal environmental footprint, and compatibility with roll-to-roll processing [45]. They also allow for the fabrication of materials with customized morphologies and surface functionalities, critical for optimizing specific electrocatalytic reactions. However, these methods often produce metastable phases that require careful stabilization, particularly under prolonged electrochemical cycling.

A related study highlighted the need for a deeper understanding of the synthesis-structure–property relationship in these materials [46]. It was emphasized that tailoring synthesis conditions to suit the final application is crucial; for instance, parameters that yield materials with enhanced HER activity may not be optimal for OER application. Thus, customizing the synthetic route based on the target reaction is a key challenge that demands multidisciplinary collaboration between materials scientists, chemists, and process engineers. Another critical aspect involves the post-synthesis processing and treatment steps, which include exfoliation, doping, annealing, and composite formation [47]. These steps must be designed in tandem

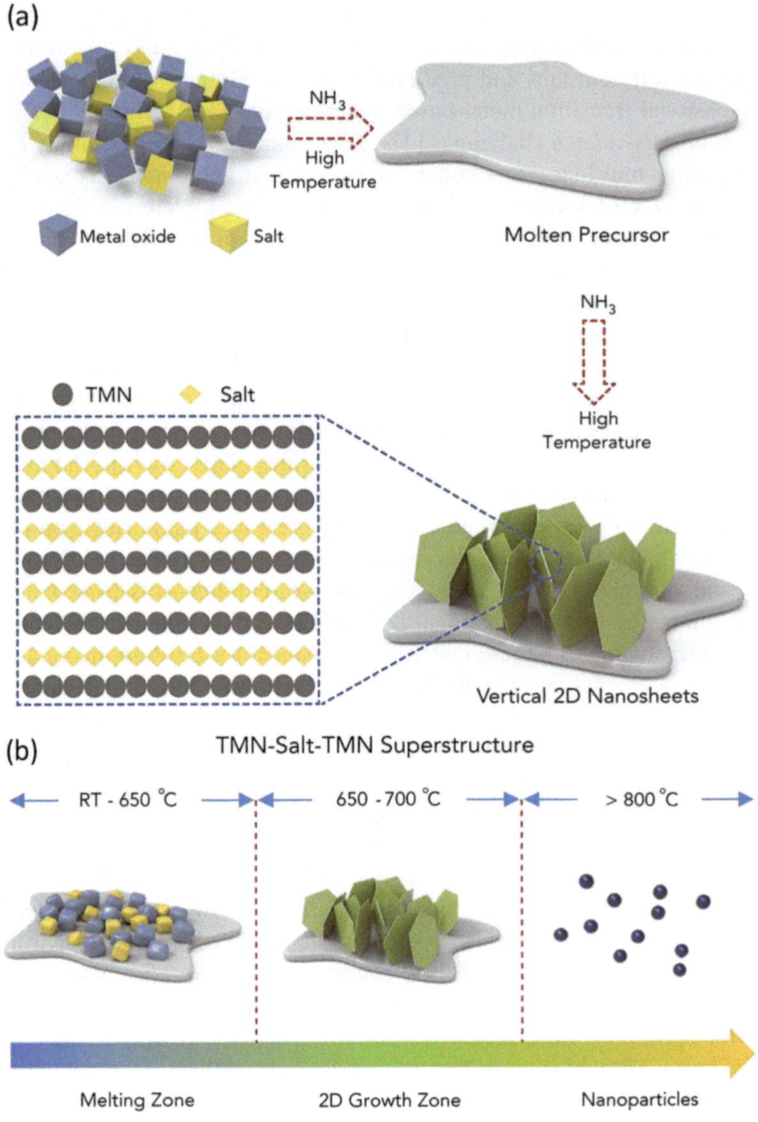

Fig. 3 **a** Schematic representation of synthesis of 2D transition-metal nitrides (TMNs) via the 'Catalytic Molten Salt Method' and **b** their morphology evolution at different temperatures. Adapted with permission from [42]. Copyright (2020) Elsevier

with the primary synthesis to ensure compatibility and retention of desired characteristics. For example, a comprehensive study discussed how metastable 2D materials, although promising in terms of electrocatalytic activity, require finely tuned synthesis conditions to preserve their structure [48]. Moreover, their large-scale production often mandates post-treatment flexibility to accommodate real-world conditions such as variable humidity, temperature fluctuations, and chemical exposure.

The synthesis of 2D material-based electrocatalysts is a complex, multi-variable process that faces significant challenges in scalability. These include ensuring uniformity in morphology and composition, preserving structural integrity during transfer or deposition, achieving desired surface functionalities, and maintaining catalytic performance under operational conditions. The quest for scalable synthesis methods is closely tied to the ongoing development of novel electrochemical systems, and the solutions to these challenges will likely come from an integrated approach combining advanced synthesis, real-time characterization, and in-situ performance evaluation [49]. As research continues to evolve, the emergence of hybrid techniques that blend traditional chemical methods with modern engineering strategies—such as machine learning-assisted synthesis optimization [50] and continuous-flow microreactor technologies [51] may offer the necessary breakthroughs to bridge the gap between laboratory synthesis and industrial application.

2.2 *Batch-to-Batch Variability*

The reliable production of 2D material-based electrocatalysts on a commercial scale necessitates the ability to consistently reproduce materials with identical physicochemical characteristics [52]. However, one of the most persistent and critical challenges in this context is batch-to-batch variability. This phenomenon refers to inconsistencies that arise in the properties of materials synthesized under identical conditions but across different batches. Such variations can include disparities in morphology, lateral size, layer thickness, defect density, surface homogeneity, and crystalline quality. These inconsistencies profoundly affect the electrocatalytic performance of the materials, influencing their activity, selectivity, and long-term operational stability.

Batch-to-batch synthetic variability in 2D electrocatalysts becomes particularly problematic when considering their application in scalable energy conversion technologies, such as hydrogen evolution, and oxygen evolution reactions [53]. The ability of a catalyst to maintain its structural and functional integrity across repeated synthesis attempts is a fundamental requirement for real-world deployment. However, due to the sensitivity of 2D materials to synthesis parameters, even marginal fluctuations can result in significant variations in electrochemical behavior. This variability can be attributed to a range of interconnected factors, the most prominent of which are the synthesis methodologies employed. As stated in the previous section (Sect. 2.1), the methods such as CVD, atomic layer deposition (ALD), hydrothermal synthesis, and liquid-phase exfoliation are among the most commonly used techniques for the preparation of 2D materials [54]. Each of these methods involves a multitude of adjustable parameters, and slight alterations in any of them can lead to pronounced changes in the final product. For instance, in the case of CVD, factors such as substrate temperature, precursor concentration, carrier gas flow rates, and growth time can directly affect the crystalline domain size, film thickness, and defect density of the synthesized material [55]. Variability introduced during CVD processes has been shown to result in MoS_2 films with fluctuating electrocatalytic efficiencies.

Similarly, liquid-phase exfoliation suffers from inconsistent control over flake size and thickness, as well as aggregation phenomena, which can result in a polydisperse suspension of nanosheets [35]. Inconsistencies in solvent choice, sonication time, centrifugation speed, and temperature can further increase batch-to-batch differences. These fluctuations impact not only the material surface area and exposure of active sites but also the electronic conductivity and chemical reactivity of the 2D layers [56]. The purity and consistency of chemical precursors are also critical contributors to batch variability. Even trace amounts of impurities in starting materials can lead to unintended doping, the formation of secondary phases, or alteration of growth dynamics. The importance of precursor purity is especially notable in vapor-phase syntheses, where impurities can influence nucleation kinetics and the eventual stoichiometry of the resulting material. Studies have highlighted that simpler and purer precursors tend to yield cleaner and more uniform 2D layered transition metal chalcogenides, suggesting that a standardized precursor supply chain is vital for minimizing batch inconsistencies (Fig. 4) [57]. In addition, environmental factors, such as ambient humidity, temperature, and exposure to air or moisture during and after synthesis, play non-trivial roles in determining the final characteristics of 2D electrocatalysts [58]. For instance, synthesis carried out under varying humidity levels can lead to differences in interlayer spacing or hydration levels of 2D materials, which directly affect their electrocatalytic performance [59]. Moreover, post-synthesis treatments such as annealing, drying, or chemical functionalization, if not standardized, can introduce further disparities between batches. Temperature gradients across the reaction chamber or inconsistencies in heat transfer can influence the uniformity of thermal treatments, impacting crystallinity and defect formation [60].

Another important dimension is the scale of synthesis. As the production of 2D materials moves from milligram-scale laboratory experiments to gram- or kilogram-scale manufacturing, the ability to maintain consistent quality becomes increasingly difficult [29]. Scale-up often introduces challenges such as non-uniform precursor distribution, inefficient heat or mass transfer, and inhomogeneous reaction environments, all of which can amplify batch-to-batch variability [61]. For example, increasing the batch size in liquid exfoliation without simultaneously optimizing sonication energy or solvent-to-material ratios may lead to inadequate exfoliation or restacking of layers, thus degrading the quality of the material. The electrocatalytic implications of batch variability are far-reaching. Differences in surface area and defect density alter the density and accessibility of active sites, affecting turnover frequency (TOF) and Tafel slopes [62]. Variations in conductivity, stemming from

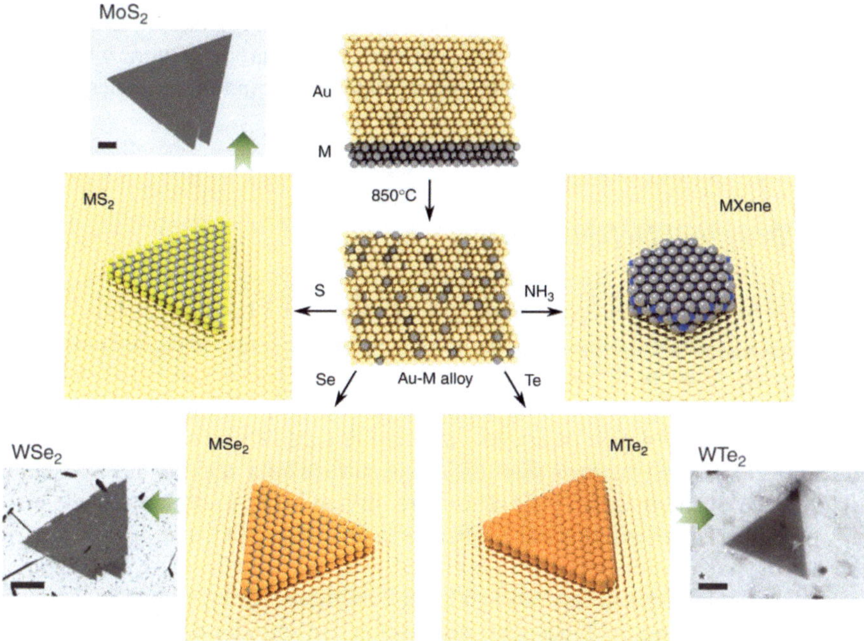

Fig. 4 Schematic representation of synthesis of 2D layered transition metal chalcogenides, where *M* and *X* vary from Mo, W, Cr, Fe, Co, Hf, Nb, V and S, Se, Te, respectively. Adapted with permission from [57]. Copyright (2019). Distributed under a Creative Commons Attribution License 4.0 (CC BY) https://creativecommons.org/licenses/by/4.0/

inconsistent crystal quality or doping levels, can hinder charge transfer during electrocatalysis. This, in turn, affects the measured overpotential, current density, and overall energy efficiency of the electrocatalytic process. In durability tests, non-uniformities in structural integrity across batches can result in drastically different stability profiles under prolonged electrochemical operation.

To address these challenges, considerable effort is being directed toward the standardization and optimization of synthesis protocols. In CVD processes, advancements have been made in the control of gas-phase precursor delivery systems and thermal profiles across substrates, facilitating more uniform film growth with reproducible properties [55]. Techniques involving in-situ monitoring of film growth via spectroscopy or mass spectrometry have emerged as valuable tools to dynamically adjust synthesis parameters in real-time, ensuring consistent material quality across batches [63]. Similarly, the development of automated synthesis platforms, where environmental parameters are tightly controlled and monitored, is playing a pivotal role in improving reproducibility [64].

The issue of batch-to-batch variability remains one of the most formidable barriers in the scale-up of 2D material-based electrocatalysts. It affects nearly every facet of catalyst performance, from initial activity to long-term durability. Tackling this challenge requires a multi-pronged approach encompassing precise control of synthesis conditions, high-purity precursor selection, environmental regulation, and real-time monitoring technologies. As progress continues in this area, a future wherein 2D electrocatalysts are produced reliably and reproducibly at industrial scales becomes increasingly tangible.

2.3 Purity and Defect Control

Achieving high purity and controlling defects are vital for optimal catalytic activity. However, large-scale synthesis often introduces impurities and structural defects that can adversely affect performance [65]. As research on 2D material-based electrocatalysts accelerates, it becomes increasingly clear that the purity and defect control in their synthesis are not just desirable—they are critical. These two aspects directly affect the electronic structure, stability, active surface area, and the electrocatalytic performance of the material [62]. Therefore, in any effort to scale up these materials for commercial or industrial applications, an uncompromising focus on purity and defect management must be prioritized. The structural integrity of 2D electrocatalysts is highly sensitive to synthesis conditions. Even minor deviations in temperature, pressure, or precursor composition can introduce unintended defects or impurities. These changes may seem negligible, but they can have pronounced effects on catalytic efficiency, durability, and overall functionality. Consequently, advanced control strategies are required during synthesis, often pushing the limits of currently available laboratory or industrial techniques.

Interestingly, while defects are often considered detrimental, strategic introduction of certain types of defects—commonly referred to as *defect engineering* (this aspect would be detailed more in next section; Sect. 3.3)—can actually enhance the catalytic activity of 2D materials [66]. In controlled scenarios, these defects create additional active sites and can tune the electronic properties of the electrocatalyst. For instance, defects introduced in Fe-doped NiS/MoS_2 heterostructures have been shown to significantly increase the electrochemical surface area (ECSA) and activate previously inert unsaturated metal sites, improving overall water splitting performance [10]. This reveals a compelling duality: defects must be controlled, not merely eliminated. The move toward intentional defect creation as a design principle rather than an unfortunate byproduct of synthesis is gaining traction. Controlled defect creation in carbon-based materials has been highlighted as a key contributor to improved electrocatalytic reactions [67]. Specific vacancies or heteroatom doping in these structures generate favorable electronic configurations that enhance their reactivity. However, this strategy demands extreme precision. Uncontrolled or excessive defects can lead to poor electrical conductivity, mechanical fragility, and a decline in long-term performance. Hence, defect density must be optimized—not maximized.

Purity plays an equally pivotal role. Impurities—whether from the precursors, or ambient environment—can act as electronic traps, alter surface chemistry, or compete with the active sites for reactants, thereby diminishing electrocatalytic efficiency [68]. Non-precious metal-based carbon electrocatalysts are particularly susceptible to such degradation [69]. The use of impure precursors introduces foreign atoms that alter the surface energy and adsorption properties of the catalyst. This vulnerability emphasizes the necessity of high-quality precursor materials and tightly controlled synthetic conditions. To address these challenges, some synthesis methodologies have adopted high-purity gel-like precursors and extremely controlled atmospheres, minimizing contamination throughout the process [70]. Such methods offer the benefit of enhanced environmental control during formation and growth stages, resulting in materials that exhibit both higher purity and improved electrocatalytic properties [71]. While these strategies are promising at the laboratory scale, translating them to industrial-scale production presents a different set of challenges, including cost, complexity, and energy consumption.

The choice of synthesis method is a critical determinant of both the purity and defect landscape of the final material [68]. Among the most explored techniques are hydrothermal synthesis, co-precipitation, and CVD. These methods offer varying degrees of control, scalability, and efficiency. Of these, CVD stands out for its ability to produce high-quality 2D materials with relatively uniform defect distribution and minimal contamination [72]. Recent advancements in CVD have focused on fine-tuning precursor flow rates, pressure, and substrate temperature to modulate both impurity levels and defect characteristics. Modifying parameters such as temperature gradients and gas flow rates in CVD systems has demonstrated the potential to not only reduce impurity content but also control the nature and distribution of defects. For example, it has been shown that adjusting the precursor concentration can influence the nucleation rate, thereby modulating the density of edge sites or grain boundaries—features that directly correlate with electrocatalytic activity [73]. However, the scalability of such finely tuned processes remains a significant hurdle, often requiring expensive equipment and precise environmental controls that are not easily replicated in bulk production settings.

Emerging computational techniques are beginning to play a pivotal role in overcoming some of these synthesis challenges [74]. High-throughput computational screening and machine learning algorithms are being employed to predict optimal synthesis conditions for targeted electrocatalytic behaviors [75]. These tools can rapidly analyze vast datasets to identify key parameters that influence defect type, concentration, and spatial distribution, as well as impurity mitigation strategies [76]. By incorporating these predictive models into experimental workflows, researchers can reduce the trial-and-error burden and move more efficiently toward optimal synthesis protocols. Such integration of data-driven approaches with experimental chemistry represents a powerful avenue for the future of electrocatalyst design [77]. For example, simulations have been used to predict how certain defect geometries in 2D materials influence adsorption energies for hydrogen or oxygen intermediates, which are crucial for reactions such as HER and OER. These insights help to rationally guide defect engineering strategies, ensuring that only beneficial imperfections are introduced during synthesis. Nonetheless, a persistent issue remains in translating lab-scale successes into commercially viable technologies. While researchers have achieved remarkable feats in small-scale synthesis—yielding materials with exceptional activity, selectivity, and durability—the methods employed are often resource-intensive, time-consuming, or environmentally unsustainable. Scaling up without compromising on purity or defect control requires innovations not only in chemical processing but also in reactor design, process automation, and real-time quality monitoring [78]. One proposed solution is the modularization of synthesis systems, allowing for the parallel production of 2D electrocatalysts under controlled micro-environments [79]. Such modular systems could maintain high precision while improving throughput, but they require significant investment in both infrastructure and operational expertise. Additionally, standardization across laboratories and production facilities will be essential to ensure reproducibility—a cornerstone of any scalable technology.

The challenges of scaling up 2D material-based electrocatalysts are deeply intertwined with the dual imperatives of purity and defect control. While defect engineering can be a powerful tool for performance enhancement, its benefits can only be realized if implemented with precision. Similarly, high-purity synthesis is non-negotiable for ensuring reproducibility and stability. The road to commercialization is paved with technical hurdles, but the convergence of advanced synthesis techniques, computational modeling, and system-level engineering offers a roadmap. Continuous innovation in these domains will be essential for transitioning 2D electrocatalysts from the laboratory bench to real-world energy conversion technologies.

3 Structural and Surface Engineering Limitations

3.1 Surface Area and Active Site Availability

2D material-based electrocatalysts have garnered significant attention due to their unique physicochemical characteristics, which include large surface areas, tunable electronic structures, and abundant exposed active sites [80, 81]. These features collectively contribute to their exceptional performance in various electrochemical applications, particularly in renewable energy conversion systems such as water splitting. However, one of the primary challenges faced during the scale-up of these materials lies in maintaining their structural and surface properties, especially the availability of surface area and active sites [81].

During large-scale synthesis or integration into device architectures, 2D materials tend to undergo aggregation, restacking, or morphological transformations [82]. Such alterations reduce the effective ECSA, directly impacting the number of accessible active sites and consequently diminishing catalytic performance. Ensuring the preservation of these features is therefore crucial in upscaling processes without compromising functionality [83]. ECSA serves as a fundamental parameter for assessing the actual catalytic area involved in electrochemical reactions. On the other hand, architectural optimization through structural manipulation also plays a significant role in maximizing active site exposure [84]. The incorporation of single-atom catalysts (SACs) within 2D materials represents a transformative strategy to maximize atomic utilization and prevent metal atom agglomeration [85]. SACs are uniformly dispersed atoms anchored on the surface of 2D substrates, which maintain catalytic activity without sacrificing stability. These configurations facilitate uniform and stable arrays of active sites that enhance selectivity and reactivity [86]. Furthermore, when embedded within 2D matrices, the coordination environment around the single atoms can be precisely modulated using heteroatom doping or functional groups, enhancing the binding strength between the catalyst surface and intermediates [87]. This synergy enables tuning of both the geometric and electronic properties of the electrocatalyst, ultimately improving catalytic kinetics and thermodynamics.

Such fine-tuning of the active site environment is instrumental in modulating reaction pathways. For example, by introducing heteroatoms like nitrogen or sulfur into the lattice of 2D materials (this aspect would be detailed more in next section; Sect. 3.2), the local charge distribution and d-band center of the metal sites can be adjusted [88]. This adjustment enables optimal adsorption energies of key intermediates, improving the overall reaction rates. These strategies have been particularly impactful in designing high-efficiency electrocatalysts for reactions such as OER and HER [80]. The intrinsic structure of 2D materials—characterized by ultrathin layers and high aspect ratios—facilitates a large number of exposed active sites. The planar architecture ensures that a majority of atoms are located on or near the surface, making them accessible for electrocatalytic interactions. This high degree of exposure, combined with the anisotropic electron transport properties of 2D materials, enables rapid charge transfer to the reaction interface, which is critical for reactions requiring fast electron exchange, such as those in electrochemical energy devices [89].

However, traditional synthesis routes often fail to preserve or enhance these favorable attributes at scale. Solvothermal methods, liquid-phase exfoliation, and CVD may introduce structural inconsistencies or uncontrollable morphologies that hinder the uniform exposure of active sites [90]. Moreover, the synthesis of MOFs and related coordination polymers into stable 2D sheets often suffers from incomplete exfoliation or stacking, reducing surface area and active site accessibility [91]. These issues necessitate the development of innovative fabrication approaches that promote stability, dispersion, and uniformity of catalytically active centers. Recent advances include the use of template-assisted growth, space-confinement strategies, and surfactant-mediated exfoliation, which offer better control over material morphology [92]. For instance, ammonia-assisted synthesis of bimetallic metal–organic framework precursors has led to the development of 2D nanosheet structures of 2D-MCo_3O_4-based material with enhanced surface area and stability [93] (Fig. 5). These materials outperform traditional precious metal-based catalysts in critical reactions such as OER due to their bifunctional nature and efficient charge transfer properties.

Another effective strategy involves the formation of heterostructures, where two or more materials are combined to create synergistic interfaces that boost electron transfer and catalytic activity [12]. These heterostructures can be engineered to facilitate electron localization or improve adsorption energies of reactive species [94]. For example, oxygen-rich cobalt-based 2D structures have demonstrated exceptional

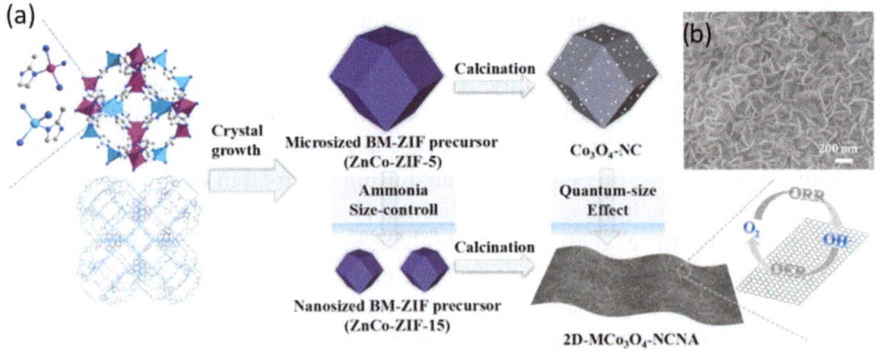

Fig. 5 **a** Schematic representation of synthesis 2D-MCo_3O_4-NCNAs via size-controlled technique from the zinc/cobalt containing ZIF precursors. **b** SEM image of 2D-MCo_3O_4-Nitrogen-doped Carbon Nanosheet Array (NCNA)-based material. Adapted with permission from [93]. Copyright (2018). John Wiley & Sons, Inc.

performance in bifunctional electrocatalysis [95] attributed to the optimized distribution of surface active sites and enhanced electrical conductivity. Electrocatalytic processes also benefit significantly from the thinness and exposed architecture of 2D materials. Their flexibility and mechanical robustness enable them to be integrated onto various substrates, including flexible and wearable electronics [96]. The layered structure not only maximizes surface reactivity but also assists in minimizing bulk diffusion barriers, thereby accelerating reaction kinetics [40]. Despite these advancements, certain persistent challenges hinder the large-scale application of 2D electrocatalysts. These include limited production yield, lack of reproducibility in defect and dopant distribution, and structural collapse during operational cycles. The aggregation and restacking of nanosheets during drying, deposition, or storage also significantly reduce their effective surface area. Addressing these issues requires a holistic approach that combines material synthesis, structural design, and device engineering.

The surface area and active site availability are two of the most critical parameters influencing the performance of 2D material-based electrocatalysts. The design strategies aimed at optimizing these parameters—such as SAC and heteroatom incorporation, and heterostructure formation—have shown great promise at the laboratory scale. However, ensuring that these features are retained or enhanced during scale-up is a complex challenge that demands innovative engineering solutions and deeper mechanistic understanding. Ongoing research focused on synthesis optimization, stability control, and interfacial tuning [97] is expected to bridge the gap between laboratory-scale breakthroughs and real-world deployment of 2D electrocatalysts in sustainable energy technologies.

3.2 Doping and Functionalization

Doping and functionalization [98] are two of the most critical approaches for enhancing the electrocatalytic properties of 2D material-based electrocatalysts. These processes aim to improve structural and electronic characteristics, increase the number and activity of catalytic sites, and improve long-term operational stability. However, while these methods have yielded significant advances at laboratory scales, their translation to industrial-scale synthesis presents considerable challenges.

One of the foremost difficulties is the uniformity of doping [99]. Uniform doping of heteroatoms into 2D materials is an essential requirement for reproducible and high-performing electrocatalysts. Nevertheless, achieving this consistently across large areas is highly complex. The spatial variation in dopant concentration often leads to a heterogeneous distribution of active sites, resulting in unpredictable or diminished catalytic performance [100]. This non-uniformity in doping not only impacts the material electronic properties but also introduces local strain and disorder within the lattice, affecting both reactivity and stability. Doping involves introducing specific atoms—often heteroatoms such as nitrogen, phosphorus, sulfur, or transition metals—into the atomic lattice of 2D materials [101]. It changes their electronic structure and can significantly enhance their catalytic behavior. For instance, nitrogen-doped carbon materials have been widely investigated due to the ability of nitrogen atoms to modify charge distribution and create high-density active sites suitable for OER and HER [102]. Nitrogen atoms introduce lone-pair electrons into the carbon matrix, improving electrical conductivity and facilitating electron transfer during electrocatalytic reactions.

The impact of nitrogen doping on the electronic and catalytic behavior of 2D materials has been systematically explored. For example, increased nitrogen content in carbon-based nanosheets enhances electrocatalytic activity due to improved electronic conductivity and a higher density of active sites [103]. These nitrogen-doped materials are often synthesized using precursors such as glucose and amide compounds, which allow for the development of oxygen-containing nitrogen-doped carbon nanosheets. The resulting materials show excellent electrochemical properties and offer a cost-effective alternative to platinum-based catalysts. Functionalization, in parallel, involves modifying the surfaces of electrocatalysts with specific chemical groups or atoms [104]. This technique aims to alter the surface chemistry to improve reactivity, selectivity, and interaction with reactants. Functionalization often enhances the adsorption behavior of reactive species and stabilizes intermediates, leading to increased reaction efficiency [105]. For example, the molybdenum carbide (Mo_2C)-based catalysts have been recognized for their outstanding performance in HER under acidic conditions. However, their limited activity in alkaline environments is often attributed to insufficient active sites and poor conductivity. A novel composite, Mo_2C embedded in Nitrogen-doped Porous Carbon nanosheets (Mo_2C@2D-NPCs), has demonstrated superior activity in alkaline HER, showing competitive energy efficiency with platinum-based benchmarks [106] (Fig. 6a). The molybdenum carbide embedded in nitrogen-doped porous carbon nanosheets with surface functional groups enhances HER activity by optimizing the binding energy of hydrogen intermediates. Another notable instance involves the fine-tuning of electronic properties through strategic surface modifications [104]. Functional groups such as –OH, $-NH_2$, or –COOH introduced during synthesis or post-treatment processes help modulate electron density on electrocatalytic centers. This modulation can significantly impact performance in applications like OER by facilitating charge transport and tuning the adsorption energy of reaction intermediates.

Despite these promising outcomes, there are inherent limitations in doping and functionalization strategies that challenge scalability. A primary concern is the lack

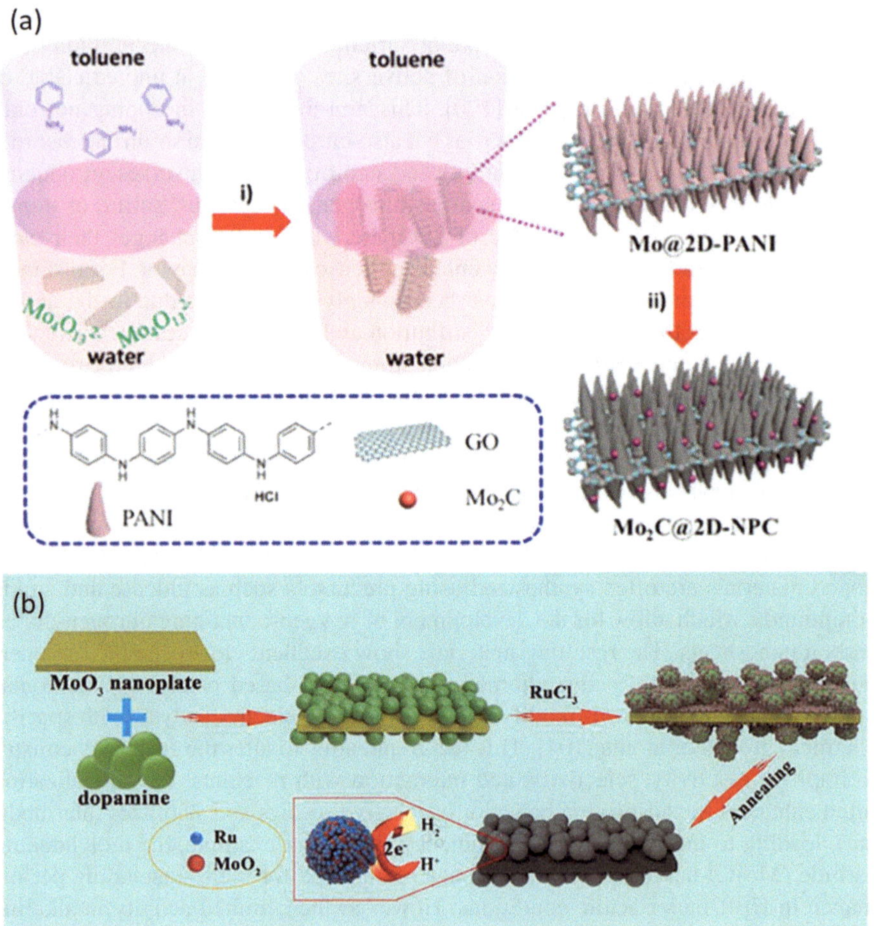

Fig. 6 **a** Synthesis of (**a**) Mo_2C/2D nanoporous carbon-based material. Adapted with permission [106]. Copyright (2017). American Chemical Society. **b** Preparation of 2D MoO_2 nanosheets with Ru nanoparticles. Adapted with permission [111]. Copyright (2020). American Chemical Society

of reproducibility due to variations in precursor materials and synthesis parameters. For example, differences in carbon or nitrogen sources can result in significant variations in morphology and surface area, leading to inconsistent electrocatalytic behavior [107]. The dependency on complex synthetic environments makes scale-up particularly difficult, as even slight deviations in temperature, pH, or precursor concentration can lead to markedly different material properties. Additionally, non-uniform distribution of dopants is a persistent problem. In many cases, the random placement of dopant atoms results in patches of high and low activity on the surface, which can lead to erratic reaction kinetics [108]. This is especially problematic for large-area electrodes needed for industrial applications, where uniform catalytic activity is essential for reliable and scalable device performance. Further, maintaining structural integrity during doping is another critical challenge. While doping can create new active sites, it may also introduce defects or structural distortions that degrade mechanical strength and reduce durability under operational stress [109]. Uncontrolled thermal treatments used to activate dopants often cause lattice disruptions or collapse porous frameworks, ultimately compromising long-term stability. For example, although intrinsic defects such as zigzag edges and pentagonal carbon arrangements can improve electrocatalytic activity, excessive or uncontrolled defect formation can lead to degradation over time [110].

Doping and functionalization also provide an opportunity to engineer synergistic interactions between active sites and reactants. For example, a hybrid electrocatalyst—2D-MoO_2/Ru—was developed using a straightforward and scalable method by the incorporation of ruthenium into MoO_2 nanosheets (Fig. 6b) [111]. This catalyst exhibited remarkable hydrogen evolution activity across a wide pH range, showcasing low overpotentials and exceptional durability by improving electronic coupling with hydrogen intermediates and maximizing active site exposure. These properties make it a compelling alternative to conventional Pt-based electrocatalysts. However, such engineered systems often require precise control over doping levels, spatial arrangement of active sites, and chemical composition, which are difficult to achieve at scale. Moreover, the thermal and chemical stability of doped or functionalized materials under real-world operational conditions, such as variable pH and temperature, remains a critical concern [112]. Many doped materials exhibit excellent performance in controlled laboratory conditions but degrade quickly in practical applications due to leaching of dopants, restructuring of active sites, or surface oxidation.

The doping and functionalization are vital strategies for addressing the structural and surface engineering limitations of 2D material-based electrocatalysts. These approaches have enabled significant advances in activity, selectivity, and stability of electrocatalytic systems. However, scaling these techniques beyond the laboratory remains challenging, primarily due to difficulties in achieving uniformity, maintaining structural integrity, and ensuring long-term durability. Advances in synthesis methodologies, in-situ characterization, and theoretical modeling will be crucial for overcoming these barriers. As the field progresses, scalable synthesis protocols and real-time quality control techniques will likely play a pivotal role in translating promising 2D electrocatalyst designs into practical energy conversion devices.

3.3 Edge Effects and Defect Engineering

Edge effects and defect engineering are critical considerations in the structural and surface engineering of two-dimensional (2D) material-based electrocatalysts [113]. These elements govern the intrinsic and extrinsic electrocatalytic properties that determine the performance of such materials in various electrochemical processes [114]. The catalytic efficiency of 2D materials, particularly in electrocatalysis for sustainable energy technologies such as water splitting, is often dictated by their edge sites and the presence of engineered or intrinsic defects. The edge atoms of 2D materials usually possess unsaturated coordination, which results in higher reactivity compared to their basal plane counterparts. Consequently, a major share of electrocatalytic activity in 2D materials arises from these edges, making their consistent engineering at scale a pivotal challenge.

Reproducing edge sites and defects in a uniform and scalable fashion is a complex task due to their stochastic nature and dependence on the synthesis method [115]. In lab-scale studies, meticulous chemical treatments and post-synthesis modifications can generate high-performance catalysts; however, these are difficult to translate to industrial-level processes. Maintaining uniformity across large production volumes without compromising performance remains one of the bottlenecks in the commercial application of these materials. One of the most studied cases is molybdenum disulfide (MoS_2), where it has been demonstrated that the edge sites exhibit significantly higher catalytic activity than the inert basal planes [116]. The edges of MoS_2 offer more favorable binding with reactants and intermediates, thereby accelerating reactions such as the HER. For example, a simple yet effective method was developed to create cost-effective and efficient electrocatalysts using iron (Fe) doped cobalt hydroxide ($Co(OH)_2$) [117] (Fig. 7). The introduction of Fe into $Co(OH)_2$ enhanced the structural features and electrocatalytic efficiency of the material for water splitting, particularly HER and OER processes. This composite electrocatalyst demonstrated robust performance and high durability, rendering it a viable candidate for large-scale hydrogen production applications.

Edge sites are generally less coordinated and thus more reactive, particularly in binding intermediates of redox reactions. The geometric and electronic structure at these edges leads to altered orbital distributions, which favor the adsorption of key species involved in electrocatalytic cycles. The capability to harness these effects is essential to optimizing reaction kinetics. For instance, the electrocatalysts with enhanced edge structures show superior performance due to the increased reactivity of their active sites and the optimized electron transfer mechanisms inherent to these structural motifs [118]. A further advancement in edge engineering was demonstrated with the creation of hierarchical 3D architectures composed of 2D layers of $MoSe_2$ and reduced graphene oxide (rGO) [119]. This strategy capitalized on the synergetic combination of high density of active edges of $MoSe_2$ with the excellent conductivity of rGO. The intercalation of rGO with $MoSe_2$ enabled an enhancement in both the number of exposed edge sites and the charge transfer properties. The improved electronic coupling at the $MoSe_2$/rGO interface facilitated lower overpotentials and

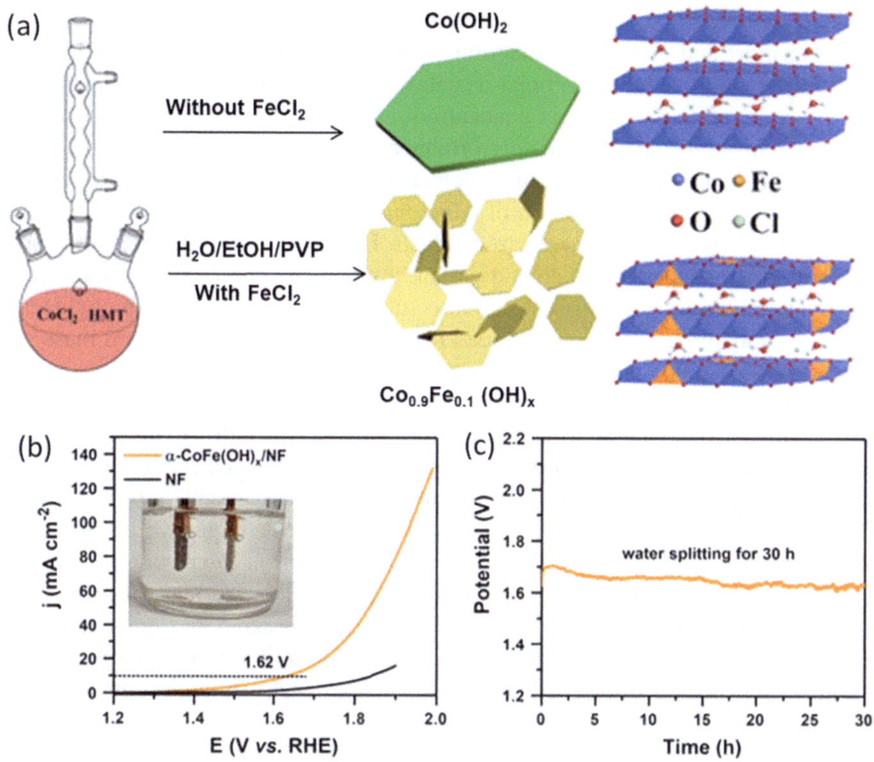

Fig. 7 **a** Schematic representation of synthesis of $Co(OH)_2$ and $Co_{0.9}Fe_{0.1}(OH)_x$ nanosheets. **b** Polarization and **c** chronopotentiometric curves of $Co_{0.9}Fe_{0.1}(OH)_x$ nanosheets. Adapted with permission from [117]. Copyright (2019). John Wiley & Sons, Inc.

higher current densities, making this hybrid structure especially suitable for high-efficiency water splitting systems. Such structures are not only efficient but also economically viable for scaling due to the reduced reliance on precious metals and the use of cost-effective carbon-based frameworks.

Beyond edge effects, defect engineering serves as an equally powerful approach to boost the performance of 2D electrocatalysts [120]. Defects, such as atomic vacancies, dislocations, heteroatom doping, and grain boundaries, introduce new active sites and can significantly modify the electronic structure of 2D materials [121]. These alterations often lead to improved charge mobility, enhanced adsorption of reactants, and facilitated electrocatalytic turnover. Intrinsic carbon defects, including zigzag edges and pentagonal ring distortions, have been shown to enhance activity, as they enable superior interaction with the intermediates [122]. The increase in electron density around these defects promotes more efficient electron transfer, which is vital for electrocatalytic processes. Further investigations revealed that specific types of defects contribute to increased catalytic activity while also providing more efficient charge transport pathways. In a study involving carbon-based materials, it was demonstrated that structural modifications such as vacancies led to improved mass transport and electron conduction within the catalyst network [123]. These structural changes, while beneficial to catalytic performance, must be introduced with precision. Excessive defect concentration can induce instability in the framework, which ultimately degrades performance and longevity. Therefore, achieving an optimal balance between beneficial defect density and structural stability remains a key challenge. The combination of edge engineering and defect control is especially promising for scaling up electrocatalysts. Hybrid materials that integrate these strategies often outperform their individual components due to synergistic effects [124]. The creation of multi-scale architectures that incorporate both engineered edge sites and strategic defects can lead to high surface areas, improved mass and charge transport, and enhanced electrochemical performance.

Despite these advantages, several hurdles remain in the translation of these lab-scale strategies to industrial-scale production. The reproducibility of defect and edge features on a large scale is limited by the complexity of synthesis routes and the inherent variability of bottom-up or top-down fabrication methods [66]. Furthermore, the characterization of edge sites and defects at scale requires advanced techniques, such as aberration-corrected transmission electron microscopy [125] and synchrotron-based X-ray absorption spectroscopy, which are often inaccessible for routine industrial quality control [126]. Both edge effects and defect engineering are indispensable in the structural and surface optimization of 2D material-based electrocatalysts. By systematically enhancing active site density through controlled edge configurations and defect structures, it is possible to substantially boost catalytic efficiencies for various energy conversion reactions. However, challenges in the scalable and reproducible synthesis of such features necessitate the development of robust, cost-effective, and scalable fabrication techniques. The integration of edge and defect engineering strategies presents a pathway toward the next generation of high-performance, durable, and scalable electrocatalysts for green energy technologies.

4 Future Perspectives and Possible Solutions

A promising direction for overcoming the existing scale-up challenges in 2D material-based electrocatalysts lies in the advancement of novel synthesis strategies. Current research demonstrates that it is possible to fabricate high-purity 2D nanomaterials under mild reaction environments, which helps retain their delicate structural and electronic properties. Avoiding harsh conditions such as extreme temperatures or corrosive chemical environments is critical to preserving the inherent catalytic properties of these materials. Such mild synthesis approaches ensure that the crystallinity, surface area, and active sites of the 2D nanosheets remain intact, which is crucial for sustained electrocatalytic activity during large-scale production.

Innovative methodologies, including self-assembly, oriented attachment, and nanoscale-confined synthesis, have enabled better control over the growth process of 2D materials. These techniques allow for precision in designing defect structures, edges, and surface functional groups, all of which are important parameters governing electrocatalytic behavior. For instance, regulating defect density can tune the local electronic structure, enhance the adsorption of key intermediates, and ultimately improve the reaction kinetics. Such strategies not only maximize the intrinsic activity of 2D catalysts but also ensure reproducibility, which is a foundational requirement for commercial deployment. Moreover, the adoption of these techniques allows for the tailoring of materials that can be reliably synthesized in batch or continuous production systems. In particular, these methods can produce large quantities of uniform nanosheets with well-defined properties, enabling the fabrication of electrode materials at scales relevant to industrial applications. As a result, advanced synthesis strategies are laying the groundwork for bridging the gap between laboratory innovation and commercial scalability. Parallel to synthetic advancements, the integration of 2D materials with other nanostructures to form hybrid or composite materials has emerged as a key strategy for overcoming both performance and scalability bottlenecks. Hybrid materials leverage the distinct properties of each component, creating synergistic systems that outperform their individual constituents. For example, combining 2D materials with conductive substrates or semiconducting components can lead to a significant enhancement in overall catalytic efficiency. The introduction of such secondary phases into 2D frameworks extends the number of catalytically active sites and improves charge transport pathways.

These composites often exhibit greater mechanical robustness and chemical stability, making them more suitable for harsh operational environments. They can also address issues related to the degradation of pure 2D materials over prolonged use. By forming strong chemical or physical bonds between different components, hybrid structures enhance stability and durability without sacrificing catalytic performance. This synergy is especially valuable in electrocatalytic systems where long-term stability under operational voltages and chemical environments is a key determinant of feasibility. Additionally, the rational design of interfaces in hybrid systems—particularly at the atomic and molecular levels—plays a vital role in enhancing electrocatalytic activity. Optimized heterostructures with engineered interfaces can facilitate rapid interfacial charge transfer, decrease overpotential, and improve the kinetics of key electrochemical reactions. The careful modulation of electronic structures at the interface allows for better orbital overlap and energy alignment, which improves the efficiency of charge transport and reaction intermediates adsorption. Two-dimensional heterostructures, especially those combining materials with complementary properties, present a unique platform for the next generation of electrocatalysts. These architectures allow for the controlled tuning of properties such as bandgap, conductivity, and catalytic site exposure. The creation of interfaces with high interfacial charge mobility and strong electronic coupling can significantly accelerate reaction kinetics. For example, multilayer heterostructures with ordered stacking and strategic interlayer bonding provide a robust platform for facilitating electrochemical reactions while maintaining mechanical and thermal integrity.

Incorporation of bridging ligands or interlayer linkers that provide electronic pathways across different domains further refines these materials. Such ligands modulate the charge distribution and electronic density at the junction, influencing the adsorption–desorption behavior of reactants and products. These engineered features enable not only enhanced activity but also selective catalysis, which is highly desirable for targeted electrochemical transformations like OER, and HER. Despite the progress in laboratory-scale innovation, the transition to commercial-scale production remains a formidable challenge. Most of the synthesis methods currently in use are limited by low throughput, high cost, and the need for specialized equipment. Therefore, scalable production strategies such as roll-to-roll manufacturing and large-area coating techniques are critical for industrial adoption.

Roll-to-roll processing, in particular, offers a promising route for producing continuous films of 2D materials or hybrid architectures on flexible substrates. This technique enables uniform deposition over large areas, making it well-suited for integrating electrocatalysts into electrochemical devices like water-splitting systems. With appropriate process control, roll-to-roll systems can ensure high reproducibility in film thickness, composition, and catalytic performance. This level of control is necessary for meeting the quality standards required in energy applications. Moreover, scalable coating techniques that utilize flexible substrates are being developed to support the deposition of 2D and hybrid materials. These include inkjet printing, spray coating, dip coating, and spin coating, each offering advantages in terms of scalability, material utilization, and adaptability to different substrate types. Importantly, these methods must be optimized to maintain the delicate structural features and surface properties that govern catalytic activity. The integration of three-dimensional (3D) architectures with 2D materials further enhances the scalability and practical applicability of these electrocatalysts. 3D-structured supports, such as foams or porous scaffolds, provide a mechanically stable platform for anchoring 2D nanosheets. This not only increases the active surface area but also facilitates mass transport and electrolyte accessibility, which are vital for efficient electrochemical reactions. When combined with scalable synthesis and deposition techniques, such hybrid structures become strong candidates for deployment in commercial devices.

In practical terms, the commercialization of 2D material-based electrocatalysts also hinges on consistent performance metrics across large batches. Quality assurance protocols must be implemented to assess structural integrity, surface area, catalytic activity, and stability over time. The adoption of standardized testing conditions and benchmarking frameworks will aid in comparing materials across research groups and industrial entities. It is equally important to consider the economic and environmental aspects of scaling up. The cost of raw materials, energy consumption during synthesis, and the use of environmentally benign solvents and precursors must be optimized. Sustainable manufacturing practices, including the reuse of reagents and recycling of byproducts, will ensure that large-scale production aligns with green chemistry principles.

The future of 2D material-based electrocatalysts in real-world applications is promising but requires a concerted effort in innovation, engineering, and industrial translation. By advancing mild and controllable synthesis techniques, engineering hybrid interfaces, and adopting scalable manufacturing approaches, it is possible to overcome the current limitations. With these strategies, the vision of deploying

2D materials in next-generation electrochemical energy systems can move closer to reality, bridging the gap between laboratory success and commercial viability.

5 Conclusion

The industrial deployment of 2D material-based electrocatalysts remains a formidable challenge despite the substantial progress made at the laboratory scale. The transition from bench-top synthesis to large-scale production is restricted by several critical factors, including synthesis scalability, structural and morphological consistency, and integration into functional electrochemical systems, long-term operational stability, and sustainability in terms of environmental and economic costs.

One of the major obstacles is the difficulty in achieving synthesis scalability while preserving the intrinsic properties of 2D materials, such as high surface area, tunable electronic properties, and active edge sites. Laboratory techniques like CVD, mechanical exfoliation, and solvothermal synthesis are typically designed for producing small quantities of material with precise control over size, thickness, and morphology. However, when these methods are scaled up, maintaining uniformity, defect control, and phase purity becomes increasingly complex. Inconsistencies in structural quality at scale can significantly impact catalytic performance, thereby reducing reliability in commercial applications.

Another pressing challenge is the engineering of 2D material structures to ensure they remain stable and active under industrial electrochemical operating conditions. In real-world applications, 2D catalysts must withstand extreme pH levels, elevated temperatures, fluctuating current densities, and prolonged reaction times. Unfortunately, many 2D materials are prone to degradation or delamination, which adversely affects their catalytic efficiency and lifecycle. Additionally, issues such as restacking of nanosheets, poor conductivity, and insufficient active site exposure further hinder their large-scale implementation. The integration of 2D materials into functional devices such as electrolyzers also presents significant hurdles. Techniques are required for uniform deposition on electrodes, maintaining intimate contact with conductive substrates, and ensuring mechanical robustness. Moreover, compatibility with current manufacturing infrastructure remains limited, leading to increased production costs and barriers to adoption. Another layer of complexity arises from the need to balance catalytic activity with selectivity, durability, and cost —that is difficult to optimize simultaneously at scale.

Environmental and regulatory concerns also play a vital role. The production of many 2D materials involves the use of hazardous solvents, high energy consumption, and generation of waste, which poses risks to both health and the environment. Thus, there is a growing demand for green and sustainable manufacturing approaches that align with industrial and environmental standards. Despite these limitations, several emerging approaches provide hope for overcoming the challenges of scaling up 2D material-based electrocatalysts. Notably, advancements in material design through hetero-structuring, doping, and defect engineering have significantly enhanced the stability and activity of 2D catalysts. Parallel progress in synthesis optimization techniques such as roll-to-roll production, electrochemical exfoliation, and hydrothermal methods have demonstrated the potential for higher throughput while maintaining material quality.

Moreover, the integration of computational and machine learning in materials research is facilitating predictive modeling of catalytic behavior and accelerating the discovery of scalable synthesis routes. Such computational tools can identify performance-property relationships and guide experimental efforts towards more efficient and scalable electrocatalysts. Looking forward, future research must prioritize the development of cost-effective and environmentally benign synthesis methods, along with efforts to streamline the integration of 2D materials into practical electrochemical systems. Establishing robust collaborations between academic researchers, and industries will be crucial in creating a conducive ecosystem for the translation of laboratory innovations into industrial products.

With sustained innovation in fabrication technologies, improved understanding of degradation mechanisms, and enhanced stability under real-world conditions, 2D material-based electrocatalysts have the potential to revolutionize the field of energy conversion. Their successful scale-up could play a transformative role in achieving global clean energy goals, particularly in applications such as hydrogen production and renewable energy systems.

References

1. Fan, F.R., Wang, R., Zhang, H., Wu, W.: Emerging beyond-graphene elemental 2D materials for energy and catalysis applications. Chem. Soc. Rev. **50**, 10983–11031 (2021). https://doi.org/10.1039/C9CS00821G
2. Anne, B.R., et al.: A review on MXene as promising support materials for oxygen evolution reaction catalysts. Adv. Func. Mater. **33**, 2306100 (2023). https://doi.org/10.1002/adfm.202306100
3. Lakshmi, K.C.S., Vedhanarayanan, B., Lin, T.-W.: Electrocatalytic hydrogen and oxygen evolution reactions: role of two-dimensional layered materials and their composites. Electrochim. Acta **447**, 142119 (2023). https://doi.org/10.1016/j.electacta.2023.142119
4. Li, W., et al.: Confinement engineering of electrocatalyst surfaces and interfaces. Adv. Func. Mater. **32**, 2207727 (2022). https://doi.org/10.1002/adfm.202207727
5. Yap, F.M., Loh, J.Y., Ng, S.-F., Ong, W.-J.: Self-supported earth-abundant carbon-based substrates in electrocatalysis landscape: unleashing the potentials toward paving the way for water splitting and alcohol oxidation. Adv. Energy Mater. **14**, 2303614 (2024). https://doi.org/10.1002/aenm.202303614
6. Zhao, Y., et al.: Recent developments in two-dimensional carbon-based nanomaterials for electrochemical water oxidation: a mini review. Catalysts **14**, 221 (2024)
7. Younis, M.A. et al.: Noble metal-free two dimensional carbon-based electrocatalysts for water splitting. BMC Mater. **1** (2019). https://doi.org/10.1186/s42833-019-0006-2
8. Quan, L., Jiang, H., Mei, G., Sun, Y., You, B.: Bifunctional electrocatalysts for overall and hybrid water splitting. Chem. Rev. **124**, 3694–3812 (2024). https://doi.org/10.1021/acs.chemrev.3c00332
9. Wang, Q. et al.: Insights into MXenes-based electrocatalysts for oxygen reduction. Energy **255** (2022). https://doi.org/10.1016/j.energy.2022.124465

10. Liu, P., Li, J., Yan, J., Song, W.: Defect-rich Fe-doped NiS/MoS_2 heterostructured ultrathin nanosheets for efficient overall water splitting. Phys. Chem. Chem. Phys. **24**, 8344–8350 (2022). https://doi.org/10.1039/d1cp05721a
11. Chen, J., et al.: 2D molybdenum compounds for electrocatalytic energy conversion. Adv. Func. Mater. **33**, 2210236 (2023). https://doi.org/10.1002/adfm.202210236
12. Wang, Y., et al.: Environmentally-friendly exfoliate and active site self-assembly: thin 2D/2D heterostructure amorphous nickel-iron alloy on 2D materials for efficient oxygen evolution reaction. Small **15**, 1805435 (2019). https://doi.org/10.1002/smll.201805435
13. Adegoke, K.A., Ogunjinmi, O.E., Adegoke, O.R., Bello, O.S.: Bifunctional two-dimensional metal organic frameworks for oxygen reaction and water splitting. Nano Energy **128**, 109897 (2024). https://doi.org/10.1016/j.nanoen.2024.109897
14. Arumugam, B., et al.: 2D titanate nanosheet-supported NiMoOx core@shell nanospikes for overall electrocatalytic water-splitting reactions. Energy Fuels **38**, 16820–16833 (2024). https://doi.org/10.1021/acs.energyfuels.4c02257
15. Lu, S., Lou, F., Yu, Z.: Recent progress in two-dimensional materials for electrocatalytic CO_2 reduction. Catalysts **12**, 228 (2022)
16. Xiong, C., Cai, C.: Co-based nanosheets with transitional metal doping for oxygen evolution reaction. Nanomaterials **12**, 1788 (2022)
17. Yi, H., et al.: Recent advance of transition-metal-based layered double hydroxide nanosheets: synthesis, properties, modification, and electrocatalytic applications. Adv. Energy Mater. **11**, 2002863 (2021). https://doi.org/10.1002/aenm.202002863
18. Niu, W.-J., He, J.-Z., Gu, B.-N., Liu, M.-C., Chueh, Y.-L.: Opportunities and challenges in precise synthesis of transition metal single-atom supported by 2D materials as catalysts toward oxygen reduction reaction. Adv. Func. Mater. **31**, 2103558 (2021). https://doi.org/10.1002/adfm.202103558
19. Wang, N., Song, S., Wu, W., Deng, Z., Tang, C.: Bridging laboratory electrocatalysts with industrially relevant alkaline water electrolyzers. Adv. Energy Mater. **14**, 2303451 (2024). https://doi.org/10.1002/aenm.202303451
20. Li, Y., et al.: Advanced electrocatalysis for energy and environmental sustainability via water and nitrogen reactions. Adv. Mater. **33**, 2000381 (2021). https://doi.org/10.1002/adma.202000381
21. Jin, H., et al.: Emerging two-dimensional nanomaterials for electrocatalysis. Chem. Rev. **118**, 6337–6408 (2018). https://doi.org/10.1021/acs.chemrev.7b00689
22. Tavar, D., et al.: Amine-functionalized graphene-rich NiFe layered double hydroxide nanostructure for enhanced electrocatalytic water and urea oxidation. ACS Appl. Nano Mater. **6**, 22517–22531 (2023). https://doi.org/10.1021/acsanm.3c03519
23. Deng, R., Zhang, B., Zhang, Q.: Electrochemical water splitting for scale hydrogen production: from the laboratory to industrial applications. ChemCatChem **16**, e202301165 (2024). https://doi.org/10.1002/cctc.202301165
24. Liu, Y., Xiao, C., Huang, P., Cheng, M., Xie, Y.: Regulating the charge and spin ordering of two-dimensional ultrathin solids for electrocatalytic water splitting. Chem **4**, 1263–1283 (2018). https://doi.org/10.1016/j.chempr.2018.02.006
25. Gao, T., et al.: Terrace-like 2D hierarchically porous iron/cobalt metal-organic framework: ambient fast synthesis and efficient oxygen evolution reaction application. Small **19**, 2207735 (2023). https://doi.org/10.1002/smll.202207735
26. Wang, X., et al.: Two-dimensional conductive metal-organic frameworks electrocatalyst: design principle and energy conversion applications. Mater. Today Energy **44**, 101652 (2024). https://doi.org/10.1016/j.mtener.2024.101652
27. Qi, Q., et al.: Two-dimensional metal-organic frameworks-based electrocatalysts for oxygen evolution and oxygen reduction reactions. Adv. Energy Sustain. Res. **2**, 2000067 (2021). https://doi.org/10.1002/aesr.202000067
28. Wang, H., et al.: Electronic modulation of non-van der Waals 2D electrocatalysts for efficient energy conversion. Adv. Mater. **33**, 2008422 (2021). https://doi.org/10.1002/adma.202008422

29. Li, J., et al.: Recent advances in the large-scale production of photo/electrocatalysts for energy conversion and beyond. Adv. Energy Mater. **14**, 2402441 (2024). https://doi.org/10.1002/aenm.202402441
30. Chen, Y., Zhao, G., Sun, W.: Strategies of engineering 2D nanomaterial-based electrocatalysts toward hydrogen evolution reaction. Mater. Renew. Sustain. Energy **9**, 10 (2020). https://doi.org/10.1007/s40243-020-00170-w
31. Cai, Z., Liu, B., Zou, X., Cheng, H.-M.: Chemical vapor deposition growth and applications of two-dimensional materials and their heterostructures. Chem. Rev. **118**, 6091–6133 (2018). https://doi.org/10.1021/acs.chemrev.7b00536
32. Jiang, Q., et al.: Two-dimensional metal-organic framework nanosheet supported noble metal nanocrystals for high-efficiency water oxidation. Adv. Mater. Interfaces **8**, 2002034 (2021). https://doi.org/10.1002/admi.202002034
33. Liu, Y., et al.: Freestanding 2D NiFe metal-organic framework nanosheets: facilitating proton transfer via organic ligands for efficient oxygen evolution reaction. Small **18**, 2201076 (2022). https://doi.org/10.1002/smll.202201076
34. Wyss, K.M., Luong, D.X., Tour, J.M.: Large-scale syntheses of 2D materials: flash joule heating and other methods. Adv. Mater. **34**, 2106970 (2022). https://doi.org/10.1002/adma.202106970
35. Zheng, W., Lee, L.Y.S.: Beyond sonication: advanced exfoliation methods for scalable production of 2D materials. Matter **5**, 515–545 (2022). https://doi.org/10.1016/j.matt.2021.12.010
36. Li, Z. et al.: Wet-chemical synthesis of two-dimensional metal nanomaterials for electrocatalysis. Natl. Sci. Rev. **9** (2021). https://doi.org/10.1093/nsr/nwab142
37. Lee, H., Heo, E., Yoon, H.: Physically exfoliating 2D materials: a versatile combination of different materials into a layered structure. Langmuir **39**, 18678–18695 (2023). https://doi.org/10.1021/acs.langmuir.3c02418
38. Hu, C., Paul, R., Dai, Q., Dai, L.: Carbon-based metal-free electrocatalysts: from oxygen reduction to multifunctional electrocatalysis. Chem. Soc. Rev. **50**, 11785–11843 (2021). https://doi.org/10.1039/d1cs00219h
39. Gao, J., Tao, H., Liu, B.: Progress of nonprecious-metal-based electrocatalysts for oxygen evolution in acidic media. Adv. Mater. **33**, 2003786 (2021). https://doi.org/10.1002/adma.202003786
40. Chia, X., Pumera, M.: Characteristics and performance of two-dimensional materials for electrocatalysis. Nat. Catal. **1**, 909–921 (2018). https://doi.org/10.1038/s41929-018-0181-7
41. Zhou, S., Shi, L., Li, Y., Yang, T., Zhao, S.: Metal-organic framework-based electrocatalysts for acidic water splitting. Adv. Func. Mater. **34**, 2400767 (2024). https://doi.org/10.1002/adfm.202400767
42. Jin, H., et al.: Molten salt-directed catalytic synthesis of 2D layered transition-metal nitrides for efficient hydrogen evolution. Chem **6**, 2382–2394 (2020). https://doi.org/10.1016/j.chempr.2020.06.037
43. Sun, Q., et al.: Molten-salt assisted synthesis of two-dimensional materials and energy storage application. Mater. Today Chem. **29**, 101419 (2023). https://doi.org/10.1016/j.mtchem.2023.101419
44. Cao, X., Tan, C., Zhang, X., Zhao, W., Zhang, H.: Solution-processed two-dimensional metal dichalcogenide-based nanomaterials for energy storage and conversion. Adv. Mater. **28**, 6167–6196 (2016). https://doi.org/10.1002/adma.201504833
45. Du, P., et al.: Roll-to-roll flash joule heating to stabilize electrocatalysts onto meter-scale Ni foam for advanced water splitting. ACS Nano **19**, 1327–1339 (2025). https://doi.org/10.1021/acsnano.4c13787
46. Bat-Erdene, M. et al.: Elemental 2D materials: solution-processed synthesis and applications in electrochemical ammonia production. Adv. Funct. Mater. **32** (2021). https://doi.org/10.1002/adfm.202107280
47. Yang, X., et al.: Review of emerging atomically precise composite site-based electrocatalysts. Adv. Energy Mater. **13**, 2301737 (2023). https://doi.org/10.1002/aenm.202301737

48. Jin, H., Song, T., Paik, U., Qiao, S.Z.: Metastable two-dimensional materials for electrocatalytic energy conversions. Acc. Mater. Res. **2**, 559–573 (2021). https://doi.org/10.1021/accountsmr.1c00115
49. Zuo, S., Wu, Z.-P., Zhang, H., Lou, X.W.: Operando monitoring and deciphering the structural evolution in oxygen evolution electrocatalysis. Adv. Energy Mater. **12**, 2103383 (2022). https://doi.org/10.1002/aenm.202103383
50. Qian, Q., et al.: Recent advancements in electrochemical hydrogen production via hybrid water splitting. Adv. Mater. **36**, 2306108 (2024). https://doi.org/10.1002/adma.202306108
51. Hu, Y., et al.: Scale-up, continuous and low-temperature production of multimetal based electrocatalysts toward water electrolysis. ACS Appl. Mater. Interfaces **16**, 47563–47570 (2024). https://doi.org/10.1021/acsami.4c08677
52. Choi, S.H., et al.: Large-scale synthesis of graphene and other 2D materials towards industrialization. Nat. Commun. **13**, 1484 (2022). https://doi.org/10.1038/s41467-022-29182-y
53. Rogdakis, K., Karakostas, N., Kymakis, E.: Up-scalable emerging energy conversion technologies enabled by 2D materials: from miniature power harvesters towards grid-connected energy systems. Energy Environ. Sci. **14**, 3352–3392 (2021). https://doi.org/10.1039/D0EE04013D
54. Kim, H.-U., et al.: Wafer-scale and low-temperature growth of 1T-WS2 film for efficient and stable hydrogen evolution reaction. Small **16**, 1905000 (2020). https://doi.org/10.1002/smll.201905000
55. Saenz, G.A., et al.: Effects of synthesis parameters on CVD molybdenum disulfide growth. MRS Adv. **1**, 2291–2296 (2016). https://doi.org/10.1557/adv.2016.484
56. Niu, L., et al.: Production of two-dimensional nanomaterials via liquid-based direct exfoliation. Small **12**, 272–293 (2015). https://doi.org/10.1002/smll.201502207
57. Shivayogimath, A. et al.: A universal approach for the synthesis of two-dimensional binary compounds. Nat. Commun. **10** (2019). https://doi.org/10.1038/s41467-019-11075-2
58. Li, H., et al.: Stability of electrocatalytic OER: from principle to application. Chem. Soc. Rev. **53**, 10709–10740 (2024). https://doi.org/10.1039/D3CS00010A
59. Martinez, U., et al.: Critical role of intercalated water for electrocatalytically active nitrogen-doped graphitic systems. Sci. Adv. **2**, e1501178 (2016). https://doi.org/10.1126/sciadv.1501178
60. Zhu, Q., Qu, Y., Liu, D., Ng, K.W., Pan, H.: Two-dimensional layered materials: high-efficient electrocatalysts for hydrogen evolution reaction. ACS Appl. Nano Mater. **3**, 6270–6296 (2020). https://doi.org/10.1021/acsanm.0c01331
61. Li, W., et al.: Unlocking efficiency: minimizing energy loss in electrocatalysts for water splitting. Adv. Mater. **36**, 2404658 (2024). https://doi.org/10.1002/adma.202404658
62. Zhao, S., Yang, Y., Tang, Z.: Insight into structural evolution, active sites, and stability of heterogeneous electrocatalysts. Angew. Chem. Int. Ed. **61**, e202110186 (2022). https://doi.org/10.1002/anie.202110186
63. Wu, B., et al.: In situ X-ray absorption spectroscopy of metal/nitrogen-doped carbons in oxygen electrocatalysis. Angew. Chem. Int. Ed. **62**, e202219188 (2023). https://doi.org/10.1002/anie.202219188
64. Shen, L., Zhou, J., Yang, T., Yang, M., Feng, Y.P.: High-throughput computational discovery and intelligent design of two-dimensional functional materials for various applications. Acc. Mater. Res. **3**, 572–583 (2022). https://doi.org/10.1021/accountsmr.1c00246
65. Zhang, Y., et al.: Insights into the dynamic evolution of defects in electrocatalysts. Adv. Mater. **35**, 2209680 (2023). https://doi.org/10.1002/adma.202209680
66. Yan, D., et al.: Defect chemistry of nonprecious-metal electrocatalysts for oxygen reactions. Adv. Mater. **29**, 1606459 (2017). https://doi.org/10.1002/adma.201606459
67. Kundu, J., Kwon, T., Lee, K., Choi, S.I.: Exploration of metal-free 2D electrocatalysts toward the oxygen electroreduction. Exploration **4** (2024). https://doi.org/10.1002/exp.20220174
68. Pumera, M.: Materials electrochemists' never-ending quest for efficient electrocatalysts: the devil is in the impurities. ACS Catal. **10**, 7087–7092 (2020). https://doi.org/10.1021/acscatal.0c02020

69. Spanos, I., Masa, J., Zeradjanin, A., Schlögl, R.: The effect of iron impurities on transition metal catalysts for the oxygen evolution reaction in alkaline environment: activity mediators or active sites? Catal. Lett. **151**, 1843–1856 (2021). https://doi.org/10.1007/s10562-020-03478-4
70. Bari, G.A.K.M.R., Jeong, J.-H.: Comprehensive insights and advancements in gel catalysts for electrochemical energy conversion. Gels **10**, 63 (2024). https://doi.org/10.3390/gels10010063
71. Wang, H., Tang, C., Chang-xin, Z., Huang, J.Q., Zhang, Q.: Emerging graphene derivatives and analogues for efficient energy electrocatalysis. Adv. Funct. Mater. **32** (2022). https://doi.org/10.1002/adfm.202204755
72. Zhang, Y., et al.: Recent progress in CVD growth of 2D transition metal dichalcogenides and related heterostructures. Adv. Mater. **31**, 1901694 (2019). https://doi.org/10.1002/adma.201901694
73. Wang, H.-F., Tang, C., Zhao, C.-X., Huang, J.-Q., Zhang, Q.: Emerging graphene derivatives and analogues for efficient energy electrocatalysis. Adv. Funct. Mater. **32**, 2204755 (2022). https://doi.org/10.1002/adfm.202204755
74. Wang, J., et al.: Non-precious-metal catalysts for alkaline water electrolysis: operando characterizations, theoretical calculations, and recent advances. Chem. Soc. Rev. **49**, 9154–9196 (2020). https://doi.org/10.1039/D0CS00575D
75. Liu, J., et al.: Toward excellence of electrocatalyst design by emerging descriptor-oriented machine learning. Adv. Funct. Mater. **32**, 2110748 (2022). https://doi.org/10.1002/adfm.202110748
76. Zhang, X., Chen, A., Chen, L., Zhou, Z.: 2D materials bridging experiments and computations for electro/photocatalysis. Adv. Energy Mater. **12**, 2003841 (2022). https://doi.org/10.1002/aenm.202003841
77. Ryu, B., Wang, L., Pu, H., Chan, M.K.Y., Chen, J.: Understanding, discovery, and synthesis of 2D materials enabled by machine learning. Chem. Soc. Rev. **51**, 1899–1925 (2022). https://doi.org/10.1039/D1CS00503K
78. Xie, Y., et al.: In situ investigation on life-time dynamic structure-performance correlation toward electrocatalyst service behavior in water splitting. Adv. Funct. Mater. **32**, 2111777 (2022). https://doi.org/10.1002/adfm.202111777
79. Ji, J., Lou, W., Shen, P.: Modular design in metal-organic frameworks for oxygen evolution reaction. Int. J. Hydrogen Energy **47**, 39443–39469 (2022). https://doi.org/10.1016/j.ijhydene.2022.09.126
80. Chen, H., et al.: Active site engineering in porous electrocatalysts. Adv. Mater. **32**, 2002435 (2020). https://doi.org/10.1002/adma.202002435
81. Cheng, R., Min, Y., Li, H., Fu, C.: Electronic structure regulation in the design of low-cost efficient electrocatalysts: from theory to applications. Nano Energy **115**, 108718 (2023). https://doi.org/10.1016/j.nanoen.2023.108718
82. Dong, R., Zhang, T., Feng, X.: Interface-assisted synthesis of 2D materials: trend and challenges. Chem. Rev. **118**, 6189–6235 (2018). https://doi.org/10.1021/acs.chemrev.8b00056
83. Wei, C., et al.: Approaches for measuring the surface areas of metal oxide electrocatalysts for determining their intrinsic electrocatalytic activity. Chem. Soc. Rev. **48**, 2518–2534 (2019). https://doi.org/10.1039/C8CS00848E
84. Zhuang, J., Wang, D.: Recent advances of single-atom alloy catalyst: properties, synthetic methods and electrocatalytic applications. Mater. Today Catal. **2**, 100009 (2023). https://doi.org/10.1016/j.mtcata.2023.100009
85. Wang, X., et al.: Electronic structure regulation of the Fe-based single-atom catalysts for oxygen electrocatalysis. Nano Energy **121**, 109268 (2024). https://doi.org/10.1016/j.nanoen.2024.109268
86. Zhang, Y., et al.: Two-dimensional electrocatalysts for efficient reduction of carbon dioxide. ChemSusChem **13**, 59–77 (2019). https://doi.org/10.1002/cssc.201901794

87. Huang, S. et al.: N-, O-, and S-tridoped carbon-encapsulated Co_9S_8 nanomaterials: efficient bifunctional electrocatalysts for overall water splitting. Adv. Funct. Mater. **27** (2017). https://doi.org/10.1002/adfm.201606585
88. Wang, Q., et al.: An overview of heteroatom doped cobalt phosphide for efficient electrochemical water splitting. Chem. Eng. J. **456**, 141056 (2023). https://doi.org/10.1016/j.cej.2022.141056
89. Zhang, J., Zhang, Q., Feng, X.: Support and interface effects in water-splitting electrocatalysts. Adv. Mater. **31**, 1808167 (2019). https://doi.org/10.1002/adma.201808167
90. Chen, C.-A., Lee, C.-L., Yang, P.-K., Tsai, D.-S., Lee, C.-P.: Active site engineering on two-dimensional-layered transition metal dichalcogenides for electrochemical energy applications: a mini-review. Catalysts **11**, 151 (2021)
91. Khan, U., Nairan, A., Gao, J., Zhang, Q.: Current progress in 2D metal–organic frameworks for electrocatalysis. Small Struct. **4** (2022). https://doi.org/10.1002/sstr.202200109
92. Li, T., et al.: Synthesis of single crystalline two-dimensional transition-metal phosphides via a salt-templating method. Nanoscale **10**, 6844–6849 (2018). https://doi.org/10.1039/C8NR01556B
93. Zhang, H., et al.: Quantum effects allow the construction of two-dimensional Co_3O_4-embedded nitrogen-doped porous carbon nanosheet arrays from bimetallic MOFs as bifunctional oxygen electrocatalysts. Chem. Eur. J. **24**, 14522–14530 (2018). https://doi.org/10.1002/chem.201802898
94. Tang, C., Zhong, L., Zhang, B., Wang, H.-F., Zhang, Q.: 3D mesoporous van der Waals heterostructures for trifunctional energy electrocatalysis. Adv. Mater. **30**, 1705110 (2018). https://doi.org/10.1002/adma.201705110
95. Zhang, W., Xu, C., Zheng, H., Li, R., Zhou, K.: Oxygen-rich cobalt–nitrogen–carbon porous nanosheets for bifunctional oxygen electrocatalysis. Adv. Funct. Mater. **32** (2022). https://doi.org/10.1002/adfm.202200763
96. Lee, G.S., et al.: 2D materials beyond post-AI era: smart fibers, soft robotics, and single atom catalysts. Adv. Mater. **36**, 2307689 (2024). https://doi.org/10.1002/adma.202307689
97. Vedhanarayanan, B., Lakshmi, K.C.S., Lin, T.-W.: Interfacial tuning of polymeric composite materials for high-performance energy devices. Batteries **9**, 487 (2023)
98. Meethale Palakkool, N., et al.: Recent advances in surface functionalized 3D electrocatalyst for water splitting. Adv. Energy Sustain. Res. **6**, 2400258 (2025). https://doi.org/10.1002/aesr.202400258
99. Li, L., et al.: Unraveling oxygen evolution reaction on carbon-based electrocatalysts: effect of oxygen doping on adsorption of oxygenated intermediates. ACS Energy Lett. **2**, 294–300 (2017). https://doi.org/10.1021/acsenergylett.6b00681
100. Rong, C., Dastafkan, K., Wang, Y., Zhao, C.: Breaking the activity and stability bottlenecks of electrocatalysts for oxygen evolution reactions in acids. Adv. Mater. **35**, 2211884 (2023). https://doi.org/10.1002/adma.202211884
101. Liu, Z., He, T., Jiang, Q., Wang, W., Tang, J.: A review of heteroatomic doped two-dimensional materials as electrocatalysts for hydrogen evolution reaction. Int. J. Hydrogen Energy **47**, 29698–29729 (2022). https://doi.org/10.1016/j.ijhydene.2022.06.306
102. Wu, B., et al.: Nitrogen-rich carbonaceous materials for advanced oxygen electrocatalysis: synthesis, characterization, and activity of nitrogen sites. Adv. Funct. Mater. **32**, 2204137 (2022). https://doi.org/10.1002/adfm.202204137
103. Li, Q., Xu, D., Ou, X., Yan, F.: Nitrogen-doped graphitic porous carbon nanosheets derived from in situ formed $G\text{-}C_3N_4$ templates for the oxygen reduction reaction. Chem. Asian J. **12**, 1816–1823 (2017). https://doi.org/10.1002/asia.201700586
104. Chen, D., Zou, Y., Wang, S.: Surface chemical-functionalization of ultrathin two-dimensional nanomaterials for electrocatalysis. Mater. Today Energy **12**, 250–268 (2019). https://doi.org/10.1016/j.mtener.2019.01.006
105. Yu, Z.-Y., et al.: Clean and affordable hydrogen fuel from alkaline water splitting: past, recent progress, and future prospects. Adv. Mater. **33**, 2007100 (2021). https://doi.org/10.1002/adma.202007100

106. Lu, C., et al.: Molybdenum carbide-embedded nitrogen-doped porous carbon nanosheets as electrocatalysts for water splitting in alkaline media. ACS Nano **11**, 3933–3942 (2017). https://doi.org/10.1021/acsnano.7b00365
107. Kanase, R.S., Arunachalam, M., Badiger, J., Das, P.K., Kang, S.H.: Highly selective and durable CO production via effective morphology and surface engineering of ZnO electrocatalysts. ACS Appl. Energy Mater. **5**, 13913–13927 (2022). https://doi.org/10.1021/acsaem.2c02553
108. Xiao, Z. et al.: Activating the basal planes in 2H-$MoTe_2$ monolayers by incorporating single-atom dispersed N or P for enhanced electrocatalytic overall water splitting. Adv. Sustain. Syst. **6** (2022). https://doi.org/10.1002/adsu.202100515
109. Zhang, A., Liang, Y., Zhang, H., Geng, Z., Zeng, J.: Doping regulation in transition metal compounds for electrocatalysis. Chem. Soc. Rev. **50**, 9817–9844 (2021). https://doi.org/10.1039/D1CS00330E
110. Jiang, Y., et al.: Significant contribution of intrinsic carbon defects to oxygen reduction activity. ACS Catal. **5**, 6707–6712 (2015). https://doi.org/10.1021/acscatal.5b01835
111. Yuan, N., Jiang, Q., Wu, Z., Tang, J.: Ru nanoparticles decorated on 2D MoO_2 nanosheets as efficient and durable electrocatalysts for the hydrogen evolution reaction in a wide pH range. J. Phys. Chem. C **124**, 10804–10814 (2020). https://doi.org/10.1021/acs.jpcc.0c01117
112. Hu, C., Dai, L.: Doping of carbon materials for metal-free electrocatalysis. Adv. Mater. **31**, 1804672 (2019). https://doi.org/10.1002/adma.201804672
113. Sun, T., et al.: Defect chemistry in 2D materials for electrocatalysis. Mater. Today Energy **12**, 215–238 (2019). https://doi.org/10.1016/j.mtener.2019.01.004
114. Ying, Y., Fan, K., Lin, Z., Huang, H.: Facing the "Cutting Edge": edge site engineering on 2D materials for electrocatalysis and photocatalysis. Adv. Mater. **37**, 2418757 (2025). https://doi.org/10.1002/adma.202418757
115. Zhang, Z., et al.: Heterostructure engineering of 2D superlattice materials for electrocatalysis. Adv. Sci. **9**, 2204297 (2022). https://doi.org/10.1002/advs.202204297
116. Cao, Y.: Roadmap and direction toward high-performance MoS_2 hydrogen evolution catalysts. ACS Nano **15**, 11014–11039 (2021). https://doi.org/10.1021/acsnano.1c01879
117. Liu, Y. et al.: Electroactive edge-site-enriched α-$Co_{0.9}Fe_{0.1}(OH)_x$ nanoplates for efficient overall water splitting. ChemElectroChem **6**, 2415–2422 (2019). https://doi.org/10.1002/celc.201900340
118. Rowley-Neale, S.J., et al.: 2D molybdenum disulphide (2D-MoS_2) modified electrodes explored towards the oxygen reduction reaction. Nanoscale **8**, 14767–14777 (2016). https://doi.org/10.1039/c6nr04073j
119. Janani, G. et al.: Aggregation induced edge sites actuation of 3D $MoSe_2$/rGO electrocatalyst for high-performing water splitting system. Aggregate **5** (2023). https://doi.org/10.1002/agt2.430
120. Liu, F., Fan, Z.: Defect engineering of two-dimensional materials for advanced energy conversion and storage. Chem. Soc. Rev. **52**, 1723–1772 (2023). https://doi.org/10.1039/D2CS00931E
121. Zhao, D., et al.: Atomic site electrocatalysts for water splitting, oxygen reduction and selective oxidation. Chem. Soc. Rev. **49**, 2215–2264 (2020). https://doi.org/10.1039/C9CS00869A
122. Zhu, J., Mu, S.: Defect engineering in carbon-based electrocatalysts: insight into intrinsic carbon defects. Adv. Funct. Mater. **30**, 2001097 (2020). https://doi.org/10.1002/adfm.202001097
123. Wu, K. et al.: Defect engineering in transition-metal (Fe, Co, and Ni)-based electrocatalysts for water splitting. Carbon Energy **6** (2024). https://doi.org/10.1002/cey2.485
124. Yun, Q., et al.: Hybridization of 2D nanomaterials with 3D graphene architectures for electrochemical energy storage and conversion. Adv. Funct. Mater. **32**, 2202319 (2022). https://doi.org/10.1002/adfm.202202319

125. Sohn, W., Kim, M., Jang, H.W.: Atomic-scale insights into the 2D materials from aberration-corrected scanning transmission electron microscopy: progress and future. Small Sci. **4**, 2300073 (2024). https://doi.org/10.1002/smsc.202300073
126. Zhu, Y., et al.: Emerging dynamic structure of electrocatalysts unveiled by in situ X-ray diffraction/absorption spectroscopy. Energy Environ. Sci. **14**, 1928–1958 (2021). https://doi.org/10.1039/D0EE03903A

Designing Electrocatalyst for Integrated HER and OER

Hani Nasser Abdelhamid

Abstract Hydrogen (H_2) gas has been considered a clean and sustainable energy source, representing a promising alternative to fossil fuels. However, the effective method for hydrogen production for large scale demand is still challenging; moreover, several production methods exist. Among these methods, water-splitting electrolysis using an electrocatalyst is an effective technique for efficient hydrogen production. Creating efficient, stable, and economical electrocatalysts is crucial for attaining high electrocatalytic hydrogen production through water splitting or electrolysis. This book chapter offers an overview of strategies and information for designing electrocatalysts with high performance in water splitting, emphasizing their activity, stability, and efficiency. It explains the key factors hydrogen evolution reaction (HER) and the oxygen evolution reaction (OER). The design developments covered nanostructured electrocatalysts utilizing noble, non-noble (noble-free) metals, and metal-free electrocatalysts. It also covers innovative approaches and insights into the structure, composition, particle size, morphology, phases, doping, defects, substrates, and active sites of electrocatalysts designed to improve the activity and stability in the HER and OER. It outlines the challenges and opportunities in advancing active and resilient electrocatalysts for the HER and OER to enhance hydrogen production through water-splitting electrolysis.

Keywords Electrocatalysts · OER · HER · Water splitting

1 Introduction

Environment and energy are urgent and critical issues that must be addressed to achieve the planet's sustained economic and social progress. Global Energy Review 2025 (IEA, International Energy Agency) shows that global energy consumption has risen significantly [1]. It reported that Global energy demand grew by 2.2% in 2024,

H. N. Abdelhamid (✉)
Department of Chemistry, College of Science, Imam Mohammad Ibn Saud Islamic University (IMSIU), Riyadh, 11623, Saudi Arabia
e-mail: hnabdelhamid@imamu.edu.sa; hani.nasserabdelhamid@gmail.com

O. P. Pandey et al. (eds.), *2D Materials for Electrochemical Water Splitting*, Engineering Materials, https://doi.org/10.1007/978-981-95-2003-9_13

driven by a 4.3% surge in electricity use due to extreme heat, electrification, and digitalization. About 80% of the world's energy consumption comes from conventional energy sources fossil fuels, including natural gas, coal, and oil. It found 38% supply growth in the renewable section, followed by gas, coal, oil, and nuclear, with an increase of 28%, 15%, 11%, and 8%, respectively. Both emerging economies (e.g., China and India) and advanced economies show significant increases in energy demand. Advanced economies also saw a rebound, particularly in the U.S. [1]. The world electricity demand rose by nearly 1,100 TWh because of new technologies, including cooling, EV adoption, industry, and data centers, representing an increase of 60%. The high consumption of fossil fuel energy increases carbon dioxide (CO_2) emissions growth, causing climate change and leading to environmental issues [1]. This concern makes renewable energy unavoidable. In 2024, renewable energy was pivotal in addressing the increasing global electricity demand, constituting 80% of the expansion in power generation [1]. Renewable sources, e.g., solar photovoltaic and wind energy, accounted for 32% of total electricity, substantially transforming the global energy composition [1]. In China, wind and solar accounted for roughly 20% of total electricity output. Renewable energy sources ensure sustainable options for increasing global energy demand, climate mitigation, and economic advancement [1].

Molecular hydrogen (H_2) gas is considered one of the most promising fuels [2, 3]. It possesses the highest gravimetric energy density, and water by-product of combustion, is environmentally advantageous, positioning it as an excellent prospective energy carrier and an alternative for future low carbon energy systems [4]. Hydrogen exhibits a substantial energy density of 120 MJ/kg, presenting a formidable substitute for fossil fuels. Currently, a significant quantity of H_2 gas is produced by steam generated by the combustion of fossil fuels [5]. This method depletes fossil fuels, creates CO_2 emissions, and exhibits a suboptimal conversion rate. Hydrogen can also be generated via other methods, such as fermentation, thermolysis, and photocatalysis [5–7]. Consequently, an efficient hydrogen generation method without environmental issues is a critical challenge.

The widespread use of hydrogen might reduce world CO_2 emissions by 830 million tonnes annually [8]. Water electrolysis is an eco-friendly, efficient, and sustainable technology for producing hydrogen and oxygen [9]. It was described for the first time in 1789. It has been considered a sustainable method due to the abundant presence of water in the universe, which serves as a precursor and a by-product of hydrogen combustion. It can be described in two half-cell reactions; typically hydrogen evolution (HER) and oxygen evolution (OER).

$$\textbf{HER}: H_2O\ (l) \rightarrow H_2 + 1/2\ O_2 \Delta E^\circ = 1.23\ V$$

$$\textbf{OER}: 2H_2O \rightarrow 4e^- + 4H^+ + O_2 \Delta E^\circ = 0\ V$$

Potential value is measured versus Normal (N) or Standard (S) hydrogen electrode (NHE/SHE).

Electrolysis facilitates the transformation of electricity into hydrogen, which can be stored and used to produce electricity using fuel cells [10–12]. Electricity can be come from renewable sources, including solar and wind. Renewable energy in electrolysis systems facilitates zero-emission hydrogen generation, improving environmental sustainability [10–12]. These systems may be tailored for various scales and are proficient for the production of hydrogen with high purity. However, the extensive implementation of electrolysis faces obstacles due to high capital and operational expenses, mainly attributed to costly catalysts and the energy-demanding characteristics of the process. Although electrolysis exhibit lower energy efficient than thermochemical processes such as steam methane reforming (SMR), its environmental impact is significantly influenced by the source of electricity. Continuous efforts in both research and development could lead to achieve high efficiency and minimize costs are crucial for electrolysis as a fundamental method for sustainable production of hydrogen and long-term energy storage [10–12].

This book chapter discusses the electrocatalyst design for water-splitting reactions involving reactions, i.e., HER and OER. It summarizes the performance assessment of various electrocatalysts in water splitting, highlighting their activity, stability, and efficiency. The design advancements encompassed nanostructured electrocatalysts employing both noble and non-noble metals, considering several key factors such as the advanced architecture or structure, composition, morphology, phase, particle size, doping and active sites of electrocatalysts aimed at enhancing electrocatalytic performance and durability in the HER and OER. We also discuss the problems and potential of advancing active and resilient electrocatalysts. Further efforts should be made to create efficient, reliable, and economical electrocatalysts, crucial for attaining high electrocatalytic hydrogen production through water splitting.

2 Hydrogen: Production and Storage

Hydrogen is increasingly acknowledged as important component in the shift towards a low-greenhouse gas (GHG) economy [13, 14]. The hydrogen value chain consists of four essential components: Hydrogen Production, Transportation, Storage, and Utilization, as depicted in Fig. 1 [8]. In response to the increasing demand for clean energy, several hydrogen production technologies have been established, including electrochemical, thermochemical, biological, chemical, and photocatalytic methods (Fig. 2) [15–17]. Effective and secure hydrogen transportation from production sites to consumers is crucial for widespread applications (Fig. 1). Transportation methods are often categorized according to the physical state of hydrogen: gaseous, liquid, or chemical hydrogen carriers. The storage of hydrogen is a challenge, increasing the cost of hydrogen usage. Storage technologies include compressed gas, cryogenic liquid, and solid-state storage devices. Hydrogen use technologies involve its incorporation into energy systems and industrial applications, e.g., fuel cells, combustion, and other techniques.

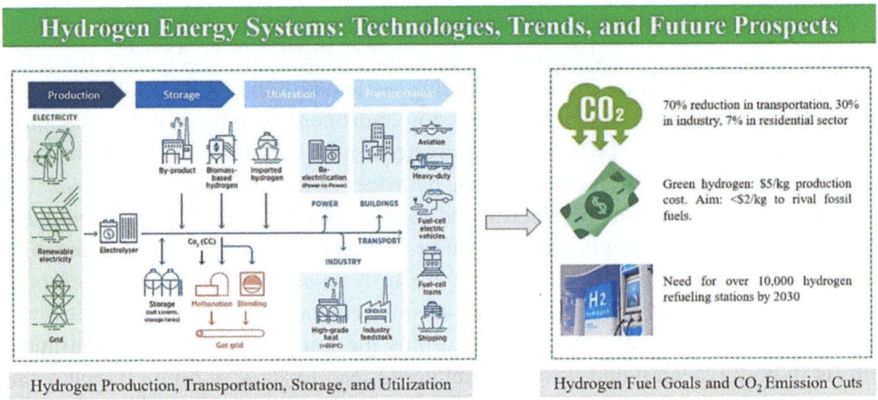

Fig. 1 Production, storage, utilization, and transportation of hydrogen. Figure reprinted with permission from Open Access Ref. [8]. © 2024 The Authors. Published by Elsevier B.V.

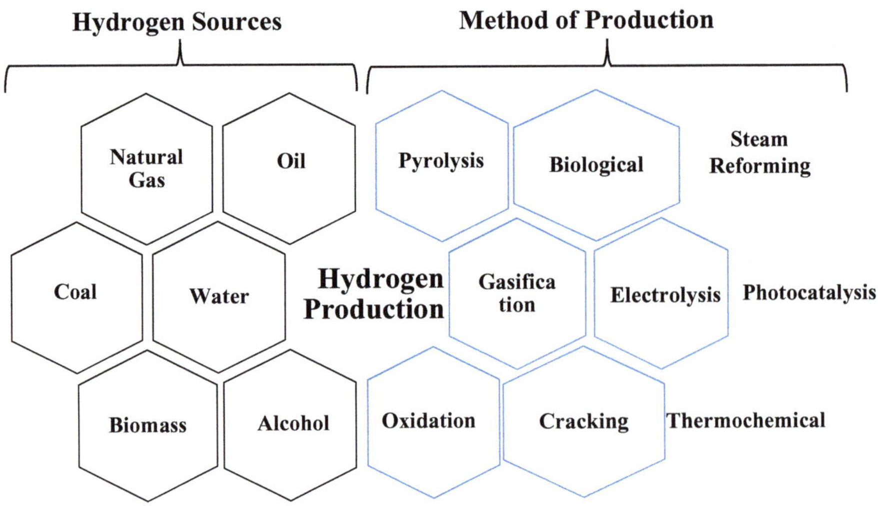

Fig. 2 Common method used for hydrogen production

Hydrogen generation can be accomplished via various methods, generally classified into non-renewable and renewable sources, differing in efficiency and technological advancement (Table 3, Fig. 2). One of the most prevalent techniques is steam methane reforming (SMR), which uses hydrocarbons, e.g., methane, as feedstock. With an efficiency of 60–75%, SMR is financially feasible yet remains non-renewable. Methanol steam reforming offers an alternative method utilizing liquid hydrocarbons at reduced reforming temperatures. Partial oxidation, a fossil fuel-based technique, achieves 60–75% efficiency by utilizing heavier hydrocarbons such as oil or coal, and it is commercially used. Auto-thermal reforming, which merges steam and oxygen, synthesizes the advantages of steam methane reforming and partial oxidation for enhanced efficiency. Solar steam reforming utilizes solar energy to facilitate the reforming process, achieving comparable efficiency (60–75%) while being designated as renewable. Electrolysis, a process that disassociates water using electricity, attains approximately 70% efficiency and is now utilized commercially. Nevertheless, its classification as renewable is contingent upon the electricity source. For example, if fueled by fossil-based electricity, it stays non-renewable. Advanced research-phase technologies encompass photoelectrolysis of water, utilizing solar energy with an efficiency of 12.4%, and photolysis with a significantly lower efficiency of 0.5%, both classified as renewable technologies. Thermochemical water splitting, utilizing heat (possibly from nuclear or concentrated solar power), achieves approximately 30% efficiency and is now under research, categorized as non-renewable. Furthermore, the synthesis of hydrogen from biomass via pyrolysis and biological processes is under investigation to enhance yield and mitigate environmental effects. The variety of feedstocks and efficiency among these technologies emphasizes the necessity of choosing hydrogen production methods according to local resource availability, sustainability objectives, and energy policy aspirations (Table 1).

3 Water Electrolysis

Electricity facilitates the electrochemical dissociation of water into oxygen and hydrogen, i.e., water electrolysis, offering a sustainable and eco-friendly method for hydrogen generation. During electrolysis, an electric current is sent through water via an electrolyzer, including an anode and a cathode separated by an electrolyte. At the anode, water molecules undergo oxidation to yield hydroxide ions (OH^-), protons (H^+), or oxygen ions (O^{2-}) in alkaline electrolysis, acidic or proton exchange memebrane (PME), and solid oxide electrolysis (SOE) or high-temperature SOE, respectively. During electrolysis, ions produced at the anode traverse the electrolyte to the cathode, where they undergo reduction to yield hydrogen gas. The generated gases of hydrogen and oxygen are further separated, and stored either compressed or liquefied.

Alkaline electrolysis uses electrolyte solutions of based such as sodium hydroxide (NaOH) or potassium hydroxide (KOH). It is economically beneficial and scalable, permitting the utilization of non-noble metal catalysts. Nonetheless, the HER in alkaline conditions often has an activity that is 2–3 orders of magnitude inferior to that in acidic settings.

PEM electrolysis takes place in acidic environments via a proton-conducting membrane. It provides high proton conductivity, small gas permeability, high efficiency, and significant hydrogen production or generated rates (HGRs). The requirement for acidic conditions limits OER catalysts to noble metals and their oxides, hence high system prices.

Table 1 Summary of hydrogen production methods

Technology	Process type	Operating temp.	Catalysts/ microorganisms	Substrates	Energy/ H_2 yield	H_2 purity	Renewable integration	Advantages	Disadvantages	Development stage	Future outlook
Alkaline electrolysis	Electrochemical	60–80 °C	Non-precious metals	Water	Moderate	High	Good	Low-cost, mature	Moderate efficiency	Mature	Increased use in hybrid renewable systems
PEM electrolysis	Electrochemical	50–80 °C	Pt-based metals		High			High efficiency, compact design	High cost, catalyst scarcity	Advanced	Enhanced efficiency, lower costs
SOE (solid oxide electrolysis)	Electrochemical	700–1000 °C	Non-precious metals	Steam	High		Limited	Efficient at high temps	Material challenges	Advanced	Coupling with waste heat, niche scaling
AEM electrolysis	Electrochemical	50–80 °C		Water	Moderate to high		Good	Balance of cost and performance	Stability and durability issues	Developmental	Potential for low-cost green H_2
Bipolar membrane electrolysis	Electrochemical	Ambient–80 °C						Low operating temp, pH flexibility	Emerging technology	Early stage	Niche integration, ongoing validation
MECs	Electrochemical (Bio)	Ambient–40 °C	Electroactive microbes	Wastewater, biomass	Low to Moderate	Moderate	Limited	Low energy input	Low yield, complex setup	Developmental	Cost-effective wastewater integration
Dark fermentation	Biological	Ambient	Anaerobic bacteria	Organic waste, biomass	Moderate	Low	Good	Waste valorization	Low efficiency, scaling	Developmental	Process optimization, hybrid systems
Photofermentation	Biological	Ambient + light	Photosynthetic bacteria	Volatile fatty acids	Low to moderate		Limited	Solar-driven	Light dependency	Early stage	Photobioreactor design improvements

(continued)

Table 1 (continued)

Technology	Process type	Operating temp.	Catalysts/ microorganisms	Substrates	Energy/ H_2 yield	H_2 purity	Renewable integration	Advantages	Disadvantages	Development stage	Future outlook
Cyanobacterial H_2 production	Biological	Ambient + light	Cyanobacteria	CO_2, water	Low		Limited	Carbon–neutral, photosynthetic	Very low yield		Biotech enhancement, strain engineering
Microbial consortia	Biological	Ambient	Diverse microbes	Biomass, wastewater	Moderate to high	Moderate	Good	Synergistic microbial use	Control complexity	Developmental	Metabolic engineering potential
BES (bioelectrochemical systems)	Electrochemical/ Bio	Ambient	Electroactive microbes	Wastewater, biomass	Low to Moderate	Moderate	Limited	Coupled waste treatment	Complex design, low yield		Resource recovery integration
Thermochemical production	Thermochemical	High	Metal catalysts	Fossil/organic feedstocks	High	High		Mature, scalable	GHG emissions	Mature	CCS + transition to renewables
Photocatalytic production	Photocatalytic	Ambient + light	Semiconductors	Water	Low	High		Solar-driven, clean	Low efficiency, material limits	Early stage	New catalysts, hybrid systems

Solid oxide electrolysis (SOE) or high-temperature SOE functions at high temperatures (700–1000 °C) with ceramic or solid oxide electrolytes, hence offering enhanced energy efficiency and the capacity to harness waste heat. Notwithstanding these benefits, its commercialization is impeded by intricate system requirements and issues related to material stability.

Several methods were considered as recent advancements in electrolysis including anion exchange membrane (AEM) electrolysis, microbial electrolysis cells (MECs), and bipolar membrane electrolysis. AEM electrolysis employs an anion-conducting membrane with cheap or noble-free metal catalysts, facilitating economical operation at reduced temperatures relative to PEM systems. Bipolar membrane electrolysis utilizes a membrane that includes a bipolar junction that improves water dissociation and hydrogen generation. MECs utilize electroactive microorganisms to transform organic substrates, including biomass or wastewater, into hydrogen, providing a pathway for bioelectrochemical hydrogen generation.

Water electrolysis involves HER and OER reactions. Table 2 summarizes the vital information for both reactions, i.e., HER and OER. HER is an electrochemical reduction of protons (acidic medium) or water (alkaline medium) to produce H_2 gas at the cathode. While OER is an electrochemical oxidation of water that produces O_2 gas at the anode, it involves multiple electron transfer steps.

In the water electrolysis process, the input energy must be sufficient to surpass the barrier, akin to other chemical reactions. In general, HER and OER require a potential of 0 and 1.23 V versus NHE. However, in reality, the reactions take place at higher values. The difference between the experimental and theoretical potential value is overpotential (denoted as η). The overpotential is a crucial factor for evaluating the performance of the electrolysis system and materials used in its cells. Appropriate electrocatalysts with low overpotential are necessary for the HER and OER to facilitate the effective generation of H_2 and O_2.

4 Factors Affecting Electrolysis

The performance of HER catalysts is governed by several interdependent factors, including Gibbs free energy of hydrogen adsorption (ΔG_H*), electronic structure, temperature, catalyst-promoter interactions, and mass transport dynamics (Table 2). Efficient HER catalysts facilitate proton adsorption and H_2 desorption, ideally possessing ΔG_H* values near zero, which can be calculated via density functional theory (DFT) methods. Transition metal catalysts, particularly those with favorable d-band centers, enable optimal binding strength and lower activation barriers. The d-band center theory is a traditional notion primarily focused on the binding energy values of adsorbed H* species (Table 2). Promoters such as Na, Fe, Co, and Ni enhance catalyst performance by modifying surface properties, stabilizing intermediates, and improving electronic interactions. They can also offer synergistic effects with base metals like Ru or MoS_2. High temperature boosts reaction kinetics, though

Table 2 Summary of key information for HER and OER

Reaction	Key equations	Comments/challenges	Evaluation methods	Catalyst examples
HER	**Acidic medium**: $2H^+ + 2e^- \rightarrow H_2$ **Alkaline medium**: $2H_2O + 2e^- \rightarrow H_2 + 2OH^-$	– Highly dependent on proton availability and electrode material – Platinum (Pt) is the benchmark catalyst due to its low overpotential – In alkaline media, HER is sluggish due to additional water dissociation step	– Linear sweep voltammetry (LSV) – Tafel slope – Electrochemical Impedance Spectroscopy (EIS) – Chronoamperometry (CA) – Chronopotentiometry (CP)	– Noble metals: Pt, Pd, Ru – Transition metal-based catalysts: MoS_2, CoP, NiFe – Carbide/nitride catalysts: WC, Mo_2C, VN
OER	**Acidic medium:** $2H_2O \rightarrow O_2 + 4H^+ + 4e^-$ **Alkaline medium:** $4OH^- \rightarrow O_2 + 2H2O$	– OER is kinetically sluggish due to multiple proton-electron transfer steps – RuO_2 and IrO_2 are the most effective catalysts, but they are expensive and unstable – Alternative transition metal-based catalysts (Ni, Co, Fe) show promise but require further optimization	– LSV – CV – Tafel – Faradaic Efficiency (FE) – CA – CP	– Noble metals: RuO_2, IrO_2 – Transition metal oxides/hydroxides: $NiFe_2O_4$, Co_3O_4, MnO_2 – Perovskite catalysts

care must be taken to avoid corrosion. Furthermore, effective mass transport is critical for maintaining reactant supply and product removal. However, effective mass transport is governed by diffusion, convection, and boundary layer dynamics. Thus, further efforts should be made to optimize flow rates, reduce polarization, and maximize catalytic efficiency, making these physical parameters crucial for designing scalable, high-performance HER systems in renewable energy applications.

Several parameters affect the electrocatalyst's performance (Fig. 3). Electrocatalyst properties affect the material's performance, including composition, phases,

Table 3 Key factors influencing HER performance

Factor	Effect on HER	Mechanism/details
ΔG_H*	Indicator of catalyst efficiency	Ideally close to 0; calculated using DFT
d-Band theory	Influences H* adsorption and binding strength	d-band center position affects M–H bond strength; lower σ* occupancy improves binding
Temperature	Enhances reaction kinetics and equilibrium	Increases reaction rate, ion mobility, and shifts equilibrium towards H_2; may cause corrosion at high T
Catalysts	Provide active sites for HER	Transition metals (e.g., Pt, Ru, Ni, Pd) lower activation energy and enhance adsorption–desorption cycles
Promoters	Improve catalyst performance without direct participation	Modify surface/electronic properties; stabilize intermediates; enhance dispersion and synergy
Synergistic effects	Amplify catalytic activity	The combined action of catalysts and promoters exceeds individual performance
Mass transport	Ensures continuous reactant supply and product removal	Affected by diffusion, convection, and boundary layer dynamics
Flow rate	Controls reactant delivery and polarization	Higher rates improve mixing and reduce polarization; excessive rates may increase energy costs

morphology, defects cocatalyst deposition, surface active sites, surface properties, doping, and atomic thickness. Electrocatalytic activity can be improved via control electronic properties by modification of morphology, defect engineering, phase transitions, cocatalyst deposition, and element doping (Fig. 3). The synthesis of the nanoarchitecture of more than one electrocatalyst offers the synergistic effect of the system structure, and the possible mutual interaction among diverse chemical components may impede further enhancements in electrocatalysis. Electocatalysis can be improved via external physical fields e.g., magnetic, electric, and light fields, generating thermal effects, decreasing charge transfer resistance, and bubble coverage on the catalyst surface [18]. Thus, external fields improved HER and OER efficiency. The summary of these factors are tabulated for HER (Table 3) and OER (Table 4).

The OER is an essential half-reaction in water splitting and electrochemical energy conversion systems (Table 4). It exhibits slow kinetics resulting from a complex four-electron transfer mechanism. The efficacy of OER catalysts depends on various interrelated factors, including the adsorption energy of reaction intermediates (e.g., OH*, OOH*, or O*), electronic structure (e.g., the e_g orbital occupancy in transition metal oxides), overpotential, temperature, catalyst composition, and mass transport conditions (Table 4). OER catalysts, e.g., IrO_2, RuO_2, NiFe-layered double hydroxide (LDHs), and perovskites, can stabilize critical intermediates and optimize adsorption energies to prevent rate-limiting processes. Promoters and dopants can improve the conductivity of the electrocatalysts, elevate active spots, and alter surface chemistry to facilitate charge transfer (Table 4) [19]. Raising temperature enhances

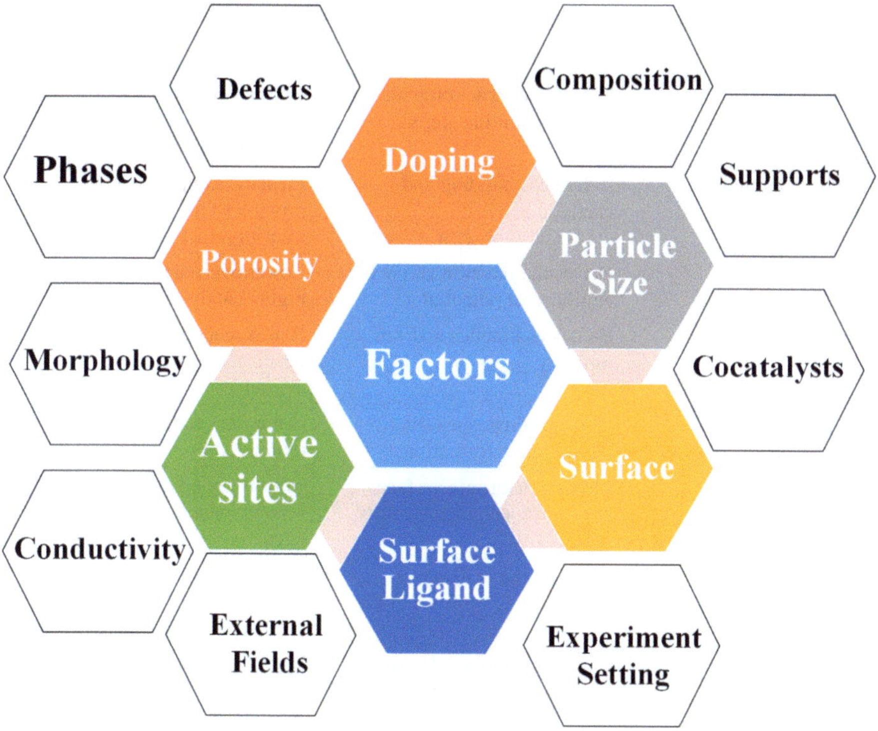

Fig. 3 Factors that should be investigated when designing electrocatalysts for HER or OER

reaction kinetics but requires regulation to prevent catalyst deterioration. Moreover, mass transfer and flow rate are essential for reducing oxygen bubble accumulation, ensuring reactant delivery, and avoiding diffusion constraints. Optimizing these variables is crucial for developing resilient, high-efficiency OER catalysts for renewable energy systems.

Dopanet and defect affect the electronic properties and conductivity of the electrocatalysts. Electrocatalytic performance of graphitic-carbon nitride (g-C_3N_4) can be improved using boron dopants and nitrogen deficiency [19]. These properties can be achieved via carbonization of the materials in the presence of sodium borohydride [19]. B and N-dopant improved electronic properties, consequently, the electrocatalytic performance of electrocatalysts [20]. Fe and Mo co-doped Ni_3S_2 were reported for OER and HER [21]. Fe-MoS_2/Ni_3S_2@nickel foam (NF) demonstrated superior HER and OER performance compared to MoS_2/Ni_3S_2@NF and Ni_3S_2@NF, achieving current densities of 10 and 50 mA cm^{-2} at low overpotentials of 74/175 and 80/160 mV for HER and OER, respectively. Furthermore, when integrated into an alkaline electrolyzer, Fe-MoS_2/Ni_3S_2@NF achieved a current of 10 mA cm^{-2} at a cell voltage of 1.60 V, demonstrating remarkable stability [21].

Table 4 Key factors influencing OER performance

Factor	Effect on OER	Mechanism/details
Adsorption	Determines reaction pathway and rate-limiting steps	Optimal catalysts balance adsorption energies to minimize overpotential
Electronic structure (e_g occupancy)	Influences bonding and catalytic activity	Ideal e_g occupancy of ~1.2 promotes OER activity in transition metal oxides
Overpotential (η)	Drives the reaction above the equilibrium potential	Lower overpotential indicates higher catalytic efficiency
Temperature	Enhances kinetics and ion mobility	Raises reaction rate; excessive heat can cause catalyst degradation
Catalysts	Provide active sites for intermediate formation	Common catalysts are metal oxides, and perovskites
Promoters/dopants	Modify electronic/surface properties, enhance stability	Improve conductivity, tune binding energies, increase active sites
Mass transport	It affects the delivery of reactants and removal of O_2	Poor mass transport can cause bubble buildup and concentration gradients
Flow rate	Controls interface dynamics and product desorption	High flow reduces O_2 accumulation and improves reactant access; it should be optimized to avoid turbulence

5 Electrocatalysts: The Evaluation Criterion for Electrocatalysts

A significant obstacle hindering the practical application of water splitting is the slow reaction kinetics of OER and HER, attributed to high overpotentials. Consequently, the electrocatalyst mainly aims to reduce the overpotentials for OER and HER to facilitate effective hydrogen and oxygen generation. Designing ideal electrocatalysts that are cost-effective, exhibit high catalytic activity, and demonstrate endurance across various media for electrolytic water splitting presents significant challenges. The design of electrocatalysts should rely on the operational parameters of the water electrolysis cell.

Water electrolysis is an endergonic reaction, according to thermodynamic values such as the positive value of Gibbs free energy (ΔG). Thus, electrocatalysts should overcome kinetic barrier. Evaluating a catalyst's performance in electrocatalytic water splitting depends on several essential properties, including activity, stability, and efficiency (Table 5). The activity is evaluated by overpotential (η), Tafel slope, and exchange current density (j_0), which may be obtained from the polarization curves. Stability is assessed by fluctuations in potential or current over time. Efficiency is characterized by faradaic efficiency and turnover frequency. These parameters are summarized in Table 5.

The thermodynamic potential for the electrochemical water-splitting reaction is 1.23 V at 25 °C and 1 atm of pressure. Nonetheless, owing to the kinetic barrier associated with the reaction, water electrolysis necessitates a potential surpassing the thermodynamic potential (1.23 V) to surmount this barrier. The additional potential is referred to as overpotential (η), principally arising from the inherent activation barriers at both the anode and cathode. Overpotential is a vital metric for evaluating the performance of electrocatalysts and measures usually at a current density of 10 mA cm^{-2}, corresponding to solar-to-hydrogen efficiency of 12.3%. Linear sweep voltammetry (LSV) is usually employed to examine the changes in overpotential before and after cyclic voltammetry (CV) cycling at a specified current density. CV is an alternate method that measures current by altering voltage at different scan rates (0.5–200 mV s^{-1}). The Tafel slope and exchange current (J_0) can also be used as metrics for evaluating activity based on the link between overpotential and kinetic current (Table 5). Tafel slope is calculated using the equation: $\eta = a + b \log j$, where η denotes overpotential and j refer to current density. The Tafel plot's provides two significant kinetic parameters. One parameter is the Tafel slope b, and the other is j, which may be determined by measuring the current at zero overpotential ($\eta = 0$). The Tafel slope, i.e., b value, is associated with the catalytic reaction mechanism. A low Tafel slope indicates an increase in current density relative to changes in overpotential, offering accelerated electrocatalytic reaction kinetics. The exchange current density characterizes the inherent charge transfer at equilibrium conditions. A high j_0 value indicates an increased charge transfer rate and a decreased reaction barrier. An improved electrocatalyst will exhibit a reduced Tafel slope and an elevated exchange current density.

The electrocatalyst's efficiency can be quantitatively evaluated using Faradaic efficiency (Table 5). Faradaic efficiency measures the effectiveness of electron transfer from the external circuit to the electrode surface during the electrochemical process. It is calculated by the ratio of the experimentally measured quantities of H_2 or O_2 to the theoretically calculated quantities of H_2 or O_2. The experimental results are derived from evaluating gas production through water–gas displacement or gas chromatography (GC). Turnover frequency (TOF) is essential for quantifying the reaction rate concerning catalytic sites, indicating the catalyst's inherent catalytic activity. It measures the quantity of reactants converted into the desired product per catalytic site per unit of time (Table 5). It is a good technique for evaluating the catalytic activity of different catalysts, especially within analogous systems or under similar conditions.

Stability is important for evaluating the catalyst's feasibility in water splitting via electrolysis (Table 5). It is assessed using potential or current measurements over time (more than 12 h) using two traditional methodologies: chronoamperometry (relation between current over time, I–t curve) or chronopotentiometry (relation between potential over time, E–t curve). The constant long constancy of the current or potential enhances the stability of the catalyst.

Table 5 Evaluation criteria for electrocatalyst performance

Property	Definition	Technique used	Equation for calculation	Comments	Key considerations
Activity	Represents the catalytic ability to drive HER and OER efficiently. It is mainly determined by overpotential (η), Tafel slope, and exchange current density (j_0)	• LSV • CV • Tafel plot	1. **Overpotential (η):** $\eta = E - E_o$ *(E = applied potential, E° = thermodynamic potential: 0 V for HER and 1.23 V for OER ($\eta_{OER} = E_{RHE} - 1.23$ V))* 2. **Tafel Slope (b):** $\eta = b \log j + a$ *(j = current density, a = intercept)* 3. **Exchange Current Density (j_0):** Obtained by extrapolating Tafel plots to $\eta = 0$ V 4. **The electrochemical active surface areas (ECSA)** ECSA = $(C_{dl}/C_s) \times A$ (where Cdl is double-layer capacitances; Cs is specific capacitance (i.e., 0.04 mF/cm^2); A represents the electrode's actual surface area	– Lower η = better catalytic performance – Lower Tafel slope = faster reaction kinetics – Higher j_0 = more excellent intrinsic catalytic activity – Mass transport limitations can affect measurement accuracy	– Electrode preparation affects activity (e.g., catalyst loading, morphology, and binder selection) – Comparison must be under identical conditions (e.g., same electrolyte, pH, and temperature)

(continued)

Table 5 (continued)

Property	Definition	Technique used	Equation for calculation	Comments	Key considerations
Stability	The durability of the electrocatalyst under prolonged operational conditions, measured by the degradation of catalytic activity over time	• Chronoamperometry (CA) (constant potential) • Chronopotentiometry (CP) (constant current) • Multi-cycle CV	1. Stability is assessed by monitoring overpotential drift or current retention (%) over time 2. Comparison of initial and final current densities after a defined test period	– Stability is influenced by catalyst dissolution, surface reconstruction, and electrode degradation – Realistic testing should include high-current conditions and long-term cycling to mimic practical applications	– Highly active catalysts often suffer from poor long-term stability due to surface oxidation or leaching – Lab-scale performance should be validated under industrial conditions

(continued)

Table 5 (continued)

Property	Definition	Technique used	Equation for calculation	Comments	Key considerations
Efficiency	Evaluate how effectively charge transfer leads to gas evolution with minimal energy loss	• Gas Chromatography (GC) • Electrochemical Impedance Spectroscopy (EIS) • Rotating Disk Electrode (RDE) studies	1. **Faradaic Efficiency (FE):** 2. $FE = \frac{\text{Experimental charge for gas evolution}}{\text{Total charge passed}} \times 100\%$ 3. **Turnover Frequency (TOF):** 4. $TOF = \frac{\text{Moles of gas produced per second}}{\text{Moles of active sites}}$ *TOF can also be calculated from CV curves at different san rates $Slope = n^2F^2A\Gamma_0/4RT$ Where n, F, A, Γ_0, R, and T represent the number of transferred electrons, Faradaic constant (96485C mol^{-1}), surface area, the concentration of active sites on the surface in units of mol cm^{-2}, absolute temperature, and the ideal gas constant, respectively $TOF = JA/nFm$ J, A, and m represent current density, the geometrical area of the electrode, and the mass of active sites (mol), respectively, while n = 4 represents the electron consumption moles for the evolution of 1 mol of O_2	– High FE (>90%) indicates minimal side reactions – TOF depends on catalyst loading, making direct material comparison difficult – Requires precise gas collection and electrode surface area normalization for accurate assessment	– Electrolyte composition affects efficiency – Side reactions can impact faradaic efficiency – Testing should simulate real-world conditions – Industrially relevant current densities should be considered

6 Evaluation Strategies of HER Catalysts

Assessing the efficacy of HER catalysts requires the consideration of many electrochemical and physicochemical criteria that collectively elucidate their activity, kinetics, stability, and inherent characteristics (Table 6). Overpotential (η) is essential, signifying the additional energy necessary to commence hydrogen evolution, with reference points such as η_1, η_{10}, and η_{100} employed at current densities values of 1, 10, and 100 mA cm^{-2}, respectively for standardized comparison. The Tafel slope (b) and j_0 provide insights into reaction kinetics, with a lower b and greater j_0 indicating enhanced catalytic efficiency. EIS facilitates the measurement of charge transfer resistance (R_{ct}), which means the efficiency of electron movement within the system. Long-term operational durability and constant potential or current electrolysis are evaluated via CV to assess catalyst stability. Faradaic efficiency (FE) quantifies the efficacy of electrical charge conversion into hydrogen gas, with values approaching 100% ensuring high catalytic selectivity (Table 6). TOF quantifies the production of hydrogen molecules per active site per second, hence assessing intrinsic catalytic activity. The hydrogen binding energy (HBE), particularly ΔG_H*, should be medium value, not excessively strong or weak, to improve the adsorption and desorption processes during HER, with optimal catalysts positioned around the apex of the Sabatier volcano plot (Table 6).

7 Evaluation Strategies of OER Catalysts

The efficacy of OER catalysts is often evaluated using various electrochemical parameters that indicate their activity, kinetics, stability, and efficiency (Table 7). Overpotential (η) denotes the extra voltage necessary beyond the theoretical potential (1.23 V vs. NHE) to facilitate the OER, with η_{10} (at 10 mA cm^{-2}) serving as a standard reference for comparison among different electrocatalysts. A low overpotential signifies a more efficient catalyst. The Tafel slope (b) indicates the kinetic characteristics of the OER; low values ensure accelerated electron transport and improved reaction rates. EIS is employed to ascertain charge transfer resistance (Rct), which indicates the facilitation of electron movement at the catalyst–electrolyte interface (Table 7). Low Rct values in OER catalysts indicate superior catalytic kinetics. Stability is a crucial property, typically assessed via long-term chronoamperometry or CV, to determine the catalyst's durability under severe oxidative conditions (Table 7). Faradaic efficiency (FE) quantifies the efficacy of electron use in producing molecular oxygen, with values approaching 100% signifying high efficiency (Table 7). TOF quantifies intrinsic activity by measuring the number of oxygen molecules produced per active site per second (Table 7).

Table 6 Summary of evaluation strategies for HER catalysts

Parameter	Definition	Ideal value/condition	Significance
Overpotential (η)	Extra voltage needed beyond thermodynamic potential (0 V vs. RHE)	η_{10} < ~100 mV	Indicates energy efficiency; η_{10} used for comparison
Tafel slope (b)	Slope of η versus log(j); related to reaction kinetics	Low b (~30–40 mV/dec)	Lower b implies faster kinetics
Exchange current density (j_0)	Current density at zero overpotential	High j_0	Reflects intrinsic catalytic activity
EIS (R_{ct}, R_{ad})	EIS; Rct = charge transfer resistance	Low R_{ct} and R_{ad}	Indicates fast charge transfer and low adsorption barrier
Stability	Durability under prolonged operation	Minimal degradation over ≥ 10 h at ≥ 10 mA cm^{-2}	Essential for practical applications
Faradaic efficiency (FE)	% of electrons used to produce H_2	Close to 100%	Measures selectivity and charge utilization efficiency
Turnover frequency (TOF)	Molecules of H_2 generated per active site per second	Higher TOF	Indicates intrinsic efficiency of catalytic sites
Hydrogen binding energy (ΔG_H)*	Free energy of H adsorption	~0 eV	Balanced adsorption/desorption, optimal HER kinetics

8 Designing an Electrocatalyst for Integrating HER and OER

There are parameters that should be considered when designing the electrocatalyst effect for HER and OER (Fig. 3). Creating an electrocatalyst that effectively combines the HER and OER necessitates meticulous attention to material composition, structural design, conductivity, substrates, doping, defects/vacancies, morphology, and porosity (Fig. 3). The selection of materials holds significant importance. There are several compositions with effective performance, including compounds of transition metals, such as layered double hydroxides, nitrides, phosphides, and oxides, as well as carbon materials doped with heteroatoms, including nitrogen, boron, and phosphorus-doped graphene or carbon nanotubes. Select or integrate materials such that one element promotes HER while the other supports OER. Carbon-based supports for dual-doped elements (such as N, S or B, P) enhance the activation of both HER and OER sites. Develop porous structures offers large surface areas, leading to high number of active sites. Interface engineering using different materials significantly improves charge transfer and efficiently separates the HER/OER zones.

Table 7 Summary of evaluation strategies for OER catalysts

Parameter	Definition	Ideal value/condition	Significance
Overpotential (η)	Extra voltage beyond 1.23 V versus RHE to initiate OER	η_{10} < ~350 mV	Indicates energy efficiency; η_{10} used for catalyst benchmarking
Tafel slope (b)	Slope of η versus log(j); reflects OER reaction kinetics	Low b (~40–60 mV/dec)	A lower slope means faster kinetics
Exchange current density (j_0)	Current at zero overpotential	High j_0	Indicates fast intrinsic activity
EIS (Rct)	EIS for charge transfer resistance	Low Rct	Better conductivity and faster charge transfer
Stability	Long-term durability under oxidative conditions	Minimal overpotential increase over ≥10 h	Key for industrial applications
Faradaic efficiency (FE)	% of electrons contributing to O_2 evolution	Close to 100%	Measures selectivity and efficient electron usage
Turnover frequency (TOF)	Molecules of O_2 evolved per site per second	High TOF	Indicates catalytic activity per active site

Utilize conductive substrates such as carbon cloth or nickel foam to facilitate optimal electron transport. Introduce vacancies, such as oxygen or nitrogen vacancies, to adjust the electronic structure and improve the adsorption of reactancs and intermediates. Incorporate heteroatoms or merge metals to modify the d-band center and influence reaction energetics. It is crucial to guarantee corrosion resistance, particularly in environments with alkaline or acidic electrolytes. Electrocatalysts can be categorized into noble-metal, noble-free metal, and metal-free electrocatalysts. The following section discusses these classes.

8.1 Noble-Based Electrocatalysts

The metals of the Platinum Group; platinum (Pt), ruthenium (Ru), palladium (Pd), iridium (Ir), and rhodium (Rh), offer the most efficient electrocatalysts for HER due to their superior characteristics, including atomic hydrogen adsorption, binding energy, and Gibbs free energy [22]. They possess low activation energies for the surface desorption of hydrogen. These catalysts can attain further properties, including high exchange current densities (j_0) values, small Tafel slopes, and almost 100% Faradaic efficiency. Pt represents the apex of the Volcanic curve depicted in Fig. 4a [23]. PGMs surpass electrocatalysts in HER due to their superior efficiency, which is characterized by a low Tafel slope and a “quasi-zero” onset overpotential when utilized with platinum. It is situated approximately at the summit of the volcano [23]. The Sabatier principle posits that platinum’s exceptional catalytic efficiency as a hydrogen evolution reaction catalyst is due to its optimum adsorption and desorption properties on its surface, as indicated by its moderate bond strength between catalyst and hydrogen. Nonetheless, noble platinum metal catalysts’ cost and maintenance demands hinder their practical application. Transition-metal nitrides, sulfides, and phosphides exemplify electrocatalysts based on earth-abundant elements that have shown considerable potential as catalysts for the HER to address this challenge. Nonetheless, the limited quantities and high costs of platinum and other associated precious metals on Earth impede the extensive application of Pt-based metal catalysts for large-scale H_2 production to meet global energy demands.

Molecular noble metal nanoclusters, with particle size less than 3 nm, are an effective category of electrocatalysts with well-defined composition and structure [22]. Whether independent or supported, atomically precise noble metal nanoclusters exhibit significant potential for various electrocatalytic reactions, especially the ORR and HER [22]. The challenges encompass the difficulty in isolating preferred products due to generating unwanted byproducts and the knowledge of reaction pathways since specific nanocluster-catalyzed reactions favor indirect multi-step pathways rather than more efficient direct routes. Significant parameters like composition, ligand environment, and surface structure remain inadequately comprehended. Doping with heteroatoms can significantly modify structure and reactivity, highlighting the necessity for further research. Furthermore, customizing ligands for reactions is essential, as seen by the performance discrepancies in HER and OER that rely on the ligand used.

Although many efficient OER catalysts, such as those based on Ir and Ru, demonstrate significant chemical stability (high dissolution resistance or low dissolution

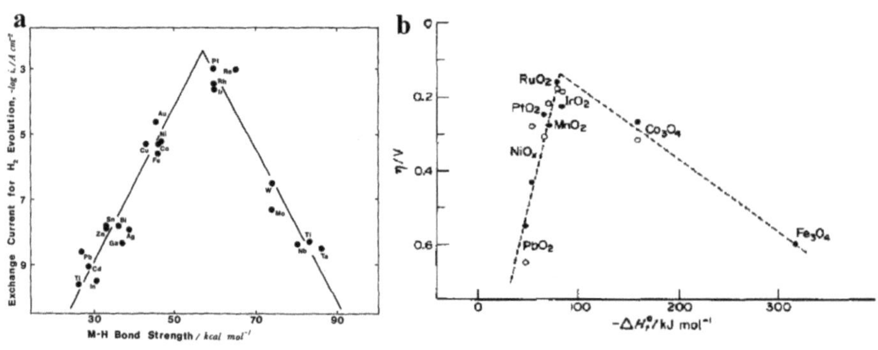

Fig. 4 Vocano plots for **a** currents for HER versus strength of the intermediate metal-hydrogen bond. Figure reprinted with permission from Ref. [23]. Copyright [24] © 1972 Published by Elsevier B.V. **b** Electrocatalytic activity in O_2 evolution at various metal oxides as a function of enthalpy, open circle for alkaline, while closed circle for acidic media. Copyright © 1984 Published by Elsevier Ltd.

properties) in acidic environments. In general, non-noble or transition metal-based materials fail to endure such conditions. Thus, it is objective to create stable and resilient OER electrocatalysts that exhibit high activity and long-term stability in acidic environments.

Noble metals are expensive electrocatalysts. Therefore, reducing the use of noble metal-based electrocatalysts or replacing them with cost-effective, non-noble or cheap metal-based electrocatalysts exhibiting high catalytic activity and stability is crucial in advancing the hydrogen economy. This aim can be achieved using abundantly available nanomaterials, particularly highlighting non-platinum electrocatalysts, including transition metal of nitrides, oxides, carbides, chalcogenides (sulfides and selenides), and nonmetallic composite catalysts.

8.2 Noble Metal-Free Electrocatalysts

Non-noble or noble metal-free metal-based electrocatalysts, including carbides, nitrides, phosphides, and chalcogenides, have garnered significant interest because of their superior performance in the OER using alkaline media. However, these catalysts experience compositional and structural conversion under OER conditions. Consequently, identifying the active sites remains a challenge, complicating the development of approaches to detect these sites to guide the design and fabrication of effective catalysts.

Efficient electrocatalysts can be achieved with superior electronic structure, surface morphology, and active sites. In this context, several materials were reported including single-atom catalysts, metal dichalcogenides, porous materials (e.g. metal–organic frameworks (MOFs), and covalent organic frameworks (COFs)), heteroatom-doped carbon materials, metal carbides, metal nitrides, as well as bimetallic and alloy catalysts have exhibited minimal overpotential. Based on Volcano plots, IrO_2 and RuO_2 are the most effective electrocatalysts for OER (Fig. 4b).

Abdelhamid*et al.* reported the synthesis of a bimetallic oxide composed of nickel (Ni) and copper (Cu)@carbon, referred to as (Ni, Cu)O@C nanocomposite [25]. They synthesized the material via the carbonization of Ni/Cu-based MOF. (Ni, Cu)O@C nanocomposite was evaluated as electrocatalyst for OER compared to CuO@C (Fig. 5) [25]. It demonstrated an overpotential of 330 mV at 10 mA cm^{-2}, markedly surpassing the performance of the CuO@C catalyst, which recorded η of 425 mV. The catalyst demonstrated a modest Tafel slope of 72 mV dec^{-1}. Additionally, the catalyst showed a TOF of 0.39 mol of O_2 per second at η of 400 mV. It exhibited remarkable stability, sustaining its performance for 24 h under alkaline conditions [25].

McAllister *et al.* investigated the optimization of redox-active immobilized molecular metal-chalcogenide materials by manipulating chalcogen or metal's stoichiometry and examining essential factors of HER [26]. They reported selective selection of chalcogen elements and their proportions (e.g., O_2^{-}/S_2^{2-}, $O^{2-}/S_2^{2-}/S_4^{2-}$) for Mo or W. At optimal conditions, the LSV curve demonstrates that stoichiometric

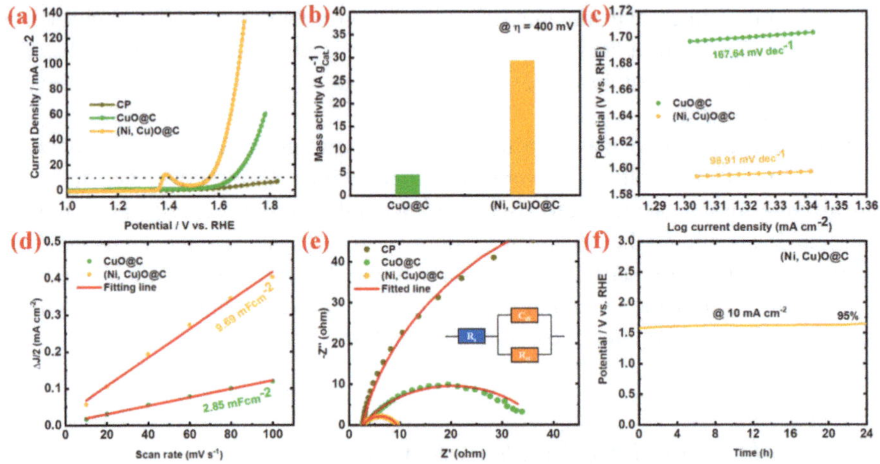

Fig. 5 Electrochemical measurements using **a** LSV curves, **b** Mass activity, **c** Tafel slopes; **d** Electrochemical double-layer capacitance; **e** Nyquist plots, and **f** Chronopotentiometry curve. Figure reprinted with permission from Ref. [25]. © 2025 Hydrogen Energy Publications LLC

and structural regulation facilitates hydrogen evolution at low overpotential without catalyst degradation over 1000 cycles [26].

A hollow nanostructure of Co_3O_4/Mo_2N heterostructures was synthesized and investigated for OER and HER [27]. The synthesis procedure involved regulated pyrolysis of Mo–Co Prussian blue analog with an appropriate Mo/Co ratio. The materials exhibit high H_2O/H* adsorption, offering a 12.9-fold and 20-fold increase in current density at a 300 mV, compared to the single-phase Co_3O_4 nanofibers for HER and OER, respectively [27].

Nickel-based materials are effective electrocatalysts for water splitting [28]. Several Ni-based catalysts were reported, including Fe-doped Ni phosphides/Ni sulfide p-p heterojunction (NiFe-PS) [29], Ni_3S_2-NiO_x/nickel foam (NF) [30], nickel-silver sulfides (Ag_2S-NiS_x) [31], Ni_3S_2 or NiS/carbon sphere [32], NiS_x/graphite [33], and Co_2C-NiTe [34]. NiFe-PS exhibits good electrocatalytic performance compared with expensive Pt-group elements such as RuO_2 (Fig. 6) [29]. Ni–Fe layered double hydroxides (NiFe-LDHs, 500 nm) are effective electrocatalysts for OER in alkaline environments [35]. The authors improved the activity of NiFe-LDH by controlling defects, doping, and exfoliating to achieve single-layer [35]. They reported controlled doping of NiFe-LDH with W by a straightforward one-step alcohothermal process to produce NiFeW-LDHs. The oxidation state of W in NiFeW-LDH was +6. The peak area of surface-adsorbed OH markedly increased with W doping, indicating improved OH adsorption on the NiFeW-LDH surface. Density functional theory (DFT) calculations demonstrated that W^{6+} promotes the adsorption of H_2O and O^{*-} intermediates, hence enhancing the activity of Fe sites. The NiFeW-LDH catalyst exhibited a low overpotential of 199 and 237 mV at 10 and 100 mA cm^{-2} in KOH (1 M) [35]. 3D Ni–V-based heterojunctions system

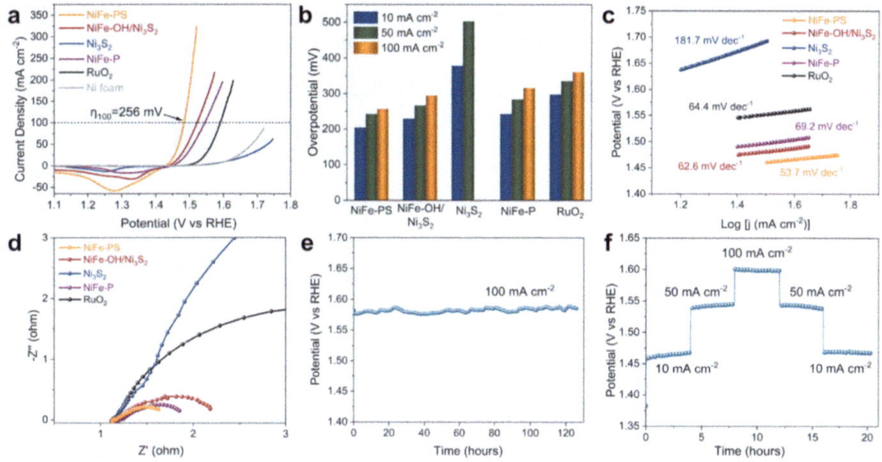

Fig. 6 **a** LSV curves, **b** overpotentials values, **c** Tafel plots, **d** Nyquist plots, **e** chronopotentiometric curve of NiFe-PS at 100 mA cm^{-2}, and **f** The staircase chronopotentiometric curve of NiFe-PS. Figure reprinted with permission from Ref. [29]. © 2022 Elsevier B.V.

supported on nickel foam (Ni_3N-VN/NF and Ni_2P-VP2/NF) were investigated for HER and OER [36]. The heterointerface in Ni_3N-VN exhibits an appropriate H* absorption energy, which is advantageous for improving HER activity, featuring an onset overpotential (η_{onset}) of zero and a Tafel slope of 37 mV dec^{-1} in KOH (1 M), comparable to that of Pt/C/NF. The combination of Ni_2P-VP_2 with oxide species for the OER demonstrates improved activity, achieving a ηonset of 220 mV and a Tafel slope of 49 mV dec^{-1}. The high electrocatalytic performance of the materials can be explained in terms of the heterointerface role in activating intermediates, the high conductivity of the material for electron transfer, and the porous structure that facilitates mass transport. The electrocatalyst offered straightforward integration for effective overall water splitting, achieving a low driving voltage of at least 1.43 V, with a measured voltage of 1.51 V, and demonstrating impressive stability over 100 h. The material can power a solar cell of 1.5 V [36].

8.3 *Metal-Free Electrocatalysts*

Metal-free electrocatalysts, including heteroatom-doped carbon materials like N, B, P, S-doped graphene [37–39], carbon nanotubes [40], COFs [41], g-C_3N_4 [19], and polymer-derived carbons [42], represent significant advancements as alternatives to conventional metal-based catalysts for OER and HER.

Carbon nanomaterials offer effective metal-free electrocatalysts. They can be easily synthesized using several cheap and abundant sources such as Glucose [43], *platycladus orientalis* tree-cone biomass [44], corn stalks [45], rice husk [46], peanut

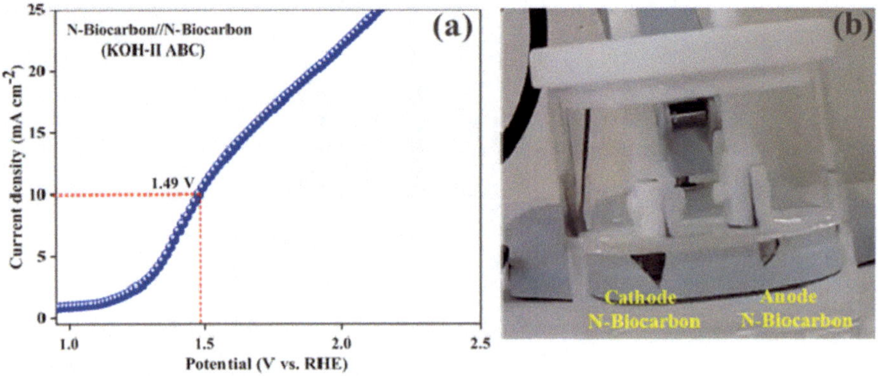

Fig. 7 **a** LSV curve for overall water splitting and **b** Digital photograph image of two-electrode setup. Figure reprinted with permission from Ref. [44] © 2023 Elsevier Ltd.

shell [47], palm waste [48], Silk cocoon [49], coconut water [50], Egg [51], and Coffee waste [52]. N-self-doped defect-rich porous carbon nanosheets were synthesized using biomass of *Platycladus orientalis* tree cones and were investigated for water splitting, i.e., HER and OER [44]. It exhibited surface areas, pore volume, and conductivity of 3369 $m^2\ g^{-1}$, 2.1 $cm^3\ g^{-1}$, and 12.69 S cm^{-1}. It showed outstanding performance in both OER and HER in H_2SO_4 (0.5 M) solution with an overpotential of 90 mV for OER and 188 mV for HER. It can be applied for overall water splitting, requiring only a cell voltage of 1.49 V to achieve current densities of 10 mA cm^{-2} (Fig. 7). The defects in the carbon, along with a higher specific and electrochemical active surface area due to N doping, significantly improve the reaction kinetics, leading to enhanced performances in both OER and HER [44].

Metal-free electrocatalysts exhibit several advantages. They can be synthesized from abundant resources offering both economical and plentiful. In contrast to noble metals such as Pt, Ru, and Ir, catalysts that do not contain metals utilize abundant elements in the Earth, lowering costs and enhancing accessibility. These materials exhibit exceptional resistance to corrosion. They demonstrate remarkable chemical stability in both acidic and alkaline environments. They provide a customizable framework. Performance enhancement can be achieved through doping, structural engineering, and surface functionalization, which adjusts electronic properties and active sites. These materials demonstrate sustainability and reduced toxicity under the principles of green chemistry. However, they demonstrate low catalytic efficiency compared to metal-based counterparts, especially under high current densities. Certain forms exhibit inadequate electrical conductivity, which decreases charge transfer efficiency. While there have been advancements, the long-term operational stability in challenging environments, particularly for OER, requires further improvement.

9 Challenges

The HER and OER encounter significant obstacles that impede their practical use in energy conversion systems (Table 8). Most of the reported electrocatalysts in HER exhibit high overpotential and slow kinetics, necessitating sophisticated catalyst design to low energy barriers and enhance proton adsorption/desorption. The adsorption of intermediates can obstruct active sites and decrease effectiveness, while the environmental reliance on acidic or alkaline medium necessitates catalysts that function across various pH levels. The expense and restricted availability of noble metal catalysts, e.g., Pt, limit scalability, necessitating the exploration of earth-abundant substitutes.

The inefficiency of existing electrolysis techniques and high production expenses hinder the advancement of hydrogen technology [8]. The electrolysis efficiency is 60–80%, with hydrogen production prices of $5 per kilogram [8]. Strategic improvements are essential to lower these costs and to achieve a high efficiency of over 80% [8]. Moreover, hydrogen storage presents distinct obstacles, necessitating up to 700 pressure settings or temperatures below −253 °C. The specified storage conditions require the advancement of materials and enhancements to infrastructure. The

Table 8 Challenges in HER and OER

Challenge	HER	OER
High overpotential and kinetics	Requires additional voltage to drive HER due to slow proton-electron transfer kinetics	OER is sluggish due to a complex four-electron transfer mechanism
Adsorption	Intermediates block active sites; strong adsorption can hinder desorption and poison catalysts	Strong adsorption of OER intermediates (e.g., OH*, OOH*, or O*) can lead to high overpotentials
pH dependency	Performance varies in acidic versus basic media; few catalysts are efficient across pH range	Catalyst stability and performance are also electrolyte-dependent
Catalyst cost and availability	Noble metals (Pt) are efficient but costly; alternatives like Ni, Co, Mo are more abundant	Ir and Ru oxides are highly active but expensive; alternatives often lack long-term stability
Catalyst stability	Surface poisoning and structural degradation over time	Deactivation due to oxidative environment and poor durability under high potentials
Mass transport issues	Poor proton supply and hydrogen removal can reduce efficiency	O_2 bubble accumulation at the electrode surface hinders electron transfer and reduces active surface area
Scalability	Requires cost-effective, abundant, and stable materials for industrial-scale hydrogen production	Scaling up demands robust OER catalysts with low overpotential and long-term operational durability

economic viability and market integration of hydrogen will critically depend on addressing these technological and infrastructural hurdles underpinned by strong regulatory frameworks [8].

The OER, which is characterized by a complicated four-electron mechanism, experiences inherently sluggish kinetics, necessitating high overpotentials and resilient catalysts capable of stabilizing various oxygenated intermediates (Table 8). Catalyst instability, low conductivity, and the buildup of oxygen bubbles further constrain OER performance.

Both HER and OER benefit from advancements in electronic structure optimization and surface engineering, as well as incorporating promoters or hybrid materials to enhance catalytic performance, decrease costs, and facilitate operation in various electrolytic conditions.

10 Conclusions

This book chapter discusses the techniques for designing and fabricating nanostructured materials for catalysis, which may provide significant insights into efficient catalyst production and the possibility for substantial future advancements in water electrolysis. It emphasizes the essential principles and strategies in designing high-performance electrocatalysts, focusing on activity, stability, and cost-effectiveness. Electrocatalysts with high efficiency toward HER and OER are challenging for enhancing hydrogen production via water-splitting electrolysis. Investigating nanostructured materials, including noble and non-noble metals and metal-free systems, while adjusting their composition, morphology, and surface characteristics can lead to substantial advancements in scalable hydrogen energy solutions. Understanding synergistic interactions, doping strategies, and defect engineering highlights the significance of comprehensive material design. Despite ongoing long-term durability and reaction kinetics challenges, the relentless advancement of innovative electrocatalysts offers a hopeful avenue for achieving sustainable, commercial-scale hydrogen production.

References

1. IEA: Global Energy Review 2025, vol. Paris http (2025)
2. Frieden, F., Leker, J.: Sustain. Energy Fuels **8**, 1806–1822 (2024)
3. Le, T.T., Sharma, P., Bora, B.J., Tran, V.D., Truong, T.H., Le, H.C., Nguyen, P.Q.P.: Int. J. Hydrogen Energy **54**, 791–816 (2024)
4. Zainal, B.S., Ker, P.J., Mohamed, H., Ong, H.C., Fattah, I.M.R., Rahman, S.M.A., Nghiem, L.D., Mahlia, T.M.I.: Renew. Sustain. Energy Rev. **189**, 113941 (2024)
5. Kamran, M., Turzyński, M.: J. Energy Storage **96**, 112601 (2024)
6. Abdelhamid, H.N.: Int. J. Hydrogen Energy **46**, 726–765 (2021)
7. Alamiery, A.: ChemPhysMater. https://doi.org/10.1016/j.chphma.2023.09.002

8. Sadeq, A.M., Homod, R.Z., Hussein, A.K., Togun, H., Mahmoodi, A., Isleem, H.F., Patil, A.R., Moghaddam, A.H.: Sci. Total Environ. **939**, 173622 (2024)
9. Qadeer, M.A., Zhang, X., Farid, M.A., Tanveer, M., Yan, Y., Du, S., Huang, Z.-F., Tahir, M., Zou, J.-J.: J. Power. Sources **613**, 234856 (2024)
10. Emam, A.S., Hamdan, M.O., Abu-Nabah, B.A., Elnajjar, E.: Int. J. Hydrogen Energy **64**, 599–625 (2024)
11. Kempler, P.A., Coridan, R.H., Luo, L.: Chem. Rev. **124**, 10964–11007 (2024)
12. Li, J., Fu, G., Sheng, X., Li, G., Chen, H., Shu, K., Dong, Y., Wang, T., Deng, Y.: Adv. Powder Mater. **3**, 100227 (2024)
13. Du, L., Yang, Y., Zhou, L., Liu, M.: Sustainability **16**, 4602 (2024)
14. Halder, P., Babaie, M., Salek, F., Shah, K., Stevanovic, S., Bodisco, T.A., Zare, A.: Renew. Sustain. Energy Rev. **199**, 114543 (2024)
15. Kumar, R., Singh, R., Dutta, S.: Energy Fuels **38**, 2601–2629 (2024)
16. Bade, S.O., Tomomewo, O.S., Meenakshisundaram, A., Ferron, P., Oni, B.A.: Int. J. Hydrogen Energy **49**, 314–335 (2024)
17. Rampai, M.M., Mtshali, C.B., Seroka, N.S., Khotseng, L.: RSC Adv. **14**, 6699–6718 (2024)
18. Qin, C., Chen, S., Gomaa, H., Shenashen, M.A., El-Safty, S.A., Liu, Q., An, C., Liu, X., Deng, Q., Hu, N.: Acta Physico-Chimica Sin. **40**, 2307059 (2024)
19. Khan, I., Khan, S., Shayan, M., Ullah, R., Rizwan, M., Khan, A., Alodhayb, A.N., Iqbal, A., Iqbal, K., Ullah, S.: Carbon N. Y. **229**, 119451 (2024)
20. Abodouh, M.M., Khedr, G.E., Allam, N.K.: Int. J. Hydrogen Energy **61**, 922–933 (2024)
21. Wang, T., Li, B., Wang, P., Xu, M., Wang, D., Wang, Y., Zhang, W., Qu, C., Feng, M.: J. Colloid Interface Sci. **672**, 715–723 (2024)
22. Qin, D.-D., Tang, Y., Ma, G., Qin, L., Tao, C.-L., Zhang, X., Tang, Z.: Int. J. Hydrogen Energy **46**, 25771–25781 (2021)
23. Trasatti, S.: J. Electroanal. Chem. Interfacial Electrochem. **39**, 163–184 (1972)
24. Trasatti, S.: Electrochim. Acta **29**, 1503–1512 (1984)
25. Alhakemy, A.Z., Elsayed, M.H., Algethami, F.K., Abdelhamid, H.N.: Int. J. Hydrogen Energy **115**, 289–298 (2025)
26. McAllister, J., Bandeira, N.A.G., McGlynn, J.C., Ganin, A.Y., Song, Y.-F., Bo, C., Miras, H.N.: Nat. Commun. **10**, 370 (2019)
27. Wang, T., Wang, P., Zang, W., Li, X., Chen, D., Kou, Z., Mu, S., Wang, J.: Adv. Funct. Mater. https://doi.org/10.1002/adfm.202107382
28. Qi, Y., Qiu, L., Ma, X., Wu, J., Xiang, J., Guo, C., Yu, J., Li, K., Tao, Z., Lv, Y.: Int. J. Hydrogen Energy **83**, 520–544 (2024)
29. Zhang, J., Yu, H., Yang, J., Zhu, X., Hu, M., Yang, J.: J. Alloys Compd. **924**, 166613 (2022)
30. Huang, Z., He, L., Zhang, W., Huang, W., Mo, Q., Yang, L., Fu, Q., Gao, Q.: J. Colloid Interface Sci. **622**, 728–737 (2022)
31. Zhang, H., Tang, H., Weng, Q., Wei, Q., Duan, M., Bo, X., Fu, F., Zan, L.: J. Solid State Chem. **316**, 123556 (2022)
32. Gao, T., Nie, M., Luo, J., Huang, Z., Sun, H., Guo, P., Xue, Z., Liao, J., Li, Q., Teng, L.: Electrochem. Commun. **129**, 107076 (2021)
33. Ma, M., Yang, G., Wang, H., Lu, Y., Zhang, B., Cao, X., Peng, D., Du, X., Liu, Y., Huang, Y.: Int. J. Hydrogen Energy **44**, 1544–1554 (2019)
34. Munawar, T., Bashir, A., El-Marghany, A., Shen, L., Tu, Z., Fatima, S., Ghafoor Abid, A., Alim Khan, S., Koc, M., Yan, C.-F., Iqbal, F.: Fuel **375**, 132445 (2024)
35. Duan, X., Sendeku, M.G., Zhang, D., Zhou, D., Xu, L., Gao, X., Chen, A., Kuang, Y., Sun, X.: Acta Physico-Chimica Sin. **40**, 2303055 (2024)
36. Yan, H., Xie, Y., Wu, A., Cai, Z., Wang, L., Tian, C., Zhang, X., Fu, H.: Adv. Mater. https://doi.org/10.1002/adma.201901174
37. Saini, R., Naaz, F., Bashal, A.H., Pandit, A.H., Farooq, U.: Green Chem. **26**, 57–102 (2024)
38. Raza, A., Hassan, J.Z., Qumar, U., Zaheer, A., Babar, Z.U.D., Iannotti, V., Cassinese, A.: Mater. Today Adv. **22**, 100488 (2024)

39. Tripathi, P., Verma, A.K., Sinha, A.S.K., Singh, S.: ChemistrySelect. https://doi.org/10.1002/slct.202403155
40. Xia, G., Habibullah, Xie, Q., Huang, Q., Ye, M., Gong, B., Du, R., Wang, Y., Yan, Y., Chen, Y., Wu, C.: Chem. Eng. J. **495**, 153405 (2024)
41. Dhanabalan, K., Sriram, G., Sannasi, V., Ajeya, K.V., Jung, S.-H., Jung, H.-Y.: Int. J. Hydrogen Energy **51**, 376–398 (2024)
42. Meng, Z., Wan, F., Wu, J., Xiong, Y., Chen, Y., Wu, M., Wang, R., Tang, H., Tian, T., Ma, J., Tang, H.: Int. J. Hydrogen Energy **102**, 1411–1420 (2025)
43. Huang, B., Liu, Y., Xie, Z.: J. Mater. Chem. A **5**, 23481–23488 (2017)
44. Murugan, N., Thangarasu, S., Bin Seo, S., Mariappan, A., Choi, Y.R., Oh, T.H., Kim, Y.A.: Renew. Energy **222**, 119801 (2024)
45. Liu, Z., Zhou, Q., Zhao, B., Li, S., Xiong, Y., Xu, W.: Fuel **280**, 118567 (2020)
46. Guchhait, S.K., Sutradhar, D., Nandi, R., Sarma, A.K.: Energy Sources Part A Recover. Util. Environ. Eff. **45**, 5957–5969 (2023)
47. Saravanan, K.R.A., Prabu, N., Sasidharan, M., Maduraiveeran, G.: Appl. Surf. Sci. **489**, 725–733 (2019)
48. Prabu, N., Saravanan, R.S.A., Kesavan, T., Maduraiveeran, G., Sasidharan, M.: Carbon N. Y. **152**, 188–197 (2019)
49. Liu, X., Zhang, M., Yu, D., Li, T., Wan, M., Zhu, H., Du, M., Yao, J.: Electrochim. Acta **215**, 223–230 (2016)
50. Raut, S.D., Shinde, N.M., Nakate, Y.T., Ghule, B.G., Gore, S.K., Shaikh, S.F., Pak, J.J., Al-Enizi, A.M., Mane, R.S.: ACS Omega **6**, 12623–12630 (2021)
51. Wu, H., Geng, J., Ge, H., Guo, Z., Wang, Y., Zheng, G.: Adv. Energy Mater. https://doi.org/10.1002/aenm.201600794
52. Pandey, K., Jeong, H.K.: Chem. Phys. Impact **6**, 100175 (2023)

Integration of 2D Materials in Water Splitting Devices

Akshidha Singla and Aman Mahajan

Abstract Two-dimensional (2D) materials are currently in limelight for water splitting applications by virtue of their exceptional physicochemical and electrochemical properties. Despite their excellent performance, interlayers restacking due to inherent Van der Waal forces which retards ionic conductivity and making them imperative for electrochemical performance in practical applications. Among various approaches, integration of 2D materials found to be promising in water splitting processes as it finely tunes the electronic band structure by synergically providing maximum surfacial active sites; facilitating swift charge transfer kinetics; high thermodynamic stability and catalytic activity as compared to their single counterparts. Accordingly, this book chapter will focus on the recent developments advanced fabrication techniques, functionalization of 2D materials by integrating them with materials of different dimensions to enhance their catalytic activity for water-splitting applications. Additionally, comprehensive understanding and in-depth analysis of electrochemical processes involved in oxygen evolution reaction (OER), hydrogen evolution reaction (HER) and photoelectrochemical (PEC) water splitting processes have been acknowledged. Finally, the chapter addresses current key challenges and proposes future directions for developing high-performing 2D material based electrocatalysts for efficient hydrogen production as a key to next-generation energy solutions.

Keywords Integration of 2D materials · Water splitting · HER · OER · Photoelectrochemical · Defect engineering

1 Introduction

Rapid pace of industrialization and globalization in twenty-first century heavily reliant on energy consumption and raises a severe energy concern due to world's dependency on fossil fuel-based resources to meet its energy demands. In 2008, the global energy consumption rate reached 15 TW and projections indicates a near

A. Singla · A. Mahajan (✉)
Department of Physics, Guru Nanak Dev University, Amritsar, Punjab, India
e-mail: aman.phy@gndu.ac.in

O. P. Pandey et al. (eds.), *2D Materials for Electrochemical Water Splitting*, Engineering Materials, https://doi.org/10.1007/978-981-95-2003-9_14

doubling of this figure by 2050 [1]. The fast depletion of these non-renewable fossil fuels possesses a global concern for energy crisis. Moreover, their utilization has led to the environmental pollution at unprecedented scale. Thus, there is a pressing need to find appropriate and viable solutions by replacing them with an alternative green, environment friendly and renewable energy sources which offer sustainable and relatively cleaner options in comparison to fossil fuels [2–4]. This growing and enormous demand for renewable and clean energy sources, sparked significant interest in hydrogen (H_2) as a potential fuel owing to its excellent energy density per mass [5] and high calorific value [6]. It has been found that demand for hydrogen will increase from the present 6–28 Mt in next 30 years as per the report of Energy and Research Institute of India [7]. An enormous demand for green hydrogen has emerged as it can decarbonize a variety of pivotal areas such as transportation, shipping and steel production. Green hydrogen has a significant potential in the transportation sector as it has the ability to replace the fossil fuels which immensely contributes towards the greenhouse gas emissions. It has the ability to power fuel cells, burnt in hydrogen turbine for electricity generation, or can be used directly as clean transportation fuel. Furthermore, hydrogen is considered as an extremely clean energy fuel as during its consumption, water is produced [8]. However, unlike other fuels derived directly from fossil-based resource, hydrogen is not readily available in its natural state. Thus, it can be generated from different primary energy sources like fossil fuel gasification, steam methane reforming, water splitting by photolysis and electrolysis, and biological biomass conversion functions. Out of all, water splitting is found to be a promising green technology for the production of clean and sustainable hydrogen, with the dissociation of water into hydrogen and oxygen producing high purity of hydrogen (>99.9%) but generally offers high cost than conventional fossil fuel gasification techniques [9]. Also, it has been expected that the energy cost for renewable hydrogen production could fall 30% by 2030 as per the report from International Energy Agency states [10]. Although water splitting technology has potential for renewable energy generation, though its implication is still in early ages [11]. In this regard, enormous research has been carried to improve the practicability by reducing the production cost and by increasing the efficacy of this technology to upscale them an industrial level [12, 13]. The primary methodologies for water splitting include photocatalytic, electrocatalytic and PEC approaches, each leveraging distinct mechanisms to achieve the desired chemical transformation.

2 Fundamentals of Water Splitting

Water electrolysis is based on the principle of splitting water molecules into oxygen at anode and hydrogen at cathode by passing an electric current [14].

At Anode

$$2H_2O \rightarrow O_2(\uparrow) + 4e^- + 4H^+, \quad E^o = 1.23\ V \tag{1}$$

At Cathode

$$4H^+ + 4e^- \rightarrow 2H_2 \uparrow, \quad E^o = 0\ V \tag{2}$$

Water electrolysis approach for hydrogen generation can be divided into two types as follows.

2.1 *Electrochemical Water Splitting*

The non-spontaneous nature of water splitting reactions is evidenced by their positive thermodynamic change, necessitating an external power source to drive the transformation. An electrolyser is used for the breakdown of water into its constituent H_2 and O_2 molecules through electrolysis, a process triggered by an external electrical potential. This device consists primarily of two electrodes: cathode and anode as a functionalized catalytic material, and an electrolyte. The water-splitting mechanism basically include two separate half-cell reactions where OER at the anode and HER takes place at the cathode [15] as shown in Fig. 1a. This electrochemical transition produces H_2 and O_2 gas, which can be used as reactants in fuel cells. Within a fuel cell, the recombination of these gases via the hydrogen oxidation and oxygen reduction reactions, respectively, produces electrical energy with water as the only waste [16]. The thermodynamic minimum potential is 1.23 V for the OER whereas HER has reduction potential of 0 V with respect to normal hydrogen electrode (NHE) as described in Eqs. (1) and (2). However, practical water electrolysis necessitates necessitate exceeding the standard potential then theoretical value for efficient water splitting to overcome the kinetic barriers, diffusion limitations, and ohmic losses.

Also, pH of the electrolytes also affects the water splitting reactions mechanisms [17] as shown in Fig. 1b.

In acidic media

At cathode electrons and hydrogen ions combines to produce hydrogen gas as shown in Eq. (4) whereas at anode process oxidation of water molecule takes place where water splits into oxygen and hydrogen ions as shown in Eq. (3):

At Anode

$$H_2O(l) \rightarrow \frac{1}{2}O_2(g) + 2H^+ + 2e^- \tag{3}$$

At Cathode

$$2e^- + 2H^+ \rightarrow H_2 \tag{4}$$

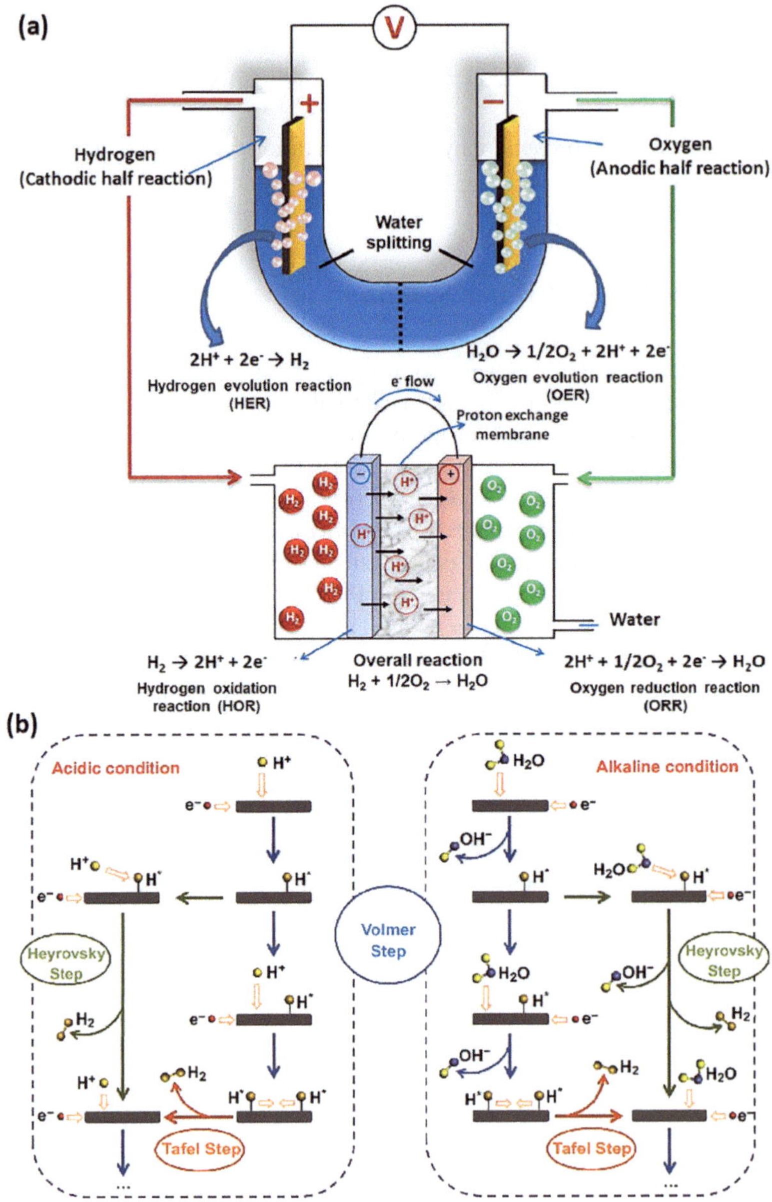

Fig. 1 **a** Schematic illustration of electrocatalytic water splitting for energy-related applications, i.e., oxygen reduction reaction in fuel cells, OER, and HER by water splitting. Adapted with permission from [16]. Copyright (2017) Vij et al., some rights reserved; exclusive licensee (American Chemical Society). Distributed under a Creative Commons Attribution License 4.0 (CC BY) https://creativecommons.org/licenses/by/4.0/; **b** schematic diagrams depicting the pathways of HER under both acidic and alkaline conditions. Adapted with permission from [17]. Copyright (2018) Wie et al., some rights reserved; exclusive licensee (Springer Nature). Distributed under a Creative Commons Attribution License 4.0 (CC BY) https://creativecommons.org/licenses/by/4.0/

In alkaline medium

The anode reaction involves OH^- ions which oxidized to produce oxygen, electrons and water as byproduct as shown in Eq. (5) whereas at cathode water and electrons reacts together to produce H_2 gas as a product as shown in Eq. (6).

At Anode

$$4OH^- \rightarrow O_2(g) + 4e^- + 2H_2O \tag{5}$$

At Cathode

$$2H_2O + 4e^- \rightarrow H_2(g) + 2OH^- \tag{6}$$

2.2 *Photoelectrochemical Water Splitting*

PEC water splitting process requires light responsive photoanode that can absorbs the photons from the light to accelerate the chemical reactions in water splitting [18] as illustrated in Fig. 2a. In PEC water splitting, a photocatalyst generates electron–hole pairs which facilitate the reduction at cathode to produce hydrogen as shown in Eq. (8) and oxidation reaction at anode necessary to split water molecules into oxygen as shown in Eq. (7). PEC water splitting is considered an eco-friendly and promising technology to harvest renewable energy sources to produce clean and green hydrogen fuel by using photon energy continuously striking the earth's surface. The reaction involve in PEC water splitting are given below:

Anode (+)

$$2H_2O \rightarrow O_2 + H^+ + 4e^- \tag{7}$$

Cathode (−)

$$4H^+ + 4e^- \rightarrow H_2 \tag{8}$$

Photocatalytic water splitting proceeds through a triphasic mechanistic pathway: (i) generation of electron–hole pairs via photonic band gap excitation; (ii) efficient charge carrier separation; and (iii) these photogenerated charge carriers participate in the surface redox reactions, i.e. HER and OER as shown in Fig. 2b. During solar illumination, absorbed photons stimulates the electron transitions from the valence band to the conduction band, creating electron–hole pairs. which are then separated by surface potentials, such that electrons in the conduction band have enough reducing power to produce hydrogen whereas holes in the valence band oxidized to generate

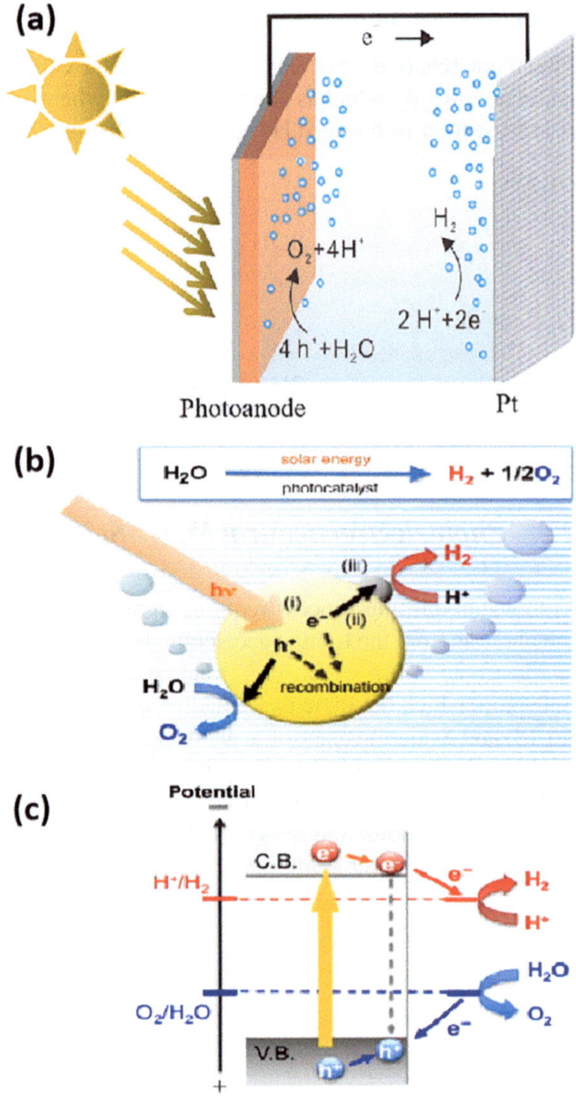

Fig. 2 Schematic illustration of **a** photoelectrochemical water splitting; **b**, **c** water splitting chemical reactions over photocatalyst. Adapted with permission from [17]. Copyright (2010) Elsevier

oxygen as shown in Fig. 2c. As a result, oxidation and reduction of water by holes and electrons at respective anode and cathode produce H_2 and O_2 gases [17].

3 Role of Catalyst in Water Splitting

There are various advantages of water splitting including HER and OER as they can be performed easily even at room temperature. Also, clean H_2 and O_2 are evolved through overall water splitting reaction and are obtained at respective cathode and anode, which can be collected easily without any gas separation technique. Despite the numerous benefits, the only limitation lies in the demand for substantial electrical energy and highly efficient and stable catalysts. Also, both HER and OER are inherently slow reactions due to their high energy the energy production efficiency. Traditionally, Pt, Ir, Ru, etc. noble metal based electrocatalysts are used and found to be highly efficient in overcoming the energy barriers for OER and HER, respectively. However, limited availability and high cost hampers their widespread scalability [19]. Therefore, significant research efforts have been dedicated to exploring low cost and efficient earth-abundant electrocatalysts for the sustainable production of H_2 and O_2.

Whereas for PEC application, the primary requirement of a photocatalyst is that the bottom of the conduction band must be more negative than the hydrogen reduction potential (0 V), while the top of the valence band should be more positive than the water oxidation potential, which is 1.23 V [20].

Further, light absorption efficiency, overpotential value, charge transport and separation characteristics, and their stability are the key factors in defining the efficiency of PEC devices. However, the recombination in charge carriers possess severe challenge in the overall performance metrices of PEC devices.

To mitigate this, extensive research efforts are devoted in surge of efficient photocatalyst or optimize the available photocatalyst to increase the photocurrent density by improving light absorption, charge separation, and catalytic activity to enhance overall performance of the device by developing efficient photocatalysts with suitable band gaps and high surface areas to maximize light absorption and catalytic activity.

Considering this, currently 2D materials like graphene [21], Borophene [22], MXene, Transition metal dichalcogenides (TMDs) [23], and Metal organic framework (MOF) [24] derived 2D materials have gained a lot of attention as prospective possibilities for electrochemical water splitting application owing to their tuneable electronic properties, high surface area, weak interlayered Van der Waal forces beneficial for easy intercalation of electrolytic ions and superior charge transport characteristics make them ideal for catalyst material improving the efficiency of electrocatalytic water splitting. Moreover, 2D materials provide enhanced light absorption, abundant active sites with less charge recombination rate; make then ideal material for catalytic reactions.

4 Importance of 2D Materials as Catalyst Materials for Water Splitting

2D materials found to be an excellent candidate for water splitting applications due to their high surface to volume ratio, abundant functional groups, unique electronic properties and abundant surface and edge enriched exposed sites and tuneable properties by couple them with other materials, make them ideal materials for various energy related applications as shown in Fig. 3.

4.1 *Structural Metrices*

2D materials are mostly arranged in a 2D planar structure composed of sp^2 hybridized atoms possessing high large surface to volume ratio surface area with high density of surface active sites; thereby allows the efficient adsorption and desorption of the electrolytic ions, resulting in enhancing the overall catalytic efficiency overall [25]. Along with this, 2D structure materials allows electrons to move freely across the plane, reducing charge transfer resistance and enhancing conductivity; beneficial in applications where fast charge transfer is important.

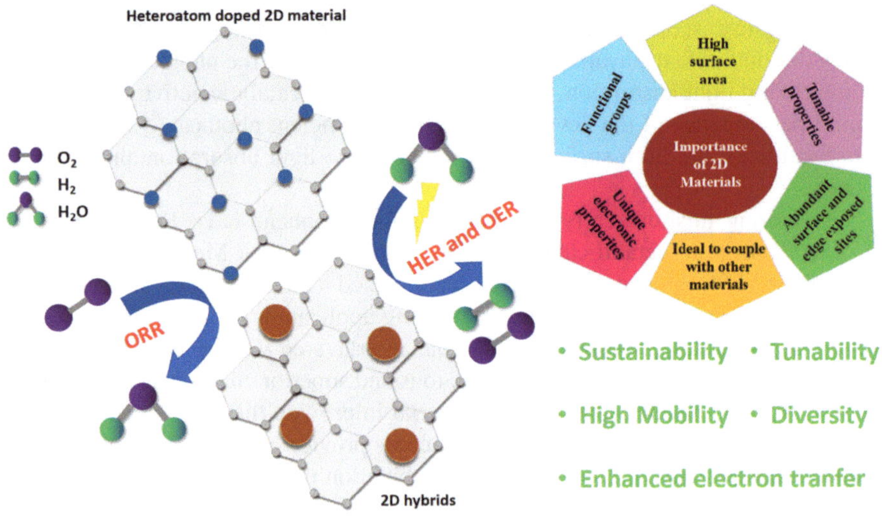

Fig. 3 Schematic illustration depicting the importance of 2D materials in water splitting applications

4.2 Mechanical Support and Fluidity

Another advantage of 2D materials is their mechanical stability and flexibility. Due to their planar structure, they can accommodate mechanical deformation. In layered material; numerous strong covalently-bonded plane is held together by weak Van der Waals forces, which gave them high mechanical strength. However, the weaker out-of-plane interactions due to Van der Waals force induces excellent mechanical flexibility in them which will offer high catalyst durability and thermal conductivity over longer duration [26].

4.3 Quantum Confinement and Band Gap Tunability

2D materials have emerged as a cornerstone of modern photonics and optoelectronics, owing to their exceptional properties, including quantum confinement effects, tunable band alignments, and pronounced surface interactions. The optical and electronic characteristics of material are intrinsically linked to its electronic band structure, which governs the behavior of charge carriers within the crystal lattice. Moreover, in 2D materials, quantum confinement effect plays a pivotal role in band gap engineering, enabling precise control over electronic transitions and carrier dynamics [27]. 2D materials exhibit a tunable band gap that can be adjusted through layer thickness, external fields, strain engineering, and heterostructure stacking. This tunability makes them highly desirable for applications in nanoelectronics, optoelectronics, and quantum technologies.

5 Synthesis of 2D Materials

To harness the advantageous merits of 2D materials like large specific surface area, tunable electronic properties, strong quantum confinement effects for energy-related applications like hydrogen generation through electrolysis, necessitates the development of innovative synthesis methodologies to achieve high density of uniformly dispersed 2D layers while overcoming the certain challenges related to material diffusion, layer aggregation, and defect formation, which directly affects the material's structural integrity and electroactive performance. Moreover, uncontrolled aggregation reduces the number of exposed active sites and limits their efficiency. Considering this, the optimal 2D materials should have high stability, excellent activity, and large number of active edge or basal plane sites. The synthesis of 2D materials involves both top-down and bottom-up approaches. Top-down methods basically employ the break down bulk materials into atomically thin layers whereas bottom-up techniques include the epitaxial growth of 2D materials from atomic or molecular precursors. The choice of synthesis technique depends on factors like scalability, material quality, and application requirements [28].

5.1 Top-Down Synthesis Approaches

Top-down synthesis method have been used to create 2D materials by exfoliating bulk stacked layered materials using mechanical exfoliation, chemical exfoliation and liquid phase exfoliation techniques. The primary advantage of top-down approach lies in the relative simplicity and scalability of the processing methods. However, challenges remain in achieving high yield defect free monolayer and controlling the flake dimensions which are critical for catalytic applications. Mechanical exfoliation has been useful in obtaining high quality graphene. Despite its effectiveness in producing pristine layers, the method suffers from scalability issues that limits the industrial applications. Chemical exfoliation using strong oxidants and intercalants, offers enhanced scalability but frequently introduces oxygenated functional groups that affects the intrinsic electronic properties of as synthesized 2D materials. Liquid phase exfoliation has emerged as a promising method which utilized solvents and ultrasonication to delaminate layered materials. Studies revealed that carefully selected solvent and surfactant facilitates the stabilization of exfoliated nanosheets, thereby minimizing agglomeration and preserving functional performance during water splitting reactions.

5.2 Bottom-Up Approaches

Bottom-up synthesis techniques offer a high degree of control over the material morphology and chemical composition. Bottom-up techniques basically include chemical vapor deposition, solvothermal synthesis and molecular self-assembly which enable the precise construction of 2D materials from the individual atoms and small molecules. Chemical vapor deposition on metal substrates such as copper or nickel is widely recognized as an effective strategy to produce large area, high quality single layered films. The process parameters including temperature, gas flow, rate, and precursor concentration can be finely tuned to modulate the crystallinity and defect density which in turn affect the charge transfer characteristics during electrochemical water splitting. Solvothermal routes, in which reactions are performed in high pressure and high temperature, autoclaves, have enabled the production of 2D nanosheets with well-defined phase and edge structures.

6 Current Challenges in Advancing 2D Materials for Water Splitting

As above mentioned 2D materials have attracted significant research interest, however still suffer from certain limitations like low hydrogen production efficiency, and inferior cycling stability due to their oxidizing nature and intersheet restacking which necessitates designing of proficient electrode/interfacial materials which can improve their performance and stability. These drawbacks can be rectified using different effective approaches such as doping, composite or heterostructure formation etc. which can efficiently help in preventing restacking issues by providing well-exposed surfacial active sites helpful in inducing catalytic properties, high selectivity with decreased band gap and increased charge transport. However, the 2D basal planes are thermodynamically unstable, creating shortage of active sites, which limits its natural catalytic activity. Further, to enhance the electrocatalytic performance, numerous attempts have been carried out to sufficiently increase the electrochemical surface area of the catalysts, exposing more active spots on the surface for electrocatalytic reactions. Although, water splitting is a viable approach for clean hydrogen production, yet its commercialization faces several fundamental challenges [29, 30] in the term of:

1. **Efficiency**: Existing water splitting technologies, especially photocatalytic and thermochemical processes, usually suffer from low energy conversion efficiencies. As a result, a large fraction of input energy is lost as heat or energy forms in other systems and thus makes the process more uneconomical.
2. **Cost implications**: Many advanced water-splitting catalysts employ expensive, precious and limited available noble metals due to their superior activity and stability like platinum, iridium, and ruthenium for efficient HER and OER which significantly increases cost of water splitting system.
3. **Durability and Stability**: Continuous operation in acidic or alkaline environments leads to corrosion, catalyst deactivation under the severe operating conditions (e.g., high temperature, electrochemical potentials) which hampers the longevity of the device. Even many lab-scale materials exhibit excellent performance towards hydrogen generation however fails to upscale at large-scale with long-term operations.

To overcoming these challenges, research developments and innovations in these areas will be crucial for making water splitting a viable and cost-effective technology for sustainable hydrogen production.

Addressing the challenges

The low cost and effective method for designing highly efficient water splitting devices using 2D materials can be achieved via strategic design and engineering of electrode/interfacial materials to enhance both performance and stability. Several approaches, including doping, composite formation, and heterostructure engineering, offer potential solutions. These methods can mitigate restacking by providing well-exposed surface active sites, thereby promoting catalytic properties, high selectivity, decreased band gaps, and increased charge transport [31]. However, their thermodynamic instability, minimal of active sites, limiting catalytic activity; necessitates strategies to expand the electrochemical surface area and having high density of active sites for the reactants to interact with catalyst surface, leading to enhanced reaction kinetics in electrocatalytic reactions. The following strategies can be implemented to enhance the device's performance:

1. **Defect Engineering**: It basically involves the introduction of defects like vacancies or edge dislocations over the 2D material structure to yield defect-induced additional active sites to improve the catalytic activity [32, 33]. These imperfections break the ideal crystalline lattice and create high-energy, more-reactive zones.
2. **Heterostructure Formation**: Heterostructures with synergistic behaviors can be produced by combining 2D materials with other materials including, but not limited to, metals, metal oxides, metal organic frameworks, or other 2D materials. The composite structures feature multiple physical and/or chemical interfaces between the individual components [34], which provider more active sites for the catalytic reaction with swift charge transfer kinetics during the catalytic activity.
3. **Surface Functionalization**: Surfacial modification of 2D materials with various functional groups can introduce new active sites and enhance their interaction with reactants. This can be achieved through chemical treatments or by grafting specific molecules onto the surface, creating more porous structures which can significantly increase the surface area and provide more active sites for catalysis [35].

By implementing these strategies, the intrinsic limitations of 2D materials can be resolved and can outperformed as an efficient and stable electrocatalysts for water splitting devices.

7 Integration of 2D Materials for Water Splitting

Among various strategies, integration strategies in 2D nanomaterials found to be an effective method as it helps in providing unique electronic structures with tuneable surface/interface micro environment; resulting in improving the overall water splitting efficiency [36]. Moreover, integrating 2D materials helps in enhancing the stability and activity of the catalytic materials during the working of water splitting application. For developing highly efficient and stable electrocatalysts for water splitting applications, this approach offers several key advantages such as:

1. **Enhanced Active Sites and Synergistic Interfaces**: The formation of heterostructures involving 2D materials creates many interfaces and abundant well exposed active sites. These interfaces, often characterized by lattice

mismatch and charge transfer, induce electronic perturbations that can lower the energy barrier for reactant adsorption and activation [37]. This synergistic effect leads to an electronic coupling between constituent materials that can create entirely new catalytic pathways.

2. **Surface Defect Engineering and enhanced catalytic activity**: Unlike bulk materials where defects can be trapped within the structure, 2D materials ensure that defects are primarily located on the surface, maximizing their impact on catalytic activity. Integrating 2D materials facilitates the formation of surface defects, such as vacancies, edges, and grain boundaries, act as preferential adsorption sites and can alter the local electronic structure, which are crucial for enhancing electrocatalytic performance.
3. **Improved Stability**: 2D heterostructures offer enhanced stability during the water splitting process due to strong interfacial interactions. The integration of 2D materials can accommodate the volumetric strain and mitigate lattice mismatch during longer operations which can act as degradation pathways. Therefore, design of robust interfaces is necessary to sustain electronic coupling towards enhanced catalytic activity.
4. **Tailoring Electronic Structures**: The electronic band structure of 2D materials is highly tuneable. Therefore, selective integration of 2D materials with other components, allows the engineering of their interfaces, band alignments, and charge transfer characteristics of the resulting heterostructures enabling fine-tuning of catalytic activity.
5. **Surface/Interface Microenvironment control**: Integrating 2D materials allows for precise control over the surface functionalization, porosity, hydrophilicity of the electrocatalyst. Therefore, precise control over the interface microenvironment allows fine-tune the interactions between the electrode and electrolyte interface, leading to improved water splitting efficiency.

Thus, integrating 2D materials with other 0D, 1D, 2D and 3D materials offers a promising pathway in designing advanced electrocatalysts for overall water splitting application. By leveraging the unique characteristics of 2D materials possessing large surface area, surface defect engineering capabilities, and tuneable electronic structures along with the integrated material can help in designing advanced catalysts with enhanced activity, stability, and control over the surface/interface microenvironment. These advancements contribute to the development of sustainable hydrogen production technologies.

7.1 2D/0D Integration

In 2D–0D heterostructures, the uniform distribution of 0D nanoparticles (NPs), quantum dots (QDs) on 2D materials found to improve the device's performance due to their synergistic interactions. A uniform distribution maximizes active sites, improves charge transfer and carrier mobility, increases stability and structural integrity by minimizing localized stress for catalysis. In this regard, graphitic carbon nitride (g-C_3N_4), with 2.7 eV bandgap has gained a significant interest for water splitting applications. However, its low electrical conductivity, limited active sites, high recombination losses; limits its photocatalytic efficacy. Various reports [38–40] have addressed this issue by creating heterojunctions containing TiO_2 for enhancing the HER activity. In this regard, He et al. [41] constructed $TiO_2/C_3N_4/Ti_3C_2$ S-scheme heterojunction photocatalyst; exhibiting higher photocatalytic activity in compared to pristine C_3N_4, TiO_2, C_3N_4/Ti_3C_2, TiO_2/C_3N_4 under visible-light irradiation. Accordingly, Jiang et al. [42] fabricated S-scheme heterojunction using 0D/2D TiO_2 nanodot/g-C_3N_4 (TOCN) via a phase-interfacial interaction. In comparison to TiO_2 nanodots and g-C_3N_4, TOCN demonstrated improved photocatalytic activity by producing 6.387 μmol of O_2 and 13.183 μmol of H_2 after 4 h of simulated sunshine irradiation. Here, the S-scheme heterojunction maintains the redox potentials of both g-C_3N_4 and TiO_2 while facilitating effective charge separation which is responsible for this improved performance. Apart from this, TMDs have been investigated as effective co-catalysts or electron transport materials for scalable H_2 production owing to their unique nanosheet morphology, high surface area, tunable band energy levels for better visible and near-infrared light responsiveness which offer significant potential for enhancing solar energy conversion efficiency. Also, in one report, the functionalized MoS_2 nanosheets were fabricated in 0D/2D hybrids using $CuInS_2$ quantum dots (CIS QDs). Here, CIS QDs connected to MoS_2 showed reduced photoluminescence lifetimes; which is beneficial in swift charge transport from photo-excited CIS to the MoS_2 nanosheets. In comparison to pristine MoS_2 and CIS-based devices, the CIS/MoS_2 hybrid device exhibits high amount of photocurrent which is basically due to efficient separation and migration of charges between CIS QDs and MoS_2 as well as from CIS QDs to MoS_2 occur [43]. Beyond this, TMDs can also function as photosensitizers as they have high ability to absorb light, generate excited states, and to accelerate charge transfer kinetics. In this context, Tajima et al. [44] fabricated water dispersible mixed dimension (2D/0D) $MoSe_2$/C_{60}-dendron nanohybrid by creating modifications in the basal plane of basal plane of $MoSe_2$ using physical modification technique which generates ($MoSe_2^+/C_{60}^-$) states through electron extraction from $MoSe_2$ exciton to C_{60}. The estimated quantum yields for HER were found to be 0.06% upon photoexcitation with A− with absorption peaks appears at $\lambda_{max} = 800$ nm and 0.27% upon photoexcitation with B− exciton with observed absorption peaks at $\lambda_{max} = 700$ nm respectively. Likewise, Vattikuti et al. [45] employed two-step hydrothermal technique to synthesize a binder-free 2D/0D-WS_2/ZrO_2 heterostructure for photocatalytic hydrogen generation. Under simulated solar radiation, this heterostructure showed enhanced photocatalytic activity, producing 1023.9 μmol h^{-1} g^{-1}, and is approximately 1.49 times more than pristine WS_2. This improvement in the performance is found due to proper band alignment in which the photosensitization effect and reduced band gap of WS_2 support the strong hydrogen evolution potential of ZrO_2. By generating photoinduced electron–hole pairs from ZrO_2 and WS_2, reduces the charge recombination and promotes effective charge separation along with their swift charge transfer. In another report, Zhang et al. [46] used a heat treatment and exfoliation technique to successfully create 2D

MoS_2 catalysts modified with 0D Mo_2C NPs to induce mechanical toughness in the heterostructure. At high current density of 1000 mA cm^{-2}; the resulting electrocatalyst demonstrates low overpotential of 412 mV with minimum Tafel slope (60 mV dec^{-1}) respectively. Along with this, thermal treatment improves the overall catalyst's mechanical durability and provides efficient electrical contact, leading to better catalytic performance. The functional groups (such as –O, –F and –OH) on the basal planes of MXenes have a major impact on their electrocatalytic performance. In particular, MXenes containing oxygen species exhibit comparative low hydrogen adsorption free energies, making them promising candidate for hydrogen generation [47–49]. Akir et al. [50] synthesized a 2D/0D V_2C/Bi composite by anchoring bismuth nanoparticles (Bi NPs) onto 2D vanadium carbide MXenes via solid-state annealing method. This V_2C/Bi catalyst showed higher HER activity in comparison to pristine V_2C in 0.5 M acidic H_2SO_4 electrolyte solution, achieving low overpotential of 663 mV at 10 mA cm^{-2} as compared to V_2C (823 mV). Also, V_2C/Bi composite showed a less Tafel slope of 173 mV dec^{-1} than V_2C (204 mV dec^{-1}), suggesting favorable HER kinetics. These results suggest that the interface between the V_2C and Bi NPs within the composite plays a vital role in enhancing HER activity due to their synergistic interactions. In another report, Panda et al. [51] synthesized high-performance Mo-MXene sheets decorated using carbon quantum dots (CQDs) followed by their surface phosphorylation to fabricate (2D/0D) based Mo-MX/C/P hybrid as illustrated in Fig. 4a. TEM image of the CQDs shows interplanar spacing of 0.21 nm corresponding to (100) plane, demonstrating a distinctive crystalline structure of carbon as shown in Fig. 4b. Also, in Mo-MX/C/P hybrid, uniformly dispersed carbon nanodots over the interfaces were observed. In Fig. 4c, d, HRTEM image shows lattice fringes with interplanar distance of 0.33 nm, confirms the presence of Mo_2CT_x crystallinity. This Mo-MX/C/P hybrid even showcase exceptional performance in both the HER and OER, requiring significantly lower onset potentials (14 mV for HER and 58 mV for OER at 0.2 mA cm^{-2}) as shown in Fig. 4e, h and minimum Tafel slope of (34 mV dec^{-1} for HER and 79 mV dec^{-1} for OER) compared to other catalysts as shown in Fig. 4f, i. Also, the Mo-MX/C/P shows enhanced stability upto 12 h during HER process as shown in Fig. 4g. The electronegativity of phosphorus and the rapid charge transfer facilitated by the CQDs within the Mo_2CT_x matrix, responsible for enhancing the catalytic performance. Specifically, hybrid achieved minimum overpotentials and Tafel slope of 53 mV and 34 mV dec^{-1} for HER and 360 mV and 79 mV dec^{-1} for OER in overall water splitting as shown in Fig. 4j. Therefore, synergistic interactions within Mo-MXene, CQDs, and electronegative phosphorus, creating numerous defect-rich sites, responsible for enhanced the catalytic activity.

In another report, Ma et al. [52] developed TMO-based 0D/2D CdS/ZnO Z-scheme heterojunction using ZnO nanosheets and CdS quantum dots. The photocatalytic hydrogen evolution rate of the resulting CdS/ZnO heterostructures was extremely high that is 22.12 mmol g^{-1} h^{-1} and is about 13 and 138 folds in comparison to pristine CdS and ZnO, respectively. Thus, development of a Z-scheme photocatalytic system and the close interaction among 2D ZnO NSs and 0D CdS QDs is

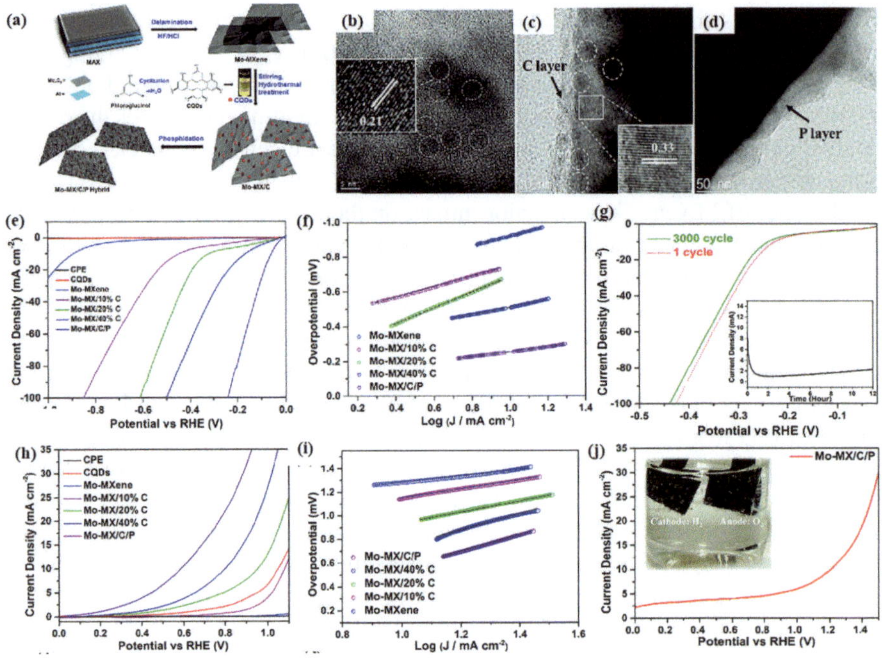

Fig. 4 **a** Schematic for the synthesis, **b–d** HRTEM images for CQDs; Mo-MX/C hybrid and MX/C/P hybrid; **e**, **h** polarization curves, **f–i** Tafel plots, with 75% iR correction for HER and OER; **g** durability test of Mo-MX/C/P hybrid for HER and **j** overall water splitting polarization plot for Mo-MX/C/P respectively. Adapted with permission from [51]. Copyright (2009) Royal Society of chemistry

responsible for this increased activity, which minimizes recombination and promotes effective charge separation and transfer.

7.2 2D/1D Integration

One of the promising approaches is 1D/2D heterostructures which offer a potential route to enhanced catalytic activity through synergistic interactions between the two components. The 1D component, typically include nanowires, nanotubes, and nanorods like morphology which provide abundant active sites and high surface to volume ratio and wide spaces between them facilitate electrolyte access and mass movement whereas the 2D component include thin sheets or flakes, facilitates charge separation and transport. In this regard, MXenes have been extensively explored with various 1D heterostructures for enhancing the catalytic activity. Likewise, Huang et al. [53] synthesized Schottky heterojunction using 1D CdS nanorod/2D Nb_2CT_x via one-step solvothermal approach. The composite showed increased photocatalytic activity, producing hydrogen at a rate of 5.3 mmol g^{-1} h^{-1}, which is about 1.7 times higher than pristine CdS. This could be attributed to well matched band energy structure and interfacial contact; facilitating charge separation and transfer. Likewise, Jiang et al. [54] synthesized Schottky heterojunction between 1D $Mn_{0.2}Cd_{0.8}S$ nanorod/2D Ti_3C_2 MXene via in situ solvothermal method; which helps in facilitating photogenerated carrier separation, produced efficient hydrogen evolution rate of 15.73 mmol g^{-1} h^{-1} with 6.72-fold improvement in HER activity in comparison to pure $Mn_{0.2}Cd_{0.8}S$ nanorods. Beyond MXenes, various other 2D materials based 1D/2D heterostructures have been extensively explored. Likewise, Du et al. [55] explored the 1D g-C_3N_4 nanotube/2D BiOI heterojunction synthesized using hydrothermal treatment. The in-situ grown BiOI on g-C_3N_4 nanotubes offers lower charge transfer resistance and facilitate effective electronic transmission pathways for enhancing the PEC activities. Here, the BiOI/g-C_3N_4 nanotubes exhibit high photocurrent density of 8.15×10^{-5} A cm^{-2} with respect to the reference (Ag/AgCl) electrode, which is approximately 1.8- and 2-times folds than that of pristine g-C_3N_4 nanotubes and BiOI respectively. Also, 2D reduced graphene oxide (rGO) based heterostructures has also been widely investigated in water splitting applications. As Samajdar et al. [56] has synthesized 1D $LaFeO_3$/2D rGO nanocomposite synergically providing the charge-separating properties of rGO to enhance the photocatalytic activity of $LaFeO_3$. $LaFeO_3$ has been selected due to cost effective and high chemical stability approach with narrow band gap perovskite oxide (ABO_3) type nature whereas the rGO sheets act as charge separators in photocatalytic water splitting process. The nanocomposite exhibited significant enhancement in photocurrent density and increases hydrogen generation rate (82 mmol g^{-1} h^{-1}) by 20 folds as compared to bare $LaFeO_3$. In another report, Bera et al. [57] reported highly active 1D/2D hybrid by decorating 1D Ag nanoparticles on $AgVO_3$ nanorods integrated with 2D MoS_2 nanosheets. This heterostructure based photocatalyst which exhibited high H_2 production of 38.6 mmol g^{-1} and is 20-fold increased in the rate of H_2 generation as compared to bare MoS_2. Also, the current density of pristine MoS_2 is ~25 μA cm^{-2} and is found to be increased from Ag-$AgVO_3$ (~35 μA cm^{-2}) to (~64 μA cm^{-2}) MoS_2/Ag-$AgVO_3$ heterostructure. The specific arrangement of $AgVO_3$ and MoS_2 band positions enables a Z-scheme charge transfer pathway which efficiently separate photoinduced charge carriers and hence reducing the recombination losses, resulting in enhancing the overall efficiency for Hydrogen generation via photocatalysis using visible light illumination. In another report, Liu et al. [58] synthesized 2D nanosheets of $ZnIn_2S_4$ with 1D nanorods arrays of TiO_2 using hydrothermal treatment. These heterojunction arrays were used as photoelectrodes in a PEC, found to enhance the photocurrent density by 3 folds. Optical characteristics for pristine TiO_2 and $ZnIn_2S_4/TiO_2$ was analysed using reflectance and absorption and spectra. It has been observed that in comparison to TiO_2 (absorption at lower wavelengths than 410 nm); $ZnIn_2S_4/TiO_2$ heterostructure shows an enhanced absorption in the region below 425 nm, therefore overall nanocomposite based photoelectrode exhibits overall high absorption over larger wavelength range. Further, from Tauc plots, it was observed that optical band-gap values were 2.91 eV and 3.02 eV for $ZnIn_2S_4$ and TiO_2 respectively as shown in Fig. 5a. In Fig. 5b the reflectance spectra show that high reflectance of TiO_2 nanorod arrays in comparison to $ZnIn_2S_4$/TiO_2 hybrid which is due to the fact that light scattering occurs multiple times within the hybrid, results in effective enhancement in effective pathway for light absorption and increases the reaction sites and hence improves the photocurrent generation during water splitting process. From Mott–Schottky in Fig. 5c the flat band potential was examined and calculated values were 0.13 and 0.29 V (vs. RHE) for $ZnIn_2S_4$/TiO_2 nanorod arrays and pristine TiO_2 which suggests heterojunction offers high charge carrier density and their efficient charge transfer with the interfaces of the photoanode with minimal recombination losses.

Figure 5d represents the proposed charge transfer mechanism towards in $ZnIn_2S_4$/TiO_2 hybrid. When light falls on $ZnIn_2S_4$/TiO_2 heterojunction, electrons and holes are generated and migrate from $ZnIn_2S_4$ to TiO_2 for in case of electrons and hole migrate from TiO_2 to $ZnIn_2S_4$ and reduce themselves to generate H_2 respectively. Further, in Fig. 5e, it has been found that $ZnIn_2S_4$/TiO_2 photoelectrode exhibit high photocurrent density possessing lower overpotential of −0.53 V to 1.2 mA cm^{-2} achieve the current density at 1.5 V. Further, photocurrent curve at 0.4 V (vs. RHE) in Fig. 5f demonstrating numerous illuminations on–off cycles in which the photocurrent density found to be 0.6 mA cm^{-2} during illumination while photocurrent decay instantly to baseline upon the removal of the light source; suggesting rapid charge carrier transport within the heterojunction. In addition, photocurrent stability measurements were performed (Fig. 5g) describing the high durability of the $ZnIn_2S_4$/TiO_2 photoelectrode upto 2400 s and the photoactivity of the electrodes was assessed by calculating incident photon-to-current conversion efficiency (IPCE) (Fig. 5h) where the value of IPCE was high for $ZnIn_2S_4$/TiO_2 than that of TiO_2 electrodes suggesting that effective separation of electron–hole pairs reducing recombination losses at $ZnIn_2S_4$/TiO_2 interface which ultimately enhanced the photocatalytic water splitting performance.

7.3 2D/2D Integration

The integration of 2D–2D materials offers large contact area, high surface area and charge carrier mobility. Generally, in 2D/2D heterostructures an electric field is generated at an interface which foster quick separation of charges separation with swift transfer kinetics, resulting in increased water splitting activity. Also, the 2D/2D heterojunctions creates proper "face contact" and "line contact" at an interface based on their synthetic approach which makes them ideal catalyst for catalytic applications. In this regard, Li et al. [59] developed low cost and advanced multiphasic 1T/2H phase of $MoSe_2$ over the MXene nanosheet as an electrocatalyst for overall water splitting application. Here, 1T/2H $MoSe_2$ provided abundant active sites, while the MXene provide highly conductive pathway for facilitating swift charge transfer kinetics. This heterostructure demonstrated excellent HER and OER activity, such as in order to achieve the current density of 10 mA cm^{-2}; possessing low overpotential (340 and 95 mV) and Tafel slope of 90 and 91 mV dec^{-1} for OER and HER

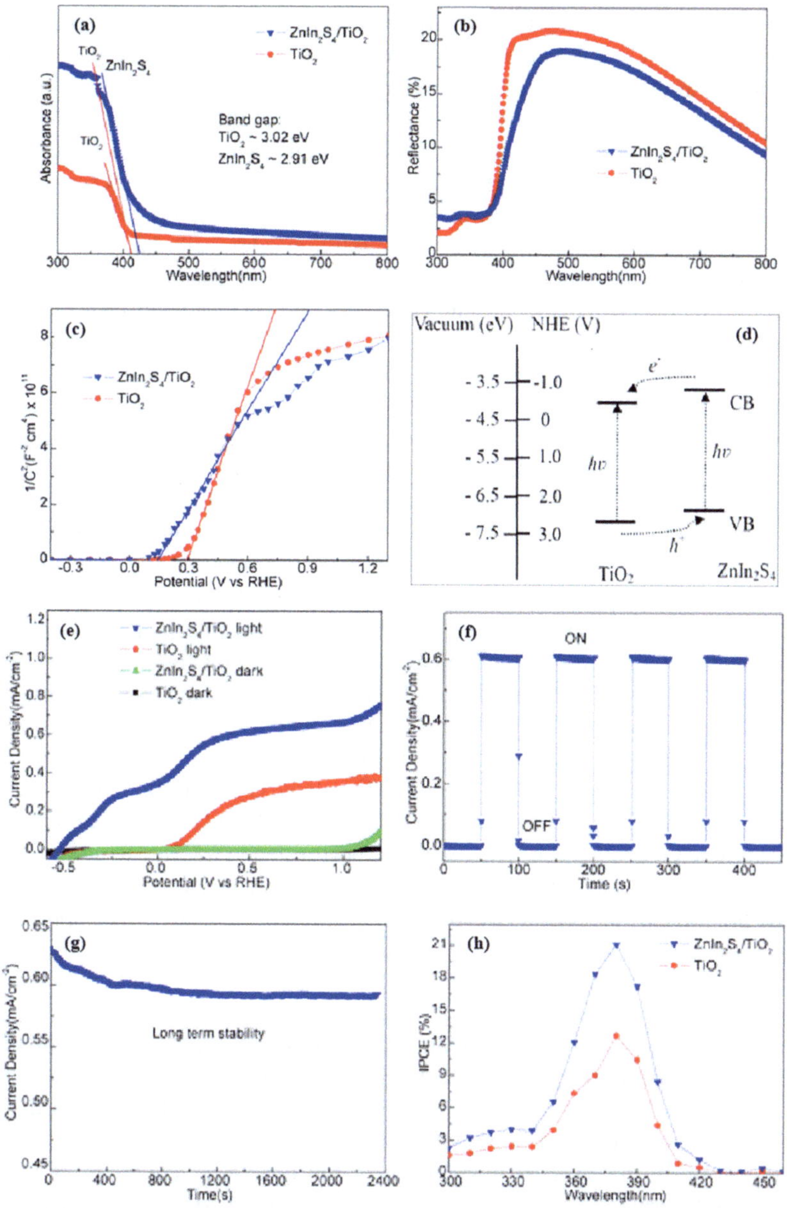

Fig. 5 **a** Absorption and **b** reflectance spectra, **c** Mott–Schottky plots of pristine TiO_2 and $ZnIn_2S_4$ nanosheet/1D TiO_2 nanorod heterojunction arrays, **d** Schematic mechanism of carrier transferring in $ZnIn_2S_4$ nanosheet/1D TiO_2 nanorod heterojunction system, **e** I–V characteristics for pristine TiO_2 and $ZnIn_2S_4$ nanosheet/1D TiO_2 nanorod heterojunction electrodes. **f** Amperometric I–t curve of $ZnIn_2S_4$ nanosheet/1D TiO_2 nanorod heterojunction electrode with light on/off cycles. **g** Photocurrent stability measurement. **h** IPCE spectra. Adapted with permission from [58]. Copyright (2014) Liu et al., some rights reserved; exclusive licensee (American Chemical Society). Distributed under a Creative Commons Attribution License 4.0 (CC BY) https://creativecommons.org/licenses/by/4.0/

and full cell demonstrating 1.64 V. Likewise, Li et al. [60] synthesized advanced 2D/2D $BiVO_4/Ti_3C_2$ composite via electrostatic interaction. The composite exhibits high interfacial contact area and shorter ion diffusion pathway; leading to an efficient separation and transfer of photoexcited carrier, resulting in exceptional activity towards photocatalytic water splitting. The amount of gas evolved using $BiVO_4/Ti_3C_2$ catalyst is 15.7 μmol which is 4 folds than that of pristine $BiVO_4$ nanosheets; this remarkable performance is pronounced due to Ti_3C_2 matrix, which effectively minimizes the overpotential for hydrogen evolution while simultaneously boosting charge carrier transfer and separation. In another report, Rasool et al. [61] synthesized WS_2 nanopetal-like sheets embedded on $Ti_3C_2T_x$ MXene layers achieving low overpotential value and minimum Tafel slope of 66 mV and 46.7 mV dec^{-1} to achieve current density of −10 mA cm^{-2} in a 1 M alkaline KOH electrolyte. Here, the interaction between $Ti_3C_2T_x$ and WS_2 creative active centre which reduces the energy barrier as predicted by theoretical calculations. Further, MXenes also found to be with layered double hydroxides (LDHs): NiFe-LDH/$Ti_3C_2T_x$ heterostructure was synthesized via hydrothermal treatment [62]. To attain the current density of 10 mA cm^{-2}, the as synthesized heterostructure exhibited minimum overpotential value of 269 mV for OER process. Furthermore, combining MXenes with metal oxides has also been proven effective. Likewise, Madhushree et al. [63] synthesized 2D $Cr_2CT_x/NiFe_2O_4$ composite using hydrothermal treatment, and to achieve the current density of 10 mA cm^{-2}, the electrocatalyst demonstrate low overpotential of 159 and 144 mV at a for OER and HER with 1.69 V cell potential in overall water splitting. Along with this, rGO was extremely investigated as components in 2D/2D heterostructures by virtue of their excellent conductivity and large surface area. As Ke et al. [64] prepared a 2D/2D WO_3 plate/rGO heterojunction. The interaction between rGO nanosheets with (0 0 2) faceted of WO_3 resulted in remarkable photocatalytic O_2 production. The improved performance was due to increased electronegative oxygen atoms, which facilitated photo-induced charge carrier separation. Beyond photocatalysis, rGO has also shown promise in electrocatalytic applications. In one of the report, Sahoo et al. [65] reported a simple microwave-assisted synthesis of a nanocomposite based on MoS_2, g-C_3N_4 and rGO. Figure 6a indicates nanoflake morphology of MoS_2 nanoflakes whereas more ordered and hexagonal shape was observed for GM (Fig. 6b). The exfoliated rGO nanosheets provided a suitable surface to regulate the synthesis of hexagonal MoS_2 nanoflakes.

In the NSGM composite, however, the MoS_2 nanoflakes displayed a distorted morphology and formed an interconnected network (Fig. 6c). HRTEM images in Fig. 6d–f confirms the successful synthesis of MoS_2 with flat edges and rGO sheets decorated with laminar-structured g-C_3N_4. Also, cluster of g-C_3N_4 coupled to surface of MoS_2 was observed which could be due agglomeration of the rGO nanosheets. This ternary composite exhibits low overpotentials of 172 and 380 mV (Fig. 6j, g) with Tafel slopes of 97 and 47 mV dec^{-1} for OER and HER to achieve current density of 20 mA cm^{-2} as shown in Fig. 6h, k. Also, CNGM electrocatalyst shows an extended stability of 40 h for HER (Fig. 6i) and 30 h for OER at 20 mA cm^{-2} current density as shown in Fig. 6l. The choice of materials in the 2D/2D heterostructure plays a crucial role in determining its performance. Zhang et al. [66] investigated the use of layered

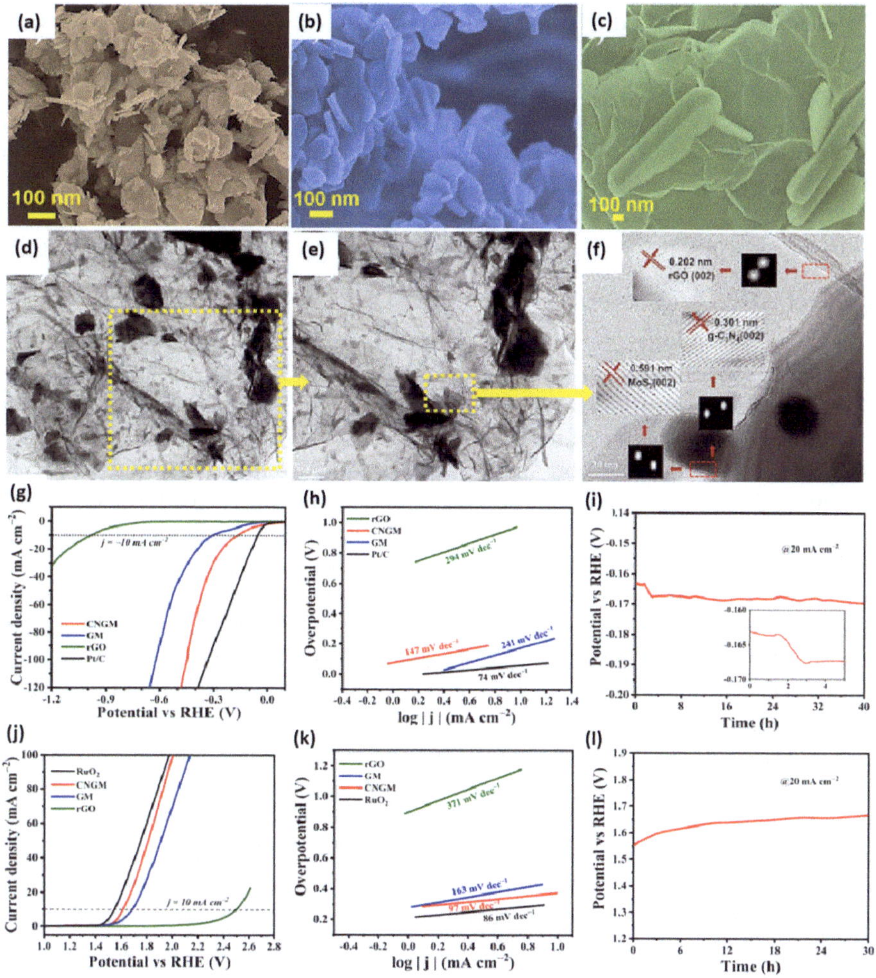

Fig. 6 SEM images of **a** commercial MoS_2, **b** GM, and **c**; TEM and HRTEM images of CNGM **d-f**; experimental results in 1 M KOH of the electrocatalysts: LSV curves, Tafel plots and chronopotentiometric stability analysis curve of CNGM towards **g–i** HER and **j–l** OER. Adapted with permission from [65]. Copyright (2025) Sahoo et al., some rights reserved; exclusive licensee (Elsevier). Distributed under a Creative Commons Attribution License 4.0 (CC BY) https://creativecommons.org/licenses/by/4.0/

perovskite oxides combined with rGO for photocatalytic H_2 production. By varying the A-site cations in the perovskite ($LaNb_2O_7$, $PrNb_2O_7$, and $NdNb_2O_7$), fine tuning in the electronic properties can be done. Here, highly exfoliated Dion-Jacobson perovskite layers with rGO layers (rGO/$NdNb_2O_7$) were assembled possessing large surface area nanocomposite and high HER activity, producing 835 μmol g^{-1} of hydrogen in sunlight which is due to an optimized synergistic effects along with cation charge density in influencing photocatalytic performance. Also, g-C_3N_4, has also been extensively studied in 2D/2D heterostructures. Mittal et al. [67] synthesized type-II heterojunction using SnS_2/g-C_3N_4 nanocomposite facilitate efficient charge separation and for reducing the recombination losses for improving overall PEC water splitting. The optimized SnS_2/g-C_3N_4 catalyst exhibit tremendous increase in photocurrent density of ~1.0 mA cm^{-2}, which is approximately ~fivefold more than that of SnS_2 photoelectrode.

7.4 2D/3D Integration

The mixed-dimensional nanostructures, particularly 2D/3D heterostructures, provide a versatile platform for designing high-performance electro/photo catalysts for water splitting. The combination of different dimensionalities allows for the optimization of active sites, charge transport, and light absorption, leading to enhanced catalytic activity and stability. Their well-defined structures provide a platform for understanding structure–activity relationships and for engineering catalysts with enhanced performance. Specifically, MOF-derived 2D materials have demonstrated significant potential in photo/electro catalysis for sustainable hydrogen production using water splitting. In this regard, Chen et al. [68] developed a 2D/3D zeolitic imidazolate framework (ZIF-67) derived catalyst in which cobalt nanoparticles were encrusted within 2D carbon nanosheets doped with nitrogen and 3D framework (Co@N-CS/N-HCP@CC) which found to exhibit remarkable catalytic performance achieving least overpotential of 248 mV for OER and 66 mV for HER with cell voltage of 1.545 V to obtain 10 mA cm^{-2} current density. The hierarchical structure provided hydrophilicity facilitates swift charge transfer kinetics by creating excessive active sites. Similarly, Tian et al. [69] synthesized 2D/3D binder-free bifunctional P-doped MOF on Ni foam catalyst for water splitting. The incorporation of phosphorus and the hybrid structure contributed to improved electronic properties and enhanced charge transfer, demonstrating excellent HER and OER activity. Here the P@Mn_1Co_3-BDC(NH_2)/NF exhibited low overpotential of 102 mV for HER and 310 mV for OER with 1.69 V cell voltage to achieve current density of 10 mV cm^{-2}. These examples indicate the importance of MOF-derived mixed-dimensional structures in achieving high electrocatalytic performance. Beyond MOF-derived materials, other mixed-dimensional chalcogenides nanostructures have also shown promising performance in water splitting applications. Sadqat et al. [70] reported 3D polyhedral nanostructures using Fe-doped manganese-based chalcogenides incorporated with amorphous carbon. The 2D/3D hierarchical structure combined with Fe doping, resulted in lower overpotentials of 210 mV and Tafel slopes of 48 mV dec^{-1} for OER, due to synergistic interaction of amorphous carbon which provide increased surface area and efficient charge transfer kinetics. Apart from this, Yang et al. [71] designed a unique 2D/3D hierarchical bifunctional electrocatalyst, by decorating Prussian blue analogue (PBA) nanocubes over CoNiFe-PS (synthesized via phosphating and sulfuration of (CoNi-LDH)); exhibited excellent performance in overall water splitting with overpotentials of 263 mV and 196 mV for OER, and 235 mV and 63 mV for HER, are at current densities of 100 mA cm^{-2} and 10 mA cm^{-2} with minimum Tafel slope of 35.81 mV dec^{-1} and 110.58 mV dec^{-1} of OER and HER as shown in Fig. 7a–f, respectively. Furthermore, a full water splitting cell (CoNiFe-PS||CoNiFe-PS) was fabricated in alkaline conditions, requiring minimal voltages of 1.693 V and 1.479 V to achieve 100 mA cm^{-2} and 10 mA cm^{-2} current density (Fig. 7g, h) respectively. The incorporation of phosphorus and sulfur creates a multi-heterostructure that modifies the catalyst's electronic structure, leading to improved conductivity and enhanced catalytic activity. The assembled two-electrode electrolyzer (CoNiFe-PS||CoNiFe-PS) outperformed a commercial RuO_2||Pt/C electrolyzer, requiring lower cell voltages for overall water splitting. Moreover, Fig. 7i represent the high stability of CoNiFe-PS//CoNiFe-PS which is found to be 100 h for overall water splitting.

In another report, Sansudin et al. [72] fabricated rGO coupled 2D/3D g-C_3N_4/$BiVO_4$ composite for the production of hydrogen using a PEC water splitting process.

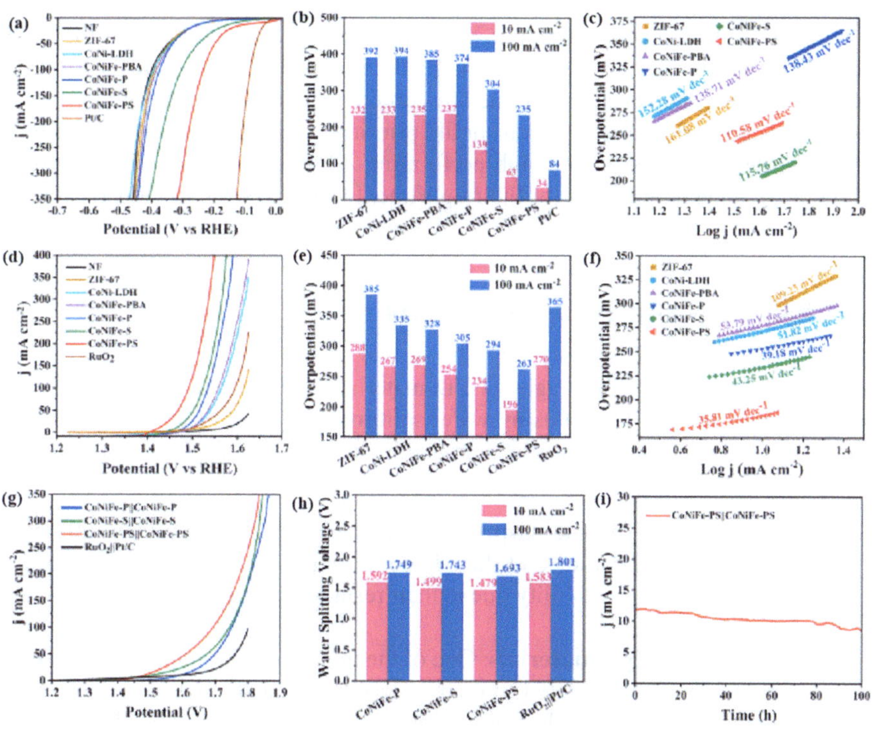

Fig. 7 **a–c** HER performance, **d–f** OER performances of ZIF-67, CoNi-LDH, CoNiFe-PBA, CoNiFe-P, CoNiFe-S and CoNiFe-PS: LSV curves, comparison of the overpotentials at the current density of 10 and 100 mA cm^{-2}, Tafel plots and **g**, **h** the LSV curves of CoNiFe-P||CoNiFe-P, CoNiFe-S||CoNiFe-S, CoNiFe-PS||CoNiFe-PS and RuO_2||Pt/C. and the comparison of the potentials at the current density of 10 and 100 mA cm^{-2}. **i** The stability test of CoNiFe-PS in electrolyzed water at a current density of 10 mA cm^{-2} for 100 h. Adapted with permission from [71]. Copyright (2024) Elsevier

Here, the rGo@g-C_3N_4/$BiVO_4$ photocatalyst produced 63.5 mmol h^{-1} of hydrogen and achieved a high photocurrent density of 14.44 mA cm^{-2}. The rGO facilitated charge transfer and the 2D/3D heterostructure promoted efficient charge separation, leading to enhanced hydrogen evolution. Further, the integration of 2D materials like MXenes with 3D nanostructures has emerged as a promising strategy. In this regard, Ullah et al. [73] designed $CoSnO_3$@MX/NF composite, where $CoSnO_3$ nanostructures were integrated into a 2D/3D MXene/Ni foam substrate. The synergistic effect between $CoSnO_3$ and MXene, along with the enhanced conductivity and interfacial coupling, beneficial in water splitting applications. For achieving a current density of 10 mA cm^{-2}, the electrocatalyst requires a minimal overpotential of 274 mV for OER and 108 mV for HER respectively. Also, Wei et al. [74] synthesized a 2D/3D WO_3/$BiVO_4$ heterostructure where worm-like $BiVO_4$ grafted onto the 2D WO_3 underlayer promoted optimized charge separation with minimized charge recombination, contributing to improved photocatalytic activity; demonstrating an excellent stability upto 3 h with remarkable photocurrent (3.85 mA cm^{-2}) at 1.23 V with respect to RHE.

In nutshell, advancement and innovation in the nanostructural engineering of 2D material is rapidly flourishing for designing durable and efficient catalyst for water splitting systems. However, constrained by certain bottlenecks necessitates advanced research. Therefore, future research should prioritize in developing innovative techniques for designing robust and high-performing 2D based catalysts with controlled morphology and composition for their upliftment from lab scale to industry level.

8 Challenges and Future Prospects

Despite the promising advantages of 2D materials which hold immense promise for producing clean hydrogen using water splitting devices, still faces several challenges which hinder their large-scale application in terms of efficiency, stability and scalability with existing technologies.

1. **Low cost and Efficient approach**: High-performing 2D materials with controlled defect densities and their large-scale production remain a technical challenge in their upscalability due to their restacking, limiting ion diffusion and catalytic activity.
2. **Instability of catalyst materials**: One of the major concerns related to these 2D integrated materials is related to their instability in harsh environmental conditions as the defects and dislocations created at the interface get easily collapsed and hence destroys the overall hierarchy of the structure which overall hampers the stability and catalytic activity of the materials.
3. **Difficulty in uncontrolled heterostructural engineering**: Currently used synthetic methodologies for integrating 2D materials with different materials into heterostructure like hydrothermal treatment often take a long time for their

synthesis and still don't provide precise control over the growth of hybrid material. The lack of uncontrolled engineering in composition and morphology poses a severe challenge in the depth analysis of their structure–property correlations and underlying mechanisms within the catalyst governing water-splitting activity.
4. **Lack of Advanced characterization techniques**: Comprehensive understanding of catalytic and structure–property correlations require advanced theoretical modeling and calculations to gain crucial insights for the optimization of catalytic mechanism within 2D materials-based catalysts.

To surmount these limitations, future research prospect should explore novel synthesis approaches and cost effective ways for developing effective 2D material based water-splitting catalysts. Therefore, future research and development should prioritize the following avenues:

1. **Scalable and Precision-Controlled Synthesis**: Research should be focused on the exploration of innovative synthesis methods which allow precise control over thickness, layer density and defect density to realize their effective potential in water splitting applications.
2. **Design and Engineering of Advanced Hybrid Materials**: Research should be focused on creation of advanced multidimensional composites with tailored interfaces could help in effective separation of charge carriers to reduce recombination losses and in improving the stability of the catalysts over longer operational duration. Also, it could help in modulating the defect density and their respective band alignments by strategic selection of constituent materials to leverage synergistic effects.
3. **Controlled Defect Engineering for Enhanced Catalytic Activity**: Research should be focused on developing advanced methods for introducing controlled defects, such as oxygen vacancies and strategically induced active sites which could helps in the effective modulation of electronic properties towards appropriate binding energies, necessary of water splitting application.
4. **Development of Earth-Abundant and Cost-effective Catalysts**: Numerous earth-abundant and cost effective materials need to be explored as substitutes for noble metal catalysts. Further, research should be focused on improving the stability and catalytic activity of these materials via compositional tuning, morphology control, and surface modification.
5. **Integration of Machine Learning and Computational Modeling**: Various theoretical computational methods such as machine learning algorithms and density functional theory should be employed to analyze large datasets and predict optimal material combinations and synthesis parameters to simulate the electronic structure and catalytic activity of materials.
6. **Integration with other Renewable Energy Sources**: Research should focus on optimizing the integration of water-splitting systems with another renewable energy resources like solar and other bio-energy sources for designing sustainable hydrogen generation systems. Optimization of coupling between the energy source and the electrolyzer is crucial that can handle the fluctuations in input energy and water quality for maximizing efficiency.

Thus, advancement in research and development in this direction can elevate the development of efficient, sustainable, and economically viable 2D material-based water-splitting catalysts, to meet the clean energy demand of the nation.

9 Conclusion

This chapter explores the integration of 2D materials for water splitting application as a frontier in sustainable energy research particularly in electrocatalytic and PEC water splitting applications owing to their exceptional physicochemical properties, tunable surface chemistry, high surface to volume ratio, high electrical conductivity; make them promising candidates. However, challenges such as interlayer restacking and limited intrinsic activity necessitate the advancement in the 2D materials to overcome their inherent limitations. With ongoing advancements in material engineering and nanotechnology, these drawbacks can be rectified. Among all, integration strategies found to improves catalytic performance by facilitating rapid charge transfer, providing active site accessibility, improve thermodynamic stability and structural integrity for long term catalytic applications and hence improving overall efficiency of water splitting devices to revolutionize sustainable hydrogen production for a clean energy future.

References

1. Podobińska Staniec, M., Wiktor Sułkowska, A., Kustra, A., Lorenc Szot, S.: Copper as a critical resource in the energy transition. Energies **18**, 969 (2025)
2. Kazemi, A., Manteghi, F., Tehrani, Z.: Metal electrocatalysts for hydrogen production in water splitting. ACS Omega **9**, 7310 (2024)
3. Molaei, M.J.: Recent advances in hydrogen production through photocatalytic water splitting: a review. Fuel **365**, 131159 (2024)
4. Sarmah, M.K., Singh, T.P., Kalita, P., Dewan, A.: Sustainable hydrogen generation and storage—a review. RSC Adv. **13**, 25253 (2023)
5. Züttel, A., Remhof, A., Borgschulte, A., Friedrichs, O.: Hydrogen: the future energy carrier. Philos. Trans. R. Soc. A: Math., Phys. Eng. Sci. **368**, 3329 (2010)
6. Bai, J., Huang, Y., Fan, X., Cui, J., Chen, B., Chen, Y., Guo, L.: Production of high calorific value hydrogen-rich combustible gas by supercritical water gasification of straw assisted by machine learning. Biores. Technol. **410**, 131275 (2024)
7. Kumar, S., Regassa Jufar, S., Kumar, S., Foroozesh, J., Kumari, S., Bera, A.: Underground hydrogen storage and its roadmap and feasibility in India toward net-zero target for global decarbonization. Fuel **350**, 128849 (2023)
8. Hosseini, S.E., Wahid, M.A.: Hydrogen production from renewable and sustainable energy resources: promising green energy carrier for clean development. Renew. Sustain. Energy Rev. **57**, 850 (2016)
9. Nnabuife, S.G., Hamzat, A.K., Whidborne, J., Kuang, B., Jenkins, K.W.: Integration of renewable energy sources in tandem with electrolysis: a technology review for green hydrogen production. Int. J. Hydrogen Energy (2024)

10. Marzouk, O.A.: Expectations for the role of hydrogen and its derivatives in different sectors through analysis of the four energy scenarios: IEA-STEPS, IEA-NZE, IRENA-PES, and IRENA-1.5 °C. Energies **17**, 646 (2024)
11. Imran, S., Hussain, M.: Emerging trends in water splitting innovations for solar hydrogen production: analysis, comparison, and economical insights. Int. J. Hydrogen Energy **77**, 975 (2024)
12. Li, X., Zhao, L., Yu, J., Liu, X., Zhang, X., Liu, H., Zhou, W.: Water splitting: from electrode to green energy system. Nano-Micro Lett. **12**, 131 (2020)
13. Ma, X., Chu, W., Wang, Y., Li, Z., Yang, J.: Increasing the efficiency of photocatalytic water splitting via introducing intermediate bands. J. Phys. Chem. Lett. **14**, 779 (2023)
14. Singh, S., Jain, S., Ps, V., Tiwari, A.K., Nouni, M.R., Pandey, J.K., Goel, S.: Hydrogen: a sustainable fuel for future of the transport sector. Renew. Sustain. Energy Rev. **51**, 623 (2015)
15. Pandian, A., Rameesha, L., Boobalan, C.: Exploring the mechanisms associated with electrocatalytic reactions for oxygen and hydrogen evolution. Sustain. Chem. One World **5**, 100046 (2025)
16. Vij, V., Sultan, S., Harzandi, A.M., Meena, A., Tiwari, J.N., Lee, W.G., Yoon, T., Kim, K.S.: Nickel-based electrocatalysts for energy-related applications: oxygen reduction, oxygen evolution, and hydrogen evolution reactions. ACS Catal. **7**, 7196 (2017)
17. Wei, J., Zhou, M., Long, A., Xue, Y., Liao, H., Wei, C., Xu, Z.J.: Heterostructured electrocatalysts for hydrogen evolution reaction under alkaline conditions. Nano-Micro Lett. **10**, 75 (2018)
18. Kumar, M., Meena, B., Subramanyam, P., Suryakala, D., Subrahmanyam, C.: Recent trends in photoelectrochemical water splitting: the role of cocatalysts. NPG Asia Mater. **14**, 88 (2022)
19. Hossain, M.N., Zhang, L., Neagu, R., Sun, S.: Exploring the properties, types, and performance of atomic site catalysts in electrochemical hydrogen evolution reactions. Chem. Soc. Rev. (2025)
20. Thakur, A., Ghosh, D., Devi, P., Kim, K.-H., Kumar, P.: Current progress and challenges in photoelectrode materials for the production of hydrogen. Chem. Eng. J. **397**, 125415 (2020)
21. Singh, J., Tyagi, P.K., Singh, V.: Potential application of reduced graphene oxide in electrocatalyst for hydrogen/oxygen evaluation reaction in water-splitting. Diam. Relat. Mater. **154**, 112113 (2025)
22. Singla, A., Dhiman, R., Mahajan, A.: A bimetallic-doped boron nanosheet electrocatalyst for efficient hydrogen evolution reaction. J. Electron. Mater. **53**, 4947 (2024)
23. Ma, J., Kuang, X.Y., Pang, J., Wanying, X., Mao, A.J., Yang, J.: Prediction of high photoconversion efficiency and photocatalytic water splitting in vertically stacked TMD heterojunctions MX_2/WS_2 and $MX_2/MoSe_2$ (M = Cr, Mo, W; X = S, Se, Te). J. Mater. Chem. A (2025)
24. Liu, S., Li, L., Yang, T., Wang, E., Yu, X., Hou, Y., Du, Z., Cao, S., Chou, K.C., Hou, X.: Enhanced overall water splitting by morphology and electronic structure engineering on pristine ultrathin metal-organic frameworks. J. Mater. Sci. Technol. **220**, 92 (2025)
25. Heo, E., Lee, S., Yoon, H.: Heterostructured 2D material-based electro-/photo-catalysts for water splitting. Mater. Chem. Front. **7**, 6154 (2023)
26. Akinwande, D., Brennan, C.J., Bunch, J.S., Egberts, P., Felts, J.R., Gao, H., Huang, R., Kim, J.-S., Li, T., Li, Y., Liechti, K.M., Lu, N., Park, H.S., Reed, E.J., Wang, P., Yakobson, B.I., Zhang, T., Zhang, Y.W., Zhou, Y., Zhu, Y.: A review on mechanics and mechanical properties of 2D materials-graphene and beyond. Extrem. Mech. Lett. **13**, 42 (2017)
27. Chen, S., Shi, G.: Two-dimensional materials for halide perovskite-based optoelectronic devices. Adv. Mater. **29**, 1605448 (2017)
28. Alam, S., Asaduzzaman Chowdhury, M., Shahid, A., Alam, R., Rahim, A.: Synthesis of emerging two-dimensional (2D) materials—advances, challenges and prospects. FlatChem **30**, 100305 (2021)
29. Acar, C., Dincer, I., Naterer, G.F.: Review of photocatalytic water-splitting methods for sustainable hydrogen production. Int. J. Energy Res. **40**, 1449 (2016)
30. Bie, C., Wang, L., Yu, J.: Challenges for photocatalytic overall water splitting. Chem **8**, 1567 (2022)

31. Shifa, T.A., Wang, F., Liu, Y., He, J.: Heterostructures based on 2D materials: a versatile platform for efficient catalysis. Adv. Mater. **31**, 1804828 (2019)
32. Singla, A., Singh, I., Mahajan, A.: The effect of transition metal ion implantation on the structural and optical properties of few-layered borophene. Nucl. Instrum. Methods Phys. Res., Sect. B **553**, 165417 (2024)
33. Mishra, R.K., Verma, K., Singh, D.S.: Defect engineering in nanomaterials: impact, challenges, and applications. Smart Mater. Manuf. **2**, 100052 (2024)
34. Li, M.Y., Chen, C.-H., Shi, Y., Li, L.J.: Heterostructures based on two-dimensional layered materials and their potential applications. Mater. Today **19**, 322 (2016)
35. Wieszczycka, K., Staszak, K., Woźniak-Budych, M.J., Litowczenko, J., Maciejewska, B.M., Jurga, S.: Surface functionalization—the way for advanced applications of smart materials. Coord. Chem. Rev. **436**, 213846 (2021)
36. Li, W., Wang, C., Lu, X.: Integrated transition metal and compounds with carbon nanomaterials for electrochemical water splitting. J. Mater. Chem. A **9**, 3786 (2021)
37. Su, T., Shao, Q., Qin, Z., Guo, Z., Wu, Z.: Role of interfaces in two-dimensional photocatalyst for water splitting. ACS Catal. **8**, 2253 (2018)
38. Zhong, X., Jin, M., Dong, H., Liu, L., Wang, L., Yu, H., Leng, S., Zhuang, G., Li, X., Wang, J.G.: TiO_2 nanobelts with a uniform coating of g-C_3N_4 as a highly effective heterostructure for enhanced photocatalytic activities. J. Solid State Chem. **220**, 54 (2014)
39. Hao, J., Zhang, S., Ren, F., Wang, Z., Lei, J., Wang, X., Cheng, T., Li, L.: Synthesis of TiO_2@g-C_3N_4 core-shell nanorod arrays with Z-scheme enhanced photocatalytic activity under visible light. J. Colloid Interface Sci. **508**, 419 (2017)
40. Zhao, T., Xing, Z., Xiu, Z., Li, Z., Yang, S., Zhu, Q., Zhou, W.: Surface defect and rational design of TiO_{2-x} nanobelts/g-C_3N_4 nanosheets/CdS quantum dots hierarchical structure for enhanced visible-light-driven photocatalysis. Int. J. Hydrogen Energy **44**, 1586 (2019)
41. He, F., Zhu, B., Cheng, B., Yu, J., Ho, W., Macyk, W.: 2D/2D/0D $TiO_2/C_3N_4/Ti_3C_2$ MXene composite S-scheme photocatalyst with enhanced CO_2 reduction activity. Appl. Catal. B **272**, 119006 (2020)
42. Jiang, Y., Sun, Z., Chen, Q., Cao, C., Zhao, Y., Yang, W., Zeng, L., Huang, L.: Fabrication of 0D/2D TiO_2 nanodots/g-C_3N_4 S-scheme heterojunction photocatalyst for efficient photocatalytic overall water splitting. Appl. Surf. Sci. **571**, 151287 (2022)
43. Ahmad, R., Srivastava, R., Yadav, S., Singh, D., Gupta, G., Chand, S., Sapra, S.: Functionalized molybdenum disulfide nanosheets for 0D–2D hybrid nanostructures: photoinduced charge transfer and enhanced photoresponse. J. Phys. Chem. Lett. **8**, 1729 (2017)
44. Tajima, T., Matsuura, T., Efendi, A., Yukimoto, M., Takaguchi, Y.: $MoSe_2$-sensitized water splitting assisted by C_{60}-dendrons on the basal surface. Chem. Eur. J. **30**, e202402690 (2024)
45. Vattikuti, S.V.P., Devarayapalli, K.C., Nagajyothi, P.C., Shim, J.: Binder-free WS_2/ZrO_2 hybrid as a photocatalyst for organic pollutant degradation under UV/simulated sunlight and tests for H_2 evolution. J. Alloy. Compd. **809**, 151805 (2019)
46. Zhang, C., Luo, Y., Tan, J., Yu, Q., Yang, F., Zhang, Z., Yang, L., Cheng, H.-M., Liu, B.: High-throughput production of cheap mineral-based two-dimensional electrocatalysts for high-current-density hydrogen evolution. Nat. Commun. **11**, 3724 (2020)
47. Shuai, T.-Y., Zhan, Q.N., Xu, H.M., Zhang, Z.J., Li, G.R.: Recent developments of MXene-based catalysts for hydrogen production by water splitting. Green Chem. **25**, 1749 (2023)
48. Biswal, L., Mohanty, R., Nayak, S., Parida, K.: Review on MXene/TiO_2 nanohybrids for photocatalytic hydrogen production and pollutant degradations. J. Environ. Chem. Eng. **10**, 107211 (2022)
49. Sherryna, A., Tahir, M.: Role of Ti_3C_2 MXene as prominent Schottky barriers in driving hydrogen production through photoinduced water splitting: a comprehensive review. ACS Appl. Energy Mater. **4**, 11982 (2021)
50. Akir, S., Azadmanjiri, J., Antonatos, N., Dekanovsky, L., Roy, P.K., Mazánek, V., Lontio Fomekong, R., Regner, J., Sofer, Z.: Atomic-layered V_2C MXene containing bismuth elements: 2D/0D and 2D/2D nanoarchitectonics for hydrogen evolution and nitrogen reduction reaction. Nanoscale **15**, 12648 (2023)

51. Panda, A., Kim, H.: Phosphorus-decorated Mo-MXene/CQD hybrid: a 2D/0D architecture for bifunctional electrochemical water splitting. Nanoscale **13**, 14795 (2021)
52. Ma, D., Shi, J.W., Zou, Y., Fan, Z., Ji, X., Niu, C.: Highly efficient photocatalyst based on a CdS quantum dots/ZnO nanosheets 0D/2D heterojunction for hydrogen evolution from water splitting. ACS Appl. Mater. Interfaces **9**, 25377 (2017)
53. Huang, J., Tao, J., Liu, G., Lu, L., Tang, H., Qiao, G.: In situ construction of 1D CdS/2D Nb_2CT_x MXene Schottky heterojunction for enhanced photocatalytic hydrogen production activity. Appl. Surf. Sci. **573**, 151491 (2022)
54. Jiang, Z., Chen, Q., Zheng, Q., Shen, R., Zhang, P., Li, X.: Constructing 1D/2D Schottky-based heterojunctions between $Mn_{0.2}Cd_{0.8}S$ nanorods and Ti_3C_2 nanosheets for boosted photocatalytic H_2 evolution. Acta Physico Chimica Sinica **37**, 2010059 (2020)
55. Du, Y., Ma, R., Wang, L., Qian, J., Wang, Q.: 2D/1D BiOI/g-C_3N_4 nanotubes heterostructure for photoelectrochemical overall water splitting. Sci. Total Environ. **838**, 156166 (2022)
56. Samajdar, S., Bera, S., Das, P.S., Finch, H., Dhanak, V.R., Chakraborty, S., Maiyalagan, T., Annapurna, K., Ghosh, S.: Exploration of 1D–2D $LaFeO_3$/rGO S-scheme heterojunction for photocatalytic water splitting. Int. J. Hydrogen Energy **48**, 17838 (2023)
57. Bera, S., Ghosh, S., Basu, R.N.: Silver as solid-state electron mediator in MoS_2/Ag-$AgVO_3$ Z-scheme heterostructures for photocatalytic H_2 generation. J. Alloy. Compd. **830**, 154527 (2020)
58. Liu, Q., Lu, H., Shi, Z., Wu, F., Guo, J., Deng, K., Li, L.: 2D $ZnIn_2S_4$ nanosheet/1D TiO_2 nanorod heterostructure arrays for improved photoelectrochemical water splitting. ACS Appl. Mater. Interfaces **6**, 17200 (2014)
59. Li, N., Zhang, Y., Jia, M., Lv, X., Li, X., Li, R., Ding, X., Zheng, Y.Z., Tao, X.: 1T/2H $MoSe_2$-on-MXene heterostructure as bifunctional electrocatalyst for efficient overall water splitting. Electrochim. Acta **326**, 134976 (2019)
60. Li, Y., Liu, Y., Xing, D., Wang, J., Zheng, L., Wang, Z., Wang, P., Zheng, Z., Cheng, H., Dai, Y., Huang, B.: 2D/2D heterostructure of ultrathin $BiVO_4$/Ti_3C_2 nanosheets for photocatalytic overall water splitting. Appl. Catal. B **285**, 119855 (2021)
61. Rasool, F., Pirzada, B.M., Talib, S.H., Alkhidir, T., Anjum, D.H., Mohamed, S., Qurashi, A.: In situ growth of interfacially nanoengineered 2D–2D WS_2/$Ti_3C_2T_x$ MXene for the enhanced performance of hydrogen evolution reactions. ACS Appl. Mater. Interfaces **16**, 14229 (2024)
62. Li, Z., Wang, X., Ren, J., Wang, H.: NiFe LDH/$Ti_3C_2T_x$/nickel foam as a binder-free electrode with enhanced oxygen evolution reaction performance. Int. J. Hydrogen Energy **47**, 3886 (2022)
63. Madhushree, R., Sunaja Devi, K.R.: Structural investigation of Cr_2CT_x/$NiFe_2O_4$ MXene composite as a bifunctional electrocatalyst for water splitting. Surf. Interfaces **52**, 104849 (2024)
64. Ke, J., Zhou, H., Liu, J., Zhang, Z., Duan, X., Wang, S.: Enhanced light-driven water splitting by fast electron transfer in 2D/2D reduced graphene oxide/tungsten trioxide heterojunction with preferential facets. J. Colloid Interface Sci. **555**, 413 (2019)
65. Sahoo, S., Al Mahmud, A., Sood, A., Dhakal, G., Tiwari, S.K., Zo, S., Kim, H.M., Han, S.S.: Microwave-assisted facile synthesis of graphitic-C_3N_4/reduced graphene oxide/MoS_2 composite as the bifunctional electrocatalyst for electrochemical water splitting. J. Sci.: Adv. Mater. Devices **10**, 100843 (2025)
66. Zhang, H., Wang, Y., Niu, W., Yoskamtorn, T., Luo, M., Tayler, R., Day, S., Edman Tsang, S.C.: Tuning 2D perovskite–graphene layered composite for photocatalysis. Sustain. Energy Fuels **8**, 3362–3371 (2024)
67. Mittal, S., Khosya, M., Singh, M., Khare, N.: 2D/2D heterojunction of SnS_2/g-C_3N_4 nanocomposite photoelectrode for improved photoelectrochemical water splitting performance. Appl. Surf. Sci. **677**, 161016 (2024)
68. Chen, Z., Ha, Y., Jia, H., Yan, X., Chen, M., Liu, M., Wu, R.: Oriented transformation of Co-LDH into 2D/3D ZIF-67 to achieve Co–N–C hybrids for efficient overall water splitting. Adv. Energy Mater. **9**, 1803918 (2019)
69. Tian, R., Wang, F., Zou, C., Pei, Z., Guo, X., Yang, H.: Modulating organic ligands to construct 2D–3D-hybrid porous P-doped metal-organic frameworks electrocatalyst for overall water splitting. J. Alloy. Compd. **933**, 167670 (2023)

70. Sadaqat, M., Manzoor, S., Aman, S., Gouadria, S., Usman, M., Joya, K.S., Najam-Ul-Haq, M., Hassan, H.M.A., Ashiq, M.N., Taha, T.A.: Mn-based hierarchical polyhedral 2D/3D nanostructures MnX_2 (X = S, Se, Te) derived from Mn-based metal-organic frameworks as high-performance electrocatalysts for the oxygen evolution reaction. Energy Fuels **36**, 10327 (2022)
71. Yang, N., Yang, W., Yang, X., Xiao, X., Zhang, L., Zhang, L.: 2D/3D hierarchical and multi-heterostructured Co/Ni/Fe phosphosulfide as a highly efficient bifunctional electrocatalyst for overall water splitting. Int. J. Hydrogen Energy **56**, 396 (2024)
72. Samsudin, M.F.R., Sufian, S.: Hybrid 2D/3D g-C_3N_4/$BiVO_4$ photocatalyst decorated with rGO for boosted photoelectrocatalytic hydrogen production from natural lake water and photocatalytic degradation of antibiotics. J. Mol. Liq. **314**, 113530 (2020)
73. Ullah, S., Ahmed, W., Zeeshan, M., Hu, S., Zhang, X.: Integral 2D/3D structured $CoSnO_3$@MXene/NF as a highly active and stable bifunctional electrocatalyst for alkaline water splitting. Int. J. Hydrogen Energy **70**, 448 (2024)
74. Wei, P., Wen, Y., Lin, K., Li, X.: 2D/3D WO_3/$BiVO_4$ heterostructures for efficient photoelectrocatalytic water splitting. Int. J. Hydrogen Energy **46**, 27506 (2021)

MIX
Papier aus verantwortungsvollen Quellen
Paper from responsible sources
FSC® C105338

If you have any concerns about our products,
you can contact us on
ProductSafety@springernature.com

In case Publisher is established outside the EU,
the EU authorized representative is:
Springer Nature Customer Service Center GmbH
Europaplatz 3, 69115 Heidelberg, Germany

Printed by Libri Plureos GmbH
in Hamburg, Germany